Fibronectin in Health and Disease

Editor

Steven E. Carsons, M.D.
Chief, Division of Rheumatology
and Immunology
Winthrop University Hospital
Mineola, and
Assistant Professor of Medicine
S. U. N. Y.
Stony Brook, New York

CRC Press, Inc.
Boca Raton, Florida

Library of Congress Cataloging-in-Publication Data

Fibronectin in health and disease / editor, Steven E. Carsons.
p. cm.
Includes bibliographies and index.
ISBN 0-8493-5064-6
1. Fibronectins--Physiological effect. 2. Fibronectins-
-Pathophysiology. I. Carsons, Steven E., 1950- .
[DNLM: 1. Fibronectins. QU 55 F443]
QP552.F53F54 1989
612'.01575--dc19
DNLM/DLC
for Library of Congress

89-817
CIP

Direct all inquiries to CRC Press, Inc., 2000 Corporate Blvd., N.W., Boca Raton, Florida, 33431.

International Standard Book Number 0-8493-5064-6

Library of Congress Card Number 89-817
Printed in the United States

PREFACE

A little more than a decade ago, fibronectin was recognized to be a significant component of mammalian extracellular matrix and blood. Multiple diverse and important biological functions, some previously known by other identities, were realized to reside on the fibronectin molecule. Certain key observations — namely, that fibronectin was intimately associated with cell-substrate adhesion and cell-particle recognition — sparked tremendous interest in a wide array of biological disciplines ranging from embryology to immunology. A decade later, we now have gathered a tremendous amount of information regarding the role of fibronectin in basic biological processes. The use of monoclonal antibodies, protein and cDNA sequencing, and of synthetic peptides has led to the identification of the mechanism responsible for fibronectin-mediated cell adhesion at the tripeptide level.

With recognition of the significance of fibronectin in basic biological processes came the realization of its potential importance in biomedical applications. Fibronectin cell-matrix interactions have important implications with regard to oncogenesis and metastases, wound healing, and embryonic development. Fibronectin cell-particle interactions clearly are important in terms of immune adherence of microorganisms, immune complex physiology, and phagocytosis of tissue debris in trauma. Fibronectin protein-protein interactions appear to play important roles in connective tissue structure and remodeling and in blood coagulaiton.

Accordingly, over the past several years, investigators have sought to define the role of fibronectin in human development, health, and disease. Basic approaches have included (1) measurement of plasma fibronectin levels in normal and pathological states; (2) examination of fibronectin deposition patterns in normal and diseased tissues; (3) *in vitro* study of fibronectin synthesis and turnover in cell culture systems; (4) molecular characterization of fibronectin isolated from pathological plasma and body fluid; and (5) the use of animal models to evaluate fibronectin as a measure of disease activity and to monitor the effect of pharmacologic intervention.

This volume appears at a time when many of the initial approaches described above have been examined; however, many additional questions regarding the role of fibronectin in normal homeostasis and disease remain unanswered. It is hoped that this volume will serve to compile and summarize the state of current knowledge in the field and serve to stimulate further investigation. This work is not intended to be an encyclopedic compendium of fibronectin research but rather to highlight areas that have been the object of intense investigation and where fibronectin has been demonstrated to play a significant role. Such areas include human development and aging, host defense, coagulation, carcinogenesis, inflammation, and connective tissue metabolism.

It appears that the next decade of fibronectin research should prove to be as interesting and informative as the past one. The development of newer reagents such as monoclonal antibodies, molecular probes, and synthetic peptides should aid in the identification and characterization of additional fibronectin species. It is likely that some will be specific for a particular tissue or pathological state. Characterization of the structural variations among these fibronectins will lead to a more complete understanding of cell-matrix, cell-particle, and protein-protein interactions in normal and abnormal physiology.

Perhaps the most significant recent development in extracellular matrix research has been the characterization of the fibronectin-receptor superfamily now commonly known as the integrins. Investigation into the dynamic interplay between fibronectins and their receptors in normal and abnormal states should further advance our knowledge of important biological processes, including development and aging, neoplasia, inflammation, and wound healing.

THE EDITOR

Steven Carsons, M.D. is currently chief of the Division of Rheumatology and Immunology at Winthrop University Hospital and Assistant Professor of Medicine at the State University of New York at Stony Brook.

Dr. Carsons received his A.B. degree in chemistry from University College of New York University in 1972 and his M.D. degree from New York Medical College in 1975. During his fellowship training at the State University of New York at Downstate he was awarded the Renee Carhart Amory fellowship by the New York Arthritis Foundation and a National Arthritis Foundation postdoctoral fellowship. He remained on the faculty of SUNY Downstate until 1982, during which time he was a NIH Clinical Associate Physician. He directed the Division of Clinical Immunology at Long Island Jewish Medical Center from 1982 to 1989. He assumed his current position in 1989.

Dr. Carsons is a Founding Fellow of the American Rheumatism Association and is a member of the American Federation for Clinical Research, The Society for Experimental Biology and Medicine, and the New York Academy of Sciences. Currently he is president of the New York Rheumatism Association.

Dr. Carsons has been the recipient of grants from the National Institutes of Health, the Lupus Foundation, and the National and New York Arthritis Foundations. He is the author of more than 50 articles, abstracts, and chapters. His current research interest is the role of fibronectin and matrix proteins in inflammatory rheumatic disease.

To my wife, Lesley
and my daughter Cara

CONTRIBUTORS

Byron E. Anderson, Ph.D.
Professor
Department of Molecular Biology
Northwestern University
Chicago, Illinois

Edwin H. Beachey, M.D.
ACOS for Research and Development
VA Medical Center and
Professor of Medicine and of Microbiology and Immunology
University of Tennessee, Memphis
Memphis, Tennessee

André D. Beaulieu, M.D., F.R.C.P.(C.)
Co-Director
Inflammation and Immunology-Rheumatology Research Unit
Centre Hospitalier de l'Université Laval
Ste-Foy, Québec, Canada

Celso Bianco, M.D.
Director for Science and Technology
The New York Blood Center
New York, New York

Nancy Burton-Wurster, Ph.D.
Senior Research Associate
Department of Veterinary Microbiology, Immunology, and Parasitology
J. A. Baker Institute for Animal Health
Cornell University
New York State College of Veterinary Medicine
Ithaca, New York

Steven E. Carsons, M.D.
Chief, Division of Reumatology and Immunology
Winthrop University Hospital
Mineola, and
Assistant Professor of Medicine
S. U. N. Y.
Stony Brook, New York

Richard A. F. Clark, M.D.
Head
Division of Dermatology
Departments of Medicine and Pediatrics
National Jewish Center for Immunology and Respiratory Disease
Denver, Colorado

M. La Fleur
Inflammation and Immunology-Rheumatology Research Unit
Centre Hospitalier de l'Université Laval
Ste-Foy, Québec, Canada

Beverly B. Lavietes, Ph.D.
Assistant Professor
Department of Medicine
State University of New York at Stony Brook
Stony Brook, New York

George Lust, Ph.D.
Professor of Physiological Chemistry
Department of Veterinary Microbiology, Immunology, and Parasitology
J. A. Baker Institute for Animal Health
Cornell University
New York State College of Veterinary Medicine
Ithaca, New York

David M. Mann, Ph.D.
Postdoctoral Research Associate
La Jolla Cancer Research Foundation
La Jolla, California

Paul Mapp
Research Fellow
Bone and Joint Research Unit
The London Hospital
London, England

Albert J. T. Millis, Ph.D.
Professor
Center for Cellular Differentiation
Department of Biological Sciences
State University of New York at Albany
Albany, New York

Dudley G. Moon, Ph.D.
Research Assistant Professor
Department of Physiology
Albany Medical College
Albany, New York

Kenneth Olden, Ph.D.
Director
Cancer Center
Department of Oncology
Howard University School of Medicine
Washington, D.C.

Kevin M. Connolly, Ph.D.
Department of Chemotherapy
Glaxo Research Division
University of North Carolina
Chapel Hill, North Carolina

Harry S. Courtney, Ph.D.
Research Microbiologist
V.A. Medical Center, and
Instructor
Department of Medicine
University of Tennessee, Memphis
Memphis, Tennessee

Ruth A. Entwistle, Ph.D.
Postdoctoral Fellow
Department of Medicine
Jewish Hospital
St. Louis, Missouri

Jiro Fujita, M.D.
Assistant Professor
Kagawa Medical School
Kagawa, Japan

Mark H. Ginsberg, M.D.
Associate Member
Committee on Vascular Biology
Research Institute of Scripps Clinic
La Jolla, California

Patricia Greenwel
Liver Research Center, and
Department of Pathology
Albert Einstein College of Medicine
Bronx, New York

David L. Hasty, Ph.D.
Associate Professor
Department of Anatomy and
Neurobiology
University of Tennessee, Memphis
Memphis, Tennessee

K. E. Herbert
Research Fellow
Rheumatology Department
Homerton Hospital
London, England

Edward F. Plow, Ph.D.
Head, Committee on
Vascular Biology
Research Institute of Scripps Clinic
La Jolla, California

Stephen I. Rennard, M.D.
Chief
Pulmonary Section
Department of Internal Medicine
University of Nebraska Medical Center
Omaha, Nebraska

Marcos Rojkind, M.D., Ph.D.
Professor
Department of Medicine and Pathology
Liver Research Center
Albert Einstein College of Medicine
Bronx, New York

Thomas M. Saba, Ph.D.
Professor and Harold C. Wiggers
Chairman
Department of Physiology
Albany Medical College
Albany, New York

Frank R. Schmid, M.D.
Chief, Arthritis-Connective Tissue
Diseases Section, and
Professor
Department of Medicine
Northwestern University Medical School
Chicago, Illinois

David L. Scott
Senior Lecturer and
Consultant Physician
Department of Rheumatology
St. Bartholemew's Hospital
London, England

W. Andrew Simpson, Jr., Ph.D.
Research Biologist
V. A. Medical Center, and
Associate Professor of Medicine and of
Microbiology and Immunology
University of Missouri School of
Medicine
Columbia, Missouri

Martin J. Humphries, Ph.D.
School of Biological Science
Department of Biochemistry
University of Manchester
Manchester, England

John E. Kaplan, Ph.D.
Professor
Department of Physiology
Albany Medical College
Albany, New York

C. Kreis
Inflammation and Immunology-
Rheumatology Research Unit
Centre Hospitalier de l'Université Laval
Ste-Foy, Québec, Canada

Vera J. Stecher, Ph.D.
Associate Scientific Director
Department of Medical and Scientific
Affairs
Winthrop Pharmaceuticals
New York, New York

J. Jay Weiss, Ph.D.
Manager, Allergy Research and
Development
Kallestad Diagnostics
Chaska, Minnesota

Kenneth M. Yamada, Ph.D.
Chief
Membrane Biochemistry Section
Laboratory of Molecular Biology
National Cancer Institute
National Institutes of Health
Bethesda, Maryland

TABLE OF CONTENTS

Chapter 1

THE STRUCTURE AND FUNCTION OF FIBRONECTINS

Steven E. Carsons

TABLE OF CONTENTS

I. INTRODUCTION

Once thought to be primarily an inert structural scaffolding, recent studies have shown that the extracellular matrix (ECM) is a complex structure containing many glycoproteins. Matrix glycoproteins mediate multiple interactions among other ECM components as well as cell membrane receptors. These interactions result in profound alterations in cell phenotype and function.[1]

A. MODULAR GLYCOPROTEINS OF THE EXTRACELLULAR MATRIX

Biochemical characterization of ECM components has resulted in the identification of a series of proteins that to a certain degree share structural and functional features. The major common feature of these proteins is a general structure represented by a series of relatively homologous modular elements (domains) interconnected by flexible regions. One or more domains binds to extracellular macromolecules while separate domains interact with cells. This modular domain arrangement allows the glycoprotein to serve as a molecular bridge between cells and the ECM. The general arrangement of a modular matrix glycoprotein is schematized in Figure 1.

Domain A represents a binding site for extracellular components, i.e., collagen, whereas domain B represents a cell-binding domain. Within the large cell-binding domain exists a discrete cell-binding sequence (open and closed circles). The cross-hatched area indicates potential regions of sequence variability adjacent to the cell-binding sequence. Such variable regions may modulate the cell-binding specificity and affinity of the molecule (see Chapter 11). Domain C represents an additional binding domain for extracellular components. Domain C, therefore, may mediate binding to heparin, or to cell surface heparin sulfate. The overall domain structure may be modified by the presence of glycosylation sites (CHO), substrates for cross-linking enzymes (XL), and sulfhydryl groups (S-S). The latter may mediate the formation of inter- and intrachain linkages. These modifications may affect the subunit structure, solubility, and protease susceptibility of the matrix glycoprotein.

Fibronectin is the most extensively studied matrix glycoprotein. Therefore, a great deal of information is available concerning its primary structure, posttranslational modification, and tissue distribution. Structural and functional information is also rapidly amassing on other matrix glycoproteins. The major characteristics of these matrix glycoproteins are reviewed in References 2 and 3 and are summarized in Table 1.

Several proteins that participate in the coagulation process, such as fibrinogen, von Willebrand factor, and thrombospondin, also have been found to share cell- and collagen-binding properties with ECM proteins.[4,5]

It is apparent that ECM proteins play a major role in several important biologic processes including coagulation, wound healing, tumor-host interactions, and inflammation. The existence of extensive structural information including primary sequence data for fibronectin

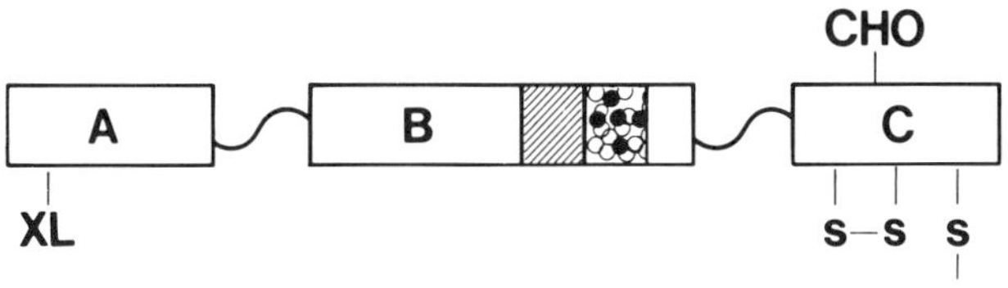

FIGURE 1. Schematic diagram of a typical, hypothetical cell-binding ECM glycoprotein. Capital letters (A,B,C) represent globular domains, whereas wavy lines indicate flexible, protease-sensitive interdomain regions. The cross-hatched area represents regions of sequence variation adjacent to an invariant cell-binding sequence (open and closed circles). CHO indicates carbohydrate; XL indicates regions involved in protein crosslinking; S-S and S indicate intra- and interchain disulfide bonds.

TABLE 1
Features of Cell-Binding Matrix Proteins[2,3]

Matrix Protein	Collagen	GAG[a]	Fibrin	Cell	Circ[b]	Matrix
Fibronectin	I,III	Yes	Yes	Yes	Yes	Yes
Collagens	Fibril formation	—	—	Yes	No	Yes
Laminin	IV	Heparin	—	Yes	Tr[d]	Yes
Vitronectin	Yes	Yes	—	Yes	Yes	Yes
Chondronectin	II	CS[c]	—	Yes	Yes	Yes

[a] Glycosaminoglycan binding.
[b] Circulating form.
[c] Chondroitin sulfate.
[d] Trace.

has enabled investigators to compare fibronectins from normal developmental stages and from various pathological conditions. Such studies have added to our understanding of the role of fibronectin and the ECM in human development and disease.

II. FIBRONECTIN — STRUCTURAL CONSIDERATIONS

A. GENERAL CHARACTERISTICS

The fibronectin monomeric subunit chain has a molecular weight of approximately 250 kDa. Estimates derived from polyacrylamide gel electrophoresis in the presence of reducing agents range from 210 to 290 kDa.[6] The plasma form circulates as a dimer having a nominal molecular weight of 450 to 500 kDa,[7] whereas fibronectins extracted from tissue and cell layers may exist as higher molecular weight multimers. All fibronectins contain approximately 4 to 8% carbohydrate.[9] The presence and type of carbohydrate may contribute to variations in molecular weight estimations and charge heterogeneity.

B. DOMAIN ORGANIZATION

1. Investigational Approach

Limited proteolytic digestion of fibronectin has been exceedingly useful in elucidating the domain organization of the molecule. Fractionation of fibronectin proteolytic digests on ligand affinity columns followed by sodium dodecyl sulfate-polyacrylamide gel (SDS-PAGE) analysis has enabled investigators to construct models depicting the linear array of domains comprising the entire fibronectin subunit. A composite domain model of fibronectin demonstrating the major ligand-binding domains and important proteolytic cleavage sites is shown in Figure 2.

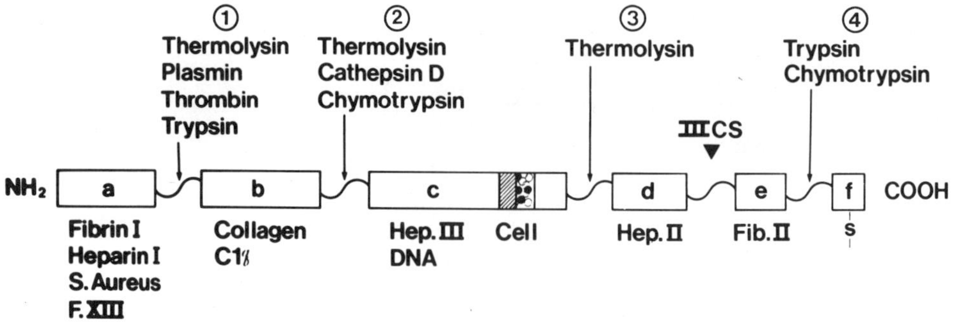

FIGURE 2. Domain model of a fibronectin subunit. Lower case letters (a to f) identify discrete domains proceeding from amino- to carboxyterminal. Ligand-binding affinities are listed beneath each domain. The open and solid circles indicate the approximate location of the Arg-Gly-Asp (RGD) cell-binding sequence. The cross-hatched region indicates a potential area of sequence variability due to insertion of a domain via differential mRNA splicing. IIICS (arrowhead) is a differentially spliced region connecting domains d and e. S is the location of the interchain disulfide link. Arrows indicate protease-sensitive sites (1 to 4) where limited proteolysis results in the formation of major domains.

Studies performed by Hahn and Yamada[10] used limited chymotryptic digestion of cellular fibronectin at site 2 (Figure 2), to establish that the collagen-binding and cell-binding regions were located on separate fragments. Ruoslahti et al.[11] obtained limited aminoterminal sequences of plasma fibronectin fragments and examined their cell- and gelatin-binding activity. They were able to conclude that the collagen-binding domain was located near the aminoterminal and the cell-binding domain was located toward the carboxyterminal. Balian et al.[12] reached a similar conclusion regarding the aminoterminal localization of the collagen-binding domain using preferential metabolic labeling of the carboxyterminal followed by digestion with cathepsin D.

Thus, it is apparent that the basic domain organization of the fibronectin molecule is responsible for its major functional characteristic — the ability to link cells to collagenous substrata via separate and distinct binding sites.

2. The Aminoterminal — Multiple Ligand-Binding and Cross-Linking Activites

Treatment of fibronectin with several proteolytic enzymes, inducing trypsin, plasmin, and thermolysin (Site 1, Figure 2), produces fragments with molecular weights ranging from 24 to 32 kDa.[13,14] These fragments represent the aminoterminal domain (domain a, Figure 2). This domain binds to affinity matrices composed of fibrin and heparin.[13,15,16] In addition, at least one other fibrin-binding domain and at least two other heparin-binding domains have been localized to other regions of the molecule.[15-17] The fibrin-binding affinity of the aminoterminal domain is relatively high, whereas the heparin-binding affinity of this domain is relatively low.[16,18]

Fibronectin binding to fibrinogen/fibrin can be covalently stabilized by transglutaminase-mediated cross-linking.[18] The transglutaminase-reactive sites appear to be glutaminyl residues located in the aminoterminal domain.[19] These sites may also participate in the cross-linking of fibronectin to itself and to collagen.[20]

Another important biological activity of fibronectin is its interaction with bacteria, particularly *Staphylococcus aureus*[21] (see Chapter 6). The *S. aureus*-binding site has also been localized to the aminoterminal domain.[21] Fibronectin may also become cross-linked to *S. aureus* by transglutaminase.[20]

3. The Collagen-Binding Domain

The collagen-binding domain of fibronectin (domain b, Figure 2) is located immediately carboxyterminal to aminoterminal domain a. Several proteases, including chymotrypsin and thermolysin (site 2, Figure 2), cleave fibronectin so as to produce collagen-binding fragments ranging in size from 40 to 60 kDa[10,13] These domains are easily purified on collagen-agarose columns. Cathepsin D treatment results in the generation of a slightly larger 70-kDa collagen-binding fragment which is composed of the collagen-binding domain joined to the ~ 30-kDa aminoterminal domain (domains a and b; Figure 2).[18]

Fibronectin has demonstrated the ability to bind to collagen types I, II, III, IV, V,[1] and VI;[22] however, fibronectin does not bind equally well to all collagens, and binding affinity appears to be determined by the collagen molecular form and genetic type. Fibronectin binds denatured collagen with much higher affinity than it binds native collagen especially with regard to type I.[23] Fibronectin appears to have greater affinity for native collagen III than for other types.[23] A major fibronectin-binding site on collagen has been localized to the (α1) CB7 cyanogen bromide fragment on the basis of binding-inhibition studies.[24]

The collagen-binding domain also contains two additional important biochemical features. The domain appears to contain a large proportion of the carbohydrate of the molecule. Carbohydrate seems to protect the collagen-binding region from proteolysis.[25] Fibronectin also binds to the C1q component of the complement system via the collagen-binding domain (see Chapter 5). This interaction probably occurs at the triple helical collagenous domain of the C1q molecule.[25a]

4. The Cell-Binding Domain — A Tetrapeptide Mediates Cell Attachment to Fibronectin

Initial studies localized the cell-binding region of the fibronectin molecule to the carboxyterminal.[11] Limited proteolytic digestion patterns that had been useful in the localization of fibrin, heparin, and collagen domains to reasonably discrete regions of the molecule could only approximate the cell-binding site to large carboxyterminal fragments ranging from 140 to 200 kDa (domain c, Figure 2).[10,11] Subsequently, Pierschbacher et al. localized cell attachment activity to a 15,000-Da fragment. These investigators then utilized a monoclonal antibody directed against the cell attachment region (3E3) to purify a sufficient quantity of this fragment to enable them to obtain the sequence of the 108 amino acids comprising the fragment.[27] Cell-attachment activity was subsequently demonstrated for a 3400-Da subfragment and eventually for a synthetic tetrapeptide — Arg - Gly - Asp - Ser (RGDS), located 20 residues from the carboxyterminal of the 108-amino acid fragment[28] (RGD, Figure 2). It is remarkable that the entire cell-binding activity of the 250-kDa subunit is expressed by this tetrapeptide (RGDS). This sequence may also play a role in the cell-binding activity of other matrix adhesive proteins such as vitronectin, interstitial collagen, and thrombospondin.[4]

5. The Carboxyterminal — Repeat Ligand-Binding Affinities and the Interchain Link

Early structural studies revealed the presence of a heparin-binding site on the large cell-binding carboxyterminal fragment.[11] Limited digestion with thermolysin and trypsin have proven to be very useful in demonstrating the presence of additional heparin- and fibrin-binding domains at the carboxyterminus of the molecule.[13,15,17] These domains (d and e, Figure 2) were found to possess affinity for heparin and fibrin. Thus, the binding affinities of the carboxyterminus essentially mirror the major binding affinities of the aminoterminal. Examination of the complete amino acid sequence of fibronectin reveals the structural basis for such functional repetition. This will be discussed in more detail in Section II.C below.

Finally, a small (3 to 6 kDa) domain located at the carboxyterminal (domain f, Figure 2) has been found to contain the interchain disulfide bond that joins the subunits of the fibronectin dimer.[17]

C. THE STRUCTURAL BASIS FOR FIBRONECTIN FUNCTION

The availability of large quantities of bovine plasma fibronectin has allowed amino acid

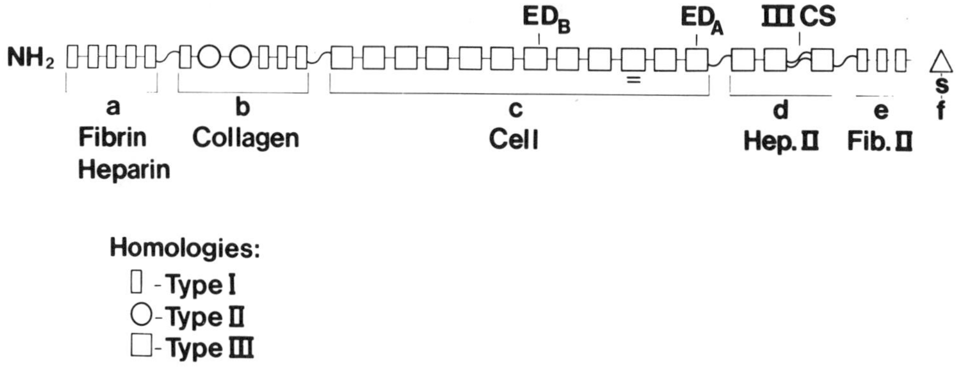

FIGURE 3. Model fibronectin subunit according to morphologic units. The major binding domains as indicated by lower case letters (a to f) are described in terms of their sequence homology and correspond to domains shown in Figure 2. ED refers to extra type III domains inserted by mRNA splicing. IIICS refers to the nontype III flexible region connecting type III units in the heparin-binding domain (domain d). Differential splicing at the IIICS region results in size heterogeneity at domain d. The A subunit contains additional IIICS sequences and is 3 to 5 kDa larger than the B subunit.

sequencing at the protein level to be completed for several large fragments spanning 90% of the individual subunit chain.[29] A complementary approach, the sequencing of c-DNA clones from rat hepatocytes and several human cell lines, has added additional information concerning the primary sequence and areas of sequence variation.[30]

Examination of the 2260-amino acid sequence reveals that the array of binding domains described above in Section II and Figure 2 corresponds to a series of homologous repeats (Figure 3).[31] This also suggests that the fibronectin gene arose via gene duplication during evolution.

Three types of repeats have been identified. Type I repeats are approximately 40 amino acids in length and contain intrachain disulfide bonds. The entire N-terminal fibrin-binding domain (domain a, Figure 2, Figure 3) is composed of type I homologies.[31] The collagen-binding domain (b, Figure 2 and 3) is made up of four type I homologies in addition to two type II regions. Type II regions are comprised of a stretch of approximately 60 amino acids and also contain intrachain disulfide bonds. The large central domain of fibronectin is composed of type III repeating units, each 90 amino acids long. The active cell attachment peptide (RGD) is probably located on the tenth type III repeat.[32]

Three important variable areas have been identified in the cell-binding domain. One type of variation occurs as a result of the insertion or deletion of an entire type III repeat following the seventh and/or the eleventh type III segment (EDA/B, Figure 3).[33, 33a] The presence of these "extra domains" referred to as ED is a feature of cellular fibronectins whereas plasma fibronectins lack the ED segments. An additional region of structural variation has been localized to a nontype III region within the C-terminal heparin-binding domain. This region connects two type III segments (III_{14} and III_{15})[29,32,34] and this has been called IIICS (type III connecting segment). The presence of additional amino acid sequences (up to 120) in the IIICS region distinguishes the larger (A chain) from the smaller (B chain) of the fibronectin heterodimer. This sequence variation results in size heterogeneity of carboxyterminal fragments containing heparin-binding domains (d, Figure 2 and 3). This region is also important since it has recently been found to contain additional cell attachment signal sequences (Reference 34 and Chapter 11). It has been determined that both regions of diversity (ED and IIICS) arise from differential splicing at the mRNA level.[32,34] Thus, all fibronectins appear to be derived from a single gene.

The string of 15 type III homology segments is followed by three additional type I segments.[29] These type I segments comprise the C-terminal fibrin-binding domain (domain e, Figure 2 and 3). It appears that type I homology regions occur where fibrin-binding domains are functionally identified. The association between type I homology and fibrin binding is apparently not restricted to fibronectin. Recently, a type I domain homologous to fibronectin type I domains has been identified in tissue plasminogen activator (tPA) which also binds fibrin.

The 12th and final type I repeat is followed by the small, protease-sensitive carboxy-terminal domain containing the interchain disulfide bond (domain f, Figure 2 and 3).[29]

III. FIBRONECTIN STRUCTURAL HETEROGENEITY

In 1974, cell matrix fibronectin was found to be immunologically identical to a circulating plasma protein,[37] subsequently identified as cold-insoluble globulin (CIg).[38] Thus, two forms of fibronectin — a circulating (or plasma) form and a cell matrix (or cellular form) — were described. During the past decade, structural, immunological, and genetic investigations have pinpointed differences between these forms of the protein (see Section II.C). Important structural and functional differences between plasma and cellular forms of fibronectin will be discussed in more detail below.

Structural differences among fibronectins have expanded to include not only plasma and cellular forms but, in addition, forms of the protein produced in embryonal, malignant, and inflammatory tissues. These fibronectins are characterized by additional subtle structural variations. Despite their subtlety, regions of structural heterogeneity appear to be strategically located in order to maximize their functional impact.

Thus it appears that there exists a ''family'' of fibronectins. Members of this family are related structurally and immunologically; however, each member (or isoform) appears to possess a unique structural feature. Although all fibronectin variants have probably not yet been identified, several generalities can be stated:

1. Structure appears to be ontogenically regulated.
2. Some structural variants may be tissue and/or organ specific.
3. The protein may be modified transcriptionally and posttranslationally in specific disease states.

A. PLASMA FIBRONECTIN

Plasma fibronectin was originally described as a nonthrombin-coagulable by-product of fibrinogen-rich plasma fractions.[39] The protein, then known as CIg was extensively characterized by Mosesson and colleagues in the 1970s.[40] Fibronectin is a significant plasma protein in vertebrates. In man, plasma fibronectin concentrations range from approximately 200 to 700 μg/ml;[41] age, sex, and weight may influence levels.[42-44] Fibronectin has been identified in cryofibrinogens obtained from the plasma of patients with malignancies,[45] and in cryoglobulins from patients with autoimmune or connective tissue disorders.[46,47] These observations suggest that plasma fibronectin forms cold-insoluble complexes with fibrinogen, immune complexes, and other plasma components (see also Chapters 5, 16). Fibronectin containing cryoprecipitates can be formed from normal plasma by the addition of heparin.[48] Formation of the heparin-precipitable fraction (HPF) is dependent on the presence of plasma fibronectin demonstrating that fibronectin plays a significant role in the formation of cryoprotein by binding to other plasma macromolecules. The participation of fibronectin in these molecular complexes has implications with regard to the metabolism and clearance of fibrin complexes in coagulation disorders[45] and immune complexes in inflammatory disease.[46,47,49]

In 1978, Saba et al. demonstrated immunological identity between plasma fibronectin and a protein known as α_2SB opsonic glycoprotein[50] (see Chapter 4). Depletion of circulating α_2SB glycoprotein could be accomplished by saturation of the reticuloendothelial system (mononuclear phagocyte system) with gelatin, a fibronectin-binding ligand. This suggested a role for plasma fibronectin in the clearance of circulating tissue debris. Levels of fibronectin-(α_2SB glycoprotein), as measured by immunoassay or bioassay, decline in experimental animals or patients experiencing severe trauma, shock, or burns. Levels return toward baseline with clinical improvement. Thus, it appears that plasma fibronectin is an important nonimmunoglobulin, noncomplement opsonin.

B. CELLULAR FIBRONECTIN

Cellular fibronectin refers to the form of the protein present in tissue and in the pericellular matrix of cultured fibroblasts. Significant physiocochemical and functional differences between cellular and plasma fibronectin have been demonstrated.

The most striking difference between these forms of the protein is their solubility. Plasma fibronectin is readily soluble in neutral Tris buffer at concentrations of 1 to 2 mg/ml. Cellular fibronectin, on the other hand, is extremely insoluble, a factor which has impeded its characterization at the protein level. Atherton and Hynes[52] produced a monoclonal antibody which selectively recognized a site located in the carboxyterminal region of cellular fibronectin. More recently, as mentioned in Section II.C, sequencing of cDNA clones for fibroblast fibronectin has demonstrated the presence of extra type III segment (the ED or extra domain) in the carboxyterminal region of cellular but not plasma fibronectin (Figure 3).[33] It does not appear that the presence of the ED accounts for the physicochemical differences between plasma and cellular fibronectin. Antibodies specific for the ED region, however, have proven to be useful in the identification of cellular fibronectin in experimental models of disease processes.[53]

In addition to physicochemical differences, several functional differences between plasma and cellular fibronectins have been identified. Cellular fibronectin mediates the agglutination of fixed sheep RBC whereas plasma fibronectin is inactive.[54] Cellular fibronectin appears to be more active in mediating the attachment and spreading of fibroblasts.[54] Cellular fibronectin binds hyaluronic acid while even high concentrations of the plasma form do not.[55] The tendency of cellular fibronectin to exist as insoluble multimers may result in an enhanced interaction with the cell surface.

C. THE RELATIONSHIP OF PLASMA TO CELL SURFACE FIBRONECTIN

Studies demonstrating the incorporation of plasma fibronectin into the extracellular matrix of cultured cells suggest that plasma and cellular fibronectin may be in equilibrium.[56-58]

Investigators have utilized biochemical manipulations of the fibronectin molecule in order to elucidate the mechanism of its insolubilization and matrix incorporation. An examination of fibronectin structure reveals several areas that are likely to be involved in multimerization and matrix incorporation (Figure 4a).

Treatment of fibronectin with heparin in the cold (see Section III.A),[48] polyamines,[59] and denaturation with guanidine[6] all mediate insolubilization of the protein. These insoluble fibronectins formed *in vitro* share many characteristics with naturally occurring matrix fibronectin. Polyamine-induced fibronectin polymers appear as linear filaments on electron microscopy and resemble naturally occurring fibronectin matrix fibers.[59] Induction of polymerization by modification of disulfides and treatment with heparin indicates that the aminoterminal region may be involved in this process. This is consistent with the localization of a fibronectin self-association site to this region of the molecule.[60] An additional carboxyterminal self-association site has been postulated to exist on horse serum fibronectin.[60]

Mosher and colleagues have studied the mechanism of incorporation of exogenous

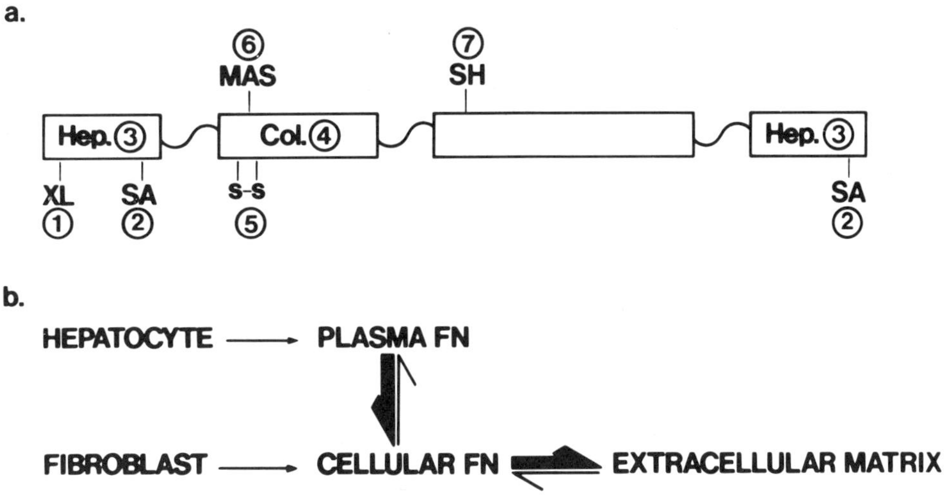

FIGURE 4. (a) Potential mechanisms of fibronectin insolubilization and matrix incorporation. Fibronectin may be covalently cross-linked by transglutaminase to itself, fibrin, or to collagen via glutaminyl residues at site 1 (XL). Fibronectin may self-associate via interactions at site 2(SA). Interaction with heparin at aminoterminal site 3 (weak) or carboxyterminal site 3 (strong) mediates co-precipitation from solution phase with fibrinogen at 4°C. Interaction with heparan sulfate may in part mediate interaction with cell surface. Fibronectin binds insolubilized collagen via the collagen-binding domain (site 4). Disulfides participating in intrachain loops (site 5) appear to be important in matrix assembly and may be part of the 70-kDa aminoterminal matrix assembly receptor (MAS; site 6). Free sulfhydryls (site 7) have been postulated to play a role in the fibronectin matrix formation. (b) Possible equilibrium between circulating and matrix fibronectins. Plasma fibronectin is secreted by the hepatocyte whereas "cellular" fibronectin is secreted by the fibroblast. Plasma fibronectin may insolubilize and become incorporated into the cellular fibronectin pool. Multimerization of cellular fibronectin appears to accompany its conversion into matrix fibers. A shift of the equilibrium toward insolubilization of circulating fibronectin and enhancement matrix incorporation may accompany tissue remodeling in wound healing and pathological states.

fibronectin into the cell layer of cultured fibroblasts.[58, 61-64] Their studies have made significant inroads toward the understanding of this complex process. Exogenous fibronectin binds to the cell layer in at least two distinct pools.[58] Binding to pool I is reversible and saturable, and bound fibronectin can be resolubilized in deoxycholate (DOC). Binding to pool II is irreversible and DOC resistant. A fraction of bound pool II molecules has been recovered as multimer. Exogenous fibronectin does not bind to isolated cell free matrix, indicating that the presence of living cells is necessary for matrix incorporation.[62]

The putative matrix assembly receptor is distinct from the cell adhesion receptor which recognizes the Arg-Gly-Asp cell adhesion sequence.[62] The 70-kDa aminoterminal cathepsin D fragment (domains a and b, Figure 2) is able to block exogenous fibronectin incorporation into matrix and thus probably contains the site that is interactive with the matrix assembly receptor.[62] Type I homology repeats (Figure 3) which contain intrachain disulfide loops and are concentrated in this region of the molecule appear to participate in the process of matrix assembly.[61]

When labeled with colloidal gold and visualized by high voltage electron microscopy, exogenous fibronectin localizes to discrete sites along the edge of the fibroblast and on retraction fibers.[63] Eventually, labeled exogenous fibronectin organizes into long fibrillar networks on the cell surface.

One of the most significant physiological alterations in fibronectin distribution is its local accumulation at sites of wound healing, inflammation, and tumor growth. Understanding the relationship between circulating and tissue-bound fibronectin will be crucial to elucidating the mechanism of fibronectin accumulation at these sites (Figure 4b).

D. VARIABILITY IN PRIMARY STRUCTURE

1. Subunit Variation

Initial characterization of plasma fibronectin revealed the presence of subunit heterogeneity. Chen, Amrani, and Mosesson described three fibronectin species corresponding to zones of migration in polyacrylamide gels.[65] Zone I material had a molecular weight of 450,000 Da and consisted of dimeric fibronectin. Zone II was composed of a series of monomeric subunits with molecular weights ranging from 200 to 235,000 Da. Zone III contained species of somewhat lower molecular weight (146 to 190,000). Subsequent SDS-PAGE studies in slab gels performed under reducing conditions showed that plasma fibronectin usually migrated as a closely spaced doublet. The higher (220 to 230 kDa) and lower (210 to 215 kDa) molecular weight chains have been designated A and B (or α and β), respectively. Isolated A and B chains were found to yield nearly identical peptide maps following extensive proteolytic digestion; however, Kurkinen et al. noted that A and B chains exhibited different kinetics with respect to plasmin digestion.[14] Subsequent use of limited proteolytic digestion with trypsin, thermolysin, and cathepsin has revealed significant subunit variations between the A and B chains. A major site of domain size heterogeneity has been localized to the heparin II-fibrin II region (Figure 2, d and e, and Section II.C).[13,15,17,66-69]

Paul and Hynes compared the subunits of plasma and cellular fibronectins using two-dimensional gel electrophoresis (IEF-PAGE) and studied the role of glycosylation, phosphorylation, and sulfation in the production of heterogeneity.[70] They concluded that observed subunit variation could not be entirely explained by posttranslational modification and predicted that primary sequence differences were in part responsible. Indeed, protein and cDNA sequencing studies have revealed the presence of the IIICS splice variant (Figure 3) which is present on the A chain but deleted from the B chain.[30] Differential splicing of the IIICS region probably accounts for the size differences between the A and B chains and their heparin II - fibrin II domains. The functional significance of these variable regions with respect to cell adhesion is discussed in Chapter 11.

2. Primary Structural Variation and Malignant Transformation

Structural variation in proximity to the cell-binding region might explain reported changes in fibronectin distribution in malignant tissue and in cultured, transformed fibroblasts (see Chapter 11).

Sekiguchi et al. compared limited proteolytic digestion patterns of fibronectins isolated from normal WI-38 fibroblasts and their SV40 transformants.[71] Constituent domains were identified using domain-specific antibodies. Fragments derived from the carboxyterminal heparin (Hep 2) domains demonstrated size heterogeneity as did fragments representing the gelatin-binding domain and the large (165 to 210 kDa) tryptic central domain. In this study, domains released from the transformed cells were generally larger, suggesting that they might contain additional sequences. Borsi et al. demonstrated the presence of a 43-kDa cathepsin D fragment in fibronectins isolated from three different malignant cell lines.[72] This 43-kDa fragment was not formed from normal fibroblast fibronectin or from plasma fibronectin, and it reacted with a monoclonal antibody specific for a site near the Hep 2 domain. These data suggest that alternative mRNA splicing is likely to be responsible for the production of this fragment in malignant cell lines. Castellani et al.[73] and Borsi et al.[74] have subsequently shown that fibronectins from tumor-derived cell lines and SV40-transformed cells contain a higher percentage of IIICS and ED-bearing subunits than do fibronectins from normal plasma or fibroblasts. These studies demonstrate that normal- and tumor-derived fibronectins display significant structural differences at the cell-binding region. These differences have been shown to be in part due to alterative RNA splicing, although differences in glycosylation and other posttranslational modifications may play an important role.

IV. POSTTRANSLATIONAL MODIFICATIONS

A. PROTEOLYTIC FRAGMENTS

As has been demonstrated elsewhere in this chapter, fibronectin is sensitive to proteases that cleave the protein at flexible regions connecting protease-resistant domains. The potential existence of fibronectin fragments *in vivo* is of extreme importance since fragments can theoretically serve as self-modulators of fibronectin function.[75] Fragments would be predicted to be of particular importance in normal and pathological situations where proteases play active roles in tissue remodeling. These include (1) interference with opsonic function in overwhelming infection, especially where the equilibrium between proteases and protease inhibitors might be in relative imbalance; (2) modulation of fibroblast chemotactic activity and matrix formation in wound healing; (3) mediation of mononuclear cell chemotaxis and the modulation of immune complex clearance in inflammation; (4) modulation of tumor cell adherence and invasion in metastasis; and (5) control of tissue migration during embryogenesis.

The identification of fibronectin fragments *in vivo* is also of practical importance in the interpretation of fibronectin measurement in various clinical situations.

1. Effect of Fragmentation on the Measurement of Fibronectin

Standard quantitative measurements of fibronectin are based on immunochemical assays. Other assays have been developed based upon functional activity, usually gelatin-binding;[76-77] however these have not been widely employed. Extensive proteolytic digestion of fibronectin results in falsely elevated values when electrophoretic and solid-phase-binding immunoassays are employed.[76-78] This may be due to alterations in electrophoretic mobility of fragments or exposure of cryptic immunogenic sites by proteolytic cleavage. On the other hand, fragmentation generally results in a reduction of bioassayable fibronectin.[76]

Most *in vitro* studies have examined extensively degraded samples; however more limited degradation appears to occur in biological fluids (see below, Section IV.A.3). Thus, the effect of *in vivo* fragmentation on fibronectin measurement is less pronounced. To examine this issue, we utilized electroimmunoassay to determine fibronectin levels of a highly inflammatory, septic synovial effusion after storage under the following conditions: (1) immediate removal of leukocytes, addition of protease inhibitors, and prompt freezing to $-30°C$; (2) removal of leukocytes followed by storage at 23°C overnight in the absence of protease inhibitors; and (3) storage at 37°C overnight in the absence of protease inhibitors following a freeze-thaw cycle to disrupt neutrophils. The results of this experiment are summarized in Table 2. Synovial fluid processed and stored under conditions purposely designed to yield maximum fragmentation of fibronectin did not result in significant alterations in measured fibronectin. It is likely that the presence of other plasma proteins including protease inhibitors modulates proteolysis in biological fluids.

2. *In Vitro* Effects of Fibronectin Fragments

Proteolytic cleavage of the fibronectin molecule at domain-specific sites results in predicted functional alterations (i.e., loss of the ability to bind cells to collagen). In some instances, proteolytic fragmentation of fibronectin results in the expression of functions not ordinarily associated with the intact protein. Table 3 lists functions specifically generated by fibronectin fragments.

3. Assessment of the Role of Fibronectin Fragments *In Vivo*

It is apparent that fragments may lack functions associated with intact fibronectin or may acquire *de novo* functional activites. Because fragments may dramatically alter the effect of fibronectin on cell behavior, their identification and characterization is of extreme biological importance.

TABLE 2
Effect of Sample Processing on Fibronectin Measurement[a]

Synovial fluid sample	Measured fibronectin[b] (μg/ml)
Centrifuged and frozen, protease inhibitors	328
Centrifuged, room temp 16 h, no inhibitors	291
Not centrifuged, 37°C, 16 h, no inhibitors	304

[a] Carsons, S. E. and Ciangarra, C., unpublished results.
[b] Measured by electroimmunoassay.

TABLE 3
Functions of Fibronectin Fragments[a]

Activity	Active Fragment	Ref.
Nodule formation in smooth muscle cells	180 — 200 kDa (carboxyterminal)	79
Transformation enhancement factor	Gelatin binding	80
Mitogenic activity	Heparin binding (35K)	81
Endothelial cell growth inhibition	Heparin binding (29K)	82
Enhanced neutrophil adhesion	n.c.[b]	83
Augmentation of monocyte phagocytes	180 kDa Gel binding	84
Macrophage agglutination factor	50 — 60 kDa Gel binding	85
Monocyte chemotaxis	60 — 200 kDa	86

[a] Activity not associated with intact fibronectin.
[b] Not characterized.

There are several difficulties inherent in the study of fragments. Isolation of the protein may result in artifactual degradation during purification. Storage of fibronectin preparations, especially with subsequent freezing and thawing, results in artifactual degradation perhaps via activation of a cryptic proteolytic site recently found to be present on fibronectin itself.[87]

Several approaches have been adopted by investigators in attempts to minimize artifactual proteolysis. Storage and isolation buffers are routinely supplemented with multiple protease inhibitors. Freeze-thaw cycles are minimized by immediate allocation of the original sample into separate aliquots. Protein in biological fluid samples can be examined for size heterogeneity by direct separation on polyacrylamide gels followed by immunofixation in order to avoid degradation which might occur secondary to purification procedures. Even under the most carefully controlled conditions, it is difficult to conclude that a fragment is actually present *in vivo*.

Fibronectin fragments are potentially of great significance in the pathogenesis of inflammation. Therefore, we utilized the techniques described above to characterize fibronectin molecular forms present in inflammatory rheumatoid synovial fluid.[88] We found that approximately 75% of immunoreactive synovial fluid fibronectin could be recovered as large 180 to 200-kDa gelatin-binding fragments. Another 10% of immunoreactive fibronectin was recovered in the gelatin-Sepharose 1*M* NaCl wash fraction as loosely bound 50 to 100-kDa fragments. The functional properties of these two classes of fragments are listed in Table 4. These functional data (Table 4) were combined with the known domain organization of fibronectin (Figure 2), in order to construct structural models of inflammatory synovial fluid fibronectin fragments. The 180- to 200-kDa species displayed diminished binding to fibrin and collagen, but normal binding to heparin. This class of fragment also demonstrated intact fibroblast chemotactic function. These data suggest that the 180- to 200-kDa species lack domain a (Figure 2) which contains the high-affinity fibrin-binding site and a low-affinity

TABLE 4
Functional Properties of Synovial Fluid Fibronectin Fragments[88]

	Molecular Weight	
	180 — 200 kDa	50 — 100 kDa
Immunoreactive Fn	75%	10%
Collagen-binding affinity[a]	33%[b]	N.D.
Fibrin-binding affinity[c]	17%[b]	N.D.
Heparin-binding affinity	100%	N.D.
Monocyte attachment to collagen[d]	100%	Inhibits
Monocyte attachment of fibrin[d]	67%	Inhibits
Synovial cell chemotaxis[e]	100%	100%

[a] Measured by solid phase gelatin-binding assay.
[b] Expressed relative to intact plasma Fn which equals 100%.
[c] Measured by solid phase fibrin-binding assay.
[d] Measured by microwell attachment assay.
[e] Measured by modified Boyden Chamber assay.

heparin-binding site. Proteolytic attack at site 1 (Figure 2) immediately N-terminal to the collagen-binding domain may have been responsible for the reduced gelatin-binding affinity observed. Since the gelatin and cell-binding domains (b and c, Figure 2) of the 180- to 200-kDa fragments were still linked in the usual linear arrangement, these fragments retained the ability to mediate monocyte attachment to gelatin. The loss of the aminoterminal fibrin-binding domain (a, Figure 2), however, abrogated the ability of these fragments to promote monocyte attachment to fibrin. The functional properties of these fragments are consistent with the existence of chronic mononuclear cell infiltrates within the collagenous extracellular matrix of pannus and with the chronic deposition of fibrin in rheumatoid synovium, the latter perhaps due to the inability of these fragments to mediate clearance of fibrin (see Chapter 16).

Although the 50- to 100-kDa fragments represented only 10% of immunoreactive fibronectin in inflammatory synovial fluid, they may have special functional significance. These fragments were not only unable to mediate cell attachment to gelatin or fibrin but competitively inhibited fibronectin-dependent cell attachment to these substrates. On the other hand, these 50- to 100-kDa fragments were entirely capable of mediating synoviocyte chemotaxis. Since these small chemotactic fragments are likely to be soluble and diffuse more readily than intact fibronectin, they may be well suited to the establishment of chemotactic gradients at inflammatory sites.

In summary, our studies of fibronectin in inflammatory synovial fluid demonstrate that: (1) fibronectin degradation at sites of inflammation is not necessarily extensive and may be productive of a restricted number of limited degradation fragments. (2) Degradation may be controlled by the local concentration of protease inhibitors and by molecular modification of fibronectin, i.e., glycosylation or binding to other macromolecules. (3) Fragments may lack functional properties of the intact protein, inhibit the function of intact fibronectin, or retain specific functional properties. It appears that a carefully controlled equilibrium among fibronectin molecular forms serves to modulate cellular function *in vivo*.

B. GLYCOSYLATION

Fibronectins are glycoproteins that contain 4 to 9% sugar by weight.[9,89,90] Most of the carbohydrate is in the form of complex oligosaccharide chains in N-glycosidic linkage to asparagine.[91-95] Recently, the presence of O-glycosidically linked carbohydrate on fibronectin has been described.[95] Despite these generalities, the study of fibronectins from different

FIGURE 5. N-linked oligosaccharides of fibronectins. The diagram represents a typical biantennary oligosaccharide. X indicates the position of the terminal sialic acid. In plasma fibronectin, X_1 and X_2 would be sialic acid residues in α 2-6 linkage. In cellular fibronectins, one chain (X_1 or X_2) would usually not be sialyated and the sialyated unit would be in α 2-3 linkage. In embryonal fibronectin, X would represent a polylactosamine unit of the form (Gal [β 1-4] Glc NAc [β 1-3])$_n$. Y represents the location of fucose which is in α 1-6 linkage to the proximal GlcNAc residues of cellular fibronectin.[91]

sources has demonstrated the presence of multiple variations in carbohydrate structure. These variations appear useful as markers for developmental and tissue-specific fibronectin isoforms, and they may modulate critical molecular functions.

1. The N-Linked Carbohydrate of Plasma and Cellular Fibronectin

The N-linked carbohydrate of plasma and cellular fibronectin is a biantennary unit of the general form displayed in Figure 5.[91-95] Consistent differences in the sialyation pattern of plasma and cellular fibronectins have been documented. Cellular fibronectins contain less sialic acid which is linked α 2-3 to galactose in contrast to the α 2-6 linkage found in plasma fibronectin.[94] Cellular fibronectin also contains fucose linked to the proximal *N*-acetylglucosamine units. Fucose is not detected in plasma fibronectin.[89,94,95]

2. Embryonal (Fetal) Fibronectin

Fibronectin can be readily isolated from amniotic fluid and placental tissue by gelatin-Sepharose chromatography and urea extraction, respectively.[90-91] The carbohydrate composition of this fibronectin is markedly different from that of plasma fibronectin.[9,90,95,97,98] Because amniotic fluid fibronectin is locally produced, it was initially referred to as a cellular fibronectin. This has resulted in some confusion in terminology; however, it is now clear that the glycosylation of amniotic fluid (embryonal) fibronectin differs significantly from that of cellular fibronectin produced by cultured fibroblasts.

Embryonal fibronectin contains approximately twice as much carbohydrate as does plasma fibronectin.[9,90,95,97,98] Glycopeptides produced from embryonal fibronectin are larger than those produced from plasma fibronectin indicating that at least part of the increased carbohydrate is present in larger oligosaccharide units.[95,97] These larger units have been found to be composed of tri- and tetraantennary N-linked chains. A small proportion of N-linked chains also contain covalently linked polylactosamine units of the form (Gal~[β1-4] Glc NAc [β1-3]). Similar polylactosamine structures have also been identified on other embryonal glycoproteins. The oligosaccharide chains isolated from embryonal fibronectins bear some similarity to the N-linked chains found on cellular fibronectin in that they are partially desialyated and contain fucose.

Hakomori and colleagues have developed a series of monoclonal antibodies that react with specific glycan structures on glycoproteins and glycolipids. One of these antibodies, FDC-6, recognizes fibronectins from amniotic fluid and fetal tissue but not from adult plasma and tissue.[99] The FDC-6 epitope is a unique structure requiring the presence of a hexapeptide Val-Thr-His-Pro-Gly-Tyr and an O-linked *N*-acetylgalactosamine at Thr.[100] This O-Linked structure has been localized to a thermolysin-sensitive site in the carboxyterminal region of fibronectin (Figure 2, site 3). In addition, other O-linked oligosaccharide units have been identified on amniotic fluid fibronectin.[95]

3. Alterations in Glycosylation Secondary to Malignant Transformation: Relationship to Embryonal Fibronectin

Distinct changes in fibronectin glycosylation have been shown to occur upon viral transformation of cultured cells. Transformation of NIL 8 cells with HSV and BHK21 C13 cells with HSV results in the synthesis of fibronectins with a greater proportion of tri- and tetraantennary chains and a higher degree of sialyation.[101,102] Using a monoclonal antibody (C6), Nichols et al. demonstrated the presence of a binary sialosyl chain on WI38 cellular and transformed fibronectins but not on plasma fibronectin.[96] This glycan was translocated from the collagen-binding domain to the aminoterminal of the central domain upon viral transformation. Thus, it appears that increased branching of N-linked chains accompanies malignant transformation and suggests a relationship between transformed fibronectin and embryonal fibronectin. Additional evidence for this relationship comes from (1) the finding of shared FDC-6 reactivity among fibronectins from malignant hepatoma lines, tumor tissue extracts, and amniotic fluid;[99] (2) the identification of lactosaminoglycan units on fibronectin from teratocarcinoma cell lines;[103] and (3) biochemical similarity between amniotic fluid fibronectin and fibronectin purified from germ cell tumors.[104]

Hakomori and colleagues have proposed an ontogenic classification of fibronectins based on glycosylation profiles.[99,105] Normal fibronectins (Nor Fn) resemble plasma fibronectin and are also found in adult tissues. Oncofetal fibronectins (Onf Fn) represent a gradual series of alterations in glycosylation beginning with the appearance of fucosylation and moderate branching seen in cellular fibronectin and progressing to extensive chain branching, lactosaminoglycosylation, and O-glycosylation seen in embryonal and malignant fibronectins.

4. Glycosylation of Inflammatory Fibronectin

Inflammatory processes generally involve some degree of nonmalignant proliferation of tissue. Fibronectin is actively produced at sites of inflammation (see Chapter 16). The presence of appreciable quantities of fibronectin in inflammatory effusions has permitted the isolation and biochemical analysis of this form of the protein. The most extensively characterized inflammatory fibronectin has been that which has been isolated from the synovial fluid of patients with inflammatory arthritis. The glycosylation of synovial fibronectin differs from that of plasma fibronectin in several respects.[106] Synovial fibronectin has a greater carbohydrate content (6.9% vs. 5.5%) but contains less sialic acid (0.7% vs. 1.2%) than does plasma fibronectin. These forms of the protein also display heterogeneity with respect to domain-specific oligosaccharide distribution (Figure 6). Synovial fibronectin demonstrates enhanced sialyation of an N-linked oligosaccharide located on a 27-kDa subfragment of the B chain. In addition, synovial fibronectin contains PNA reactive desialyated O-linked glycans on the collagen-binding domain. This is consistent with the finding of biochemically detectable *N*-acetylgalactosamine in synovial fibronectin. Recently, the presence of the FDC-6 reactive epitope (Section IV.B.3) has been localized to synovial fibronectin. Analysis of synovial fibronectin glycopeptides, however, has not revealed the presence of the large N-linked tetraantennary and lactosaminoglycan chains found on embryonal fibronectin. Thus, inflammatory synovial fibronectins differ from the plasma form and share certain but not all characteristics of the embryonal (oncofetal) form. It is not yet known if the specific glycosylation pattern of inflammatory synovial fibronectin is characteristic of fibronectins isolated from other inflammatory sites; however, it appears to represent a distinct fibronectin glycosylation variant associated with the inflammatory process.

5. Functional Consequences of Fibronectin Glycosylation Heterogeneity

Given the organization of functional domains on the fibronectin molecule, modification of glycan structure at critical locations is likely to result in functional alterations. Several oligosaccharide-dependent functional alterations have been described.

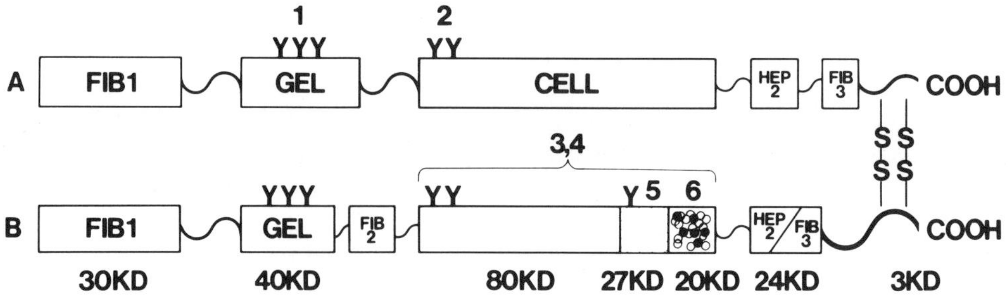

FIGURE 6. Model of fibronectin demonstrating major structural features and glycosylation heterogeneity between plasma and synovial fibronectins. FIB, GEL, DNA, CELL, HEP indicate major domains. The approximate domain size is shown under each domain in kilodaltons (KD). Y = N-linked oligosaccharide. (1) Approximate location of conformation-dependent inhibition of WGA reactivity in plasma fibronectin; plasma fibronectin gel-binding domains acquire WGA positivity following chymotryptic cleavage. (2) Approximate location of conformation-dependent PNA reactive site on synovial fluid fibronectin. This chymotrypsin-sensitive region (residue 601-602)[36] is located in a 22-amino acid sequence containing 6 serine residues as potential O-glycosylation sites. (3 and 4) 140-kDa nonheparin-binding chymotryptic fragment localized to fibronectin "B" chain. (5) 27-kDa WGA reactive synovial fibronectin fragment. A sialyated N-linked oligosaccharide is potentially located at residue 847, 977, or 1214.[37] (6) Nonglycosylated 20-kDa domain containing the RGD sequence. (Reproduced from Carsons, S., Lavietes, B. B., Slomiany, A., Diamond, H. S., and Berkowitz, E., *J. Clin. Invest.*, 80, 1342, 1987. With copyright permission of Rockefeller University Press.)

An important function of the carbohydrate moiety of a glycoprotein is protection of the protein from proteolytic attack.[25] Bernard, Yamada, and Olden demonstrated that the heavily glycosylated collagen-binding domain of chick embryo fibroblast fibronectin was selectively resistant to a broad variety of proteases.[25] When N-linked glycosylation was inhibited by tunicamycin treatment, the collagen-binding domain became increasingly susceptible to proteolysis. While deglycosylated fibronectin is not known to occur *in vivo*, naturally occurring fibronectin forms also demonstrate differential protease susceptibility on the basis of glycosylation heterogeneity. Zhu et al. showed that the gelatin-binding domain from placental (embryonal) fibronectin was more resistant to proteolysis than the corresponding domain from plasma fibronectin.[97] It has been postulated that the increased carbohydrate content and oligosaccharide chain length of embryonal and inflammatory fibronectins allow them to function in areas of active proteolysis and tissue remodeling.[105,106]

Recent studies have shown that glycosylation can also modulate the interaction of fibronectin with collagen and with the cell membrane. Jones et al. compared glycosylated and nonglycosylated human skin fibroblast fibronectins and found that the deglycosylated form had an increased affinity for gelatin and enhanced adhesion-promoting properties.[107] Zhu and Laine were able to separate a polylactosaminyl-collagen-binding fragment from those fragments containing only biantennary, complex N-linked units. The polylactosaminyl unit displayed decreased gelatin-binding affinity.[108]

C. OTHER POSTTRANSLATIONAL MODIFICATIONS

Phosphorylation and tyrosine sulfation of fibronectins have been well documented.[20,109-112] The effect of these modifications on fibronectin function has not been described.

V. FUTURE DIRECTIONS

In this chapter, I have attempted to provide an overview of the structural basis for fibronectin function. Chapters that follow will describe in detail the role of fibronectin structure-function relationships in normal cell and tissue physiology and in pathological states. Fibronectin molecular heterogeneity will be addressed in relation to several areas of investigation. It is becoming apparent that the determination of fibronectin structure is complex, involving multiple transcriptional and posttranslational events. Future studies should address whether specific fibronectin molecular forms are capable of identifying a particular developmental stage, tissue form, or pathologic process. In addition it will be important to determine whether those molecular alterations impart specific functional attributes to the protein, such as cell or tissue specific adherence.

The recent identification of a discrete set of cell membrane glycoproteins as the integrin fibronectin receptor family[113] will further facilitate our understanding of fibronectin-cell surface interactions and their consequences. The integrins appear to be products of a cell adhesion receptor gene superfamily. Each member of this group consists of an α-β heterodimer with subunit molecular weights of approximately 150 kDa and 100 kDa. Integrins mediate cell attachment to fibronectin, vitronectin, laminin, and fibrinogen. Within this superfamily, subgroups exist whose members share a common β chain but have nonhomologous α chains. It is felt that the matrix-protein specificity displayed by each protein is conferred by the α chain. Studies examining the cell and tissue distribution of integrins and their relation to fibronectin expression in development and disease will add an important dimension to our understanding of extracellular matrix function.

ACKNOWLEDGMENTS

I would like to thank Drs. Deane Mosher, Mike Mosesson, and Ken Yamada for many helpful discussions. I also would like to thank Dr. Herb Diamond for support and encouragement. Original work cited in the chapter has been supported by the NIH 3MO1 RR00318-14AISI and AM 20625-06, the Kroc Foundation, the SLE Foundation, and the Arthritis Foundation, New York chapter. Thanks to Ann Marchesiello for preparation of the manuscript and Dennis Marchesiello for preparation of the illustrations.

REFERENCES

1. **Yamada, K. M.,** Cell surface interactions with extracellular materials, *Annu. Rev. Biochem.,* 52, 761, 1983.
2. **Yamada, K. M., Akiyama, S. K., Hasagawa, T., Hasagawa, E., Humphries, M. J., Kennedy, D. W., Nagata, K., Urushihara, H., Olden, K., and Chen, W-T.,** Recent advances in research on fibronectin and other cell attachment proteins, *J. Cell. Biochem.,* 28, 79, 1985.
3. **Hewitt, A. T., Varner, H. H., Silver, M. H., and Martin, G. R.,** The role of chondronectin and cartilage proteoglycan in the attachment of chondrocytes to collagen, *Limb Dev. Regenerat.,* Part B, 23, 1983.
4. **Ruoslahti, E. and Pierschbacher, M. D.,** New perspectives in cell adhesion: RGD and integrins, *Science,* 238, 491, 1987.
5. **Pareti, F. I., Fujimura, Y., Dent, J. A., Holland, L. Z., Zimmerman, T. S., and Ruggeri, Z. M.,** Isolation and characterization of a collagen binding domain in human von Willebrand factor, *J. Biol. Chem.,* 261, 15310, 1986.
6. **Mosher, D. F. and Johnson, R. B.,** In Vitro formation of disulfide-bonded fibronectin multimers, *J. Biol. Chem.,* 258, 6595, 1983.
7. **Mosesson, M. W., Chen, A. B., and Huseby, R. M.,** The cold-insoluble globuline of human plasma: studies of its essential structural features, *Biochim. Biophys. Acta,* 386, 509, 1975.
8. **McConnell, M. R., Blumberg, P. M., and Rossow, P. W.,** Dimeric and high molecular weight forms of the large external transformation-sensitive protein on the surface of chick embryo fibroblasts, *J. Biol. Chem.,* 253, 7522, 1978.
9. **Balian, G. B., Crouch, E., Click, E. M., Carter, W. G., and Bornstein, P.,** Comparison of the structures of human fibronectin and plasma cold-insoluble globulin, *J. Supramol. Struct.,* 12, 505, 1979.
10. **Hahn, L-H. E. and Yamada, K. M.,** Isolation and biological characterization of active fragments of the adhesive glycoprotein fibronectin, *Cell,* 18, 1043, 1979.

11. **Ruoslahti, E., Hayman, E. G., Engvall, E., Cothran, W. C., and Butler, W. T.,** Alignment of biologically active domains in the fibronectin molecule, *J. Biol. Chem.*, 256, 7277, 1981.
12. **Balian, G., Click, E. M., and Bornstein, P.,** Location of a collagen-binding domain in fibronectin, *J. Biol. Chem.*, 255, 3234, 1980.
13. **Sekiguchi, K. and Hakomori, S.,** Domain structure of human plasma fibronectin differences and similarities between human and hamster fibronectins, *J. Biol. Chem.*, 258, 3967, 1983.
14. **Kurkinen, M., Vartio, T., and Vaheri, A.,** Polypeptides of human plasma fibronectin are similar but not identical, *Biochim. Biophys. Acta*, 624, 490, 1980.
15. **Sekiguchi, K., Fukuda, M., and Hakomori, S.,** Domain structure of hamster plasma fibronectin isolation and characterization of four functionally distinct domains and their unequal distribution between to subunit polypeptides, *J. Biol. Chem.*, 256, 6452, 1981.
16. **Seidl, M. and Hormann, H.,** Affinity chromatography on immobilized fibrin monomer. IV. Two fibrin-binding peptides of a chymotryptic digest of human plasma fibronectin, *Hoppe-Seyler's Z. Physiol. Chem.*, 364, S83, 1983.
17. **Hayashi, M. and Yamada, K.,** Domain structure of the carboxyl-terminal half of human plasma fibronectin, *J. Biol. Chem.*, 258, 332, 1983.
18. **Richter, H., Seidl, M., and Hormann, H.,** Location of Heparin-binding sites of fibronectin. Detection of a hitherto unrecognized transamidase sensitive site, *Hoppe-Seyler's Z. Physiol. Chem.*, 362, 399, 1981.
19. **Mosher, D. F. and Johnson, R. B.,** Specificity of fibronectin-fibrin cross-linking, *Ann. N.Y. Acad. Sci.*, 408, 583, 1983.
20. **Mosher, D. F.,** Cross-linking of fibronectin to collagenous proteins, *Mol. Cell. Biochem.*, 58, 63, 1984.
21. **Proctor, R. A., Mosher, D. F., and Olbrantz, P. J.,** Fibronectin binding to *Staphylococcus aureus, J. Biol. Chem.*, 257, 14788, 1982.
22. **Carter, W. G.,** The role of intermolecular disulfide bonding in deposition of GP140 in the extracellular matrix, *J. Cell Biol.*, 99, 105, 1984.
23. **Engvall, E., Ruoslahti, E., and Miller, E. J.,** Affinity of fibronectin to collagens of different genetic types and to fibrinogen, *J. Exp. Med.*, 147, 1548, 1978.
24. **Kleinman, H. K., Wilkes, C. M., and Martin, G. R.,** Interaction of fibronectin with collagen fibrils, *Biochemistry*, 20, 2325, 1981.
25. **Bernard, B. A., Yamada, K. M., and Olden, K.,** Carbohydrates selectively protect a specific domain of fibronectin against proteases, *J. Biol. Chem.*, 257, 8549, 1982.
25a. **Sorvillo, J., Gigli, I., and Pearlstein, E.,** Fibronectin binding to complement subcomponent C1q localization of their respective binding sites, *Biochem. J.*, 226, 207, 1985.
26. **Pierschbacher, M. D., Hayman, E. G., and Ruoslahti, E.,** Location of the cell-attachment site in fibronectin with monoclonal antibodies and proteolytic fragments of the molecule, *Cell*, 26, 259, 1981.
27. **Pierschbacher, M. D., Ruoslahti, E., Sundelin, J., Lind, P., and Peterson, P. A.,** The cell attachment domain of fibronectin determination of the primary structure, *J. Biol. Chem.*, 257, 9593, 1982.
28. **Pierschbacher, M. D. and Ruoslahti, E.,** Cell attachment activity of fibronectin can be duplicated by small synthetic fragments of the molecule, *Nature (London)*, 309, 30, 1984.
29. **Skorstengaard, K., Jensen, M. S., Sahl, P., Petersen, T. E., and Magnusson, S.,** Complete primary structure of bovine plasma fibronectin, *Eur. J. Biochem.*, 161, 441, 1986.
30. **Hynes, R.,** Molecular biology of fibronectin, *Annu. Rev. Cell Biol.*, 1, 67, 1985.
31. **Petersen, T. E., Thogersen, H. C., Skorstengaard, K., Vibe-Pedersen, K., Sahl, P., Sottrup-Jensen, L., and Magnusson, S.,** Partial primary structure of bovine plasma fibronectin: three types of internal homology, *Proc. Natl. Acad. Sci. U.S.A.*, 80, 137, 1983.
32. **Kornblihtt, A. R., Umezawa, K., Vibe-Pedersen, K., and Baralle, F.,** Primary structure of human fibronectin: differential splicing may generate at least 10 polypeptides from a single gene, *EMBO J.*, 4, 1755, 1985.
33. **Kornblihtt, A. R., Vibe-Pedersen, K., and Baralle, F. E.,** *EMBO J.*, 3, 221, 1984.
33a. **Balza, E., Borsi, L., Allemanni, G., and Zardi, L.,** Transforming growth factor β regulates the levels of different fibronectin isoforms in normal human cultured fibroblasts, FEBS, 228 (1), 42, 1988.
34. **Schwarzbauer, J. E., Tamkun, J. W., Lemischka, I. R., and Hynes, R. O.,** Three different fibronect mRNAs arise by alternative splicing within the coding region, *Cell*, 35, 421, 1983.
35. **Humphries, M. J., Komoriya, A., Akiyama, S. K., Olden, K., and Yamada, K.,** Identification of two distinct regions of the type III connecting segment of human plasma fibronectin that promote cell type-specific adhesion, *J. Biol. Chem.*, 262, 6886, 1987.
36. **Banyai, L., Varadi, A., and Patthy, L.,** Common evolutionary origin of the fibrin-binding structures of fibronectin and tissue-type plasminogen activator, *FEBS Lett.*, 163, 1983.
37. **Ruoslahti, E. and Vaheri, A.,** Novel human serum protein from fibroblast plasma membrane, *Nature (London)*, 248, 789, 1974.
38. **Ruoslahti, E. and Vaheri, A.,** Interaction of soluble fibroblast surface antigen with fibrinogen and fibrin. Identity with cold insoluble globulin of human plasma, *J. Exp. Med.*, 141, 497, 1975.

39. **Edsall, J. T.,** Some early history of cold-insoluble globulin, *Ann. N.Y. Acad. Sci.,* 312, 1, 1978.
40. **Mosesson, M. W.,** Structure of human plasma cold-insoluble globulin and the mechanism of its precipitation in the cold with heparin or fibrin-fibrinogen complexes, *Ann. N.Y. Acad. Sci.,* 312, 11, 1978.
41. **Mosher, D. F., Proctor, R. A., and Grossman, J. E.,** Fibronectin: role in inflammation, in *Advances in Inflammation Research,* Vol. 2, Weissmann, G., Ed., Raven, New York, 1981, 187.
42. **Labat-Robert, J., Potazman, J. P., Derouette, J. C., and Robert, L.,** Age-dependent increase of human plasma fibronectin, *Cell Biol. Int. Rep.,* 5, 969, 1981.
43. **van Hejden, W. C. H., Kok-Verspuy, A., Harff, G. A., and van Kamp, G. J.,** Rate-nephelometric determination of fibronectin in plasma, *Clin. Chem.,* 31, 1182, 1985.
44. **Gluud, C., Dejgaard, A., and Clemmensen, I.,** Plasma fibronectin concentrations in patients with liver diseases, *Scand. J. Clin. Lab. Invest.,* 43, 533, 1983.
45. **Stathaskis, N. E., Mosesson, M. W., Chen, A. B., and Galanakis, D, K,** Cryoprecipitation of fibrin-fibrinogen complexes induced by the cold-insoluble globulin of plasma, *Blood,* 51, 1211, 1978.
46. **Beaulieu, A. D., Valet, J. P., and Strevey, J.,** The influence of fibronectin on cryoprecipitate formation in rheumatoid arthritis and systemic lupus erythematosus, *Arthritis Rheum.,* 24, 1383, 1981.
47. **Anderson, B., Rucker, M., Entwistle, R., Schmid, F. R., and Wood, G. W.,** Plasma fibronectin is a component of cryoglobulins from patients with connective tissue and other diseases, *Ann. Rheum. Dis.,* 40, 40, 1981.
48. **Stathakis, N. E. and Mosesson, M. W.,** Interactions among heparin, cold-insoluble globulin, and fibrinogen in formation of the heparin-precipitable fraction of plasma, *J. Clin. Invest.,* 60, 855, 1977.
49. **Carsons, S. E., Lavietes, B. B., and Diamond, H. S.,** The role of fibronectin in rheumatic disease, in *Fibronectin,* Mosher, D. F., Ed., Academic Press, New York, 1989, 327.
50. **Saba, T. M., Blumenstock, F. A., Weber, P., and Kaplan, J. E.,** Physiologic role of cold-insoluble globulin in systemic host defense: implications of its characterization as the opsonic a_2-surface-binding glycoprotein, *Ann. N.Y. Acad. Sci.,* 312, 43, 1978.
51. **Saba, T. M., and Jaffe, E.,** Plasma fibronectin (opsonic glycoprotein): its synthesis by vascular endothelial cells and role in cardiopulmonary integrity after trauma as related to reticuloendothelial function, *Am. J. Med.,* 68, 577, 1980.
52. **Atherton, B. T. and Hynes, R. O.,** A difference between plasma and celullar fibronectins located with monoclonal antibodies, *Cell,* 25, 133, 1981.
53. **Peters, J. H., Ginsburg, M. H., Bohl, B. P., Sklar, L. A., and Cochrane, C. G.,** Intravascular release of intact cellular fibronectin during oxidant-induced injury of the in vitro perfused rabbit lung, *J. Clin. Invest.,* 78, 1596, 1986.
54. **Yamada, K. M. and Olden, K.,** Fibronectins — adhesive glycoproteins of cell surface and blood, *Nature (London),* 275, 179, 1978.
55. **Laterra, J. and Culp, L. A.,** Differences in hyaluronate binding to plasma and cell surface fibronectins, *J. Biol. Chem.,* 257, 719, 1982.
56. **Murray, B. A., Ansbacher, R., and Culp, L. A.,** Adhesion sites of murine fibroblasts on cold insoluble globuline-adsorbed substrata, *J. Cell Physiol.,* 104, 335, 1980.
57. **Oh, E., Pierschbacher, M., and Ruoslahti, E.,** Deposition of plasma fibronectin in tissues, *Proc. Natl. Acad. Sci. U.S.A.,* 78, 3218, 1981.
58. **Meckeown-Longo, P. J. and Mosher, D. F.,** Binding of plasma to cell layers of human skin fibroblasts, *J. Cell Biol.,* 97, 466, 1983.
59. **Vuento, M., Vartio, T., Saraste, M., Von Bonsdorff, C-H., and Vaheri, A.,** Spontaneous and polyamine-induced formation of filamentous polymers from soluble fibronectin, *Eur. J. Biochem.,* 105, 33.
60. **Ehrismann, R., Roth, D. E., Eppenberger, H. M., and Turner, D. C.,** Arrangement of attachment-promoting, self associaton, and heparin-binding sites in horse serum fibronectin, *J. Biol. Chem.,* 257, 7381, 1982.
61. **McKeown-Longo, P. J. and Mosher, D. F.,** Mechanism of formation of disulfide-bonded multimers of plasma fibronectin in cell layers of cultured human fibroblasts, *J. Biol. Chem.,* 259, 12210, 1984.
62. **McKeown-Longo, P. J. and Mosher, D. F.,** Interaction of the 70,000-mol-wt amino-terminal fragment of fibronectin with the matrix-assembly receptor of fibroblasts, *J. Cell Biol.,* 100, 364, 1985.
63. **Pesciotta, D. M., Peters, P., and Mosher, D. F.,** Localization of cell surface sites involved in fibronectin fibrillogenesis, *J. Cell Biol.,* 104, 121, 1987.
64. **Mosher, D. F. and McKeown-Longo, P. J.,** Assembly of fibronectin-containing extracellular matrix: a glimpse of the machinery, *Biopolymers,* 24, 199, 1985.
65. **Chen, A. B., Amrani, D. L., and Mosesson, M. W.,** Heterogeneity of the cold-insoluble globulin of human plasma (C1q), a circulating cell surface protein, *Biochim. Biophys. Acta,* 493, 310, 1977.
66. **Click, E. M. and Balian, G.,** Domain structure of human plasma and cellular fibronectin. Use of a monoclonal antibody affinity to identify three different subunit chains, *Biochemistry,* 24, 6685, 1985.
67. **Sekiguichi, K. and Hakomori, S.,** Identification of two fibrin-binding domains in plasma fibronectin and unequal distribution of these domains in two different subunits: a preliminary note, *Biochem. Biophys. Res. Commun.,* 97, 709, 1980.

68. **Garcia-Pardo, A., Rostagno, A., and Frangione, B.,** Primary structure of human plasma fibronectin, characterization of a 38 kDa domain containing the C-terminal heparin-binding site (Hep III site) and a region of molecular heterogeneity, *Biochem. J.*, 241, 923, 1987.
69. **Pande, H., Calaycay, J., Lee, T. D., Legesse, K., Shively, J. E., Siri, A., Borsi, L., and Zardi, L.,** Demonstration of structural differences between the two subunits of human-plasma fibronectin in the carboxy-terminal heparin-binding domain, *FEBS Lett.*, 162, 403, 1987.
70. **Paul, J. I. and Hynes, R. O.,** Multiple fibronectin subunits and their post-translational modifications, *J. Biol. Chem.*, 259, 13477, 1984.
71. **Sekiguchi, K., Siri, A., Zardi, L., and Hakomori, S.,** Differences in domain structure between human fibronectins isolated from plasma and from culture supernatants of normal and transformed fibroblasts, *J. Biol. Chem.*, 260, 5105, 1985.
72. **Borsi, L., Allemanni, G., Castellani, P., Rosellini, C., Di Vinci, A., and Zardi, L.,** Structural differences in the cell binding region of human fibronectin molecules isolated from cultured normal and tumor-derived human cells, *FEBS Lett.*, 192, 71, 1985.
73. **Castellani, P., Siri, A., Rosellini, C., Infusini,E., Borsi, L., and Zardi, L.,** Transformed human cells release different fibronectin variants than do normal cells, *J. Cell Biol.*, 103, 1671, 1986.
74. **Borsi, L., Carnemolla, B., Castellani, P., Rosellini, C., Vecchio, D., Allemanni, G., Chang, S. E., Taylor-Papadimitriou, J., Pande, H., and Zardi, L.,** Monoclonal antibodies in the analysis of fibronectin isoforms generated by alternative splicing of mRNA precursors in normal and transformed human cells, *J. Cell Biol.*, 104, 595, 1987.
75. **Brown, R. A.,** Failure of fibronectin as an opsonin in the host defense system: a case of competitive self inhibition?, *Lancet,* 2, 1058, 1983.
76. **Cotton, G. and Brown, R. A.,** The effect of proteolytic degradation of plasma fibronectin on the responses of functional and immunometric assays for intact fibronectin, *Clin. Chim. Acta,* 153, 73, 1985.
77. **Selmer, J., Eriksin, H., and Clemmensen, I.,** Native and degraded fibronectin: new immunological methods for distinction, *Scand. J. Clin. Lab. Invest.*, 44, 57, 1984.
78. **Kottgen, E., Hoeft, S., Muller, C., and Hell, B.,** Functional analysis of plasma fibronectin with special consideration of binding interference, *J. Clin. Chem. Clin. Biochem.*, 24, 541, 1986.
79. **Brennan, M. J., Millis, A. J. T., Mann, D., and Fritz, K. E.,** Structural alterations in fibronectin correlated with morphological changes in smooth muscle cells, *Dev. Biol.*, 97, 391, 1983.
80. **De Petro, G., Barlati, S., Vartio, T., and Vaheri, A.,** Transformation-enhancing activity of gelatin-binding fragments of fibronectin, *Cell Biol.*, 78, 4965, 1981.
81. **Savill, C. M. and Ayad, S. R.,** The mitogenic activity of a heparin-binding fibronectin fragment (Mr 35,000) produced by cathepsin D digestion, *Anticancer Res.*, 6, 321, 1986.
82. **Homandberg, G. A., Kramer-Bjeke, J., Grant, G. C., and Eisenstein, R.,** Heparin-binding fragments of fibronectin are potent inhibitors of endothelial cell growth: structure-function correlations, *Biochim. Biophys. Acta,* 874, 61, 1986.
83. **Vercellotti, G. M., McCarthy, J., Furcht, L. T., Jacob, H. S., and Moldow, C. F.,** Inflamed fibronectin: an altered fibronectin enhanced neutrophil adhesion, *Blood,* 62, 1063, 1983.
84. **Czop, J. K., Kadish, J. L., and Austen, F.,** Augmentation of human monocyte opsonin-independent phagocytosis by fragments of human plasma fibronectin, *Proc. Natl. Acad. Sci. U.S.A.*, 78, 3649, 1981.
85. **Godfrey, H. P., Angadi, C. V., Wolstencroft, R. A., and Bianco, C.,** Localization of macrophage agglutination factor activity to the gelatin-binding domain of fibronectin., *J. Immunol.*, 133, 1417, 1984.
86. **Norris, D. A., Clark, R. A. F., Swigart, L. M., Huff, J. C., Weston, W. I., and Howell, S. E.,** Fibronectin fragment(s) are chemotactic for human peripheral blood monocytes, *J. Immunol.*, 129, 1612, 1982.
87. **Keil-Dlouha, V. and Planchenault, T.,** Potential proteolytic activity of human plasma fibronectin, *Proc. Natl. Acad. Sci. U.S.A.*, 83, 5377, 1986.
88. **Carsons, S., Lavietes, B. B., Diamond, H. S., and Kinney, S. G.,** The immunoreactivity, ligand and cell binding characteristics of rheumatoid synovial fluid fibronectin, *Arthritis Rheum.*, 28, 601, 1985.
89. **Fukuda, M. and Hakomori, S.,** Proteolytic and chemical fragmentation of galactoprotein a, a major transformation-sensitive glycoprotein released from hamster embryo fibroblasts, *J. Biol. Chem.*, 254, 5442, 1979.
90. **Ruoslahti, E., Engvall, E., Hayman, E. G., and Spiro, R. G.,** Comparative studies on amniotic fluid and plasma fibronectins, *Biochem. J.*, 193, 295, 1981.
91. **Fukuda, M. and Hakomori, S.,** Carbohydrate structure of galactoprotein a, a major transformation sensitive glycoprotein released from hamster embryo fibroblasts, *J. Biol. Chem.*, 254, 5451, 1979.
92. **Takasaki, S., Yamashita, K., Suzuki, K., Iwanaga, S., and Kobata, A.,** The sugar chains of cold insoluble globulin. A protein related to fibronectin, *J. Biol. Chem.*, 254, 8548, 1979.
93. **Fisher, S. J. and Laine, R. A.,** Carbohydrate structure of the major glycopeptide from human cold-insoluble globulin, *J. Supramol. Struct.*, 11, 391, 1979.

94. **Fukuda, M., Levery, S. B., and Hakomori, S.,** Carbohydrate structure of hamster plasma fibronectin evidence for chemical diversity between cellular and plasma fibronectins, *J. Biol. Chem.*, 257, 6856, 1982.
95. **Krusius, T., Fukuda, M., Dell, A., and Ruoslahti, E.,** Structure of the carbohydrate units of human amniotic fluid fibronectin, *J. Biol. Chem.*, 260, 4110, 1985.
96. **Nichols, E. J., Fenderson, B. A., Carter, W. G., and Hakomori, S.,** Domain-specific distribution of carbohydrates in human fibronectin and the transformation-dependent translocation of branched type 2 chain defined by monoclonal antibody C6, *J. Biol. Chem.*, 261, 11295, 1986.
97. **Zhu, B. C-R., Fisher, S. F., Pande, H., Calaycay, J. Shively, J. E., and Laine, R. A.,** Human placental (fetal) fibronectin: increased glycosylation and higher protease resistance than plasma fibronectin, *J. Biol. Chem.*, 259, 3962, 1984.
98. **Pande, H., Corkill, J., Sailor, R., and Shively, J. E.,** Comparative structural studies of human plasma and amniotic fluid fibronectins, *Biochem. Biophys. Res. Commun.*, 101, 265, 1981.
99. **Matsuura, H. and Hakomori, S.,** The oncofetal domain of fibronectin defined by monoclonal antibody FDC-6: its presence in fibronectins from fetal and tumor tissues and its absence in those from normal adult tissues and plasma, *Proc. Natl. Acad. Sci. U.S.A.*, 82, 6517, 1985.
100. **Matsuura, H., Takio, K., Titan, K., Green, T., Levery, S. B., Sayan, M. E. K., and Hakomori, S.,** The oncofetal structure of human fibronectin defined by monoclonal antibody FDC-6, *J. Biol. Chem.*, 263, 2666, 1988.
101. **Wagner, D. D., Ivatt, R., Destree, A. T., and Hynes, R. O.,** Similarities and differences between the fibronectins of normal and transformed hamster cells, *J. Biol. Chem.*, 206, 11708, 1981.
102. **Delannoy, P., Debray, H., and Montreuil, J.,** Modifications de la structure des glycannes de las fibronectine provoquees par la transformation de cellules BHK21/C13 induite par le virus du sarcome du hamster, *C. R. Acad. Sci. Paris,* 301, 767, 1985.
103. **Cossu, G. and Warren, L.,** Lactosaminoglycans and heparin sulfate are covalently bound to fibronectins synthesized by mouse stem teratocarcinoma cells, *J. Biol. Chem.*, 258, 5603, 1983.
104. **Ruoslahti, E., Jalanko, H., Comings, D. E., Neville, A. M., and Raghavan, D.,** Fibronectin from human germ-cell tumors resembles amniotic fluid fibronectin, *Int. J. Cancer,* 27, 763, 1981.
105. **Sekiguchi, K., Klos, A. M., Hirohashi, S., and Hakomori, S.,** Human tissue fibronectin: expression of different isotypes in the adult and fetal tissues. *Biochem. Biophys. Res. Commun.*, 141, 1112, 1986.
106. **Carsons, S., Lavietes, B. B., Slomiany, A., Diamond, H. S., and Berkowitz, E.,** Carbohydrate heterogeneity of fibronectins synovial fluid fibronectin resembles the form secreted by cultured synoviocytes but differs from the plasma form, *J. Clin. Invest.*, 80, 1342, 1987.
107. **Jones, G. E., Arumugham, R. G., and Tanzer, M. L.,** Fibronectin glycosylation modulates fibroblast adhesion and spreading, *J. Cell Biol.*, 103, 1663, 1986.
108. **Zhu, B. C. R. and Laine, R. A.,** Polylactosamine glycosylation on human fetal placental fibronectin weakens the binding affinity of fibronectin to gelatin, *J. Biol. Chem.*, 260, 4041, 1985.
109. **Teng, M. H. and Rifkin, D. B.,** Fibronectin from chicken embryo fibroblasts contains covalently bound phosphate, *J. Cell. Biol.*, 80, 784, 1979.
110. **Ledger, P. W., Uchida, N., and Tanzer, M. L.,** Immunocytochemical localization of procollagen and fibronectin in human fibroblasts: effects of the monovalent ionophore, monensin, *J. Cell. Biol.*, 87, 663, 1980.
111. **Iu, A. and Hunter, T.,** Structural comparison of fibronectins from normal and transformed cells, *J. Biol. Chem.*, 256, 7671, 1981.
112. **Etheredge, R. E., Han, S., Fossel, E., Tanzer, M. L., and Glimcher, M. J.,** Identification and quantitation of O-phosphoserine in human plasma fibronectin, *FEBS,* 186, 259, 1985.

Chapter 2

FIBRONECTIN IN DEVELOPMENT

Beverly B. Lavietes

TABLE OF CONTENTS

I. INTRODUCTION

Morphogenesis involves interactions between cells and their extracellular environment such that cells begin to move, progress in a directed fashion, stop at the correct location, sort out, and differentiate appropriately. Fibronectin is only one of the cell surface and/or extracellular matrix molecules involved in these processes. We will discuss the role of fibronectin as a representative of the kinds of cell-matrix interactions that occur and their functional sequelae.

We will consider, first, general questions regarding the role of extracellular matrix in developmental processes. Next, we will discuss *in vitro* models and their relevance to *in vivo* phenomena. Finally, we will examine specific tissue systems where fibronectin has been shown to play a role.

II. MORPHOGENESIS

Morphogenesis involves translocations of cells in three dimensions over time. Development of form in a multicellular organism begins with cell division to produce a population of cells from the zygote. This is followed by gastrulation, movement of sheets of cells to form the three major germ layers, ectoderm, endoderm, and mesoderm, from which all tissue types derive.

Directed migration of cells to specific locations in the embryo occurs at later stages. Neural crest cells migrate from the top of the neural fold at precise developmental times to different locations to give rise to specific cell types, i.e., neurons, glia, pigment, skeleton, muscle, hormone producing cells.[1] Such cell relocation involves both locomotion and anchorage. Locomotion mechanistically implies repetitive detachment from and reattachment of cells to the substratum. Initiation of locomotion requires loss of adhesion to both cells and substratum in the location of origin; cessation of locomotion, a reestablishment of anchorage in the target location.

Within specific regions of the embryo, types of cells sort out and separate from one another. In the limb buds, chondrogenic mesenchyme condenses to form skeletal elements, separating from myogenic mesenchyme. These changes are associated with changes in composition of extracellular matrix.[2]

All of these phenomena involve changing patterns of adhesion between cells and their substrata. Full understanding of morphogenesis, therefore, requires knowledge of both cell-cell and cell-substratum adhesion factors. See Edelman[3,4] for reviews of specific cell adhesion molecules. The complex domain structure of fibronectin (see Chapter 1) allows it to mediate both kinds of interactions.

Extracellular matrix provides the substratum upon which cells migrate. The chemical composition of matrix may provide migrating cells the information that tells them where they are in the embryo. Cells have specific receptors to recognize matrix molecules.[5,6] Qualitative and quantitative variation in those receptors with developmental time could modulate detachment, locomotion, and reattachment of cells.

Extracellular matrix is synthesized and deposited by cells and, therefore, its composition may change as a result of the interaction of cells with it. Migrating cells leave "footprints", remnants of adhesion complexes, on the substrata over which they move. Cells may also release enzymes which digest the matrix through which they move.[7] Some polypeptide growth factors which play a role in development may act through alterations in composition of extracellular matrix. Transforming growth factor β, for example, increases accumulation of fibronectin and type I collagen.[8-10] Similarly, interaction of cells with extracellular matrix components, presumably mediated by specific receptor complexes, can alter the pattern of synthesis of matrix components.[11]

III. *IN VITRO* MODELS AND *IN VIVO* PHENOMENA

In vitro models have been developed to test hypotheses about the role of extracellular matrix components in development and differentiation. The models attempt to isolate and examine specific aspects of cell-matrix interactions.

Adherence and/or spreading of cells on various substrata has been used to obtain information on specificity and mechanism of cell-matrix interactions.[5] Adherence of different cell types to a given substratum or adherence of one cell type to different substrata provide evidence for qualitative and quantitative differences in cellular affinities for specific matrix components. Inhibition of adherence by soluble matrix molecules or fragments of matrix molecules have led to identification of the protein sequences to which cells bind.[5,12] These sequences, in turn, have been used to identify and isolate the receptor complexes on cell surfaces.[6,13] Antibodies specific to cell surface receptors can promote cell attachment and differentiation.[14] Cell adhesion and spreading on fibronectin and other matrix molecules are affected by alterations in the carbohydrate residues on these molecules.[15,16]

Phenotypic responses of cells to different substrata *in vitro* have clearly demonstrated that matrix can modulate differentiation. This is best shown by pluripotential cells which express different specific phenotypes on different substrata. Neural crest cells will be discussed in detail below. Cultured F9 embryonal carcinoma cells, another well-described development model system, show variable synthesis of fibronectin, type IV collagen, and laminin as a function of their state of differentiation.[17]

Significant differences in cell behavior occur when cells are cultured in a three-dimensional gel compared to on a two-dimensional surface. Avian lens epithelium embedded in collagen gels undergoes mesenchymal transformation.[18] Dedifferentiated flattened chondrocytes, when suspended in agarose gels, resume a rounded shape and begin to synthesize type II collagen and cartilage-specific proteoglycan.[19] In organ cultures of differentiating mouse salivary gland, extracellular matrix components stabilize the branching lobular morphology.[20]

In vitro studies with a variety of systems have shown that cells interact with extracellular matrix through specific receptors on their surfaces. Changes in receptor composition with developmental time could affect cellular adhesion to the substratum and/or adjacent cells promoting shape, change, and/or migration. A migrating cell could recognize a new environment and engagement of more surface receptors with matrix components could stop locomotion (Figure 1). For example, the cell in Figure 1A acquires more surface receptors with developmental time and will stop moving as the receptors bind ligand in the environment. On the other hand, changes in matrix composition could have similar effects on cells with an invariant set of receptors. The cell in Figure 1B has two species of receptors. It moves across a substratum sparsely populated with one type of ligand and stops when it reaches an environment with two types of ligand complementary to its receptor composition. If the cell in Figure 1A acquired the circle type receptors instead of square type receptors, it would also halt in the new environment on the right side of Figure 1B. Inductive soluble factors such as hormones may act indirectly through modulation of extracellular matrix composition secondary to changes in cellular anabolic or catabolic metabolism, which creates or destroys ligands.

Let us now turn to specific examples of developmental change involving fibronectin.

IV. FIBRONECTIN IN SPECIFIC SYSTEMS

A. EARLY EMBRYONIC CELL MIGRATION IN SEA URCHINS AND AMPHIBIANS

The role of fibronectin in cell migration appears to be highly conserved through evolution. Developmental stage-specific, cell-specific immunofluorescent staining for fibronectin has

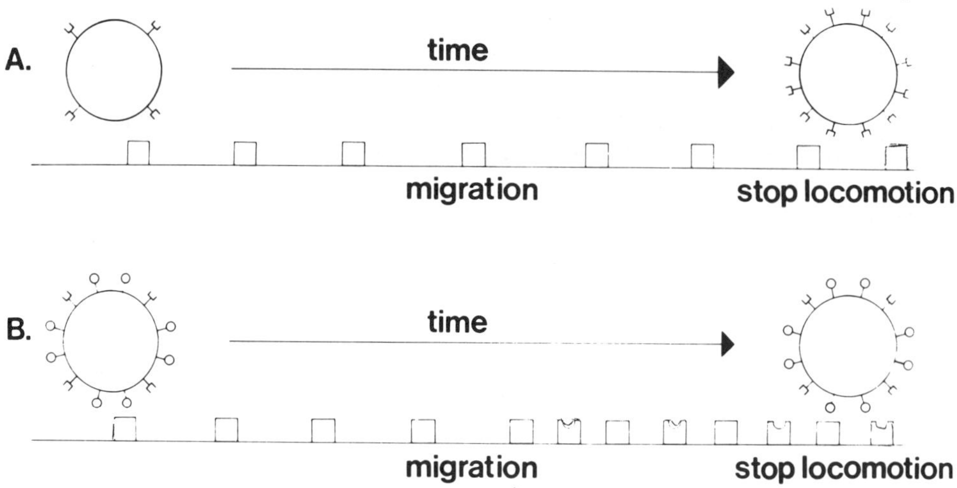

FIGURE 1. Receptor-extracellular matrix interactions regulate cell behavior. (A) A cell with receptors for square ligand in substratum migrates, making and breaking receptor-ligand complexes. With developmental time it acquires more receptors. When a greater number of receptor-ligand complexes form, the cell stops locomotion. (B) A cell with receptors for two different ligand species migrates over a substratum containing only one type of ligand. When it reaches an environment where both ligands are present, more receptor-ligand complexes form, and the cell stops moving. Similarly, if the cell in A acquired the second type of receptor instead of more of the square type, it, too, would recognize and stop in the environment seen in the right side of B.

been demonstrated in the sea urchin embryo.[21] Early in sea urchin embryogenesis, at the blastula stage, primary mesenchyme cells begin to migrate through the blastocoel, an environment rich in sulfated glycosaminoglycans. Specific fibronectin staining was detected only on primary mesenchyme cells at the midblastula stage when they are migratory; they do not stain at earlier or later developmental stages.[21] These results suggest that fibronectin mediates attachment of the primary mesenchyme cells to the substratum upon which they migrate.[21]

In a subsequent study, Katow and Hayashi[22] showed that isolated primary mesenchyme cells show dose dependence for fibronectin to support migration *in vitro*. The cells migrate in a nondirected, zig-zag fashion which may be consistent with fibronectin localization on the cells rather than the substratum.[22] In amphibian gastrulation, mesodermal cells migrate in a directed fashion on a fibronectin-containing extracellular matrix covering the roof of the blastocoel.[23,24] Antibodies to fibronectin or synthetic peptides representing the cell binding and sequence of fibronectin inhibit this directed migration.[25,26]

B. NEURAL CREST

The neural crest is a transient structure arising in vertebrate embryos when the neural folds appose to form the neural tube. Crest cells subsequently migrate into different regions of the embryo and form a multitude of different cell types. These include pigment cells in the skin, neurons of autonomic and sensory ganglia, glia, adrenomedullary cells, skeletal and connective tissue of the head. This system has been used by a number of laboratories to explore the mechanisms by which stable restrictions on differentiative potential arise in a pleuripotential population. Recent reviews of neural crest differentiation include References 27 to 30.

Grafting experiments have shown that neural crest cells from one site placed into a new site differentiate according to their new location in the embryo rather than express their original potential fate (see reviews listed above). Crest cells from older embryo transplanted into younger hosts seem to reacquire the characteristics of the earlier stage. These observations suggested that environmental cues guide crest cell differentiation.

Histochemical and immunochemical analyses of embryos at different development stages have identified a rich diversity of molecular environments through which crest cells migrate. These include varying compositions of fibronectin, collagens, laminin, and an assortment of glycosaminoglycans, e.g., hyaluronic acid, chondroitin sulfates, and heparin sulfate.[29] Since these different environments induce specific behavioral responses in neural crest cells, e.g., initiation or cessation of locomotion, pigment formation, catecholamine synthesis, the responses of neural crest cells in culture to substrata of varying composition have been studied (see reviews). We will concentrate here on studies involving fibronectin.

Electron micrographic studies showed that the cell free space which neural crest cells enter after detaching from the neural tube contains a random meshwork of collagen fibrils upon which the migrating cells appear to align.[31,32] Since fibronectin mediates cell attachment to collagen and fibronectin was localized in the same matrix,[33] Greenberg et al.[34] determined *in vitro* that fibronectin mediated the attachment of crest cells to collagen. They found first that calf serum increased attachment of crest cells to collagen types I to V and that serum could be replaced by purified cellular fibronectin. Moreover, serum depleted of fibronectin did not promote attachment of cells to type I collagen. In addition, these authors showed that fibronectin increased movement of neural crest cells in modified Boyden chambers.[34] Checkerboard analysis showed increases in both directed and random movement of crest cells in response to increasing concentrations of fibronectin; the cells responded similarly to 160-kDa cell-binding fragment of fibronectin but not to collagen (40 kDa)- or heparin (50 kDa)-binding domains.[34]

Because glycosaminoglycans are also present in the cell free space and because interaction between matrix components can alter their binding and adhesive properties, Erickson and Turley[35] studied changes in neural crest cell morphology and motility on experimental substrata of varying compositions. Collagen and chondroitin sulfate, but not hyaluronate, were permissive substrata for adherence and motility, but fibronectin was the most effective in promoting directional neural crest cell movement. Addition of glycosaminoglycans to culture medium altered the movement of cells on preformed fibronectin-collagen substrata. Chondroitin sulfate caused cells to round up, suggesting decreased cell-substratum adhesion, but did not alter speed of movement of the adherent cells. Hyaluronate, on the other hand, markedly reduced speed of movement but had little effect on morphology. These results show that molecules bound to other extracellular matrix molecules may act to promote attachment, but the same molecules binding to a cell from solution may block binding sites and prevent attachment. If fibronectin is the primary molecule to which neural crest cells attach and move, the direction and/or speed of migration of the cells may be modified by changes in fibronectin concentration or concentrations of other matrix components.[34,35]

The specificity of the requirement for fibronectin in neural crest cell migration was demonstrated by Rovasio et al.[36] using anti-fibronectin antibodies, cell-binding fragments of fibronectin, and providing a choice of substrata to the same cell population. Neural crest cells, placed on culture dishes with alternating stripes of glass and fibronectin, migrated exclusively on the fibronectin stripes. Antibody (Fab′) to 160-kDa cell-binding fragment of fibronectin reversibly inhibited directed migration on the fibronectin stripes. Individual cells displayed more random movement while groups of cells moved in a more directed fashion (Figure 2). The presence of surrounding cells appeared to prevent changes in direction. Examination by immunofluorescence of the extracellular matrix under migrating crest cells showed a realignment of fibronectin-containing fibrils. With time in culture, crest cells showed a decreased adhesion to fibronectin and increased binding to laminin which may correlate with the shift in morphology to epithelial type structures which occurs at the end of migration.

Together these data show the complexity of cell-matrix interactions and demonstrate the variety of mechanisms by which specificity and information transfer may be accomplished, e.g.,

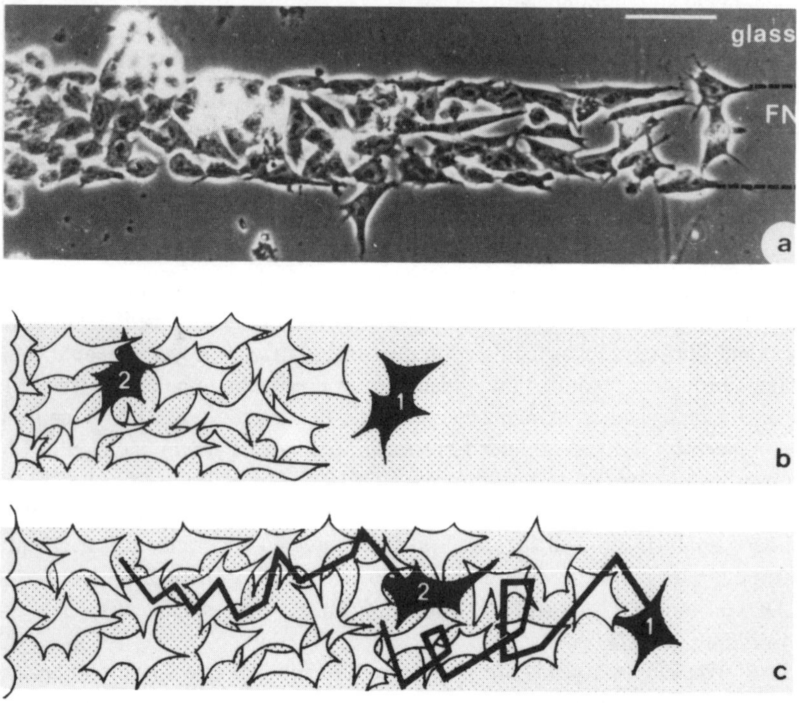

FIGURE 2. Video analysis of neural crest cell migration on fibronectin stripes. (a) Crest cells migrated strictly along fibronectin stripes after 24 h of culture; most crest cells remained in close contact with partial overlapping of cell bodies. (b and c) Tracks of cell surrounded by other cells (2); it showed a persistence of movement as compared to a pioneer cell (1) which changed directions much more frequently and even occasionally migrated backwards. Dotted region: plasma fibronectin stripes; undulating line at left corresponds to neural tube edge. Bar, 50μm; magnification × 380. (Reproduced from Rovasio, R. A. et al., in *Journal of Cell Biology,* 1983, vol. 96, 470, by copyright permission of the Rockefeller University Press.)

1. Cells have specific receptors for matrix molecules.
2. Changes in cell receptor populations change cellular affinity for matrix components.
3. Cellular activity modifies the extracellular matrix.

More recently, antibodies to cell surface receptors and synthetic peptides which copy the cell-binding sequence of fibronectin have been used to confirm *in vivo* the conclusions from the *in vitro* model systems. Boucaut et al.[26] showed that the decapeptide Arg-Gly-Asp-Ser-Pro-Ala-Ser-Lys-Pro, which contains the cell-binding sequence of fibronectin, when injected into chick embryos at the 5-somite stage, inhibited the subsequent onset of neural crest migration. Instead of spreading laterally, neural crest cells accumulated as a compact projection from the doral neural tube (Figure 3).[26]

Early avian embryos have been mapped by immunofluorescence for the distribution of the 140-kDa cell receptor for fibronectin and laminin.[37,38] Although a general codistribution of receptor and ligands was seen, at no stage were either distributed in an exclusive pattern mimicking known pathways of neural crest migration. Levels of both receptor and ligands changed at different states supporting the notion that initiation of migration and relocalization of crest cells results from a balance of cell-cell and cell-substratum adhesions.[38] Analysis of receptor distribution on cells *in vitro* suggested that migration is associated with a diffuse distribution of receptors mediating labile adhesions, whereas stationary cells are anchored at specific sites having a high local receptor concentration.[37]

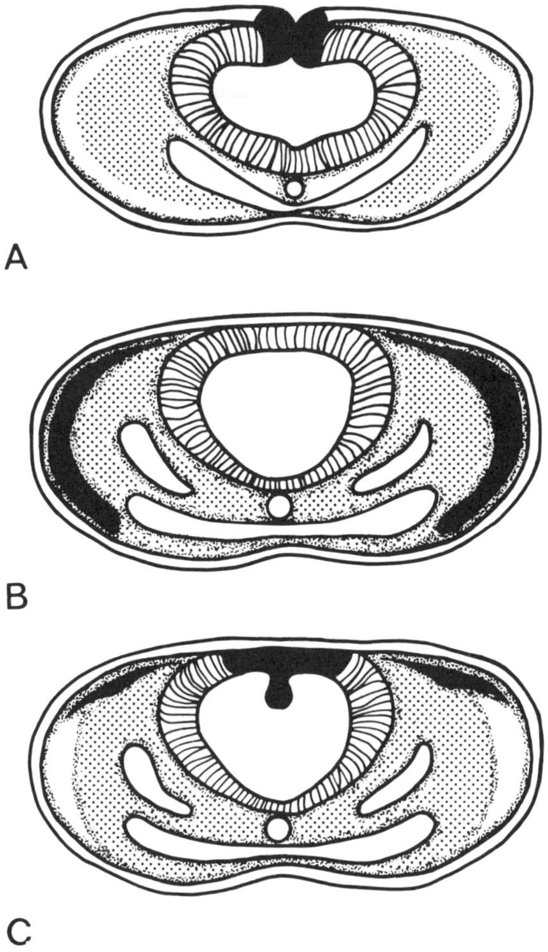

FIGURE 3. Normal and perturbed pattern of neural crest migration in avian embryos. (A) Diagram of a five-somite chick embryo, cross-sectioned at the level of the mid-mesencephalon, neural crest cells (solid black areas) originate at the junction of the fusing neural folds. A cell-free space has already formed laterally under the ectoderm. A fibronectin-rich basement membrane (finely stippled areas) lies under the ectoderm adjacent to the cell-free space, which is the presumptive crest cell pathway. (B) An 11-somite chick embryo, cross-sectioned at the level of the mid-mesencephalon. Neural crest cells have reached their furthest lateral extent occuping the previously cell-free space, but will continue to migrate ventrally. (C) An 11-somite embryo injected at the five-somite stage with the peptide Arg-Gly-Asp-Ser-Pro-Ala-Ser-Ser-Lys-Pro sequence from the cell binding region of fibronectin. The majority of the migrating neural crest cells have been forced into the neural tube lumen. Some cells are also found within the mesenchymal mesoderm. (Reproduced from Boucaut, J. C., et al. in the *Journal of Cell Biology,* 1984, vol. 99, 1829, by copyright permission of the Rockefeller University Press.)

The neural crest system poses questions beyond those of guidance along migratory pathways. Because the cells that take different pathways have different developmental fates, does the environment through which the cells travel affect differentiation? Are developmental restrictions imposed on pleuripotential cells by environmental cues, or do subpopulations with limited potential exist in the original population which are selected by different extracellular environments? Can neural crest cells cultured on different substrata express different phenotypes? Is developmental potential limited by exposure to different environments?

Trunk neural crest dispersed on tissue culture plastic tend to differentiate into melanocytes. Exposure of the same cell population to cellular fibronectin promotes expression of catecholamine synthesis rather than melanin, i.e., differentiation into neuronal rather than pigment cells.[39,40] These data support the hypothesis that matrix components like fibronectin influence differentiation, but the mechanism by which this occurs remains to be elucidated.[30]

C. LIMB BUD

The limb bud is another system in which fibronectin is associated with selection between potential phenotypes.[2] The mesodermal mesenchyme of the limb gives rise to cartilage, bone, muscle, and connective tissue. The first stage in chondrogenesis is condensation of the mesenchyme. Dessau et al.[41] reported homogeneous codistribution of fibronectin and type I collagen throughout limb bud mesenchyme from avian embryo stages 19 to 23. At stage 24 when condensation begins, fibronectin and type I collagen stain increases in intensity in the center of the limb bud. Later, when cartilage matrix is being synthesized, fibronectin and type I collagen staining disappears and is replaced by type II collagen stain. Dessau et al.[41] interpreted these observations to suggest that fibronectin mediated both cell-collagen I interactions and cell-cell interactions, effecting condensation.

The transient requirement for fibronectin and its subsequent loss in differentiated cartilage were mimicked *in vitro* in cultures of dissociated chondrocytes.[42,43] The cells make fibronectin and type I collagen until the cell mass is great enough to reestablish rounded chondrocyte morphology and specific cartilage matrix synthesis. Fibronectin is not seen in differentiated cartilage matrix unless the samples are treated with hyaluronidase to remove proteoglycan.[44,45]

Fibronectin synthesis is high in the prechondrogenic proliferation phase of the bone matrix-induced endochondral bone-forming model in rats.[45,46] Fibronectin synthesis declines during the cartilage phase of development but rises with the onset of bone formation and hematopoiesis.[45] Endochondral bone formation could be blocked by local injection of specific anti-fibronectin antibodies at the time of implantation of the bone matrix graft. The antibody apparently blocks coating of the bone matrix by endogenous plasma fibronectin thus preventing the initial colonization of the bone matrix by fibroblasts (Figure 4).[47]

Together, the above results suggest that fibronectin is required for cell-cell and cell-substratum interactions which allow for mesenchymal fibroblasts to achieve either a critical density or three-dimensional arrangement that is permissive for cartilage-specific matrix synthesis. Fibronectin synthesis declines in phenotypic chondrocytes. If exogenous fibronectin is added to chondrocyte cultures, the cells spread extensively on the substratum become fibroblastic in appearance and switch back to type I collagen synthesis.[48,49] Vitamin A, when added to chondrocyte cultures, also causes similar morphological changes which are accompanied by enhanced fibronectin retention by the cells.[50] The fibronectin synthesized by treated cultures has a slightly higher molecular weight[50] which may be a function of differential glycosylation of the molecule.[51]

While at avian stages 22 to 23 fibronectin becomes concentrated in the condensing early skeletal primordia,[41,44] fibronectin is somewhat reduced in the premyogenic regions of the limb.[52] Fibronectin-positive cells accumulate in the furrows as the muscle anlage separate. Since myoblasts in culture tend to align along fibronectin deposits, the spatial arrangement of muscles bundles *in vivo* may be determined by fibronectin-containing matrix laid down

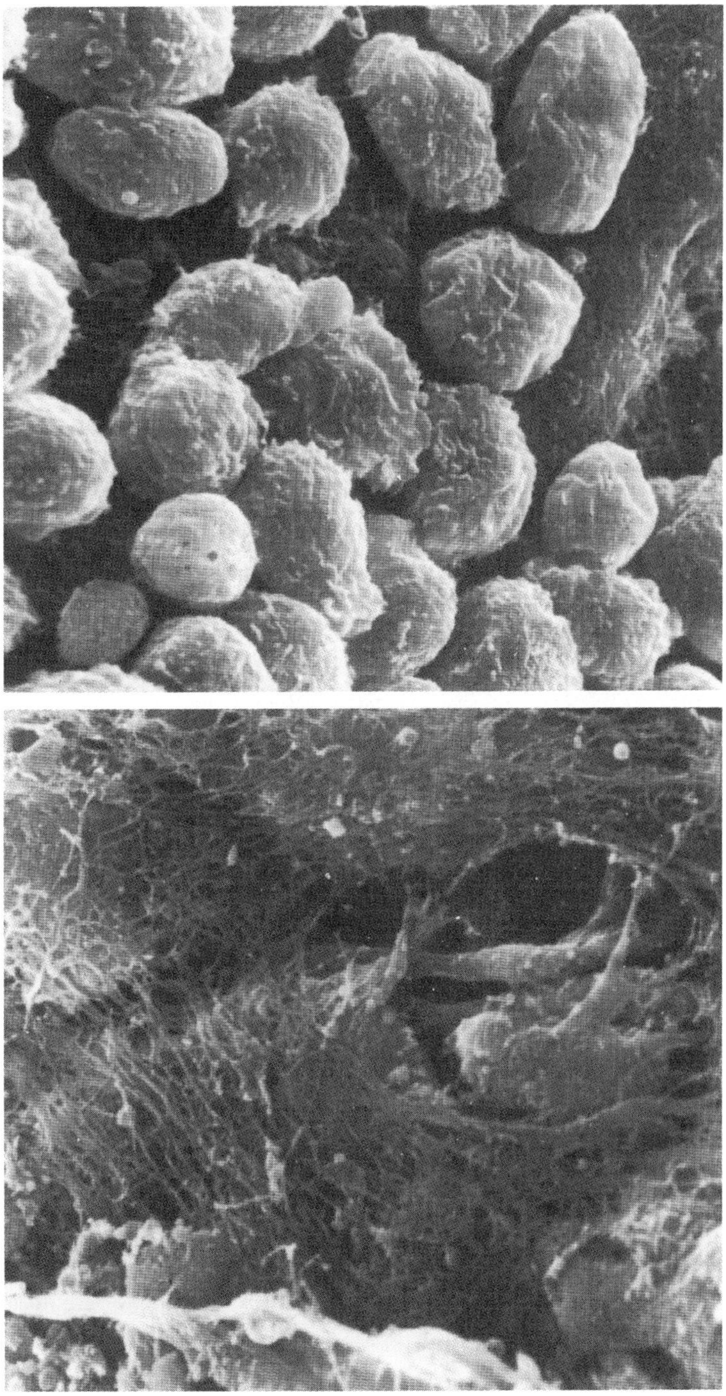

FIGURE 4. Scanning electron micrographs of day 3 bone matrix implants treated with anti-rat albumin (top), or anti-rat fibronectin (bottom). Note the reduced number of cells attached to the matrix treated with anti-fibronectin. ×3000. (From Weiss, R. E. and Reddi, A. H., *Exp. Cell Res.*, 133, 247, 1981. With permission.)

by connective tissue cells.[52] The morphology and adherence of embryonic skeletal myoblasts in culture are significantly altered by CSAT, the monoclonal antibody directed against the 140-kDa cell surface receptor for fibronectin and laminin.[53] Analysis of extracellular matrix biosynthesis as a function of proliferative and differentiative stage in a cultured myoblast line showed significant changes in amount and type of collagen.[54] Parallel studies on a nonfusing mutant subline showed depressed collagen synthesis and enhanced laminin synthesis. Fibronectin synthesis was unchanged during differentiation and equal in the mutant.[54]

Differentiation of bone and muscle, then, involves responses to complexes of extracellular matrix elements, not one particular molecule. The precursor mesenchymal cells interact with extracellular matrix elements, and their ultimate developmental fate is thus determined. The instructional role of matrix elements is most clearly demonstrated by the bone matrix implant model in which cartilage and bone differentiation can be elicited from embryonic skeletal muscle as well as mesenchymal fibroblasts.[55,56]

Extracellular matrix, in general, and fibronectin, in particular, serve not only as indicators of developmental changes in tissue architecture[57] but also as effectors of developmental change by several mechanisms. Changes in matrix composition both reflect and elicit further changes in cell behavior and tissue organization.

While early studies concentrated on the distribution of extracellular matrix molecules in developing embryos, more recent work attempts to correlate changes in cell receptors with the distribution of these molecules. Shifts in number and/or distribution of 140-kDa receptor complex has been related to changes in adhesion and other responses of embryonic cells to matrix components. Such changes appear to be involved in lung differentiation,[58] neurite extension,[14] and erythroid differentiation.[59]

In other developing systems, matrix components such as fibronectin and laminin seem to play a secondary but significant role in stabilizing changes induced by other mechanisms. In avian somitogenesis, cell-cell adhesive molecules, such as N-cadherin and neural cell adhesion molecule (N-CAM), control tissue remodeling but fibronectin and laminin appear later and stabilize the newly formed epithelial structures.[60] This stabilizing function of matrix molecules is common in the epithelial-mesenchymal interactions of gland formation.[20]

Soluble factors may effect changes in tissue by altering cellular biosynthesis of extracellular matrix. Transforming growth factor B, which has been implicated in a number of regulatory events, increases fibronectin and collagen synthesis by fibroblasts.[8] Drugs such as bleomycin[61] and dexamethasone[62-64] alter extracellular matrix synthesis and deposition.

V. SUMMARY

The interactions that occur between extracellular matrix and cells in development are multifaceted and ongoing. The encounter of a cell with a matrix component can lead to many outcomes as a function of cellular receptors, processing of information, altered cell biosynthesis, and altered matrix structures with which other cells may react. The implications for this multiplicity of interactions and responses for disease processes are discussed in the following chapters.

REFERENCES

1. **LeDouarin, N. M.,** Ontogeny of the peripheral nervous system from the neural crest and the placodes. A developmental model studied on the basis of the quail-chick chimaera system, *Harvey Lect.*, series 80, 137, 1986.
2. **Solursh, M.,** Cell and matrix interactions during limb chondrogenesis in vitro, in *The Role of Extracellular Matrix in Development,* Trelstad, R. L., Ed., Alan R. Liss, New York, 1984, chap. 12.

3. **Edelman, G. M.,** Cell adhesion molecules, *Science,* 219, 450, 1983.
4. **Edelman, G. M.,** Cell adhesion and the molecular processes of morphogenesis, *Annu. Rev. Biochem.,* 54, 135, 1985.
5. **Yamada, K. M.,** Cell surface interactions with extracellular materials, *Annu. Rev. Biochem.,* 52, 761, 1983.
6. **Hynes, R. O.,** Integrins: a family of cell surface receptors, *Cell,* 48, 549, 1987.
7. **Krystose, K. A. and Seeds, N. W.,** Plasminogen activator release at the neuronal growth cone, *Science,* 213, 1532, 1981.
8. **Ignotz, R. A. and Massague, J.,** Transforming growth factor-B stimulates the expression of fibronectin and collagen and their incorporation into the extracellular matrix, *J. Biol. Chem.,* 26, 4337, 1986.
9. **Ignotz, R. A., Endo, T., and Massague, J.,** Regulation of fibronectin and type I collagen mRNA levels by transforming growth factor-β, *J. Biol. Chem.,* 262, 6443, 1987.
10. **Raghow, R., Postlethwaite, A. E., Keski-Oja, J., Moses, H. L., and Kang, A. H.,** Transforming growth factor-β increases steady state levels of type I procollagen and fibronectin messenger RNA's posttranscriptionally in cultured human dermal fibroblasts, *J. Clin. Invest.,* 79, 1285, 1987.
11. **Sugrue, S. P. and Hay, E. D.,** The identification of extracellular matrix (ECM) binding sites on the basal surface of embryonic corneal epithelium and the effect of ECM binding on epithelial collagen production, *J. Cell Biol.,* 102, 1907, 1986.
12. **Ruoslahti, E. and Pierschbacher, M. D.,** Arg-gly-asp: a versatile cell recognition signal, *Cell,* 44, 517, 1986.
13. **von der Mark, K., Mollenhauer, J., Kuhl, V., Bee, J., and Lesot, H.,** Anchorins: a new class of membrane proteins involved in cell-matrix interactions, in *The Role of Extracellular Matrix in Development.,* Trelstad, R. L., Ed., Alan R. Liss, New York, 1984, chap. 4.
14. **Hall, D. E., Neugebauer, K. M., and Reichardt, L. F.,** Embryonic neural retina cells response to extracellular matrix proteins: developmental changes and effects of the cell substratum attachment antibody (CSAT), *J. Cell Biol.,* 104, 623, 1987.
15. **Hughes, R. C. and Mills, G.,** Functional differences in the interactions of glycosylation-deficient cell lines with fibronectin, laminin, and type IV collagen, *J. Cell. Physiol.,* 128, 402, 1986.
16. **Jones, G. E., Arumugham, R. G., and Tanzer, M. L.,** Fibronectin glycosylation modulates fibroblast adhesion and spreading, *J. Cell Biol.,* 103, 1663, 1986.
17. **Grover, A. and Adamson, E. D.,** Roles of extracellular matrix components in differentiating teratocarcinoma cells, *J. Biol. Chem.,* 260, 12252, 1985.
18. **Hay, E. D.,** Cell-matrix interactions in the embryo: cell shape, cell surface, cell skeletons, and their role in differentiation, in *The Role of Extracellular Matrix in Development,* Trelstad, R. L., Ed., Alan R. Liss, New York, 1984, chap. 1.
19. **Benya, P. D. and Shaffer, J. D.,** Dedifferentiated chondrocytes reexpress the differentiated collagen phenotype when cultured in agarose gels, *Cell,* 30, 215, 1982.
20. **Bernfield, M., Banerjee, S. D., Koda, J. E., and Rapraeger, A. C.,** Remodeling of the basement membrane as a mechanism of morphogenetic tissue interaction, in *The Role of Extracellular Matrix in Development,* Trelstad, R. L., Ed., Alan R. Liss, New York, 1984, chap. 23.
21. **Katow, H., Yamada, K. M., and Solursh, M.,** Occurrence of fibronectin on the primary mesenchyme cell surface during migration in the sea urchin embryo, *Differentiation,* 22, 120, 1982.
22. **Katow, H. and Hayashi, M.,** Role of fibronectin in primary mesenchyme cell migration in the sea urchin, *J. Cell Biol.,* 101, 1487, 1985.
23. **Boucaut, J. C. and Darribere, T.,** Fibronectin in early amphibian embryos, *Cell Tiss. Res.,* 234, 135, 1983.
24. **Lee, G., Hynes, R., and Kirschner, M.,** Temporal and spatial regulation of fibronectin in early *Xenopus* development, *Cell,* 36, 729, 1984.
25. **Boucaut, J. C., Darribere, T., Boulekbache, H., and Thiery, J. P.,** Prevention of gastrulation but not neurulation by antibodies to fibronectin in amphibian embryos, *Nature,* 307, 364, 1984.
26. **Boucaut, J. C., Darribere, T., Poole, T. J., Aoyama, H., Yamada, K. M., and Thiery, J. P.,** Biologically active synthetic peptides as probes of embryonic development: a competitive peptide inhibitor of fibronectin function inhibits gastrulation in amphibian embryos and neural crest cell migration in avian embryos, *J. Cell Biol.,* 99, 1822, 1984.
27. **Le Douarin, N. M., Cochard, P., Vincent, M., Duband, J. L., Tucker, G. C., Teillet, M. A., and Thiery, J. P.,** Nuclear, cytoplasmic, and membrane markers to follow neural crest cell migration: a comparative study, in *The Role of Extracellular Matrix in Development,* Trelstad, R. L., Ed., Alan R. Liss, New York, 1984, chap. 17.
28. **Bronner-Fraser, M.,** Latex beds with defined surface coats as probes of neural crest migratory pathways, in *The Role of Extracellular Matrix in Development,* Trelstad, R. L., Ed., Alan R. Liss, New York, 1984, chap. 18.

29. **Weston, J. A., Ciment, G., and Girdlestone, J.,** The role of extracellular matrix in neural crest development: a reevaluation, in *The Role of Extracellular Matrix in Development,* Alan R. Liss, New York, 1984, chap. 19.
30. **Weston, J. A.,** Phenotypic diversification in neural crest-derived cells: the time and stability of commitment during early development, *Curr. Top. Dev. Biol.,* 20, 195, 1986.
31. **Tosney, K. W.,** The early migration of neural crest cells in the trunk region of the avian embryo: an electron microscopic study, *Dev. Biol.,* 62, 317, 1978.
32. **Lofberg, J., Ahlfors, K., and Fallstrom, C.,** Neural crest cell migration in relation to extracellular matrix organization in the embryonic axolotl trunk, *Dev. Biol.,* 75, 148, 1980.
33. **Mayer, B. W., Jr., Hay, E. D., and Hynes, R. O.,** Immunocytochemical localization of fibrinectin in embryonic chick trunk and area vasculosa, *Dev. Biol.,* 82, 267, 1981.
34. **Greenberg, J. H., Seppa, S., Seppa, H., and Hewitt, A. T.,** Role of collagen and fibronectin in neural crest cell adhesion and migration, *Dev. Biol.,* 87, 259, 1981.
35. **Erickson, C. A. and Turley, E. A.,** Substrata formed by combinations of extracellular matrix components alter neural crest motility in vitro, *J. Cell Sci.,* 61, 299, 1983.
36. **Rovasio, R. A., Delouvee, A., Yamada, K. M., Timpl, R., and Thiery, J. P.,** Neural crest cell migration: requirements for exogenous fibronectin and high cell density, *J. Cell Biol.,* 96, 462, 1983.
37. **Duband, J. L., Rocher, S., Chen, W. T., Yamada, K. M., and Thiery, J. P.,** Cell adhesion and migration in the early vertebrate embryo: Location and possible role of the putative fibronectin receptor complex, *J. Cell Biol.,* 102, 160, 1986.
38. **Krotoski, D. M., Domingo, C., and Bronner-Fraser, M.,** Distribution of putative cell surface receptor for fibronectin and laminin in the avian embryo, *J. Cell Biol.,* 103, 1061, 1986.
39. **Sieber-Blum, M., Sieber, F., and Yamada, K. M.,** Cellular fibronectin promotes adrenergic differentiation of quail neural crest cells in vitro, *Exp. Cell Res.,* 133, 285, 1981.
40. **Loring, J., Glimelius, B., and Weston, J. A.,** Extracellular matrix materials influence quail neural crest cell differentiation in vitro, *Dev. Biol.,* 90, 165, 1982.
41. **Dessau, W., von der Mark, H., von der Mark, K., and Fischer, S.,** Changes in the patterns of collagens and fibronectin during limb bud chondrogenesis, *J. Embryol. Exp. Morphol.,* 57, 51, 1980.
42. **Dessau, W., Sasse, J., Timpl, R., Jilek, F., and von der Mark, K.,** Synthesis and extracellular deposition of fibronectin in chondrocyte cultures. Response to the removal of extracellular cartilage matrix, *J. Cell Biol.,* 79, 342, 1978.
43. **Dessau, W., Vertel, B. M., von der Mark, H., and von der Mark, K.,** Extracellular matrix formation by chondrocytes in monolayer culture, *J. Cell Biol.,* 90, 78, 1981.
44. **Kosher, R. A., Walker, K. H., and Ledger, P. W.,** Temporal and spatial distribution of fibronectin during development of the embryonic chick limb bud, *Cell Differ.,* 11, 217, 1982.
45. **Weiss, R. E. and Reddi, A. H.,** Synthesis and localization of fibronectin during collagenous matrix-mesenchymal cell interaction and differentiation of cartilage and bone in vivo, *Proc. Natl. Acad. Sci. U.S.A.,* 77, 2074, 1980.
46. **Weiss, R. E. and Reddi, A. H.,** Appearance of fibronectin during the differentiation of cartilage, bone and bone marrow, *J. Cell Biol.,* 88, 630, 1981.
47. **Weiss, R. E. and Reddi, A. H.,** Role of fibronectin in collagenous matrix-induced mesenchymal cell proliferation and differentiation in vivo, *Exp. Cell Res.,* 133, 247, 1981.
48. **Pennypacker, J. P., Hassell, J. R., Yamada, K. M., and Pratt, R. M.,** The influence of an adhesive cell surface protein on chondrogenic expression in vitro, *Exp. Cell Res.,* 121, 411, 1979.
49. **West, C. M., Lanza, R., Rosenbloom, J., Lowe, M., and Holtzer, H.,** Fibronectin alters the phenotypic properties of cultured chick embryo chondroblasts, *Cell,* 17, 491, 1979.
50. **Hassell, J. R., Pennypacker, J. P., Kleinman, H. K., Pratt, R. M., and Yamada, K. M.,** Enhanced cellular fibronectin accumulation in chondrocytes treated with vitamin A, *Cell,* 17, 821, 1979.
51. **Bernard, B. A., DeLuca, L. M., Hassell, J. R., Yamada, K. M., and Olden, K.,** Retinoic acid alters the proportion of high mannose to complex type oligosaccharides on fibronectin secreted by cultured chondrocytes, *J. Biol. Chem.,* 259, 5310, 1984.
52. **Chiquet, M., Eppenberger, H. M., and Turner, D. C.,** Muscle morphogenesis. Evidence for organizing function of exogenous fibronectin, *Dev. Biol.,* 88, 220, 1981.
53. **Decker, C., Greggs, R., Duggan, K., Stubbs, J., and Horwitz, A.,** Adhesive multiplicity in the interaction of embryonic fibroblasts and myoblasts with extracellular matrices, *J. Cell Biol.,* 99, 1398, 1984.
54. **Nusgens, B., Delain, D., Senechal, H., Winard, R., Lapierre, Ch.M., and Wahrmann, J. P.,** Metabolic changes in extracellular matrix during differentiation of myoblasts of the L6 line and of a myo $^{-}$ non-fusing mutant, *Exp. Cell Res.,* 162, 51, 1986.
55. **Sampath, T. K., Nathanson, M. A., and Reddi, A. H.,** In vitro transformation of mesenchymal cells derived from embryonic muscle into cartilage in response to extracellular matrix components of bone, *Proc. Natl. Acad. Sci. U.S.A.,* 81, 3419, 1984.

56. **Syftestad, G. T., Triffitt, J. T., Urist, M. R., and Caplan, A. I.,** An osteo-inductive bone matrix extract stimulates the in vitro conversion of mesenchyme into chondrocytes, *Calcif. Tissue Int.,* 36, 625, 1984.
57. **Paranko, J., Pelliniemi, L. J., Vaheri, A., Foidart, J. M., and Lakkala-Paranko, T.,** Morphogenesis and fibronectin in sexual differentiation of rat embryonic gonads, *Differentiation,* 23 (Suppl.), 572, 1983.
58. **Chen, W. T., Chen, J. M., and Mueller, S. C.,** Coupled expression and colocalization of 140K cell adhesion molecules, fibronectin, and laminin during morphogenesis and cytodifferentiation of chick lung cells, *J. Cell Biol.,* 103, 1073, 1986.
59. **Patel, V. P. and Lodish, H. L.,** The fibronectin receptor on the mammalian erythroid precursor cells: characterization of developmental regulation, *J. Cell Biol.,* 102, 449, 1986.
60. **Duband, J. L., Dufour, S., Hatta, K., Takeichi, M., Edelman, G. M., and Thiery, J. P.,** Adhesion molecules during somitogenesis in the avian embryo, *J. Cell Biol.,* 104, 1361, 1987.
61. **Raghow, R., Kang, A. H., and Pidikiti, D.,** Phenotypic plasticity of extracellular matrix gene expression in cultured hamster lung fibroblasts, *J. Biol. Chem.,* 262, 8409, 1987.
62. **Lavietes, B. B., Carsons, S., Diamond, H. S., and Laskin, R. S.,** Synthesis, secretion and deposition of fibronectin in cultured human synovium, *Arthritis Rheum.,* 28, 1016, 1985.
63. **Oliver, N., Neuby, R. F., Furcht, L. T., and Bourgeois, S.,** Regulation of fibronectin biosynthesis by glucocorticoid in human fibrosarcoma cells and normal fibroblasts, *Cell,* 33, 287, 1983.
64. **Raghow, R., Gossage, D., and Kang, A. H.,** Pretranslational regulation of type I collagen, fibronectin, and a 50-kilodalton noncollagenous extracellular protein by dexamethasone in rat fibroblasts, *J. Biol. Chem.,* 261, 4677, 1986.

Chapter 3

FIBRONECTIN AND AGING

Albert J. T. Millis and David M. Mann

TABLE OF CONTENTS

I. INTRODUCTION

A. AGING, A PERSPECTIVE

The biological phenomenon of organismal aging involves a complex series of progressive and irreversible changes. Aging is associated with a diminution in the efficiencies of many physiological activities including muscle output, cardiac output, and sensory perception.[1] Common manifestations of aging include atrophy of organs and tissues; decrease in the resistance to infection; the onset of certain diseases (cardiovascular and arthritic); and trauma. There is no simple definition or adequate comprehensive description of the aging process because each organ and tissue system expresses specific manifestations of the process. In short, aging appears to be a composite of the physiological changes occurring in various tissue systems.[2] The timing of the onset of age-associated changes shows considerable variation among individuals and apparently reflects differences in genetic and environmental factors. Those factors can influence the rate of aging, including the onset and severity of age-related diseases.[3]

Experimental investigations into the aging process of humans and other animals have utilized either whole animal systems, isolated tissues, or *in vitro* model systems which appear to mimic limited aspects of the aging process.[4-6] In recent years, gerontologists have begun to utilize each of these systems to focus on the molecular basis of aging.[7-9]

Investigations of the molecular basis of aging have been facilitated by the development and refinement of cell culture methodology. Current studies of *in vitro* cellular aging are based on the important discovery that cultures of human fibroblasts have a finite proliferative lifespan, *in vitro*. During the 1960s, Hayflick and his collaborators proposed that the progressive retardation of growth and the concomitant decrease in population doubling potential of nontransformed human fibroblasts was an expression of aging at the cellular level.[10] Subsequently, those initial observations were confirmed using different strains of human fibroblasts and other types of "differentiated cells".[6]

There is additional evidence to support the contention that *in vitro* cellular aging represents a valid model system with which to investigate aging, *in vivo*. For example, there is a significant correlation between the *in vitro* lifespan of animal cells growing in tissue culture and the age of the donor organism.[11-13] Cells cultured from young donors proliferate more extensively than cells derived from older donors. In addition, cells cultured from species with relatively short *in vivo* lifespans have less proliferative capacity, *in vitro*, than cells cultured from longer-lived species. The maximum number of population doublings achieved, *in vitro*, is proportional to the mean maximal life span of the donor species.[3,14,15] Finally, the proliferative capacity of fibroblasts cultured from patients suffering from premature aging conditions such as Werner's syndrome, Down's syndrome, diabetes, Hutchinson-Gilford syndrome, or progeroid syndromes[3] is less than that of age-matched control cells.[5-7,15] The subject of cell aging and senescence has been the focus of several recent review articles.[5,6,16,17]

Many age-related phenomena appear to involve changes in connective tissue properties such as cell adhesion, cell recognition, and tissue organization. Therefore, investigators have begun to analyze age-related changes in connective tissue macromolecules including collagens, elastin, proteoglycans, and glycoproteins.[18,19] The objective of this article is to review the relationship between the connective tissue glycoprotein, fibronectin, and aging and to discuss the kinds of mechanisms that can generate age-associated changes in fibronectin structure and deposition. *In vitro* studies of the relationship between fibronectin and aging have focused on fibroblasts as the experimental model.

B. FIBRONECTINS

Fibronectins are high molecular weight glycoproteins present on cell surfaces, in connective tissue matrices, in basement membranes, and in extracellular fluids including plasma.

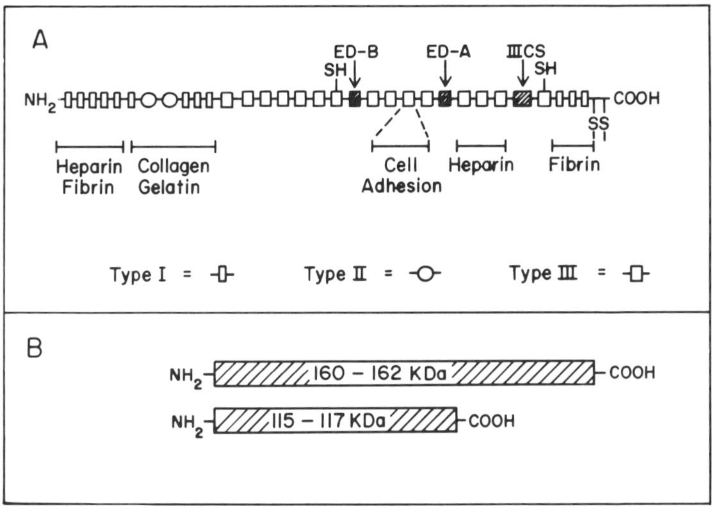

FIGURE 1. Model of the fibronectin polypeptide. Panel A shows each of the three classes of sequence homology units (types I, II, and III) and the sites for interaction with heparin, collagen, fibrin, and the cell adhesion receptor. Alternative splicing of mRNA has been shown to generate different isoforms in the sites indicated as ED-B, ED-A, or IIICS regions. Figure is redrawn and modified from Zardi, L. et al.[53] (Panel B) Proposed location of fibronectin polypeptide fragments whose relative mobilities differ in fibronectin preparations isolated from early and late passage cell culture media. The assignment of the positions of the fragments in the fibronectin polypeptide was based on apparent molecular weights and demonstrated affinities for either gelatin or heparin. As shown, the 160- to 162-kDa fragment is positioned (in relation to the model in Panel A) to show an affinity for heparin and the absence of affinity for gelatin. The 115- to 117-kDa fragment is positioned to demonstrate an absence of affinity for either gelatin or heparin.

Different forms of fibronectin comprise a group of structurally related, large glycoproteins with similar biological activities.[20-31] The fibronectin polypeptide interacts with cells and extracellular matrix macromolecules through specific sites on the polypeptide chain.[30-36] Figure 1A shows a representation of a fibronectin polypeptide (monomer) emphasizing regions of peptide sequence homology (types I, II, and III) and the molecular domains that interact with cells, collagens, glycosaminoglycans, and fibrin.[23-26] The location of peptide regions believed to be modified as a consequence of mRNA splicing are also indicated (ED-A, ED-B, and IIICS). Their significance will be discussed later in this article.

II. FIBRONECTIN IN CULTURED CELLS, AGING *IN VITRO*

A. CELLULAR SENESCENCE

In vitro aging ("cellular senescence") has been studied using cells explanted and cultured from young and old donors; from donors with age-associated diseases; and cells that were initially derived from young donors but subsequently aged (via serial subcultivation) *in vitro*. Studies of fibronectin production and aging have been restricted to fibroblast cultures derived from donors with age-associated diseases and to cultures aged *in vitro*.

The "*in vitro* lifespan" of a strain of human fibroblasts can be defined as the maximum number of population doublings attainable by the culture. For embryonic human fibroblasts, this number is approximately 50 ± 10 population doublings.[37] The "*in vitro* age" of a culture refers to the percentage of the *in vitro* lifespan completed at the time of the assay. In practice, cells are seeded into culture vessels, grown to confluency, harvested, counted,

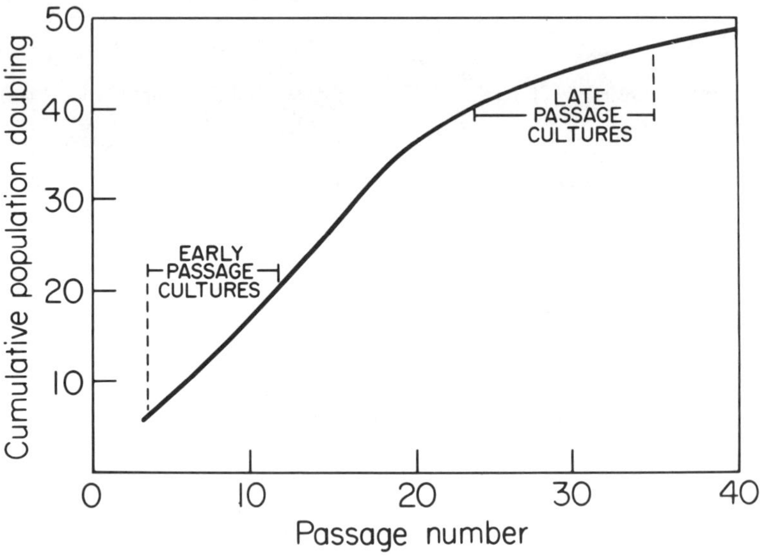

FIGURE 2. *In vitro* lifespan of cultured human fibroblasts. The relationship between passage number and the cumulative number of population doublings is shown for a strain of nontransformed fibroblasts. In general, the early and late passage cultures, referred to in the text, had completed less than 40% or more than 80% of the total *in vitro* lifespan. The data presented are from strain RIG,[54,82] a strain derived from a newborn foreskin biopsy, and represents cells grown in medium 199 supplemented with 20% fetal bovine serum. This figure was provided by Dr. Jane Sottile, Department of Biological Chemistry, University of Wisconsin School of Medicine, Madison.

and reseeded at a subconfluent density (usually less than 50% of the confluent density). This process is referred to as cell passage, and the passage number associated with a culture indicates the number of times that specific cell lineage has been harvested and reseeded in this manner.

The number of population doublings between each passage can be calculated by knowing the number of cells previously seeded in the vessel, the plating efficiency, and the number of cells present at the time of harvesting. A plot showing the relationship between cumulative populations doublings and passage number is shown in Figure 2. Conventionally, early passage cultures have completed less than 40% of the *in vitro* lifespan and are considered to be composed of "young cells". Similarly, late passage cultures have completed more than 80% of the *in vitro* lifespan and are composed of "old cells". As Figure 2 demonstrates, early passage cultures proliferate more rapidly than late passage cells.

In contrast to early passage cell cultures, late passage cell cultures exhibit reductions in cell motility, the rate of cell spreading upon the substrate, and plating efficiency.[9,38] All of those parameters can be mediated by fibronectin.

B. FIBRONECTIN IN CONDITIONED MEDIA

During the course of *in vitro* aging, human fibroblasts continue to synthesize and secrete fibronectin into the culture medium ("conditioned medium"). However, reports differ as to whether the amount synthesized is influenced by the *in vitro* age of the culture.[9,38,40] One report indicated that in comparison to early passage cultures, late passage cells secreted twice as much fibronectin into the culture medium.[39] In contrast, however, recent kinetic studies indicated that over a 24-h period, confluent cultures of early and late passage cultures secreted similar levels of fibronectin into the medium.[40] It is likely that this apparent contradiction results from differences in the procedures used to calculate the amount of fibronectin rather than from age-associated changes in the amount produced. For example, when normalized to cell number, it appeared that late passage cells secreted more fibronectin than was secreted by early passage cells, but when normalized to total cellular protein (milligrams per culture) both early and late passage cultures secreted similar levels.[41]

Diabetes mellitus represents an age-associated disease and the onset of diabetes a physiological marker for aging.[42] Analysis of cells cultured from genetically diabetic mice indicated that they secreted twice as much fibronectin per cell as control cultures.[43] Analysis of the fibronectin synthesized and secreted by late passage fibroblasts indicated that it was functionally and structurally distinct from fibronectin produced by the same cell strain, at earlier passage levels. For example, the decrease in cell to substrate adhesion, characteristic of late passage fibroblasts, was partially reversed by providing the culture with fibronectin isolated from early passage cultures but not with fibronectin isolated from late passage cultures.[9] The adhesive contacts produced in response to early and late passage cell fibronectins were analyzed via interference reflection microscopy, and it was reported that the two fibronectins differed in their abilities to promote the formation of adhesive sites known as focal contacts.[44] Fibronectin isolated from aged cell cultures did not fully support either the formation of focal contacts or a normal cell morphology by early or late passage cells.[9,44] Because aged cells formed focal contacts in response to other fibronectins, those findings were interpreted as evidence that the aged cell fibronectin was defective in promoting cell-substrate adhesion. However, the possibility remains that the differential effect was produced by an undetected component that copurified with the secreted fibronectin.

The ability of soluble cellular fibronectins to interact with collagens, heparin, and the cell surface was also evaluated. The aged cell fibronectin showed reduced binding to native interstitial collagens (types I and II) even in the presence of heparin.[45] This suggests the possibility of defective interactions between fibronectin and other extracellular matrix components, *in vivo*. The functional differences appear to be subtle, however, as binding to denatured collagens (gelatin) was similar for both fibronectins. Further, in a Boyden-chamber assay of cell migration, fibronectins from both early and late passage cell cultures were equally active.[45] The basis for the apparent functional deficiency of aged cell fibronectin was not established, but it was postulated that it may have resulted from a structural alteration in one or more of the domains necessary for maintaining the characteristic phenotype of young fibroblasts in culture.

Fibronectin structure was evaluated using polyacrylamide gel electrophoresis (PAGE). Comparison of secreted fibronectins established that although both were present as dimers, the aged cell fibronectin migrated more slowly than young cell fibronectin, suggesting that the two may differ in molecular weight.[45,46] Figure 3 shows those differences in a comparison of the electrophoretic mobilities of reduced preparations of human fibronectins. A similar comparison of fibronectins extracted from the cell layer, with 1 *M* urea, also indicated that the fibronectin synthesized by late passage cells had a greater apparent molecular weight.[46]

A combination of the techniques of limited proteolysis followed by affinity chromatography and PAGE revealed structural differences in the fibronectins isolated from early and late passage cultures. Two of the polypeptide fragments derived from late passage cell fibronectin migrated more slowly than the corresponding fragments from early passage cell fibronectin. Both sets of fragments lacked affinity for gelatin and were localized to the central region of the intact fibronectin polypeptide (as represented in Figure 1B). One of those polypeptide fragments was a heparin-binding fragment of 160 to 162 kDa and the other had an apparent molecular weight of 115 to 117 kDa. It is possible that these differences were due to primary sequence differences, since the central portion of the fibronectin molecule has been shown to contain several sites of sequence variation (Figure 1A).[47-53] However, differences in the posttranslational modifications of the fibronectins produced by early and late passage cells were not ruled out.

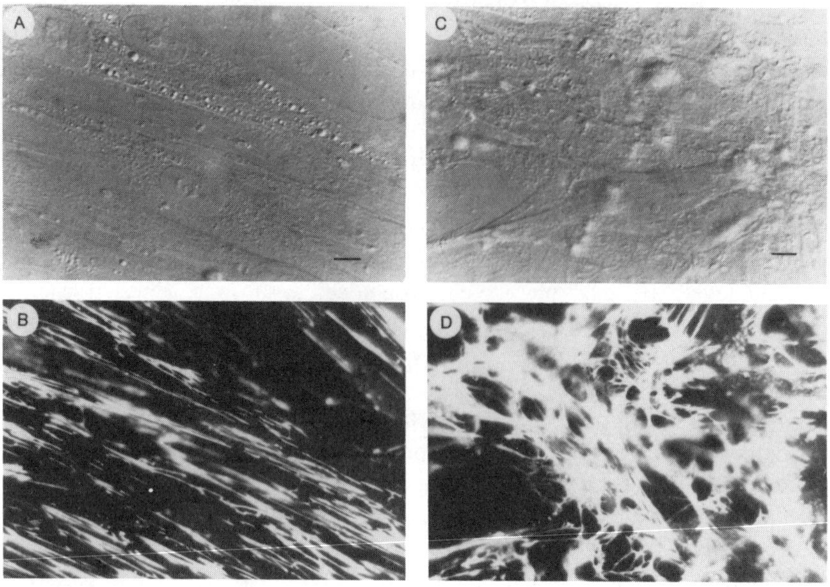

FIGURE 3. Relative mobilities of human fibronectins. Fibronectins isolated from human plasma (A), bovine plasma (B), conditioned medium produced by an early passage culture of human fibroblasts (C), and conditioned medium produced by a late passage culture of human fibroblasts (D) were transferred from a polyacrylamide slab gel to nitrocellulose paper and probed with a monoclonal antifibrinonectin antibody specific for human fibronectin. (From Sorrentino, J. A. and Millis, A. J. T., *Mech. Age. Dev.*, 28, 83—97, 1984. With permission of Elsevier Scientific Publishers Ireland Ltd.)

C. FIBRONECTIN IN THE CELL LAYER

Two early studies demonstrated that the level of an iodinated cell surface component of chick embryo fibroblasts was modified as a function of culture age.[54,55] Although both of those proteins had molecular weights of approximately 220 kDa and are now presumed to be fibronectin, there was no further characterization. More recently, immunofluorescence microscopy was used to visualize fibronectin in human fibroblast cell layers, and it was reported that late passage cells retained only minimal levels of immunoreactive protein.[39] In contrast, another study demonstrated that confluent late passage cells were associated with fibronectin fibrils that were thicker and more disorganized than the fibers associated with early passage cells (Figure 4).[56] The basis for the differences in those reports has not been established, but it may reflect differences in culture conditions. One set of cultures was incubated in culture medium for several weeks after passaging and then assayed[39] while the other set was examined just 7 d after passaging. Differences in cell density can also contribute to the distribution patterns of cell surface fibronectin.

In addition, a detailed study of fibronectin synthesis, in human fibroblasts, concluded that late passage cells synthesized and retained in the cell layer more fibronectin per cell than early passage cultures, but when expressed as fibronectin per milligram of protein, a difference was not observed. They also established that the rates of fibronectin degradation were similar in cultures of both ages.[41] An immunoassay was used to quantitate the levels of fibronectin retained in the cell layers of confluent cultures at equivalent cell densities of early and late passage cells. Those results established that twice as much fibronectin was retained in the late passage cell layers than in early passage cell layers.[40] The relationship between fibronectin production and incorporation in cells cultured from diabetic mice also showed an increase in the relative incorporation of fibronectin into the matrix.[43]

Fibronectin can interact with cultured cells either by providing an insoluble substrate

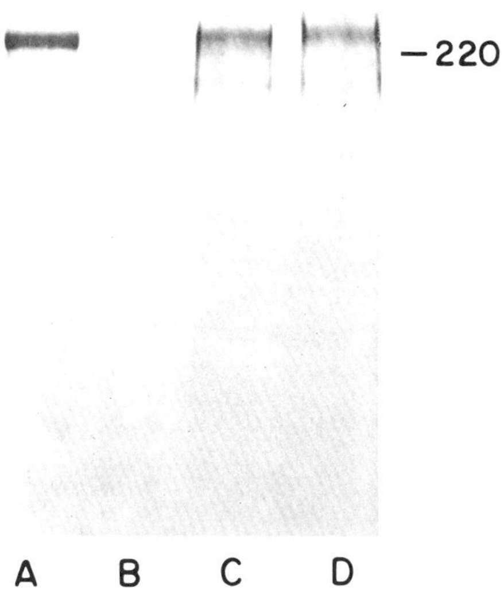

FIGURE 4. Nomarski and fluorescent images of human fibroblasts. Early (A,B) and late passage (C,D) human fibroblasts (strain RIG) were stained with a polyclonal antifibronectin antibody. Nomarski (A,C) and fluorescent (B,D) images are shown. Bar = 10 μM. (From Edick, G. F. and Millis, A. J. T., *Mech. Age. Dev.*, 27, 249—256, 1984. With permission of Elsevier Scientific Publishers Ireland Ltd.)

for the attachment of suspended cells or through the binding of soluble fibronectin to cells already attached to a substrate. In the first example, cells attach to a fibronectin-coated substrate via receptor-mediated binding to a site in the carboxyterminal region of the cell adhesion domain of fibronectin (indicated in Figure 1A).[57-61] Adhesion appears to require interaction between the amino acid sequence Arg-Gly-Asp-Ser (RGDS) of fibronectin and a member of the integrin family of cell surface receptors.[25,61-63] Recent studies, using antibodies to the human fibronectin receptor, demonstrated that early and late passage cells express immunoprecipitable integrin. However, the actual number of receptors per cell was not quantified.[89]

There is also evidence that fully spread cells interact with fibronectin by binding at a site in the aminoterminal region of the soluble fibronectin molecule.[64,65] A cell surface receptor molecule for this ''matrix organizing domain'' of fibronectin has not been isolated; however, recent studies implicate an 18-kDa cell surface molecule as a candidate for the receptor molecule.[66] A comparative study of the binding of soluble (plasma-derived) fibronectin to confluent cell layers established that although both cultures bound iodinated fibronectin, there were differences in the amount bound and the fate of the bound protein. Scatchard analysis of fibronectin bound into the deoxycholate-soluble fraction indicated that late passage cells had 1.42×10^6 and early passage cells had 0.63×10^6 binding sites per cell.[67] Further, the two cell cultures processed the bound fibronectin differently. The late passage cells transferred more of the bound fibronectin into the deoxycholate-insoluble fraction (believed to represent incorporation into the extracellular matrix) than did early passage cells.[42,67-71] The turnover of cell layer fibronectin was similar in early and late passage cells; with a $T_{1/2}$ of 4 to 5 d.[40,41,67]

III. FIBRONECTIN IN PLASMA AND AMNIONIC FLUID

In general, the concentration of human plasma fibronectin appears to increase with age. In one study, the immunoreactive plasma fibronectin concentration was increased from 105 to 210 μg/ml in serum samples from donors ranging in age from 1 month to 3 years. Subsequently the level remained relatively constant from 3 to 15 years.[72] Among older donors, further significant increases were observed.[73] Comparison of structure or function of those plasma-derived samples was not reported.

Levels of plasma fibronectin in patients affected with diabetes were also measured and compared to age-matched control samples. Interestingly, the age-associated increase observed in the control samples was present but at greatly attenuated levels in the samples derived from donors with diabetes.[74]

Structural comparison of fibronectins derived from fetal and adult human plasma indicated that although the two proteins were very similar, there were differences in carbohydrate composition. In contrast to adult plasma, fetal plasma fibronectin contained a substantial amount of fucosylated biantennary glycans and an increased level of glycopeptides that were not bound by concanavalin A.[75] Other glycosylation differences between amnionic fluid fibronectin and adult plasma fibronectin[76-78] and between fetal placental fibronectin and plasma fibronectin were also reported.[79,80]

IV. FIBRONECTIN IN TISSUES

Fibronectin distribution was analyzed in tissues obtained from donors afflicted with disease conditions associated with advancing age. In tissue sample from diabetics, there was an increase in the level of immunoreactive fibronectin.[81] The most significant increases were in dermoepidermal and capillary basement membranes. Increased tissue-associated fibronectin was also reported to occur in atherosclerotic lesions.[82] In addition, it was reported that skin biopsy samples from patients with Werner's syndrome, a condition of premature aging,[83] displayed increased immunoreactive fibronectin in the papillary dermis.[81]

Studies with mouse skin explants demonstrated that radiolabeled methionine was incorporated more efficiently into fibronectin by skin from a strain with inherited diabetes than by control skin samples from nonaffected mice.[74] It was not established whether the fibronectin in the skin and vascular tissues were derived from plasma or produced by the tissue cells located in that region. Cells and tissues can incorporate both types of fibronectin into the extracellular matrix and knowledge of the source of the fibronectin may be important.[84,85]

V. SUMMARY AND CONCLUSIONS

The results presented in this review imply age-associated changes in the relationship between fibronectin and connective tissue cells. Current data suggest that those changes may result from structural differences in the fibronectin protein as well as differences in the way fibronectin is incorporated and organized in the extracellular matrix. The concentration of fibronectin in human plasma is increased during human aging; the association of fibronectin with connective tissue or connective tissue cells is altered in diabetic tissue and as a function of "*in vitro* aging"; and the fibronectin produced by *in vitro* aged cells appear to have diminished capability to promote cell adhesion and to bind to native collagens. It is essential to emphasize that these conclusions are preliminary as they are drawn from a relatively small number of studies. Investigations using other cell types as well as tissue samples and materials from donors with syndromes of premature senescence are necessary to substantiate these relationships. However, these studies of fibronectin and aging do bring into focus two areas that merit discussion and further study; the structure of the fibronectin polypeptides and the interaction of fibronectin with the cell surface or extracellular matrix.

A. FURTHER PROSPECTS, FIBRONECTIN STRUCTURE

There is evidence that differences in fibronectin structure can result either from post-translational modifications or from differences in the primary amino acid sequence of the fibronectin polypeptide.[86] For example, it is well established that changes in carbohydrate composition affect molecular weight, sensitivity to proteolysis, and function.[30-35] This suggests the possibilities that the age-associated increase in plasma fibronectin may result from a posttranslational modification affecting turnover, from translation of different sets of fibronectin mRNAs, or from increased synthesis and secretion. These questions should be investigated.

The results of nucleic acid and amino acid sequencing studies of fibronectin and the fibronectin coding region of the genome provide convincing evidence for the existence of multiple isoforms of fibronectin.[26] The consensus, derived from Southern blotting studies of the human, chick, and rat genomic DNAs, is that each species contains a single gene for fibronectin. However, experiments using nucleotide probes derived from fibronectin cDNAs revealed the possibility that each cell line or species can produce multiple species of fibronectin mRNA and protein.[47-52,87] The different fibronectin mRNAs can be generated by alternative splicing, between exons and within an exon, in the regions indicated as ED-A, ED-B, and IIICS on Figure 1A.[53] In addition, fibronectin mRNAs ranging in size from 7.7 to 8.6 kb[88] have been identified in RNA preparations isolated from human cells, further supporting the concept that cells synthesize multiple isoforms of fibronectin mRNA. Presently it is not known whether proteins translated from the different mRNAs are functionally distinct.

Analysis of fibronectin mRNAs isolated from aged cells or tissues has not been reported, but the technology to facilitate that analysis is available. It will be important to know whether the mRNAs used to produce fibronectin in aged cells and tissues are similar or different from those used by young cells and tissues.

B. FIBRONECTIN FUNCTION IN AGING

Considered collectively, the studies reviewed here indicate that late passage cells and aged tissues may be more effective than early passage cells and young tissues in promoting fibronectin fibrillogenesis. It is possible that the increase in the level of plasma fibronectin contributes to the increased deposition of fibronectin observed in tissue from older individuals. Given the potential significance of matrix fibronectin, the mechanism of fibronectin fibrillogenesis should be examined in detail.

Undoubtedly the fibronectins are important for maintaining tissue homeostasis. Age-related changes in the localization, structure, and function of fibronectins have been reported. Understanding the basis and effects of these changes is likely to contribute to our understanding of the complex series of events referred to as aging. Continued studies in this field may yield new information regarding age-related pathologies including tissue and joint degenerative diseases, cardiovascular diseases, and defective wound healing.

ACKNOWLEDGMENTS

The authors' research was supported by grants received from the National Institutes of Health (National Institute on Aging) and the National Science Foundation.

REFERENCES

1. **Burnet, F. M.,** in *Immunology, Aging, and Cancer,* W. H. Freeman, San Francisco, 1976, 81.
2. **Cutler, R. G.,** in *Aging and Cell Function,* Johnson, J. E., Ed., Plenum Press, New York, 1984, 1.
3. **Sacher, G. A. and Hart, R. W.,** in *Genetic Effects on Aging,* Bergsma, D., Harrison, D. E., and Paul, N. W., Eds., Alan Liss, New York, 1977, 73.
4. **Hayflick, L.,** in *Handbook of the Biology of Aging,* 2nd ed., Finch, C. E. and Hayflick, L., Eds., Van Nostrand-Reinhold, New York, 1985, 159.
5. **Norwood, T. H. and Smith, J. R.,** in *The Handbook of the Biology of Aging,* 2nd ed., Finch, C. E. and Hayflick, L., Eds., Van Nostrand-Reinhold, New York, 1985, 291.
6. **Smith, J. R. and Lincoln, D. W.,** in *International Reviews in Cytology,* Vol. 89, Academic Press, New York, 1984, 151.
7. **Woodhead, A. D., Blackett, A. D., and Hollaender, A.,** *Molecular Biology of Aging,* Plenum Press, New York, 1985, 1.
8. **Lumpkin, C. K., McClung, J. K., Pereira-Smith, O. M., and Smith, J. R.,** Existence of high abundance antiproliferative mRNA's in senescent human diploid fibroblasts, *Science,* 232, 393, 1986.
9. **Chandrasekhar, S. and Millis, A. J. T.,** Fibronectin from aged fibroblasts is defective in promoting cellular adhesion, *J. Cell. Physiol.,* 103, 47, 1980.
10. **Hayflick, L.,** The limited in vitro lifetime of human diploid cell strains, *Exp. Cell Res.,* 37, 614, 1965.
11. **Schneider, E. L. and Mitsui, Y.,** The relationship between in vitro cellular aging and in vivo human aging, *Proc. Natl. Acad. Sci. U.S.A.,* 73, 3584, 1976.
12. **Martin, G. M.,** Cellular aging-clonal senescence. A review (part I), *Am. J. Pathol.,* 89, 484, 1977.
13. **Martin, G. M., Sprague, C. A., and Epstein, C. J.,** Replicative life-span of cultivated human cells, *Lab. Invest.,* 23, 86, 1970.
14. **Rohme, D.,** Evidence for a relationship between longevity of mammalian species and life spans of normal fibroblasts in vitro and erythrocytes in vivo, *Proc. Natl. Acad. Sci. U.S.A.,* 8, 5009, 1981.
15. **Cristofalo, V. J. and Stanulis-Praeger, B. M.,** in *Advances in Cell Culture,* Vol. 2, Maramorosch, K., Ed., Academic Press, New York, 1982, 1.
16. **Macieria-Coelho, A.,** Changes in membrane properties associated with cellular aging, *Int. Rev. Cytol.,* 83, 183, 1983.
17. **Stanulis-Praeger, B. M.,** Cellular senescence revisited: a review, *Mechan. Age. Dev.,* 38, 48, 1987.
18. **Labat-Robert, J.,** in *Structural Glycoproteins in Cell-Matrix Interactions,* Labat-Robert, J., Timpl, R., and Robert, L., Eds., Karger, Basel, 1986, 17.
19. **Aizawa, S., Mitsui, Y., Kurimoto, F., and Nomura, K.,** Cell surface changes accompanying aging in human diploid fibroblasts, *Exp. Cell Res.,* 127, 143, 1980.
20. **Yamada, K. M.,** Cell surface interactions with extracellular materials, *Annu. Rev. Biochem.,* 52, 761, 1983.
21. **Hynes, R. O.,** in *Cell Biology of the Extracellular Matrix,* Hay, E. D., Ed., Plenum Press, New York, 1981, 95.
22. **Yamada, K. M.,** in *Cell Biology of the Extracellular Matrix,* Hay, E. D., Ed., Plenum Press, New York, 1981, 95.
23. **Ruoslahti, E., Engvall, E., and Hayman, E. G.,** Fibronectin: current concepts of its structure and function, *Collagen Rel. Res.,* 1, 95, 1981.
24. **Vaheri, A., Salonen, E. M., and Vartio, T.,** in *Fibrosis (Ciba Foundation Symposium,* Vol. 114), Evered, D., and Whelan, J., Eds., Pitman, London, 1985, 111.
25. **Hakomori, S., Fukuda, M., Sekiguchi, K., and Carter, W. G.,** in *Extracellular Matrix Biochemistry,* Piez, K. A. and Reddi, A. H., Eds., Elsevier, New York, 1984, 229.
26. **Ruoslahti, E.,** Fibronectin and its receptors, *Annu. Rev. Biochem.,* 57, 375, 1988.
27. **Yamada, K. M., Olden K., and Hahn, L. H. E.,** in *The Cell Surface: Mediator of Developmental Processes,* Subtelny, S. and Wessells, N. K., Eds. Academic Press, New York, 1980, 1.
28. **Yamada, K. M., Hayashi, M., Hirano, H., and Akiyama, S. K.,** in *The Role of Extracellular Matrix in Development,* Alan Liss, New York, 1984, 37.
29. **Yamada, K. M., Humphries, M. J., Hasegawa, T., Hasegawa, E., Olden, K., Chen, W., and Akiyama, S. K.,** in *The Cell in Contact. Adhesions and Junctions as Morphogenetic Determinants,* Edelman, G. M. and Thiery, J., Eds., John Wiley & Sons, New York, 1985, 303.
30. **Yamada, K. M. and Olden, K.,** Fibronectins—adhesive glycoproteins of cell surface and blood, *Nature (London),* 275, 179, 1978.
31. **Ruoslahti, E., Hayman, E. G., Pierschbacher, M. D., and Engvall, E.,** in *Methods in Enzymology,* Vol. 82, Academic Press, New York, 1982, 803.
32. **Mosesson, M. W., Chen, A. B., and Huseby, R. M.,** The cold insoluble globulin of human plasma: studies of its essential structural features, *Biochem. Biophys. Acta,* 386, 509, 1975.

33. **Yamada, K. M., Scheslinger, D. H., Kennedy, D. W., and Pastan, I.,** Characterization of a major fibroblast cell surface protein, *Biochemistry,* 16, 5552, 1977.
34. **Yamada, K. M. and Kennedy, D. W.,** Fibroblast cellular and plasma fibronectins are similar but not identical, *J. Cell Biol.,* 80, 492, 1979.
35. **Ruoslahti, E., Engvall, E., Hayman, E. G., and Spiro, R. G.,** Comparative studies on amnionic fluid and plasma fibronectins, *Biochem. J.,* 193, 295, 1981.
36. **Ruoslahti, E., Hayman, E. G., Engvall, E., Cothran, W. C., and Butler, W. T.,** Alignment of biologically active domains in the fibronectin molecule, *J. Biol. Chem.,* 256, 7277, 1981.
37. **Hayflick, L.,** Cell biology of aging, *Fed. Proc.,* 38, 1847, 1979.
38. **Kelley, R. O., Vogel, K. G., Crissman, H. A., Lujan, C. J., and Skipper, B. E.,** Development of the ageing cell surface: reduction of gap junction-mediated metabolic cooperation with progressive subcultivation of human embryo fibroblasts (IMR-90), *Exp. Cell Res.,* 119, 127, 1979.
39. **Vogel, K. G., Kelley, R. O., and Stewart, C.,** Loss of organized fibronectin matrix from the surface of aging diploid fibroblasts (IMR-90), *Mech. Age. Dev.,* 16, 295, 1981.
40. **Mann, D. M., McKeown-Longo, P. J., and Millis, A. J. T.,** Binding of soluble fibronectin and its subsequent incorporation into the extracellular matrix by early and late passage human skin fibroblasts, *J. Biol. Chem.,* 263, 2756, 1988.
41. **Shevitz, J., Jenkins, C. S. P., and Hatcher, V. B.,** Fibronectin synthesis and degradation in human fibroblasts with aging, *Mechan. Age. Dev.,* 35, 221, 1986.
42. **Goldstein, S. and Harley, C. B.,** In vitro studies of age-associated diseases, *Fed. Proc.,* 38, 1862, 1979.
43. **Phan-Thanh, L., Robert, L., Derouette, J. C., and Labat-Robert, J.,** Increased biosynthesis and processing of fibronectin in fibroblasts from diabetic mice, *Proc. Natl. Acad. Sci. U.S.A.,* 84, 1911, 1987.
44. **Chandrasekhar, S., Norton, E., Millis, A. J. T., and Izzard, C. S.,** Functional changes in cellular fibronectin from late passage fibroblasts in vitro, *Cell Biol. Int. Rep.,* 7, 11, 1983.
45. **Chandrasekhar, S., Sorrentino, J. A., and Millis, A. J. T.,** Interactions of fibronectin with collagen: age-specific defect in the biological activity of human fibroblast fibronectin, *Proc. Natl. Acad. Sci. U.S.A.,* 80, 4747, 1983.
46. **Sorrentino, J. A. and Millis, A. J. T.,** Structural comparisons of fibronectins isolated from early and late passage cells, *Mech. Age. Dev.,* 28, 83, 1984.
47. **Kornblihtt, A. R., Vibe-Pedersen, K., and Baralle, F. E.,** Human fibronectin: molecular cloning evidence for two mRNA species differing by an internal segment coding for a structural domain, *EMBO J.,* 3, 221, 1984.
48. **Kornblihtt, A. R., Vibe-Pedersen, K., and Baralle, F. E.,** Human fibronectin: cell specific alternative mRNA splicing generates polypeptide chains differing in the number of internal repeats, *Nucleic Acids Res.,* 12, 5853, 1984.
49. **Kornblihtt, A. R., Umezawa, K., Vibe-Pedersen, K., and Baralle, E.,** Primary structure of human fibronectin: differential splicing may generate at least 10 polypeptides from a single gene, *EMBO J.,* 4, 1755, 1985.
50. **Schwarzbauer, J. E., Tamkun, J. W., Lemischa, I. R., and Hynes, R. O.,** Three different fibronectin mRNAs arise by alternative splicing within the coding region, *Cell,* 35, 421, 1983.
51. **Schwarzbauer, J. E., Paul, J. I., and Hynes, R. O.,** On the origin of species of fibronectin, *Proc. Natl. Acad. Sci. U.S.A.,* 81, 1424, 1985.
52. **Schwartzbauer, J. E., Patel, R. S., Fonda, D., and Hynes, R. O.,** Multiple sites of alternative splicing of the rat fibronectin gene transcript, *EMBO J.,* 6, 2573, 1987.
53. **Zardi, L., Carnemolla, B., Siri, A., Petersen, T. E., Paolella, G., Sebastio, G., and Baralle, F. E.,** Transformed human cells produce a new fibronectin isoform by preferential alternative splicing of a previously unobserved exon, *EMBO J.,* 6, 2337, 1987.
54. **Fry, M. and Weissman-Shomer, P.,** Surface proteins of young and senescent cultured avian fibroblasts, *Cell Biol. Int. Rep.,* 1, 399, 1977.
55. **Courtois, Y. and Hughes, R. C.,** Surface labeling of senescent chick fibroblasts by lactoperoxidase-catalysed iodination, *Gerontology,* 22, 371, 1976.
56. **Edick, G. F. and Millis, A. J. T.,** Fibronectin distribution on the surfaces of young and old human fibroblasts, *Mech. Age. Dev.,* 27, 249, 1984.
57. **Klebe, R. J.,** Isolation of a collagen dependent cell attachment factor, *Nature (London),* 250, 248, 1974.
58. **Pearlstein, E.,** Plasma membrane glycoprotein which mediates adhesion of fibroblasts to collagen, *Nature (London),* 262, 479, 1976.
59. **Rubin, K., Johansson, S., Pettersson, I., Ocklink, C., Obrink, B., and Hook, M.,** Attachment of rat hepatocytes to collagen and fibronectin: a study using antibodies to cell surface components, *Biochem. Biophys. Res. Commun.,* 81, 86, 1979.
60. **Ruoslahti, E. and Hayman, E. G.,** Two active sites with different characteristics in fibronectin, *FEBS Lett.,* 97, 221, 1979.

61. **Grinnell, F.,** Fibroblast receptor for cell-substratum adhesion: studies on the interaction of baby hamster kidney cells with latex beads coated by cold-insoluble globulin (plasma fibronectin), *J. Cell Biol.,* 86, 104, 1980.
62. **Pierschbacher, M. D., Hayman, E. G., and Ruopslahti, E.,** Synthetic peptide with cell attachment activity of fibronectin, *Proc. Natl. Acad. Sci. U.S.A.,* 80, 1224, 1983.
63. **Hynes, R. O.,** Integrins: a family of cell surface receptors, *Cell,* 48, 549, 1987.
64. **McKeown-Longo, P. J. and Mosher, D. F.,** Binding of plasma fibronectin to cell layer of human skin fibroblasts, *J. Cell Biol.,* 97, 466, 1983.
65. **McKeown-Longo, P. J. and Mosher, D. F.,** Interaction of the 70,000 molecular weight amino-terminal fragment of fibronectin with the matrix-assembly receptor of fibroblasts, *J. Cell Biol.,* 100, 364, 1985.
66. **McKeown-Longo, P. J.,** Identification of an 18 kilodalton cell surface protein which binds to an amino-terminal fragment of fibronectin, *J. Cell Biol.,* 106, 241a, 1987.
67. **Mann, D. M.,** Assembly of Fibronectin into the Extracellular Matrix of Early and Late Passage Human Skin Fibroblasts, Ph.D. thesis, State University of New York, Albany, 1987, 1.
68. **Hedman, K., Kurkinen, M., Alitalo, K., Vaheri, A., Johansson, S., and Hook, M.,** Isolation of the pericellular matrix of human fibroblast cultures, *J. Cell Biol.,* 81, 83, 1979.
69. **Selmin, O., Bressan, G., Bortolussi, M., and Volpin, D.,** Cell surface-collagen interaction: binding sites for type 1 collagen, on human and bovine fibroblasts, *Cell Biol. Int. Rep.,* 10, 160, 1986.
70. **Choi, M. G. and Hynes, R. O.,** Biosynthesis and processing of fibronectin in NIL.8 hamster cells, *J. Biol. Chem.,* 254, 12050, 1979.
71. **Goldberg, B.,** Binding of soluble type I collagen to the fibroblast plasma membrane, *Cell,* 16, 265, 1979.
72. **McCaferty, M. H., Lepow, M., Saba, T. M., Cho, E., Meuwissen, H., White, J., and Zuckerbrod, S. F.,** Normal fibronectin levels as a function of age in the pediatric population, *Pediatr. Res.,* 17, 482, 1983.
73. **Labat-Robert, J., Potazman, J. P., Derouette, J. C., and Robert, L.,** Age-dependent increase of human plasma fibronectin, *Cell Biol. Int. Rep.,* 5, 969, 1981.
74. **Labat-Robert, J. and Robert, L.,** Tissue and plasma fibronectin in diabetes, *Monogr. Atheroscler.,* 13, 164, 1985.
75. **Yamaguchi, Y., Isemura, M., Kosakai, M., Sato, A., Suzuki, M., Kan, M., and Yosizawa, Z.,** Characterization of fibronectin from fetal human plasma in comparison with adult plasma fibronectin, *Biochem. Biophys. Acta,* 790, 53, 1984.
76. **Ruoslahti, E., Engvall, E., Hayman, E. G., and Spiro, R. G.,** Comparative studies on amnionic fluid and plasma fibronectins, *Biochem. J.,* 193, 295, 1981.
77. **Krusius, T., Fukuda, M., and Ruoslahti, E.,** Structure of the carbohydrate units of human amnionic fluid fibronectin, *J. Biol. Chem.,* 260, 4110, 1986.
78. **Yamaguchi, Y., Isemura, M., Yosizawa, Z., Kan, M., and Sato, A.,** A weaker gelatin-binding affinity and increased glycosylation of amnionic fluid fibronectin than plasma fibronectin, *Int. J. Biochem.,* 18, 437, 1986.
79. **Zhu, B.C-R., Fisher, S. R., Pande, H., Calaycay, J., Shively, J. E., and Laine, R. A.,** Human placental (fetal) fibronectin: increased glycosylation and higher protease resistance than plasma fibronectin, *J. Biol. Chem.,* 259, 3962, 1984.
80. **Zhu, B. C-R. and Laine, R. A.,** Polylactosamine glycosylation on human fetal placental fibronectin weakens the binding affinity of fibronectin to gelatin, *J. Biol. Chem.,* 260, 4041, 1985.
81. **Labat-Robert, J. and Robert, L.,** Modifications of fibronectin in age-related diseases: diabetes and cancer, *Arch. Gerontol. Geriatr.,* 3, 1, 1984.
82. **Labat-Robert, J., Szendroi, M., Godeau, G., and Robert, L.,** Comparative distribution patterns of type I and III collagens and fibronectin in human arteriosclerotic aorta, *Path. Biol.,* 33, 261, 1985.
83. **Salk, D., Bryant, E., Au, K., Hoehn, H., and Martin, G. M.,** Systematic growth studies, cocultivation, and cell hybridization studies of Werner syndrome cultured skin fibroblasts, *Hum. Genet.,* 58, 310, 1981.
84. **Oh, E., Pierschbacher, M. D., and Ruoslahti, E.,** Deposition of plasma fibronectin in tissues, *Proc. Natl. Acad. Sci. U.S.A.,* 78, 3218, 1981.
85. **Millis, A. J. T., Hoyle, M., Mann, D. M., and Brennan, M. J.,** Incorporation of cellular and plasma fibronectins into smooth muscle cell extracellular matrix in vitro, *Proc. Natl. Acad. Sci. U.S.A.,* 82, 2746, 1985.
86. **Hynes, R. O.,** Molecular biology of fibronectin, *Annu. Rev. Cell Biol.,* 1, 67, 1985.
87. **Paul, J. I. and Hynes, R. O.,** Multiple fibronectin subunits and their post-translational modifications, *J. Biol. Chem.,* 259, 13477, 1984.
88. **Columbi, M., Barlati, S., Kornblihtt, A. R., Baralle, F. E., and Vaheri, A.,** Analysis of fibronectin mRNAs in human normal and transformed cells, *Cell Biol. Int. Rep.,* 10, 201, 1986.
89. **Mann, D. M. and Millis, A. J. T.,** unpublished results, 1988.

Chapter 4

FIBRONECTIN: ROLE IN PHAGOCYTIC HOST DEFENSE AND LUNG VASCULAR INTEGRITY

Thomas M. Saba

TABLE OF CONTENTS

I. INTRODUCTION AND OVERVIEW

The term "reticuloendothelial system"[3,54] has been used to describe the sessile and wandering mononuclear macrophages[38] capable of "rapidly ingesting" and accumulating foreign colloidal and particulate matter (Figure 1). However, not all endothelial and reticular cells are phagocytic, and thus the term "macrophage system"[69] may be more appropriate. For this review, either the term "macrophages" or "reticuloendothelial cells" will be used to describe these mononuclear phagocytes. Macrophage or reticuloendothelial system (RES) phagocytic defense mechanisms are important to organ function and host survival in critically ill surgical, trauma, or burn patients.[58] Early studies on the RES repeatedly demonstrated the uptake of intravenously injected particles in sessile phagocytic cells of the liver, spleen, and bone marrow.[3,54] This was especially intense in Kupffer cells lining the blood sinusoids of the liver. Cellular engulfment of foreign particulate matter is intimately related to cellular "recognition" mechanisms. Thus, phagocytes can discriminate foreign macromolecules and effete autologous tissue debris from healthy indigenous matter.

One method which had been widely employed in the evaluation of Kupffer cell function has been the so-called "RE blockade" technique, in which a sustained depression of the hepatic macrophage system was "supposedly produced" by the intravenous injection of large doses of gelatin-coated foreign particles which are rapidly removed from the blood by RE cells in the liver. However, the technique of RE blockade, although advancing our knowledge of RE functions, has led to many conflicting observations. Indeed, depending on the type of colloid, the dose injected, or the postinjection time studied, one can observe a state of RE depression, RE stimulation, or normal RE behavior in terms of hepatic phagocytic removal.[54] This is, in part, related to the fact that intravenous infusion of gelatin-coated particles not only results in particle ingestion by the RE cells, but will also deplete a specific "recognition protein" from the plasma.[4,55] This protein binds to the gelatin coat on the particle and serves as an "opsonic probe" to enhance particle recognition and ingestion by RE cells. This opsonic protein is identical to plasma fibronectin[5,60] or cold-insoluble globulin (CIg) which has a high affinity for the denatured collagen or the gelatin coat on the particle. This role of fibronectin as an adhesive opsonic plasma protein that can influence phagocytosis[60] will be emphasized in this review. In addition, since the soluble pool of fibronectin in plasma may serve as a reservoir for some of the more insoluble adhesive fibronectin in tissue matrices and vascular beds,[15,45] its role in vascular permeability and tissue structural integrity will also be highlighted. Such functions are important to problems of phagocyte dysfunction,[63,64] altered vascular permeability,[8,12] and impaired wound healing.[23,24]

II. INTRAVASCULAR PHAGOCYTOSIS BY THE RES; KUPFFER CELL FUNCTION

Although a differential capacity exists for different phagocytic cells, the hepatic and splenic macrophages appear to play the predominant role in the vascular clearance process.[3,54] In the normal host, uptake by the splenic, pulmonary, and bone marrow phagocytes of blood-borne particles is not as overtly extensive as that manifested by liver Kupffer cells. However, in disease states, impairment of liver phagocytosis is usually associated with a greater extrahepatic participation. This RES clearance process defends the host against endotoxemia, bacteremia, or blood-borne microaggregate tissue debris. Such blood-filtering activity lends credence to its reference as a "systemic host defense" mechanism.[44,54] Opsonizing substances in blood are determinants of this phagocytic defense mechanism. For example, complement (C_3) as well as immunoglobulin G (IgG) are the main opsonizing substances that promote bacterial phagocytosis. In contrast, for ingestion of nonbacterial

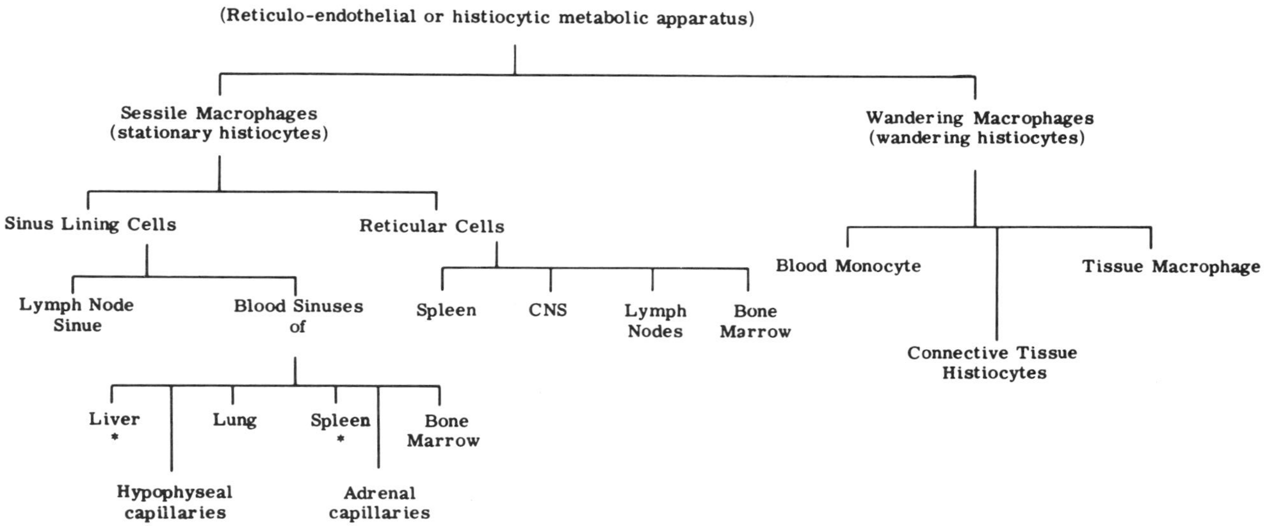

FIGURE 1. Slightly modified version of Aschoff's concept of the reticuloendothelial system. Liver and spleen (*) are major organs participating in the clearance of particulate matter from blood. Polymorphonuclear leukocyte is excluded as it is not classically part of the RES. (From Saba, T. M., *Arch. Int. Med.*, 126, 1031, 1970. Copyright 1970, American Medical Association. With permission.)

particulates an important role appears to exist for plasma fibronectin or opsonic α_2 surface-binding protein.[56,60,62,63] However, soluble fibronectin can also increase the phagocytosis of Clq-coated immune complexes as well as IgG-coated particles. This role of fibronectin, as a mediator of the phagocytic process, may be the "link" between plasma fibronectin deficiency and events such as phagocytic depression,[60] as well as microvascular embolic injury[43] often observed in patients during septic shock or disseminated intravascular coagulation.

Kupffer cells of the liver represent the single largest compartment of RE cells or mononuclear macrophages localized in one region. Various estimates suggest they may comprise up to 70 to 80% of the fixed or sessile macrophage population. Certainly, in humans, dogs, rats, and mice, they represent the predominant fraction of the RE cells which line the vascular compartment and mediate blood clearance activity.[54] Estimates suggest that up to 20 to 30% of the fixed cells in the liver in mice and rats are active Kupffer cells or dormant sinusoidal cells with phagocytic potential. In humans, approximately 15% of the hepatic cell number appears to be active Kupffer cells or potentially phagocyticly active cells within the hepatic sinusoid.

Several factors account for the intense phagocytic capability of the liver RE system with respect to removal of blood-borne particulates. First, the number of phagocytic cells involved is enormous. Second, Kupffer cells are highly responsive to opsonic stimuli which may include complement, immunoglobulin, and fibronectin. Third, their anatomical distribution provides them rapid access to the vascular compartment in an organ with a large capillary or sinusoidal network. Fourth, the liver receives a major fraction of the cardiac output (25 to 30% in man) which results in an adequate delivery of blood-borne particulates to Kupffer cells. Fifth, Kupffer cells remove various factors, i.e., bacteria, viruses, tumor cells, denatured proteins, immune complexes, cellular debris, etc., over a wide range of particulate sizes. In essence, they are a versatile and efficient group of phagocytic cells uniquely adapted for systemic host defense.

As already mentioned, Kupffer cells appear to be very responsive to an opsonic stimulus. This response can be demonstrated with isolated Kupffer cells, the isolated perfused liver, and *in vivo*. As such, opsonization with C_3 and/or IgG can rapidly accelerate bacterial clearance from the blood by Kupffer cells. Kupffer cells and other macrophages are also sensitive to the opsonic influence of plasma fibronectin coating of nonbacterial particles.[6,16,25,57] Intravenous injection of denatured protein microaggregates, gelatin-coated test colloids,[4,55] or specific immune complexes will deplete plasma fibronectin and transiently blockade Kupffer cell function. During such RES blockade, nonspecific defense against blood-borne bacteria and endotoxin, as well as hemorrhage, trauma, burn, and intravascular coagulation, is impaired.[31,75] Such findings provide experimental support for the concept that Kupffer cells in conjunction with plasma fibronectin may play a prominent role in nonspecific host defense against trauma, burn, tissue injury, and sepsis. However, it should also be noted that the postphagocytic secretory activities of Kupffer cells and other macrophages have also been implicated in potentiating cellular and microvascular injury during events such as sepsis and/or endotoxemia. Such secretory products or so-called biological mediators may include tumor necrosis factor (TNF) and prostaglandins whose excessive release can contribute to metabolic, hormonal, and molecular imbalances.

III. LEUKOCYTES AND VASCULAR INJURY

Cells of the mononuclear phagocytic system[69] are derived from the blood monocyte pool which, in turn, originates from pro-monocytes in the bone marrow. These blood monocytes take up residence in various tissues as mature and active phagocytes. Such phagocytic groupings include the alveolar macrophage, the peritoneal macrophage, and perhaps Kupffer cells. In contrast, granulocytes in the blood are derived from hematopoietic stem cells in the bone marrow. This blood granulocyte pool can be viewed as consisting of freely circulating as well as marginated cells. The total pool of mature granulocytes in the bone marrow on a per kilogram body weight basis is many times larger than that fraction in the circulation. Mature granulocytes are released from the bone marrow storage compartment under the influence of hormonal factors at a rate sufficient to maintain an apparent steady-state in terms of blood levels. The neutrophil or polymorphonuclear leukocyte (PMN) comprises approximately 50 to 70% of the total circulating blood leukocyte cell population in most species. The mature PMN is considered a terminal cell, since it does not subsequently divide. Its half-life in blood is approximately 6 to 7 h, but once it has migrated and/or become attracted into various tissue spaces, it can apparently manifest an extended half-life of approximately 24 to 72 h. PMNs are highly phagocytic and readily participate in local defense at sites of tissue injury and invasive wound infection. The mature PMN also possesses extensive bactericidal capacity and lysosomal enzymatic "machinery" to allow it to effectively digest and/or kill endocytosed bacteria.

The migration of leukocytes from the blood into tissues is a complex biologic process. Chemotactic factors are released from the site of tissue injury during the inflammatory response. These factors enhance the mobilization of blood leukocytes to the focus of injury. Neutrophil mobilization includes penetration through the vascular endothelium, traversing the subendothelial barrier, migration through the interstitial matrix, and subsequent adherence at a site of injury and/or inflammation. Accordingly, leukocytes are undoubtedly triggered to selectively adhere to a region of the endothelial barrier in close proximity to the site of injury. In terms of the current emphasis on fibronectin, it should be noted that fibronectin is highly sensitive to proteolytic cleavage[29,36] by PMN-containing enzymes, such as cathepsin G and leukocyte elastase. Moreover, selected fragments of fibronectin are chemotactic for blood monocytes, tissue macrophages, and fibroblasts. Thus, it is possible that the fragmentation of tissue matrix fibronectin by proteases released from activated leukocytes at a site of tissue injury[47] may release small molecular weight chemotactic fibronectin fragments which can modulate the inflammatory response. They may also influence ongoing leukocyte adherence to vascular endothelial surface.

However, the adherence of activated blood leukocytes or other phagocytic cells to localized regions of the vascular endothelial surface may also contribute to microvascular injury.[26] Granulocytes and macrophages release proteases as well as oxygen metabolites, such as superoxide anion (O_2^-), hydrogen peroxide (H_2O_2), and hydroxyl radical (OH^-). Thus, while phagocyte activation and margination at sites of tissue injury is an efficient defense mechanism, it may also contribute to local tissue injury, if unregulated. The association of pulmonary leukostasis with acute lung vascular injury during bacteria[12,42] or thrombin-induced lung microembolization may be mediated by this mechanism. Fibronectin fragmentation in tissues after injury could serve a useful function, since the attraction of mononuclear macrophages to sites of injury at an appropriate time interval following the initial immigration of leukocytes would optimize the removal of devitalized tissue and cellular debris. In parallel, intact plasma fibronectin sequestered into tissues after injury due to its affinity for collagen, actin, fibrin, or deposited C-reactive protein may opsonically augment the removal of nonbacterial particulates and cellular debris by phagocytic cells, i.e., wound debridement. From this perspective, fibronectin may be an essential molecule in the local macrophage defense mechanism and subsequent wound repair process after injury.[58]

IV. FIBRONECTIN AND PHAGOCYTIC RECOGNITION

Ingestion of foreign objects by phagocytic cells is a highly selective process which is linked to the "recognition capability" of the macrophages and/or leukocytes. While the basis for such recognition has yet to be delineated, plasma opsonins[54] have been implicated in this "discrimination process". The binding of opsonized particles to the phagocyte membrane may be a key step in this recognition mechanism, which is then followed by activation of the cell, cytoplasmic alterations, membrane invagination, increased adhesiveness of the plasma membrane, pseudopod extension, and particle internalization. The "adhesive" property of fibronectin may be ideal for this opsonic function. Thus, fibronectin-mediated macrophage adherence to the surface of a small particle coated with denatured collagen, which is to be endocytosed, may be somewhat analogous to the fibronectin-mediated adherence of a macrophage or other cell type to a noningestible collagenous matrix or foreign surface. In both cases, its "ligand" characteristics are manifested.

For nonbacterial particulates, i.e., cytoskeletal debris, collagenous debris, and membranous fragments, the soluble form of fibronectin appears to influence their uptake by macrophages, perhaps via a separate fibronectin receptor region on the membrane of macrophages. Most likely, the region of fibronectin that binds to the surface of the foreign particle is distinct from the domain region of fibronectin which binds to the membrane receptor. Contractile proteins such as actin and myosin exist in macrophages and granulocytes. These contractile proteins play an intimate physiologic role in pseudopod extension and subsequent particulate ingestion. The anatomical and biochemical linkage between the membrane binding of opsonized particles and the intracellular contractile machinery has not been fully clarified. Fibronectin has a high affinity for actin and the existence of a transmembranous connection between surface-bound fibronectin and intracellular actin must be considered.

V. FIBRONECTIN MOLECULE: STRUCTURE AND FUNCTION RELATIONSHIPS

Fibronectin is a high molecular weight adhesive glycoprotein found in the blood, lymph, and tissue fluid as well as in the extracellular matrix of many cells.[29,39,58,72] A soluble form of fibronectin is found in plasma and lymph, while an antigenically related fibronectin is found in connective tissues, basement membranes, and on various cell surfaces. The soluble form in plasma has been called opsonic α_2 surface-binding (SB) glycoprotein, cold insoluble globulin (CIg), humoral recognition factor, aspecific opsonin, as well as plasma fibronectin. Insoluble tissue fibronectin has been called large external transformation sensitive (LETS) protein, galactoprotein a, cell surface protein (CSP), and soluble fibroblast antigen, as well as cell surface fibronectin. In tissue culture, fibronectin can be produced by fibroblasts,[72] vascular endothelial cells,[30] macrophages,[11] and hepatocytes,[67] as well as other cell types. In culture, fibronectin is released and/or secreted into the media by such substrate-attached cells. Both plasma and tissue fibronectin appear to have a molecular weight of about 440,000 to 450,000 Da, and the intact molecule consists of two very similar disulfide-bonded monomers of 220,000 to 230,000 Da.[29,39,40,72] They are very similar, but depending on the tissue compartment, they may not be identical.[73] Thus, in the adult, the fibronectin found in tissues may be a mixture of locally synthesized and plasma-derived fibronectin. The relative composition of the tissue pool in the fetus or newborn may be very different, with tissue fibronectin being primarily locally derived, since the low plasma levels in newborns (about 30 to 40% of adult levels) may not provide an adequate plasma-to-tissue gradient essential for flux and/or incorporation into tissues *in vivo*. Fibronectin from plasma can incorporate into the tissue pool as has been documented by both isotopic[15] and immunofluorescent[45] studies. Whether fibronectin from various insoluble tissue pools normally contributes to the plasma pool remains to be determined, although it can undoubtedly be released in either an intact or fragmented form in various disease states. Perhaps a dynamic equilibrium exists as previously speculated,[58] which can be disturbed after trauma, burn, or septic shock.

Fibronectin monomer consists of at least four globular domains separated by flexible protease-sensitive areas of polypeptide chains.[29,40] Fibronectin manifests high affinity for native and denatured collagen, fibrin, fibrinogen, actin, heparin, Clq, *Staphylococcus aureus,* etc. as well as the surface of some cell types. In its intact dimeric form in blood, it does not readily bind to the luminal surface of healthy vascular endothelial cells, although it may bind to the sub-luminal surface of healthy vascular endothelial cells. Thus, fibronectin is an adhesive modular glycoprotein[29] with an affinity for many entities that can be endocytosed by phagocytic cells[55,58] (see Chapter 1). This structural organization supports the concept that fibronectin is an important molecule in the blood and tissue fluids required for macrophage clearance of actin containing cytoskeletal debris, intracellular membranous or non-membranous debris, and extracellular collagenous debris. It may also influence the blood clearance and relative liver vs. nonhepatic localization of blood-borne immune complexes as well as certain antigens such as DNA and *S. aureus* which can bind to fibronectin.[13] This latter possibility suggests that pathologically exposed vascular bed matrix-localized tissue fibronectin could potentially augment deposition of circulating immune complexes or antigens in vascular beds of organs such as the kidney.[13] Efficient hepatic Kupffer cell phagocytosis of such blood-borne immune complexes may protect the kidney from such immune complex deposition. We now know that fibronectin readily binds to Clq and can also augment the ingestion of Clq-coated immune complexes. In contrast to the opsonic role of plasma or soluble fibronectin, tissue fibronectin appears to be an adhesive glycoprotein essential for tissue integrity and cell-cell interaction. This "dual physiological role" of fibronectin (opsonic vs. adhesive) is unique and forms the basis[58] for much of the experimental work on this highly complex molecule.

VI. PLASMA FIBRONECTIN DEFICIENCY IN SEPTIC SURGICAL OR TRAUMA PATIENTS

Immunoreactive and bioassayable plasma fibronectin deficiency has been documented in septic surgical, trauma, or burn patients, as well as patients with disseminated intravascular coagulation (DIC), especially during multiple organ failure.[8,17,18,33,41,48,50,51,56,63,68] In addition, nonsurviving trauma patients have a persistent humoral opsonic fibronectin deficit, while survivors manifest an early restoration of this plasma component following injury.[62-64] The discovery[5,60] that plasma fibronectin (CIg) and opsonic α_2SB glycoprotein were identical indicated that human plasma cryoprecipitate could be used to reverse opsonic deficiency in septic injured patients. Infusion of "fresh" plasma cryoprecipitate into septic trauma, burn, or surgical patients rapidly reversed the bioassayable and immunoreactive fibronectin deficiency.[56] In initial studies, this resulted in an apparent improvement in the febrile and septic state as well as pulmonary function.[56,62,63] These patients also manifested increased alertness, decreased tachycardia, and a lowering of body temperature. A tendency to return toward deficient levels was often observed within 48 h after the infusion of only 10 U of fresh plasma cryoprecipitate, especially in patients with persistent sepsis or a focus of tissue injury and/or invasive wound infection. Such observations supported the concept that the passive administration of plasma fibronectin may potentially augment RE systemic defense in septic injured patients, if it could be administered in doses sufficient to maintain a normal or slightly elevated plasma level. However, standardizing such a therapeutic approach will not be easy, since we now recognize that the volume of fibronectin-rich cryoprecipitate infused, the age of the plasma cryoprecipitate (storage duration) which can influence its opsonic activity, the extent of fragmentation of the fibronectin molecule in cryoprecipitate, as well as its factor XIII and possibly its fibrinogen content, are all variables which may influence the opsonic as well as organ function responses observed in patients.

Pulmonary abnormalities and adult respiratory distress are common manifestations of organ failure in septic surgical and/or injured patients. Burn patients with mainly cutaneous injury can also manifest respiratory distress which may be the result of a ventilation-perfusion imbalance and interstitial pulmonary edema.[58] Capillary leaking in the lung is believed to be a major contributing factor to pulmonary edema and/or posttraumatic pulmonary insufficiency. This altered integrity of the vascular bed is accentuated during periods of sepsis and/or intravascular coagulation after surgery and/or trauma. It is quite possible that sequestration of activated blood leukocytes in the lung vascular and interstitial spaces and their release of proteases as well as oxygen metabolites contributes to this acute lung injury by initiating the fragmentation and/or release of tissue fibronectin found between lung endothelial cells and their basement membrane. We have observed that plasma fibronectin deficiency in sheep amplifies the increase in lung vascular permeability with Gram-negative sepsis following surgery[42] and that i.v. infusion of fibronectin-rich cryoprecipitate[59] or purified opsonically active plasma fibronectin[12] can limit the increase in lung capillary leaking. Thus, the plasma pool of this molecule may be a reservoir for fibronectin in tissues and its plasma concentration may be critical to integrity of the lung vascular barrier.

Saba[53] and Kaplan and Saba[31] first documented opsonic deficiency in animals after surgery or trauma and its relationship to RES function. Subsequently, Scovill, et al.[64] demonstrated a deficit of plasma opsonic activity in patients following blunt trauma. Clinically defined sepsis in these patients was preceded by a fall in plasma opsonic activity by as much as 2 d. As reviewed in 1970,[54] this opsonic factor was initially characterized as glycoprotein with a high affinity for gelatin, heparin, and foreign surfaces and clearly cryoprecipitable, especially in the presence of heparin. Subsequently, this opsonic protein was discovered to be identical to CIg or plasma fibronectin.[5,60] Its purification and the preparation of an appropriate antibody allowed for its measurement by immunoassay, which revealed that deficiencies of antigenic fibronectin in plasma or serum correlated with loss of bioassayable opsonic activity.[4,6,55,56] However, we have also come to realize that the antigenic fibronectin content of a plasma or serum sample may not always correlate with its fibronectin-mediated opsonic augmentation of phagocytosis, especially if the dimeric fibronectin molecule in such samples has undergone proteolytic fragmentation. Fragmented fibronectin will not retain full opsonic activity, and can actually competitively inhibit intact dimeric fibronectin in many of the phagocytic assays.[19] Moreover, depending on the immunoassay used (electroimmunoassay, immunoturbidimetric assay, etc.), fragmentation of fibronectin can result in a very high or very low antigen concentration. In addition, it is clear that "gelatin binding activity" is not a reliable index of opsonic or functional activity relative to stimulation of phagocytosis. The molecule can have a great loss in ability to promote phagocytosis, yet still have "gelatin binding" activity. Finally, the fragmented molecule in plasma will not manifest the normal incorporation into the tissue pool.[45]

Infusion of fresh plasma cryoprecipitate (usually less than 1-month old), known to be rich in plasma fibronectin will reverse the opsonic deficits in septic surgical and trauma patients.[57] Robbins, et al.[50] also demonstrated a drop in plasma fibronectin and a sequestration of CIg (fibronectin) in the drainage fluid of patients following mastectomy. In fact, the relative amount of fibronectin in the drainage fluid at 2, 4, and 8 h following operation was higher than circulating levels. Rubli et al.,[51] studying a group of 98 intensive care patients, 33 of which were septic with 23 manifesting acute respiratory failure, noted that fibronectin as well as antithrombin III, transferrin, and pre-albumin were significantly lower in septic as compared to nonseptic injured patients, thus reflecting a pattern of general protein depletion in septic patients. In this group of patients, they observed that patients with plasma fibronectin levels less than 200 μg/ml exhibited a mortality of 86%. However, if the levels were greater than 200 μg/ml, the mortality was only 31%. In patients undergoing elective surgery, Richards et al.[48] demonstrated a significant postoperative drop in plasma fibronectin which returned to normal levels by the third postoperative day. Patients with preexisting intra-abdominal infections also had lower plasma fibronectin levels preoperatively, which correlated with a higher incidence of multiple organ failure. In patients without multiple organ failure, Richards et al.[48] observed that when plasma fibronectin normalized by the 5th postoperative day, the mortality was 12.5%; whereas patients with sustained low levels of plasma fibronectin beyond the 5th postoperative day exhibited a mortality of 62%. Thus, it would appear that normalization of plasma fibronectin is associated with a better clinical course and greater survival rate.

After severe thermal injury, patients are often difficult to manage and require large amounts of volume support to maintain hemodynamic stability. This may be due to both loss of fluid and proteins into the area of injury as well as a generalized increase in vascular permeability. Lanser et al.[33] initially described the acute depletion of plasma fibronectin following thermal injury. This acute depletion of plasma fibronectin followed by rapid recovery of fibronectin levels after sublethal burn injury has been well characterized by Deno et al.[14] Collagenous debris may exist in plasma following burn injury which may bind fibronectin and limit its biological functions, a concept supported by the acute rise in hydroxyproline containing protein in plasma early after burn.[34] However, a second phase of fibronectin deficiency is typically observed with the development of sepsis, and resolution of the septic episode is associated with normalization of plasma fibronectin. Ekindjian et al.,[17] although unable to closely correlate the extent of burn injury with the degree of plasma fibronectin depletion, also showed a clear correlation of low fibronectin levels with the presence of sepsis in burn patients. Eriksen et al.[18] demonstrated that the plasma fibronectin will fall in patients acutely following burn injury, following excision and grafting of the burn area, as well as during periods of septicemia postburn. Indeed, the fall in plasma fibronectin preceded by 2 to 3 d the development of positive blood cultures. Such observations prompted Brodin et al.[9] to suggest that plasma fibronectin levels may be of prognostic value with respect to sepsis and ultimate mortality in adult burn patients. This response may also exist in infants, since low fibronectin levels have been documented in high risk newborns as well as septic infants with respiratory distress.[74] Fibronectin levels in such infants will normalize following resolution of the septic episode. If plasma fibronectin is sequestered in areas of subendothelial or tissue injury due to its high affinity for actin, fibrin, or denatured collagen, or if it is depleted during the process of RES removal of blood-borne immune complexes as well as cytoskeletal, intracellular, or collagenous debris, then persistent deficiency may reflect consumption in excess of synthesis and/or release. Its deficiency in the critically ill patient may be amplified and protracted if limited nutritional support exists, as emphasized by both animal[16] and patient[28] studies. Moreover, clinical intervention such as transfusion of stored blood with microaggregates[66] or cardiopulmonary bypass[65] can also lower plasma fibronectin concentrations.

VII. LUNG VASCULAR PERMEABILITY AND FIBRONECTIN LEVELS

Although numerous clinical studies[9,17,18,22,48,52,56,64] have documented that persistent low levels of plasma fibronectin points to a poor prognosis in the critically ill patient, very few studies have documented a correlation of depressed plasma fibronectin levels with dysfunction of a specific organ system, other than the reticuloendothelial systems as well as lung vascular permeability. The earliest findings which suggest such a correlation were reported by Scovill et al.[62,63] in that low plasma fibronectin levels in trauma and burn patients with severe sepsis was temporally associated with organ failure, such as a ventilation-perfusion mismatch, a blunted peripheral vascular response to ischemia, and an abnormal endogenous creatinine clearance. Fourrier et al.[20] studying patients with adult respiratory distress syndrome (ARDS) showed that low angiotension converting enzyme (ACE) levels, perhaps reflecting pulmonary endothelial injury or dysfunction, were of substantial value in predicting the presence of ARDS. Additionally, they showed that although a low plasma fibronectin level was not predictive of the presence of ARDS, it was correlated with severity of ARDS in such patients. In contrast, Maunder, et al.[35] showed low plasma fibronectin levels in septic patients, but found no correlation between the development of ARDS and a low plasma fibronectin level. Similarly, Bagge et al.,[2] studying a group of patients with both blunt and operative trauma as well as sepsis, confirmed the existence of low plasma fibronectin levels in these patients as compared to normal controls, but were unable to document a difference in plasma fibronectin levels between patients that developed and those that did not develop ARDS.

Although the etiology of ARDS is unclear, it is generally accepted that alterations in pulmonary microvascular permeability contribute to the increased fluid and protein flux in the lung. It is this aspect of the pathophysiology of pulmonary failure to which fibronectin may relate. Utilizing the lung lymph fistula model, we observed that the increase in lung transvascular protein clearance following infusion of live *Pseudomonas* appears to be amplified in fibronectin-deficient sheep.[42] The administration of fibronectin-rich cryoprecipitate, but not cryoprecipitate depleted of fibronectin,[59] resulted in a decrease in the lung vascular permeability as reflected in transvascular protein clearance determinations. The direct influence of i.v. infusion of purified human plasma fibronectin[61] on lung lymph flow and transvascular protein clearance after postoperative *Pseudomonas* bacteremia has been studied in my laboratory by Drs. Cohler and Charash (postdoctoral fellows)[12] utilizing the chronic, unanesthetized sheep lung lymph fistula model. In this study, sheep were given the bacteria after surgery at two different times, i.e., when their fibronectin levels were normal and when their endogenous fibronectin levels were markedly elevated. Then, sheep were given the same postoperative bacterial challenge after their fibronectin levels were experimentally elevated by i.v. infusion of purified human plasma fibronectin.[12,61]

The protocol for these previous studies[12] as reviewed below was as follows. At surgery, induction of general anesthesia was accomplished with 2.5% sodium pentothal (15 mg/kg). Sheep were endotracheally intubated and mechanically ventilated at a rate of 15 to 18 breaths per minute and a tidal volume of 10 to 15 ml/kg. They were maintained throughout surgery on 78% N_2O, 20% O_2, and 1 to 1.5% halothane anesthesia. They received 1 liter of 0.9% saline intravenously during the surgical preparation. The lung lymph fistula was surgically prepared under sterile conditions via the right thoracotomy through the seventh intercostal space. The efferent duct of the caudal mediastinal lymph node was cannulated with a silastic catheter (0.047 in. O.D., 0.025 in. I.D.) and the posterior portion of the node was ligated and transected to minimize contamination of lung lymph by diaphragmatic lymphatics. The silastic catheter was tunneled and secured beneath the parietal pleura before exiting the thorax at the ninth intercostal space. The lung lymph fistula cannula was then secured on the skin at a level comparable to the *in vivo* position of the caudal mediastinal lymph node when the sheep was standing in the erect position, and the thoracotomy was suture closed.

After this surgery, sheep received 2 ml/kg/h of normal saline i.v. for 36 h. Postoperative recovery of the sheep was excellent; they were soon mobile. The sheep were studied 48 h after surgical preparation of the fistula, and then 5 d later at a time corresponding to a phase of rebound elevation of plasma fibronectin. Accordingly, this protocol allowed for comparison of lung protein clearance or so-called "lung vascular permeability" with postoperative bacteremia during a period of either normal or elevated plasma fibronectin levels.[12] Sheep received an intravenous infusion of 5×10^8 live saline-washed *Pseudomonas aeruginosa* suspended in 50 ml of normal saline over a 60-min period. For every animal, anticoagulated blood and lymph samples (EDTA) were collected at 30-min intervals for 6 h postbacterial challenge, and thereafter at 12, 24, and 48 h. The identical protocol was used after either the first or second bacterial challenge. Lymph flow (Q_L) and lymph-to-plasma total protein concentration (L/P) as well as the transvascular protein clearance (TVPC),

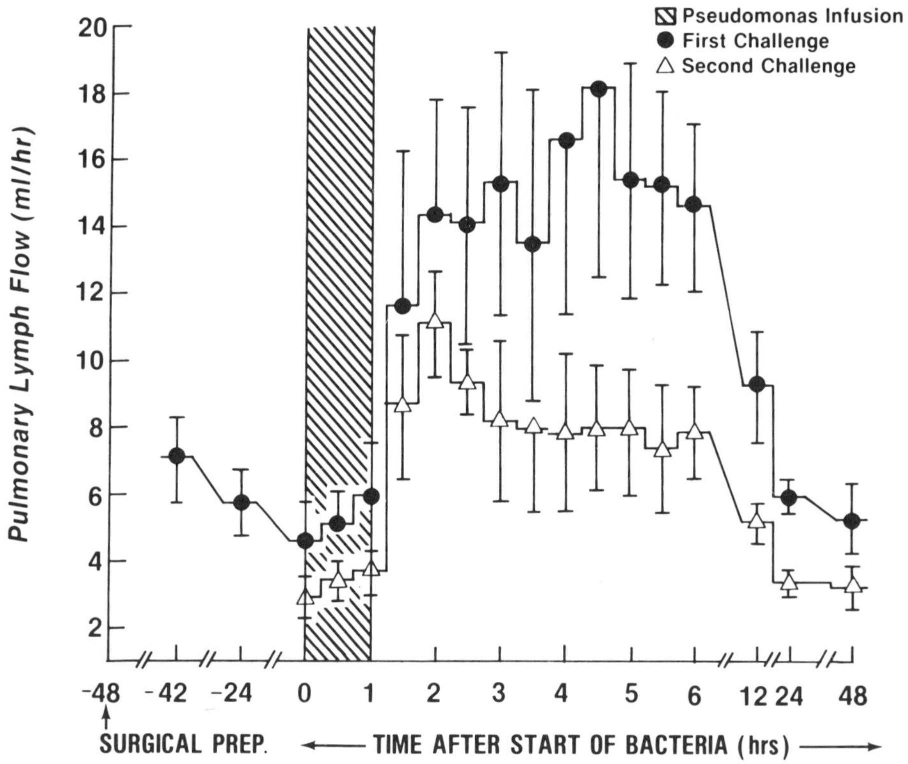

FIGURE 2. Pulmonary lymph flow (Q_L) in sheep after intravenous infusion of *Pseudomonas*. There were six sheep in the first challenge group; four of these had patent fistula catheters and were reevaluated in the second challenge group. Second challenge animals showed an attenuated increase in this parameter. Q_L in both groups was significantly ($p < 0.05$) elevated at 2 h as compared to baseline values. When both groups were also compared to each other, Q_L in the first challenge group was greater over the 3- to 5-h ($p < 0.10$) and $5^1/_2$- to 24-h ($p < 0.05$) intervals as compared to the level in the second-challenge sheep. Values are expressed as mean ± SE. (From Cohler, L. F., Saba, T. M., Lewis, E., et al., *J. Appl. Physiol.*, 63, 623, 1987. With permission.)

which is the product of lymph flow (ml/h) × total protein L/P concentration ratio, were quantified. Systemic arterial pressure, pulmonary arterial pressure, pulmonary wedge pressure, central venous pressure, and cardiac output were also measured.

As shown in Figure 2,[12] pulmonary lymph flow (Q_L) increased at 2 h after the initial bacterial infusion. The early rise in Q_L after the first bacterial challenge was sustained over the 3- to 12-h period as compared to baseline, even though pulmonary pressures had essentially returned to baseline (14.8 ± 3.1 mmHg vs. 18.5 ± 4.4 mmHg at 6 h). After the second bacterial challenge (Figure 2), the sheep also demonstrated the early marked elevation in lymph flow, but the second phase of the response over the 3 to 12 h was greatly attenuated compared to the first bacterial challenge. We observed a declining L/P ratio[12] which was at its lowest by 2 h after the start of bacterial infusion. This decline in L/P ratio coincided with the early rise in Q_L. After the first bacterial challenge, sheep demonstrated a rebound elevation of L/P ratio by 3 h that continued to rise above baseline and reached its maximum at 4 h. This elevation in L/P ratio, along with increased Q_L, reflected an increased flux of protein-rich fluid across the endothelial barrier whose permeability was evidently increased resulting in leakage of protein which would normally be restricted to a greater degree to the vascular space. While the second challenge animals showed a similar biphasic response in the L/P ratio, the rebound peak L/P ratio in this group was delayed by 2 to 3 h and was only significantly elevated above baseline by 6 h, suggesting that any increase in permeability

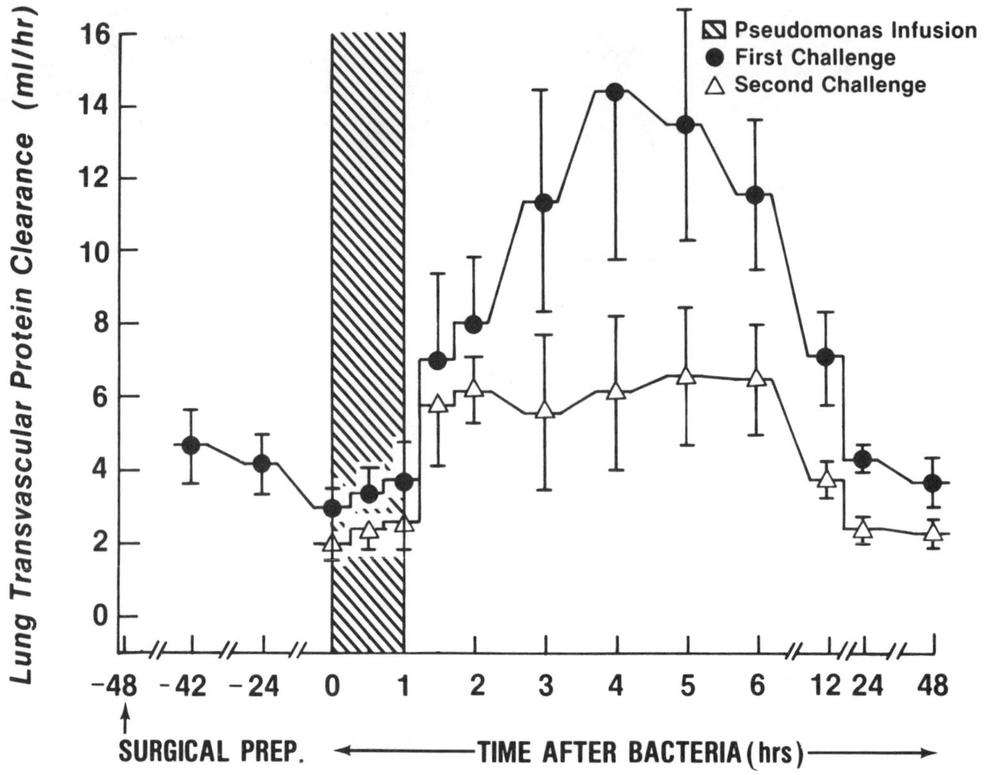

FIGURE 3. Lung transvascular protein clearance (TVCP) in sheep after intravenous infusion of Pseudomonas. There were six sheep in the first challenge group; of these, four had patent fistula catheters and were reevaluated in the second challenge group. Second challenge animals showed an attenuated increase in this parameter. TVPC in both groups was significantly ($p < 0.05$) elevated at 2 h as compared to baseline values. When both groups were also compared to each other, the TVPC in the first challenge group was greater over the 3- to 5-h ($p < 0.10$) and 5- to 24-h ($p < 0.05$) intervals as compared to the level in second challenge sheep. Values are expressed as mean ± SE. (From Cohler, L. F., Saba, T. M., Lewis, E., et al., *J. Appl. Physiol.*, 63, 623, 1987. With permission.)

that may have developed in the second group was much attenuated. The calculated lung TVPC used as an index of lung vascular permeability in these two groups of sheep is shown in Figure 3.[12] With the first bacterial challenge, the TVPC increased by 1.5 to 2.0 h postbacterial infusion. In the second phase (3 to 12 h), a progressive and sustained elevation in TVPC as compared to baseline was observed until 12 h after this first bacterial challenge. After the second bacterial challenge, the initial increment in TVPC was similar to that observed after the first bacterial infusion, but the second phase (3 to 12 h) elevation in TVPC or lung vascular permeability was attenuated.

The pulmonary arterial pressure after infusion of this sublethal dose of *Pseudomonas* is typically characterized by an early rise in mean pulmonary arterial pressure, which peaks at 1.0 to 2.0 h after the start of the 60-min infusion and gradually declines over the 3- to 6-h period to a new level slightly above baseline pressures.[12] Significant hypotension was not seen with this low dose, sublethal, bacterial infusion. The hemodynamic response in terms of pulmonary arterial pressure, systemic arterial pressure, and cardiac output was very similar after either the first or second bacterial dose. This hemodynamic pattern, coupled with the observed changes in L/P, Q_L, and TVPC,[12] suggested that while pressure changes could be a major factor mediating the early increment in lymph flow and decline in L/P; the second phase of elevated lung protein clearance over 3 to 12 h (Figure 3) could not be

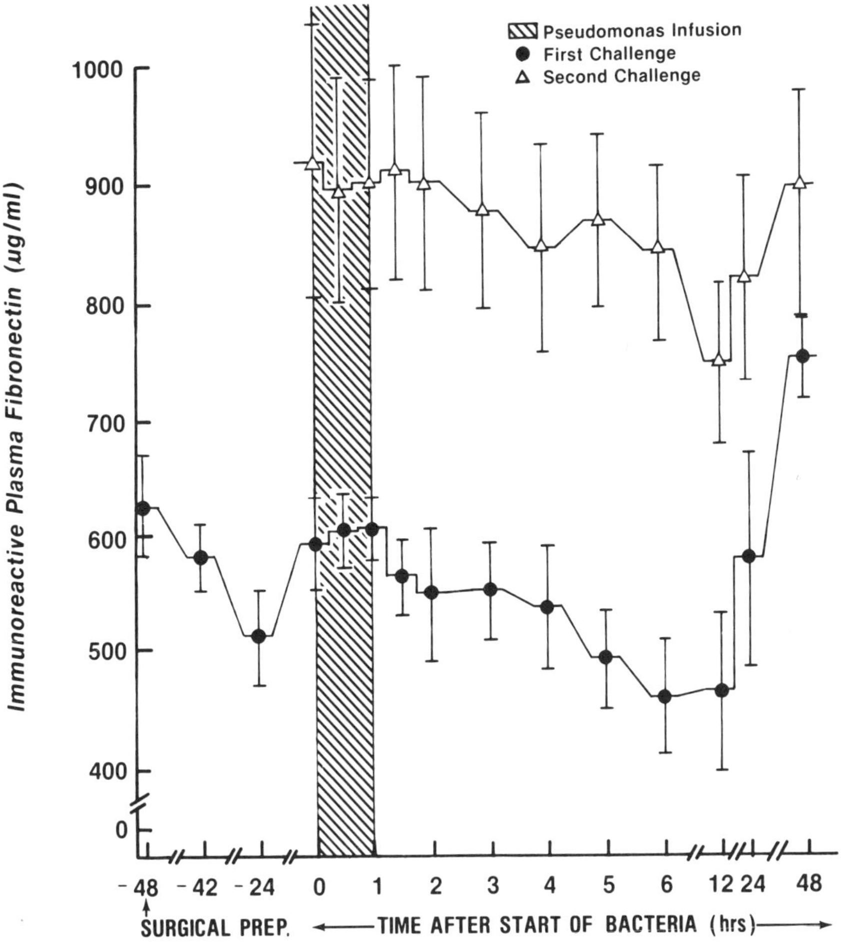

FIGURE 4. Immunoreactive plasma fibronectin levels in sheep after intravenous infusion of *Pseudomonas*. Animals challenged a second time with bacteria showed significantly ($p < 0.01$) higher baseline plasma fibronectin levels. Fibronectin levels declined slowly over the 1- to 12-h period after the first bacterial challenge; the fibronectin levels did not significantly ($p < 0.05$) decline after the second bacterial challenge. Values are expressed as mean ± SE. (From Cohler, L. F., Saba, T. M., Lewis, E., et al., *J. Appl. Physiol.*, 63, 623, 1987. With permission.)

the result of only pressure changes. This implied a definite increase in lung vascular permeability which was attenuated after the second bacterial challenge.

Plasma fibronectin determinations (Figure 4) demonstrated[12] the anticipated early drop in plasma fibronectin after surgery which normalized by 48 h (zero time) when we started the protocol. Then, plasma fibronectin slowly declined after the first bacterial infusion, and reached its lowest level by 6 h (control = 590 ± 37 µg/ml; 6 h = 459 ± 49 µg/ml). Their plasma fibronectin levels began to rise (737 ± 34 µg/ml), after the start of the first bacterial challenge in association with restoration of the permeability of the vascular barrier (Figure 4). However, their fibronectin levels were markedly elevated at the start of the second experimental protocol (first challenge baseline = 590 ± 37 µg/ml; second challenge baseline = 921 ± 114 µg/ml). Thus, the total amount of soluble fibronectin in plasma was markedly increased at the time these sheep manifested an attenuated response to the second dose of bacteria.

We then determined if i.v. therapy with purified human plasma fibronectin[12] in post-

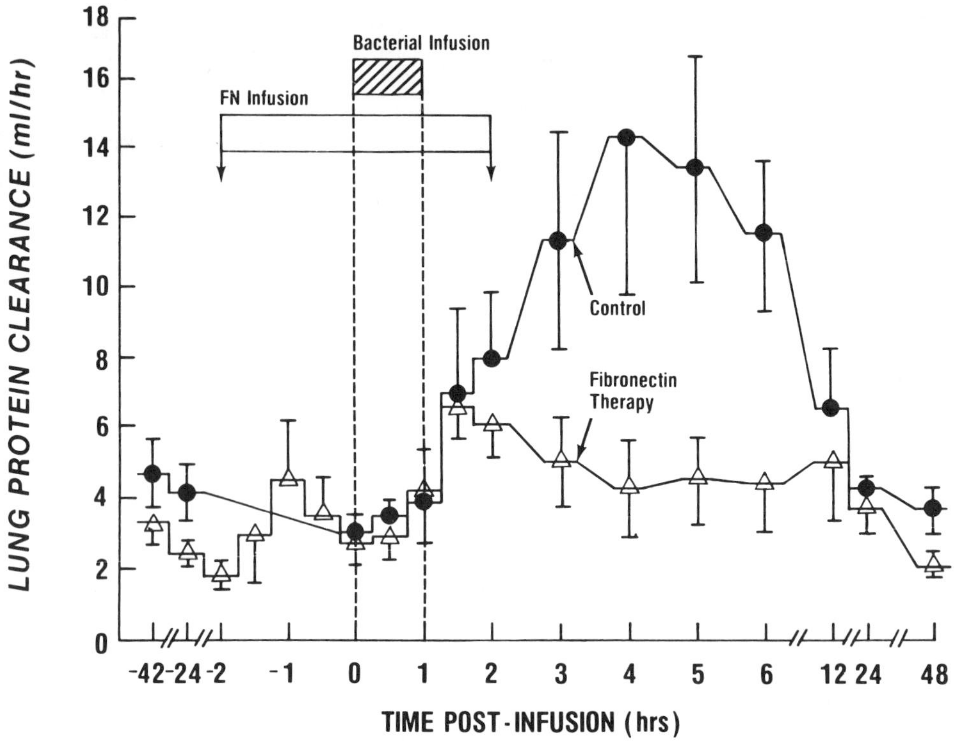

FIGURE 5. Lung protein clearance (LPC) in control (n = 6) and fibronectin-treated (n = 4) sheep with postoperative bacteremia. LPC returned to baseline by 3 to 4 h in the fibronectin-treated sheep. When both groups were also compared to each other, lung protein clearance in the controls at 3, 4, 5, and 6-h was greater ($p < 0.05$) than the lung protein clearance in the fibronectin-treated group at the same intervals. The L/P ratios used in the calculation of LPC were only determined at 60-min intervals between 2 to 6 h. Values are expressed as mean ± SE. (From Cohler, L. F., Saba, T. M., Lewis, E., et al., *J. Appl. Physiol.*, 63, 623, 1987. With permission.)

operative bacteremic sheep would elicit the same apparent protection of the lung vascular barrier. Both groups demonstrated an early rise in Q_L by $1^1/_2$ h after the start of bacterial infusion. However, the second phase response over 3 to 12 h in sheep infused with the purified fibronectin was markedly different. Control animals demonstrate a sustained elevation in Q_L with recovery toward baseline by 12 to 24 h. Sheep receiving plasma fibronectin therapy manifested a transient increase in Q_L which returned to baseline by 3 to 4 h.[12] Accordingly, while the fibronectin-treated animals also showed an early rise in lung protein clearance (LPC), their LPC used as an index of lung vascular permeability returned to a level not different from normal baseline by 3 h (Figure 5). Thus, elevation of plasma fibronectin, in normal sheep, to a concentration very similar to that observed in sheep with endogenous hyperfibronectinemia also attenuated the increase in lung protein vascular permeability with postoperative bacteremia. Such observations[12] suggest that the concentration of plasma fibronectin may not only influence phagocytic function, but may also influence the integrity or permeability of the lung vascular barrier during periods of bacteremic or septic shock after surgery or trauma.

VIII. ACUTE LUNG INJURY — ADULT RESPIRATORY DISTRESS

Sepsis, with or without bacteremia, is a frequent complication observed in postsurgical, trauma, and burn patients who often develop noncardiogenic pulmonary edema and respiratory distress. Although many mechanisms may mediate ARDS, an increase in pulmonary microvascular permeability may be a contributing factor. Such patients often demonstrate very low levels of plasma fibronectin early after injury, as well as prior to and in association with sepsis. Since plasma fibronectin enhances RES clearance of effete indigenous and foreign particulate matter from the blood, it has been reasoned that the fibronectin-mediated RES systemic defense may protect the lung microcirculation from embolic injury.[43] Such embolization could lead to a ventilation-perfusion mismatch and altered gas exchange, which are characteristics of ARDS.

From another perspective, tissue fibronectin has been localized between endothelial cells and in the subendothelial basement membrane matrix by immunofluorescence. Endothelial cells in culture show decreased adherence and impaired spreading when the media is free of fibronectin.[21,27] Normal confluent patterns of these cells are observed with addition of fibronectin. Moreover, endothelial cells secrete as well as incorporate fibronectin into their extracellular matrices. Indeed, the addition of fibronectin fragments will cause disadhesion of cultured endothelial cells from their collagenous substratum[27] suggesting that the intact fibronectin molecule may be required for cell adherence, and fragments of fibronectin may compete with the intact molecule. Since fibronectin is very sensitive to proteolysis, the margination of activated leukocytes in the lung microcirculation which has been documented with bacteremia or the activation of resident intravascular pulmonary macrophages may initiate fragmentation of tissue matrix fibronectin by the local release of proteases, thus contributing to altered endothelial cell adhesion and increased microvascular endothelial permeability. Such alterations are consistent with the intense increase in lung vascular permeability and disturbances in fibronectin kinetics reported after i.v. protease infusion.[11]

Endothelial cells, fibroblasts, hepatic parenchymal cells, as well as macrophages[1,49] can synthesize and elaborate fibronectin, and in some cases deposit this glycoprotein into their extracellular matrix. Additionally, soluble fibronectin from plasma can incorporate into the insoluble tissue fibronectin pool. Indeed, the rapid increase in fibronectin in areas of vascular injury[10] and wound repair cannot be totally accounted for by local synthesis and, in part, reflects the increased flux of plasma fibronectin into the tissue pool of fibronectin.[14-15] Such incorporation may involve plasma transglutaminase[32] with subsequent formation of fibronectin multimers in the extracellular matrix mediated by disulfide exchange in the amino-terminal portion of fibronectin.[37] Accordingly, the sequestration and long-term retention of fibronectin at sites of vascular and tissue injury may reflect a reparative role of soluble fibronectin from the plasma reservoir.[14,15]

Brigham et al.[8] initially reported that sheep given i.v. *Pseudomonas* manifest a marked increase in pulmonary vascular permeability. Niehaus et al.[42] observed that sheep with plasma fibronectin deficiency manifested a greater increase in transvascular protein clearance with bacteremia than sheep with normal plasma fibronectin levels. Part of this response may be mediated by the generation of circulating fibrin aggregates[34] and the increased extrahepatic localization of such particulates especially during RE blockage.[43] This pathophysiological response may be amplified by protease release from marginated leukocytes[26] whose lung sequestration is initiated by blood-borne bacteria or particulate debris not rapidly removed by the RES. In sheep, in particular, some of the infused bacterial or products of bacterial septicemia may be directly cleared by "intrapulmonary sessile macrophages" which reside in the lung vascular space at high concentrations in sheep. This may directly release oxygen radicals or proteases which initiate acute lung vascular injury. Our findings do not negate a role for blood leucocytes or local macrophages in the increased lung vascular permeability with sepsis, but rather emphasize a potential role for plasma fibronectin in lung microvascular integrity. If plasma fibronectin levels can influence the integrity of the lung microvascular barrier, it may have potential therapeutic value in the treatment of septic surgical or trauma patients with "capillary leaking", pulmonary edema, and ARDS.

IX. SUMMARY

Thus, a relationship may exist between reticuloendothelial system phagocytic failure and the development of lung failure in septic surgical or injured patients. Of particular importance may be fibronectin. A unique relationship may exist between the more soluble plasma fibronectin and the antigenically related insoluble cell surface fibronectin, as it appears that plasma fibronectin can be incorporated into the tissue fibronectin pool. Accordingly, its deficiency and/or fragmentation may be important to the pathophysiology of altered phagocytic host defense and altered vascular integrity.[52] It should be pointed out that cryoprecipitate infusion as a means to elevate fibronectin levels is not a proven modality of therapy for the septic injured patient. Controlled clinical studies with opsonically active fresh cryoprecipitate[57] as well as purified opsonically active plasma fibronectin[61] must be done to test the value of fibronectin infusion in the therapy of septic shock. If purified human fibronectin is used, it should be pasteurized and documented to have retained its opsonic activity after pasteurization before use in such clinical trials. If cryoprecipitate is used as a source of fibronectin, then only fresh (less than 1 to 2 months old) plasma cryoprecipitate should be used, since it has recently been shown by immunoblot analysis[7] that the fibronectin-mediated opsonic activity of plasma cryoprecipitate declines with storage. This loss of activity appears to be related to the presence of fragments of fibronectin[7] apparently formed during storage in excess of several months, even when cryoprecipitate is stored at $-80°C$. Antigen concentration and gelatin binding activity are not adequate or reliable measures of opsonic or function activity, and caution must be used in relying on these parameters.

The initial discovery[5,60] of the common identity of fibronectin (CIg) and opsonic α_2SB glycoprotein was the missing link that pointed to a relationship between fibronectin and phagocytic cell function. Since that time, many studies have directly or indirectly pointed to a role for fibronectin in phagocytic cell behavior. For example, fibronectin is a nonimmune opsonin for denatured proteins as well as cellular and/or tissue debris;[6,32,55,70] it is a ligand that in conjunction with complement can amplify *Staphylococcus* uptake by leukocytes; it can bind to Clq and increase the phagocytic removal of Clq-coated immune complexes; and it can augment the uptake of C3b-coated red blood cells by activated leukocytes as well as monocyte uptake of particles coated with IgG.[46] Macrophage binding to fibronectin also appears to increase the number of Fc receptors on the surface of macrophages as well as influence complement-mediated phagocytic ingestion of bacteria.[71] However, the additional role of fibronectin in vascular permeability is a new biologic dimension with exciting possibilities. The model presented in Figure 6 provides an overview of mechanisms by which disturbances in fibronectin kinetics and/or structure could potentially influence lung microvascular integrity. The current intense scientific interest in fibronectin will undoubtedly expand our understanding of how this molecule influences the behavior of phagocytic cells,[6,25,46,58] and the process of wound repair,[10,23,24] and the microvascular response[11,12] to tissue injury.

ACKNOWLEDGMENTS

Studies presented in this review were supported by NIH grant GM-21447. Utilization of the Albany Trauma Research Center (GM-15426) for clinical studies is also acknowledged. Mrs. Deborah Moran and Mrs. Maureen Davis provided secretarial and editorial assistance. The efforts of Larry Cohler, M.D. and William Charash, M.D., who were postdoctoral fellows (GM-07033) and Ph.D. trainees in the laboratory of the author, is especially acknowledged with respect to the sheep lung vascular permeability studies reviewed. The technical assistance of E. Cho, E. Lewis, M. Lewis, and V. Gray is also acknowledged.

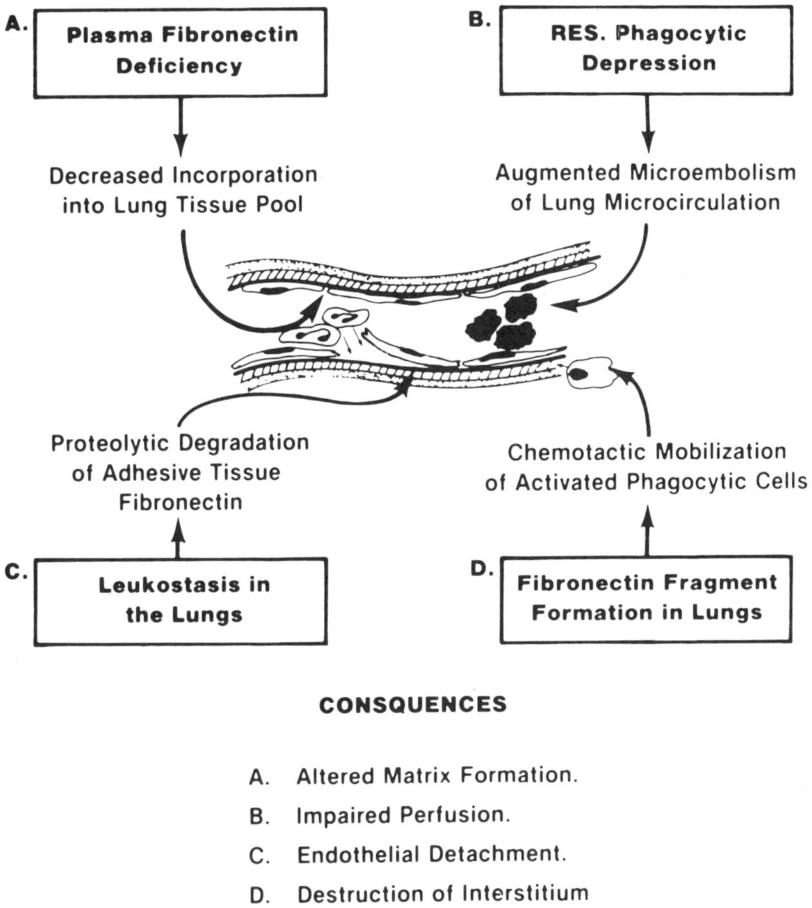

FIGURE 6. Mechanisms by which disturbances in fibronectin can contribute to lung vascular injury and altered vascular permeability. Microvascular embolization due to inefficient RES particulate removal; altered gradient for incorporation of plasma fibronectin into tissue pool; proteolytic fragmentation of lung tissue fibronectin by enzymes released from marginated leukocytes and/or activated macrophages with abnormal cell-cell contact or endothelial cell adhesion to collagenous substratum; and chemotactic attraction of active monocytes, leukocytes, and other cells by fibronectin fragments are depicted. Oxygen metabolites, such as superoxide anion (O_2^-), hydrogen peroxide (H_2O_2), and the hydroxyl radical (OH^-), released from these activated cells could further injure endothelial and nonendothelial cells. Proteolytic enzyme release can disrupt the interstitial and extracellular matrix. (From Saba, T. M., in *Critical Care: State of the Art,* Vol. 7, The Society of Critical Care Medicine, Fullerton, CA, 1986, 437. With permission.)

REFERENCES

1. **Alitalo, K., Hovi, T., and Vaheri, A.,** Fibronectin is produced by human macrophages, *J. Exp. Med.,* 151, 602, 1980.
2. **Bagge, L., Hedstrand, U., Hook, M., Johansson, S., Lind, E., Modig, J., and Saldeen, T.,** Fibrinolysis inhibition and fibronectin in the blood in patients with the delayed microembolism syndrome, *Uppsala J. Med. Sci.,* 88, 81, 1983.
3. **Benacerraf, B., Biozzi, G., Halpern, B. N., and Stiffel, C.,** Physiology of phagocytosis of particles by the reticuloendothelial system, in *Physiopathology of the Reticuloendothelial System,* Halpern, B. N., Ed., Charles C Thomas, Springfield, IL, 1957, 52.

4. **Blumenstock, F., Weber, P., Saba, T. M., and Laffin, R.,** Electroimmunoassay of alpha-2-opsonic protein during reticuloendothelial blockade, *Am. J. Physiol.*, 232, R80, 1977.
5. **Blumenstock, F. A., Saba, T. M., Weber, P., and Laffin, R.,** Biochemical and immunological characteristics of human opsonic α_2-SB-glycoprotein: its identity with cold-insoluble globulin, *J. Biol. Chem.*, 253, 4287, 1978.
6. **Blumenstock, F. A., Saba, T. M., Roccario, E., Cho, E., and Kaplan, J.,** Opsonic fibronectin after trauma and particle injection as determined by a peritoneal macrophage monolayer assay, *J. Reticuloendothelial Soc.*, 30, 61, 1981.
7. **Blumenstock, F. A., Valeri, C. R., Saba, T. M., and Cho, E.,** Progressive loss of fibronectin mediated opsonic activity in plasma cryoprecipitate with storage, *Vox Sanguinis,* 54, 129, 1988.
8. **Brigham, K. L., Woolverton, W. C., Blake, L. A., and Staub, N. C.,** Increased sheep lung vascular permeability caused by Pseudomonas bacteremia, *J. Clin. Invest.*, 54, 792, 1974.
9. **Brodin, B., von Schenck, H., Schildt, B., and Liljedahl, S.-O.,** Low plasma fibronectin indicates septicaemia in major burns, *Acta Chir. Scand.*, 150, 5, 1984.
10. **Clark, R. A. F., Dvorak, H. F., and Colvin, R. B.,** Fibronectin in delayed-type hypersensitivity skin reactions: associations with vessel permeability and endothelial cell activation, *J. Immunol.*, 126, 787, 1981.
11. **Cohler, L. F., Saba, T. M., and Lewis, E. P.,** Lung vascular injury with protease infusion: relationship to plasma fibronectin, *Ann. Surg.*, 202, 240, 1985.
12. **Cohler, L. F., Saba, T. M., Lewis, E., Vincent, P. A., and Charash, W. E.,** Plasma fibronectin and lung protein clearance with gram-negative bacteremia: response to fibronectin therapy, *J. Appl. Physiol.*, 63, 623, 1987.
13. **Cosio, F. G. and Bakaletz, A. P.,** Role of fibronectin on the clearance and tissue uptake of antigen and immune complexes in rats, *J. Clin. Invest.*, 80, 1270, 1987.
14. **Deno, D. C., McCafferty, M. N., Saba, T. M., and Blumenstock, F. A.,** Mechanism of acute depletion of plasma fibronectin following thermal injury in rats: appearance of gelatinlike ligand in plasma, *J. Clin. Invest.*, 73, 20, 1984.
15. **Deno, D. C., Saba, T. M., and Lewis, E. P.,** Kinetics of endogenously labelled plasma fibronectin: incorporation into tissues, *Am. J. Physiol.*, 245, R564, 1983.
16. **Dillon, B. C., Saba, T. M., Cho, E., and Lewis, E.,** Opsonic deficiency in the etiology of starvation-induced reticuloendothelial phagocytic dysfunction, *Exp. Mol. Pathol.*, 36, 177, 1982.
17. **Ekindjian, O. G., Marien, M., Wasserman, D., Bruxelle, J., Cazalet, C., Konter, E., and Yonger, J.,** Plasma fibronectin time course in burned patients: influence of sepsis, *J. Trauma,* 24, 214, 1984.
18. **Eriksen, H. O., Kalaja, E., Jensen, B. A., and Clemmensen, I.,** Plasma fibronectin concentrations in patients with severe burn injury, *Burns,* 10, 422, 1984.
19. **Ehrlich, M. I., Krushell, J. S., Blumenstock, F. A., and Kaplan, J. E.,** Depression of phagocytosis by plasmin degradation products of plasma fibronectin, *J. Lab. Clin. Med.*, 98, 263, 1981.
20. **Fourrier, F., Chopin, C., Wallaert, B., Mazurier, C., Mangalaboyi, J., and Durocher, A.,** Compared evolution of plasma fibronectin and angiotensin-converting enzyme levels in septic ARDS, *Chest,* 87, 191, 1985.
21. **Gold, L. I. and Pearlstein, E.,** Fibronectin-collagen binding and requirement during cellular adhesion, *Biochem. J.*, 186, 551, 1980.
22. **Gonzalez-Calvin, J., Scully, M. F., Sanger, Y., Fok, J., Kakkar, V. V., Hughes, R. D., Gimson, A. E. S., and Williams, R.,** Fibronectin in fulminant hepatic failure, *Br. Med. J.*, 285, 1231, 1982.
23. **Grinnell, F.,** Fibronectin and wound healing, *J. Cell Biochem.*, 26, 107, 1984.
24. **Grinnell, F., Billingham, R. E., and Burgess, L.,** Distribution of fibronectin during wound healing *in vivo, J. Invest. Dermatol.*, 75, 181, 1981.
25. **Gudewicz, P. W., Molnar, J., Lai, M. Z., Beezhold, D. H., Siefring, G. E., Jr., Credo, R. B., and Lorand, L.,** Fibronectin mediated uptake of gelatin-coated latex particles by peritoneal macrophages, *J. Cell Biol.*, 87, 427, 1980.
26. **Harlan, J. M., Killen, P. D., Harker, L. A., and Striker, G. E.,** Neutrophil-mediated endothelial injury *in vitro:* mechanisms of cell detachment, *J. Clin. Invest.*, 68, 1394, 1981.
27. **Hayman, E. G., Pierschbacher, M. D., and Ruoslahti, E.,** Detachment of cells from culture substrate by soluble fibronectin peptides, *J. Cell Biol.*, 100, 1948, 1985.
28. **Howard, L., Dillon B., Saba, T. M., Hofman, S., and Cho, E.,** Decreased plasma fibronectin during starvation in man, *J. Parent. Enteral Nutr.*, 8, 237, 1984.
29. **Hynes, R. O. and Yamada, K. M.,** Fibronectins: multifunctional modular glycoproteins, *J. Cell Biol.*, 95, 369, 1982.
30. **Jaffe, E. A. and Mosher, D. F.,** Synthesis of fibronectin by cultured human endothelial cells, *J. Exp. Med.*, 147, 1779, 1978.
31. **Kaplan, J. E. and Saba, T. M.,** Humoral deficiency and reticuloendothelial depression after traumatic shock, *Am. J. Physiol.*, 230, 7, 1976.

32. **Kiener, J. L., Cho, E., and Saba, T. M.,** Factor XIII_a as a modulator of plasma fibronectin alterations during experimental bacteremia, *J. Trauma,* 26, 1013, 1986.
33. **Lanser, M. E., Saba, T. M., and Scovill, W. A.,** Opsonic glycoprotein (plasma fibronectin) levels after burn injury: relationship to extent of burn and development of sepsis, *Ann. Surg.,* 192, 776, 1980.
34. **La Celle, P., Blumenstock, F. A., McKinley, C., Saba, T. M., Vincent, P. A., and Gray, V.,** Blood-borne collagenous debris complexes with plasma fibronectin after thermal injury, *Blood,* in press.
35. **Maunder, R. J., Harlan, J. M., Pepe, P. E., Paskell, S., Carrico, C. J., and Hudson, L. D.,** Measurement of plasma fibronectin in patients who develop the adult respiratory distress syndrome, *J. Lab. Clin. Med.,* 104, 583, 1984.
36. **McDonald, J. A., Baum, B. J., Rosenberg, D. M., et al.,** Destruction of a major extracellular adhesive glycoprotein (fibronectin) of human fibroblasts by neutral proteases from polymorphonuclear leukocyte granules, *Lab. Invest.,* 40, 350, 1979.
37. **McKeown-Longo, P. J. and Mosher, D. F.,** Mechanism of formation of disulfide-bonded multimers of plasma fibronectin in cell layers of cultured human fibroblasts, *J. Biol. Chem.,* 259, 12210, 1984.
38. **Metchnikoff, E.,** Sur la lutte des celluler de l'organisme contre l'invasion des microbes, *Ann. Inst. Pasteur,* 1, 321, 1887.
39. **Mosher, D. F.,** Changes in the plasma cold insoluble globulin concentration during experimental Rocky Mountain Spotted Fever infection in rhesus monkeys, *Thromb. Res.,* 9, 37, 1976.
40. **Mosesson, M. W. and Amrani, D. L.,** The structure and biologic activities of plasma fibronectin, *Blood,* 56, 145, 1980.
41. **Mosher, D. F. and Williams, E. M.,** Fibronectin concentration is decreased in plasma of severely ill patients with disseminated intravascular coagulation, *J. Lab. Clin. Med.,* 91, 729, 1978.
42. **Niehaus, G. D., Schumacker, P. T., and Saba, T. M.,** Influence of opsonic fibronectin deficiency on lung fluid balance during bacterial sepsis, *J. Appl. Physiol.,* 49, 693, 1980.
43. **Niehaus, G. D., Schumacker, P. T., and Saba, T. M.,** Reticuloendothelial clearance of blood-borne particulates; relevance to experimental lung microembolization and vascular injury, *Ann. Surg.,* 191, 479, 1979.
44. **Nolan, J. P.,** Endotoxin, reticuloendothelial function, and liver injury, *Hepatology,* 1, 358, 1981.
45. **Oh, E., Pierschbacher, M., and Ruoslahti, E.,** Deposition of plasma fibronectin in tissues, *Proc. Natl. Acad. Sci. U.S.A.,* 78, 3218, 1981.
46. **Pommier, G. C., O'Shea, J., Chused, T., et al.,** Studies on the fibronectin receptors of human peripheral blood leukocytes: morphologic and functional characterization, *J. Exp. Med.,* 159, 137, 1984.
47. **Richards, P. S., Saba, T. M., Del Vecchio, P. J., Vincent, P. A., and Gray, V. C.,** Matrix fibronectin disruption in association with altered endothelial cell adhesion induced by activated polymorphonuclear leukocytes, *Exp. Mol. Pathol.,* 45, 1, 1986.
48. **Richards, W. O., Scovill, W. A., and Shin, B.,** Opsonic fibronectin deficiency in patients with intra-abdominal infection, *Surgery,* 94, 210, 1983.
49. **Rieder, H., Birmelin, M., and Decker, K.,** Synthesis and functions of fibronectin in rat liver cells *in vitro, in Sinusoidal Liver Cells,* Knook, D. L. and Wissa, E., Eds., Elsevier Biomedical Press, 1982, 193.
50. **Robbins, A. B., Doran, J. E., Reese, A. C., and Mansberger, A. R., Jr.,** Cold insoluble globulin levels in operative trauma: serum depletion, wound sequestration, and biological activity: an experimental and clinical study, *Am. Surgeon,* 46, 663, 1980.
51. **Rubli, E., Bussard, S., Frei, E., Lundsgaard-Hansen, P., and Pappova, E.,** Plasma fibronectin and associated variables in surgical intensive care patients, *Ann. Surg.,* 197, 310, 1983.
52. **Saba, T. M.,** Plasma and tissue fibronectin: its role in pathophysiology of the critically-ill septic patient, in *Critical Care: State of the Art,* Vol. 7, The Society of Critical Care Medicine, Fullerton, CA, 1986, 437.
53. **Saba, T. M.,** Effect of surgical trauma on the clearance and localization of blood-borne particulate matter, *Surgery,* 71, 675, 1972.
54. **Saba, T. M.,** Physiology and pathophysiology of the reticuloendothelial system, *Arch. Intern. Med.,* 126, 1031, 1970.
55. **Saba, T. M. and Di Luzio, N. R.,** Reticuloendothelial blockade and recovery as a function of opsonic activity, *Am. J. Physiol.,* 216, 197, 1969.
56. **Saba, T. M., Blumenstock, F. A., Scovill, W. A., and Bernard, H.,** Cryoprecipitate reversal of opsonic α_2-surface binding glycoprotein deficiency in septic surgical and trauma patients, *Science,* 201, 622, 1978.
57. **Saba, T. M., Blumenstock, F. A., Shah, D. M., Kaplan, J. E., Cho, E., Scovill, W., Stratton, H., Newell, J., Gottlieb, M., Sedransk, N., and Rahm, R.,** Reversal of fibronectin and opsonic deficiency in patients: a controlled study, *Ann. Surg.,* 199, 87, 1984.
58. **Saba, T. M. and Jaffe, E.,** Plasma fibronectin (opsonic glycoprotein): its synthesis by vascular endothelial cells and role in cardiopulmonary integrity after trauma as related to reticuloendothelial function, *Am. J. Med.,* 68, 577, 1980.

59. **Saba, T. M., Niehaus, G. D., Scovill, W. A., Blumenstock, F. A., Newell, J. C., Holman, J., Jr., and Powers, S. R., Jr.,** Lung vascular permeability after reversal of fibronectin deficiency in septic sheep, *Ann. Surg.*, 198, 654, 1983.
60. **Saba, T. M., Blumenstock, F. A., Weber, P., and Kaplan, J. E.,** Physiologic role for cold-insoluble globulin in systemic host defense: implications of its characterization as the opsonic α_2SB glycoprotein, *N.Y. Acad. Sci.*, 312, 43, 1978.
61. **Saba, T. M., Blumenstock, F. A., Shah, D. M., Landaburu, R. H., Hrinda, M. E., Deno, D. C., Holman, J. M., Jr., Cho, E., Phelan, C., and Cardarelli, P. M.,** Reversal of opsonic deficiency in surgical, trauma, and burn patients by infusion of purified human plasma fibronectin, *Am. J. Med.*, 80, 229, 1986.
62. **Scovill, W. A., Annest, S. J., Saba, T. M., Blumenstock, F. A., Newell, J. C., Stratton, H. H., and Powers, S. R.,** Cardiovascular hemodynamics after opsonic α_2 surface binding glycoprotein therapy in injured patients, *Surgery*, 86, 284, 1979.
63. **Scovill, W. A., Saba, T. M., Blumenstock, F. A., Bernard, H. R., and Powers, S. R., Jr.,** Opsonic α_2 surface binding glycoprotein therapy during sepsis, *Ann. Surg.*, 188, 521, 1978.
64. **Scovill, W. A., Saba, T. M., Kaplan, J. E., Bernard, H., and Powers, S. R., Jr.,** Deficits in reticuloendothelial humoral control mechanisms in patients after trauma, *J. Trauma*, 16, 898, 1976.
65. **Snyder, E. L., Barash, P. G., Mosher, D. F., and Walter, S. D.,** Plasma fibronectin level and clinical status in cardiac surgery patients, *J. Lab. Clin. Med.*, 102, 881, 1983.
66. **Snyder, E. L., Mosher, D. F., Hezzey, A., and Golenwsky, G.,** Effect of blood transfusion on *in vivo* levels of plasma fibronectin, *J. Lab. Clin. Med.*, 98, 336, 1981.
67. **Tamkun, J. W. and Hynes, R. O.,** Plasma fibronectin is synthesized and secreted by hepatocytes, *J. Biol. Chem.*, 258, 4641, 1983.
68. **Todd, T. R., Glynn, M. F. X., Silver, E., and Redmond, M. D.,** A randomized trial of cryoprecipitate replacement of fibronectin deficiencies in the critically-ill, *Am. Rev. Resp. Dis.*, 129, A102, 1984.
69. **Van Furth, R.,** *Mononuclear Phagocytes in Immunity, Infection and Pathology*, Blackwell Scientific, Oxford, 1975, 1.
70. **Walton, K. W., Almond, T. J., Robinson, M., and Scott, D. L.,** An experimental model for the study of the opsonic activity of fibronectin in the clearance of intravascular complexes, *Br. J. Exp. Path.*, 65, 191, 1984.
71. **Wright, S. D., Licht, M. R., Craigmyle, L. S., and Silverstein, S. C.,** Communication between receptors for different ligands on a single cell: ligation of fibronectin receptors induces a reversible alteration in the function of complement receptors on cultured human monocytes, *J. Cell Biol.*, 99, 336, 1984.
72. **Yamada, K. M. and Olden, K.,** Fibronectins: adhesive glycoproteins of cell surface and blood, *Nature (London)*, 275, 179, 1978.
73. **Yamada, K. M. and Kennedy, D. W.,** Fibroblast cellular and plasma fibronectin are similar but not identical, *J. Cell Biol.*, 80, 492, 1979.
74. **Yoder, M. C., Douglas, S. D., Gerdes, J., Kline, J., and Polin, R. A.,** Plasma fibronectin in healthy newborn infants: respiratory distress syndrome and perinatal asphyxia, *J. Pediatr.*, 102, 777, 1983.
75. **Zweifach, B. W., Benacerraf, B., and Thomas, L.,** Relationship between the vascular manifestations of shock produced by endotoxin, trauma and hemorrhage. II. The possible role of the RES in resistance to each type of shock, *J. Exp. Med.*, 106, 403, 1957.

Chapter 5

FIBRONECTIN INTERACTIONS WITH COMPLEMENT PROTEINS: CHARACTERISTICS OF THE BINDING INTERACTIONS AND POTENTIAL BIOLOGICAL SIGNIFICANCE

Byron E. Anderson, Ruth A. Entwistle, J. Jay Weiss, and Frank R. Schmid

TABLE OF CONTENTS

I. INTRODUCTION

Fibronectin (Fn) has a number of well-characterized binding interactions with a variety of macromolecules as described elsewhere in this book. Fn also binds to two molecules of the classical complement system, C1q and C3. The Fn-C1q binding has been studied extensively by a number of laboratories and characterized with respect to ionic strength dependency, solution vs. solid phase binding systems, and the domains of each molecule potentially contributing binding sites. The binding of Fn to C1q, as an isolated molecule or bound to immune complexes has been studied. These immune complexes may be formed on a solid phase, in solution, or on cell surfaces. Many kinetic and stoichiometric parameters of the binding of Fn to C1q have also been determined. The Fn binding to the C3 component of complement is much less well documented. C3 will not bind to solid phase adsorbed Fn, whereas the converse situation will result in binding. There is also evidence that Fn will bind to C3 which is bound either to immune complexes or on cell surfaces.

The characteristics of the molecular interactions between Fn and C1q and C3 are described below, and we document the extensive evidence for the Fn-C1q interaction. The difficulties in demonstrating significant binding at physiologic ionic strengths are discussed with regard to other potential molecular interactions. In addition, we speculate on the involvement of the Fn-C1q binding in pathologic conditions whereby the interaction may effectively trap immune complexes at site of immunologic injury with resultant contributions to inflammation.

The sequence of events yielding immune complexes, the activation of complement, the subsequent binding of Fn to either or both C1q and C3 (and possibly immunoglobulin and certain antigens of the complexes), and the biologic significance of the Fn binding are summarized in Figures 1 and 2. As depicted, multiple types of interactions are possible (as indicated by the arrows). Interactions may involve more than one binding site on a single Fn molecule via the same binding site of each subunit of the Fn dimer or by utilizing different binding sites. In addition, the source of Fn may be the soluble, circulating form, or the cell surface or connective tissue matrix form. With cell- or tissue-bound Fn, there is the possibility of immune complex-complement protein interaction if appropriate binding sites on the Fn molecule are available for such interactions — this situation will be discussed later.

First, the formation of immune complexes and complement activation, and the interactions of Fn with C1q, and with C3 and immunoglobulin, will be described. The possible biologic and pathologic consequences of immune-complex-associated Fn are also discussed with regard to experimental findings and are summarized at the end of this chapter.

II. FORMATION OF IMMUNE COMPLEXES AND THE ACTIVATION OF THE CLASSICAL COMPLEMENT CASCADE

A. HETEROGENEOUS NATURE OF IMMUNE COMPLEXES

The immunoglobulins elicited in response to an immunogenic substance will be primarily of the IgM, IgG, and IgA classes, and of their respective subclasses. The relative concentrations of antibody immunoglobulin classes and subclasses reactive with an antigen is dependent on time after antigenic challenge, routes of immunization, the physical-chemical nature of antigen(s), and genetically determined factors. In addition, each class and subclass of antibody molecule consists of multiple isotypes and different binding sites (as defined by the amino acid sequences of the hypervariable regions in the variable domains of the heavy and light chains) to the same epitope of the antigen, and to different epitopes of the same antigen, or molecular complex of antigens. This heterogeneity and polyclonal nature of the antibody response, together with the relative concentrations of antigen and antibody, yield immune complexes (Ab_xAg_y) where the x:y ratio will vary in a population of complexes,

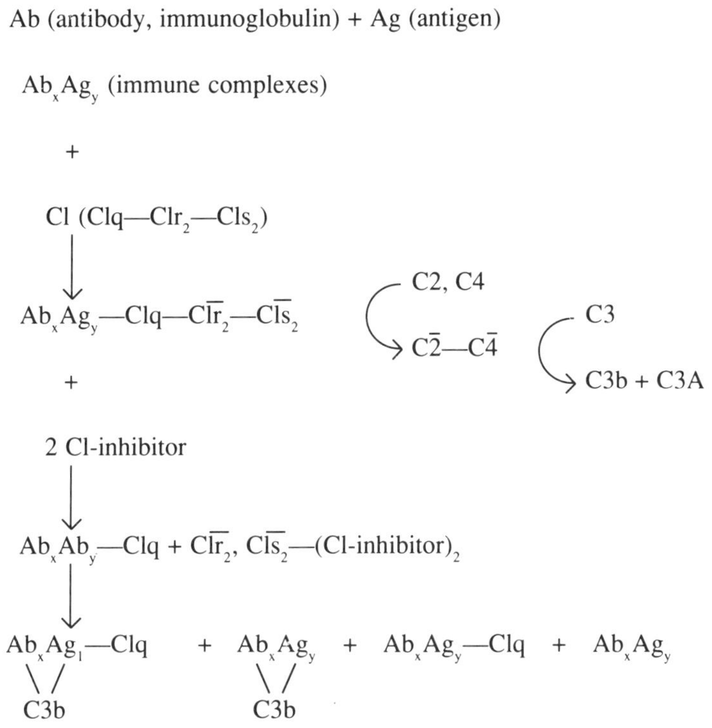

FIGURE 1. The antibodies (Ab_x), of the different immunoglobulin classes, bind to epitopes of antigens (e.g., either unassociated molecules or antigens as part of cells or molecular complexes) forming immune complexes. The immune complex composition (ratio and numbers, x and y, of Ags and Abs) and size will be dependent on Ag and Ab concentrations, size, and number of epitopes per Ag molecule, and on the class of immunoglobulins of the Abs (e.g., IgM or IgG). Immune complexes place two or more Ab molecules in close proximity for the binding of one or more C1 molecular complexes with activation of the proteolytic properties of C1r and C1s (activation denoted by the bar over the subcomponents). C1-inhibitor may bind to $C1r_2$-$C1s_2$ with subsequent dissociation of that complex leaving the immune complexes with attached C1q. C1s proteolytically activates both C2 and C4, and the resultant $C\bar{2}$-$C\bar{4}$ complex in turn activates C3 to C3a and C3b, the latter being able to form a covalent bond attachment to the immune complex proteins. The immune complexes will be heterogeneous both because of the varying numbers of Ag and Ab per complex, and because some complexes may not bind C1, C1q may dissociate from the complexes, and all immune complexes may not have bound C3b (or a degradative product of C3b).

the distribution of ratios will change over time, the sizes of the complexes (x + y) will differ, and immunoglobulin class and subclass distributions in individual immune complexes will also be different. In addition, the distributions of size, antibody: antigen ratios, and immunoglobulin types of the immune complexes will change over time as a function of antigen load, maturation of the immune response, and immune complex clearance. A single IgM class immunoglobulin bound to antigen has the ability to fix complement, and immune complexes having more than one IgG molecule in close proximity can also bind the C1 complement component with activation of the complement system.[1] However, each IgG subclass differs in its ability to activate complement. Thus, immune complexes, either as relatively soluble complexes or as part of cell surface complexes, will differ in their content of attached complement proteins.

The heterogenity of immune complexes is well documented, and several mechanisms are available in the reticuloendothelial system for recognition, binding, and clearance of the varied forms of the complexes. Because the binding of Fn to immune complexes may be partly dependent on antigen and immunoglobulin contents of the immune complexes, and on attached complement proteins (C1q and C3), as well as the spatial relationships of the

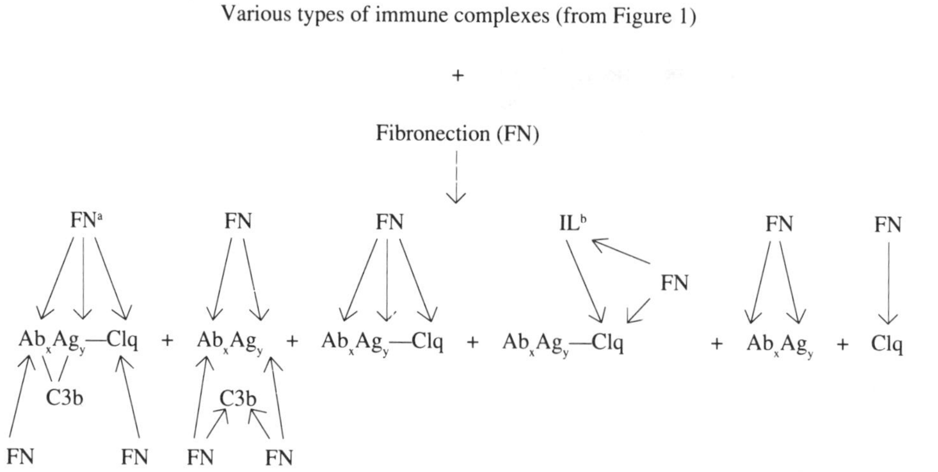

FIGURE 2. Clearance of immune complexes (with or without associated Fn, C1q, or C3b) and C1q-Fn by: (1) binding (and possible phagocytosis) by cells having Fn, C1q, C3, and Fc receptors including monocytes and macrophages, fibroblasts, and endothelial cells, epithelial cells, neutrophils, and platelets, and trapping of immune complexes in tissues by (2) binding to connective tissue matrix components such as Fn, laminin, collagen, and glycosaminoglycans. [a]Because Fn is dimeric, there are at least two ligand-binding sites per 440-kDa dimer that could bind to the same site of two of the same adjacent molecules (e.g., C1q, or Ab, or C3b) of the immune complex; or different binding domains of one Fn dimer can bind to ligand-binding sites of two different molecules (e.g., C1q and C3b) of the immune complex. There are possible a large number of such interactions as indicated by the arrows. In addition, and not indicated in the diagram, a single immune complex composed of x and y numbers of Ab and Ag, respectively, may also contain more than one bound C1q or C3b molecule, with the possibility of one Fn molecule binding to two such bound C1q or C3b molecules. IL[b] = an interacting ligand, i.e., a molecule, such as heparin, that may bind both to an immune complex component and to Fn, thus promoting binding of Fn to the immune complex.

molecules and potential Fn binding sites, immune-complex-bound Fn may also be variable in terms of amount and affinity. Furthermore, it is not yet clear whether the binding of Fn (or particular conformational states of Fn or Fn fragments) is appreciable under physiologic conditions or whether the immune complexes can bind to cell or connective tissue fixed Fn.

Thus, it is to be expected that the characteristics defined for Fn binding to immune system molecules will vary for the different systems studied.

B. THE ACTIVATION OF THE CLASSICAL COMPLEMENT SYSTEM

The classical complement cascade is a set of plasma proteins which complement defense mechanisms of the immune system by operating as chemotactic factors, inflammation factors, opsonins, and membrane attack proteins.[2] C1, the first component of this cascade, is a calcium-dependent zymogen composed of C1q and two each of the proteins C1r and C1s.[3] C1q is both the recognition and initiation protein for activating the classical complement cascade. C1q, through one or more of its six globular heads, binds to the Fc domains of the immunoglobulin of immune complexes, inducing presumably a conformational change in the C1q molecule which causes the auto-activation of C1r (to C$\overline{1r}$, the bar denoting the activated forms of the complement proteins) to exhibit proteolytic activity. As shown in Figure 1, C$\overline{1r}$ then acts on C1s yielding C$\overline{1s}$ which in turn proteolytically activates C2 and C4. The latter molecules form an active complex, C$\overline{2}$-C$\overline{4}$, which cleaves C3 into the fragments C3a and C3b. C3a is the smaller fragment which exhibits chemotactic activity for granulocytes and monocytes and causes increased permeability of capillaries. C3b can form an amide bond with amino groups of adjacent molecules using an internal thio ester mechanism. Thus, a portion of the C3b molecules may become covalently bound to cell surfaces and to immune complexes in the microenvironment surrounding the complexes with attached and activated $C\bar{1}$. C3b may be further degraded into additional fragments with other activities. C3b also cleaves C5 into C5a and C5b, the latter fragment resulting in the activation of the C6 to C9 sequence.

C1-inhibitor is a control protein of the classical complement cascade which has a greater affinity for $C\bar{1}$ than C1 and specifically binds to $C\bar{1}\bar{r}_2$-$C\bar{1}\bar{s}_2$ forming a $C\bar{1}\bar{r}_2$-$C\bar{1}\bar{s}_2$-(C1-inhibitor)$_2$ complex which dissociates from C1q leaving C1q bound to the immune complexes.[4] It has been shown that Fn does not bind to intact Cl[5] but will bind to Clq under conditions described below. It is plausible that once activated $C\bar{1}\bar{r}_2$-$C\bar{1}\bar{s}_2$ has been removed by C1-inhibitor, Fn can bind to the remaining C1q-immune complex with subsequent biologic consequences. As shown in Figure 2, the binding of Fn may be enhanced by its possible interactions with the antigen(s), immunoglobulins, and bound C3 fragments of the immune complexes. Although the interactions between Fn and C1q are the most extensively studied of the immune complex components, it cannot be presumed that Fn interacts primarily or first with C1q. Rather Fn, or Fn fragments, may initially bind to any of the immune complex components, then form a second interaction with consequent development of a high binding constant.

III. BINDING INTERACTIONS BETWEEN Fn AND C1q

Most of the studies describing the kinetic parameters and experimental conditions for Fn-C1q binding have utilized one of the two molecules adsorbed to a solid phase, such as a plastic surface or a surface coated with immune complexes. Even those studies that have suggested a solution phase interaction between Fn and C1q may be interpreted to mean that one of the components forms a liquid phase microaggregate necessary for binding of the second component. Before discussing the particulars of the Fn-C1q binding, the characteristics of Fn pertinent to its interactions with the immune system components, as well as the structure of C1q, will be described.

A. PHYSICAL-CHEMICAL AND BIOLOGICAL PROPERTIES OF Fn OF SIGNIFICANCE TO ITS BINDING TO C1q

Fibronectin has multiple domains capable of binding many different types of ligands as documented in this book. The functions of Fn in wound healing, coagulation, cell migration, and clearance of some types of bacteria and cellular debris may be in some ways similar to its interactions with complement proteins and immune complexes.

A number of post translational modifications including glycosylation, sulfation, phosphorylation, and acetylation,[6,7] yield different molecular forms of Fn. The specificities of these forms of Fn, as well as proteolytic fragments of Fn, with regard to binding to complement proteins have, for the most part, not been investigated, although it should be kept in mind that the different modifications of Fn may have differences in biologic functions.

Fn has four major domains on each of its monomers. In general, the intradomain sequences are resistant to proteolysis, while the interdomain regions are sensitive to cleavage by a number of proteinases. The *in vivo* significance of this sensitivity to proteolysis was postulated by Ruoslahti et al.[8] to be critical for different functions of Fn. For example, activation or inhibition of Fn binding might be controlled by cleaving of different domains of the protein. This has been shown to be the situation in a number of systems including the opsonic,[9] monocyte chemoattractant,[10] and lectin-like activities of Fn.[11] Likewise, the binding of Fn to certain immune system proteins may be facilitated by prior proteolytic activation of the Fn molecule although such phenomena have not been systematically investigated.

Although the secondary structure of Fn is apparently almost entirely composed of β and random types,[12] certain tertiary and quaternary structural features may be quite variable.

Diffusion, sedimentation, and electron microscopic data show two extreme conformations of Fn that vary according to pH and ionic strength conditions.[13-18] At neutral pH and physiologic ionic strength, Fn is an oblate ellipsoid with a sedimentation coefficient of 13.5S. At pH 3 or pH 11, the sedimentation value decreases to 8S indicating a much more extended, fibrillar shape. The same type of extended conformation seems to occur in salt concentrations above 0.5 *M*. Electron micrographs of Fn prepared under high and low salt conditions (or on charged or uncharged EM grids) support these findings, showing that Fn molecules are more folded and compact when prepared in low ionic strength conditions and extended in higher ionic strength conditions,[15,17] with dimensions of 51 × 32 nm (compact form) and 135 × 23 nm (extended form).

Williams et al.[16] have shown that the folded conformation of plasma Fn at neutral and physiologic ionic strength becomes more extended after binding to the collagen fragment CB7, the fragment containing the major Fn binding sequence of collagen.[19] Mosher and Johnson,[20] as well as Williams et al.,[16] postulated that Fn binding first to a high affinity and then low affinity ligands could combine to favor the more extended conformation of the protein. Variations of this idea have been suggested by a number of investigators as well.[21-24] For example, Rouslahti and Engvall[24] proposed that Fn in a cell surface matrix could be bound to a cell surface receptor, to heparin sulfate, and to collagen forming a very high affinity, and essentially structural function in the connective tissue matrices. In a similar manner, Fn binding to more than one immune complex component could yield an essentially irreversible Fn-immune complex matrix. Also, the binding between Fn and C1q apparently is greatly reduced (and not significant?) at physiologic ionic strengths, which calls into question the biologic importance of the interaction. However, Fn binding to another component of the immune complexes could result in a change of conformation which then could promote the Fn-C1q binding (evidence for a higher affinity binding is presented below). Also, another molecule, not an intrinsic part of but interacting with immune complex components, could enhance the Fn-C1q binding as has been shown with heparin.

B. PHYSICAL-CHEMICAL AND BIOCHEMICAL PROPERTIES OF C1q

C1q, the recognition subunit of the first component of the classical complement cascade, is a complex glycoprotein with a molecular weight of 460 kDa. A number of reviews describe the structure and immunoglobulin binding of C1q,[3,25-27] and the pertinent properties of C1q are summarized as follows.

C1q is composed of three kinds of polypeptide chains designated A, B, and C. A total of 18 polypeptide chains, 6 each of the A, B, and C subunits, constitute the entire protein. Approximately half of each subunit sequence resembles collagen with the typical gly-x-y sequence, the unusual amino acids hydroxyproline and hydroxylysine, and the glucosyl-galactosyl disaccharides linked to some of the hydroxylysines. The collagen-like portions of C1q are located on the aminoterminal end of the protein. Six triple helices, each composed of an A, B, and C chain, are bound together noncovalently (as are the triple helices of collagen). There is a disulfide bond between pairs of the C chains giving an increased stability of C1q as evidenced by its thermal transition temperature of 48°C vs. 39° for type I collagen.[28] The two triple helices thus held together by covalent and noncovalent bonds are referred to as a doublet. Three doublets are combined into an additional helix at the aminoterminal half of the collagen-like portion; this portion extends further as six A + B + C helices to the six globular head regions (each of the head regions composed of the carboxyterminal ends of one each of the A, B, and C chains).

The amino acid sequences for C1q are known for the A and B chains and for over half of the C chain.[29-32] The division of the collagen-like region of C1q into the super helical aminoterminal and the six A + B + C helices was predicted first from the amino acid sequence and later confirmed by electron microscopy.[33,34] In this region of the sequences, the A chain has a threonine inserted between two gly-x-y triplets, and the C chain has an alanine residue instead of a glycine residue in one of its triplets.[3,31] The noncollagen, or globular, head regions of the C1q are not unusual in their amino acid compositions. There is one stretch of amino acids in the A and B chains that is homologous to a sequence in α_2-macroglobulin. This sequence in α_2-macroglobulin is the site of cross-linking by plasma transglutaminase. Therefore, it is possible that C1q globular regions can also be cross-linked by this enzyme. The A and B chains themselves are highly homologous with each other with 40 to 73% homologies in some regions. The secondary structure of these regions has not been determined experimentally. Secondary structural predictions, according to the method of Chou and Fasman,[35] indicate approximately 35% β-sheet structure and 5% α-helical structure.

By enzymatic cleavage with either collagenase or pepsin, one can isolate, respectively, intact globular or collagen-like regions of C1q. Using these fragments, it was determined that C1r and C1s bind at the collagen-like region of C1q, and immunoglobulins IgG and IgM bind at the globular regions.[36,37] The binding constants of C1q for immunoglobulins and $C1r_2$-$C1s_2$, and the binding constants of C1q fragments for immunoglobulins and $C1r_2$-$C1s_2$ have been measured by several laboratories.[37-40] The monomeric IgG-C1q binding constant was approximately 10^4 M^{-1}.[38] The C1q globular fragments bind immune complexes (i.e., multimeric IgG) with a weak binding constant of 10^4 M^{-1}.[37] In these two reactions, one component was multimeric (i.e., C1q or the immune complexes) and the other component was monomeric (IgG or the C1q globular fragments). The binding of intact C1q and aggregated IgG was shown to be a much stronger reaction, having a binding constant of 10^7 to 10^8 M^{-1}.[39,40] The great increase in binding is due to a functional affinity interaction of more than one of the six C1q-binding sites and the multiple IgG binding sites.

C. BIOLOGICAL FUNCTIONS OF C1q

Information on the participation of C1q in different cellular processes has been increasing. As mentioned in the introduction, binding of Fn and C1q might be a mechanism for the formation of fibrosis at sites of chronic inflammation or immune complex deposition. A study by Storrs et al.[41] has shown that C1q has a high affinity for both plasma membranes and subcellular membranes. They postulated that such binding might serve to either initiate or enhance inflammatory reactions at the sites of nonimmune tissue damage. Further study will be required to establish this.

Loos[42] has shown that macrophages synthesize C1q and during C1q secretion, membrane-bound C1q might serve the macrophage as both an Fc-binding protein and a polyanion-binding protein. Bohnsack et al.[43] have measured a high affinity reaction of C1q for the basement membrane protein laminin. Intact C1 would not bind. Because activated $\overline{C1r}_2$-$\overline{C1s}_2$ are removed by C1-inhibitor, it is plausible that laminin might be the mechanism by which C1q-immune complexes are deposited on basement membranes in various diseases. Interestingly, in solid phase binding assays, laminin binds C1q-immune complexes better than Fn.[43] The deposition of C1q-immune complexes in basement membrane and connective tissue might also be mediated by proteoglycans which have been shown to bind C1q at its collagen-like region.[44]

C1q is synthesized by fibroblasts, macrophages, monocytes, and columnar intestinal epithelial cells.[45-47] *In vitro* monocytes and macrophages synthesize more C1q than C1r and C1s.[45,47] This is indirect evidence that C1q may be necessary for functions independent of $C1r_2$-$C1s_2$. After C1-inhibitor removes activated $\overline{C1r}_2$-$\overline{C1s}_2$, C1q is in fact functioning independently, if at all, so it is possible that C1q has some still undefined biological functions.

A variety of lymphoblastoid cells as well as monocytes, macrophages, fibroblasts, endothelial cells, and platelets have been shown to bind C1q.[48-52] Fibroblasts also bind C1q with two different affinities by receptors that are currently undefined.[53] It is plausible that all of the cell types that bind C1q, with the exception of platelets, might be phagocytosing either C1q-immune complexes, or effete C1q.

D. INTERACTION STUDIES ON Fn AND C1q

Based on the structural similarities of C1q and collagen and the high affinity of Fn for collagen, numerous laboratories initiated studies to determine if Fn would bind C1q. At the time of the earlier studies, the function of Fn as an opsonic protein, its affinity for complement component C3, its presence on macrophages and in platelets, and its chemotactic properties for fibroblasts were not well known. The co-presence of Fn and C1q in plasma, in cryoglobulins, and at wound repair sites was known. The questions were then (1) would Fn and C1q bind; (2) what conditions allowed binding to occur; and (3) what were the *in vivo* consequences of binding?

Determining the binding conditions of two substances involves a study of concentrations, reaction time, temperature, pH, and buffer composition. A relative affinity scale for comparing the interactions under specified conditions can then be constructed. In addition, identification of activators, inhibitors, and control processes can be obtained. From this type of information one can then postulate how these elements might fit together in *in vivo* pathways.

The following discussion will describe experiments of several laboratories and the information subsequently learned about the interaction of Fn and C1q. Comparisons of the assay systems used and the results of the studies are difficult due to experimental differences used by each laboratory.

Ingham et al.[54] used a variety of approaches to identify complex formation between Fn and C1q. These included selective precipitation by polyethylene glycol, gel filtration and affinity chromatography, sucrose density gradient analysis, and electrophoretic studies. The results of these experiments established two important criteria for Fn-C1q binding. First, binding required one of the proteins to be immobilized, for example by cyanogen bromide conjugation to Sepharose. The second criterion is that binding was enhanced by low ionic strength buffer conditions.

Isliker et al.[55] and Bing et al.[56] used Fn-coated polystyrene and either ^{125}I-C1q or alkaline phosphatase conjugated anti-C1q to measure Fn-C1q binding. In 100 m*M* sodium acetate buffer at pH 7.4, they measured the K_d of the reaction as 82 n*M*. The incubation time for the association was 30 min at 32°C. They showed, using the indirect enzyme conjugated antibody system, that an apparent saturation occurred with approximately 100 n*M* of C1q. They did not quantitate the maximum amount of C1q bound to the Fn, and the amount of Fn bound to the polystyrene was not determined, therefore the composition of the C1q-Fn complexes could not be calculated.

Bing et al.[56] compared gelatin and C1q binding to solid phase Fn. Using the same incubation temperature and buffer conditions, they determined a K_d value of 131 n*M* for Fn-gelatin. This value indicated that Fn-C1q complexes (K_d = 82 n*M*) are just slightly higher in affinity than Fn-gelatin complexes. Their finding was in contrast to earlier work by Isliker et al.[55] who showed that gelatin was a much better inhibitor of Fn-gelatin interactions than C1q by at least two orders of magnitude. More succinctly, the contradiction is that Fn-C1q complexes had a higher affinity constant than Fn-gelatin complexes, but in inhibition assays, gelatin was a better inhibitor (had an apparent higher affinity) than C1q. The construct of the binding reaction is thus seen to be very important, especially with respect to which protein is on the solid phase. Solid phase Fn was used to measure the affinity of Fn for gelatin and C1q. For the inhibition experiments with the same three proteins, gelatin was the solid phase reactant. To compare inhibition more consistently, use of solid phase Fn and liquid phase C1q and gelatin would better ensure measurement of similar binding sites.

Another example of differences in binding behavior in liquid vs. solid phase systems was described in the study by Menzel et al.[57] They showed that C1q bound native collagens by its own collagen-like portion. Using C1q, C1q fragments, and collagen linked to cyanogen bromide Sepharose, or liquid phase C1q, C1q fragments, or collagen, all the permutations of solid phase and liquid phase reactants were evaluted. Each permutation in the experimental design was needed to determine that solid phase C1q allowed a conformation favorable for binding collagen through the collagen-like portion of C1q.

Bing et al.[56] prepared C1q fragments to determine what portion of C1q was binding to Fn. They reported that the C1q globular heads caused no inhibition of C1q binding, while the collagen-like portion of C1q gave a K_i of 59 n*M*. Both fragments were used at 50-n*M* concentrations but, to be comparable to an intact C1q molecule with six globular regions, the C1q globular fragment should have been at least sixfold more concentrated. As is clearly shown by the interaction studies of intact C1q, aggregated IgG, vs. C1q or IgG fragments, weak binding interactions of isolated domains are increased many fold by the increased functional affinity possible with multivalent proteins.

Bing et al.[56] also showed that cellular Fn from hamster fibroblast culture supernatants bound C1q. The source of the Fn was unfractionated culture supernatants which were incubated with C1q-Sepharose at physiologic ionic strength. These data agree with that of Ingham et al.[54] who showed that plasma Fn could bind to C1q-Sepharose at physiologic ionic strength.

Isliker et al.[55] observed high background binding in their solid phase C1q assay using alkaline phosphatase conjugated anti-Fn. It was possible that Fn contaminants in their C1q could be causing this high background, so they absorbed the C1q on a gelatin-Sepharose column. The absorbed C1q showed decreased background binding with their alkaline phosphatase-anti-Fn antibody. They concluded from this that Fn was present in small amounts in their C1q preparation. They also observed an increased activity of this absorbed C1q in initiating complement-mediated lysis of antibody-coated red blood cells. If Fn-C1q complexes occurred *in vivo,* they postulated a similar inhibition of lytic activity could occur.

In the studies of Bing et al.[56] the C1q-Fn binding was determined at various ionic strength and pH conditions. They found that binding was just slightly higher at pH 6 than at pH 7.4, but did not vary greatly between pH 5 and 8. The binding was quite sensitive to different ionic strengths, and was inhibited 80% by addition of 100 m*M* NaCl. Their assay buffer contained 100 m*M* sodium acetate and had an ionic strength of $\mu = 0.1$. This sensitivity to the ionic strength confirmed the observations of Ingham et al.[54] that Fn-C1q interactions were enhanced at lower ionic strengths.

Pearlstein et al.[5] and Sorvillo et al.[58] used a solid phase C1q assay to measure Fn-C1q binding. They also used low ionic strength conditions, but their buffer was composed of 0.05 *M* Tris-acetate, pH 7.4, containing 2.5 mg/ml bovine serum albumin (BSA), while no BSA was present in the reaction buffer of Bing et al.[56] The K_d of the reaction, using a range of Fn concentrations from 60 to 600 n*M*, was 37 n*M*. In agreement with Bing et al.[56] and Ingham et al.[54] showed a considerable ionic strength dependency of the reaction. In addition, however, they reported that Fn binding to gelatin also had a similar ionic strength dependency. With their ionic strength conditions of $\mu = 0.05$, they measured the K_d of gelatin-Fn dissociation as 102 n*M*. The pH dependency between pH 3.5 to 9.5 for Fn binding to solid phase C1q was similar to that of C1q binding to solid phase Fn, i.e., binding was maximal at pH 6, relatively constant from pH 5 to 8, and then decreased significantly at pH values outside of this range.

Pearlstein et al.[5] also measured the dissociation of Fn-C1q complexes in order to establish that the binding reaction was reversible. After 2 h, approximately 50% of their complexes had dissociated. If the reaction is a first order process, the $k_d = 0.693/t_{1/2}$. Since $t_{1/2}$ in this case was 120 min, and the k_d is therefore equal to 5.8×10^{-3} min^{-1}. Given that their

Scatchard analysis showed a single class of binding sites with an equilibrium constant of 37 n*M*, one can calculate the k_a for the reaction by using the relationship $K_d = k_d/k_a$; k_a would be in this case 1.6×10^{-4} min^{-1} nm^{-1}. The maximal dissociation Pearlstein et al.[5] reported was 60% after 3 h. Dissociation was not monitored past 3 h to determine if all of the complexes could dissociate. If only a single class of binding sites is involved in the interaction, no further measurements should be necessary.

In order to study the relative binding of Fn and $C1r_2$-$C1s_2$ to C1q under different circumstances, Sorvillo et al.[58] and Pearlstein et al.[5] studied binding, inhibition, and fragment binding. They found that Fn bound solid phase immune complexes containing C1q, but not those containing C1. They also showed that Fn bound liquid phase immune complexes containing C1q. Increasing the amount of C1q in the complexes resulted in increased Fn binding. The increase was not directly proportional to the amount of C1q bound. Binding to C1q containing immune complexes in the liquid phase could be measured at physiologic ionic strength, but was enhanced at low ionic strength.

Activated $C\overline{1r}_2$-$C\overline{1s}_2$ inhibited binding of Fn to solid phase C1q. Maximum inhibition, though, was only 60%. Dissociation of the C1 complex by addition of ethylenediamine tetraacetic acid (EDTA) or by addition of C1-inhibitor again allowed Fn binding. Comparing the dissociation constants of Fn-C1q complexes (K_d = 82 n*M*), with the 140-n*M* K_d value for C1q-$C1r_2$-$C1s_2$, Bing et al.[56] suggested that Fn could compete for C1q binding sites after $C1r_2$-$C1s_2$ had been activated. *In vivo* activated $C\overline{1r}_2$-$C\overline{1s}_2$ is removed by the binding to C1-inhibitor. C1q then would be available for Fn binding as depicted in Figure 1 and discussed previously. Inhibition of Fn binding to C1q was also shown by liquid phase C1q. This is in sharp contrast to the data of Ingham et al.[54] which showed negligible liquid phase binding of Fn and C1q. This discrepancy has not been further documented.

In order to localize the region of C1q involved in binding to Fn, Pearlstein et al.[5] prepared collagen-like and globular fragments of C1q. They coated microtiter wells with the collagen-like fragment and observed only 15% of the binding of ^{125}I-Fn compared to the amount that bound to intact C1q. With the globular fragments of C1q, inhibition studies showed 70% inhibition of the binding of Fn to intact solid phase C1q.

With the collective results of all of these experiments, the following conclusions could be made. Fn binds C1q at or near the same site that C1r and C1s bind. C1r and C1s, even when activated, have a higher affinity than Fn for C1q, but cannot totally inhibit Fn binding; therefore, the relative affinities are apparently close in magnitude. The globular head fragments of C1q partially inhibit Fn binding to C1q, indicating that a portion of the Fn-C1q interaction is due to the globular regions of the C1q. Because only 15% as much Fn bound to the solid phase collagen-like fragment of C1q as bound to intact solid phase C1q, it would appear that the collagen-like region contributes part of the C1q binding site (additional studies are described below). Alternatively, it was not clear if the low binding to the collagen-like fragment was due to either a conformational change in the solid phase fragment or a decreased amount of solid phase fragment. Finally, in comparing Fn binding to C1q vs. Fn binding to gelatin, it should be noted that the analogy would be more accurate if Fn binding to collagen was used because the C1q collagen-like region more closely resembles collagen than gelatin.

Together the papers described here encompass a large number of permutations of binding and inhibition studies, but a few more permutations might produce some additional information about possible interactions of these proteins *in vivo*. It would, for example, be interesting to determine if collagen competes for Fn in the presence of C1q containing immune complexes; in addition, one could study C1q-immune complex and collagen (gelatin) competition for Fn. The sensitivity of either competition to changes in ionic strength could indicate different reactivities operating in different physiologic conditions. Indeed one can speculate that it might be advantageous if Fn cross-linked collagen and C1q-containing immune complexes *in vivo* to immobilize the immune complexes and allow for localized phagocytosis.

Solid phase C1q is very likely different than either solid phase or liquid phase C1q-immune complex. It would be interesting to compare equilibrium constants of the different forms of C1q in these systems (experimental results described below). One technical problem in these experiments is that to be closest to physiologic systems, both C1q and collagen should be solid phase reactants. Experimentally, this has been accomplished by coating the proteins on a polystyrene tube and having a fluid phase ligand equilibrate between the two surfaces.

To determine more precisely the binding sites on C1q and Fn, two studies have been done by Reid and Edmondson[59] and by Sorvillo et al.[60] Their approaches used fragments of C1q and Fn obtained by either collagenase, pepsin, or trypsin digestion, to measure how well solid phase fragments bound Fn. The data from these studies indicated that the site on the Fn molecule binding to C1q was localized to the 50-kDa gelatin-binding fragment. In addition, these studies showed that the binding site for Fn on C1q was located at the interface between the globular and collagen-like portions of C1q. The binding site on C1q is in a portion of the protein that remains intact after either collagenase or pepsin digestion. This was determined by finding that either fragment of C1q on solid phase could bind Fn, and in the liquid phase, C1q collagen-like fragments inhibited Fn binding to solid phase C1q. Liquid phase globular portions of C1q only produced a small amount of inhibition. This was explained by this fragment perhaps having a liquid phase conformation that masked its binding site.

These two studies still leave several questions unanswered. Liquid phase inhibitors in concentrations up to 400 n*M* caused a maximum of 80% inhibition. This included intact liquid phase Fn. Competitive inhibition controls using the same solid phase fragments as the fluid phase inhibitors (e.g., solid and liquid phase 50-kDa Fn fragments) were not done. Thus, it could not be stated that all of the binding was due to the fragment alone. It was also not shown whether liquid phase reactants inhibited Fn binding by binding the liquid phase Fn or the solid phase C1q. As in the study of Bing et al.[56] C1q globular fragment binding was compared on a molar basis to that of intact C1q and the collagen-like fragment of C1q. As mentioned before, this is not a fair comparison because a single C1q molecular has six globular domains, and when attached to intact C1q the functional affinity of six globular heads in close proximity may be greatly increased.

Another question only partially answered by these studies is why there is increased binding of Fn to heat-denatured C1q. Engvall and Ruoslahti[61] showed that Fn bound more readily to heat-denatured collagen compared to native collagens of many types. *In vivo*, they postulated that Fn binds to partially denatured collagen fibrils. If Fn-C1q binding is analogous to the Fn-collagen interactions, one would expect to see an increased binding of heat-denatured C1q. Bing et al.[56] had reported such an increase in Fn binding to C1q that had been heat-denatured for 30 min at 56°C. Reid and Edmondson[59] repeated this experiment and also heat-denatured the collagen-like fragment of C1q. They found that the solid phase-adsorbed heat-denatured C1q bound increased amounts of Fn, but the heat-denatured collagen-like fragment did not. They did not heat-denature the globular fragments of C1q, assuming the globular portions were not involved in binding. Thus, possible contributions of this fragment to the binding were not addressed.

Finally, in the various inhibition experiments reported by both Sorvillo et al.[60] and Reid and Edmondson,[59] a combination of fragments that would mimic the intact proteins was not examined. For example, a mixture of C1q globular and C1q collagen-like fragments, or Fn 29-kDa, 50-kDa, and 200-kDa fragments were not used to inhibit the intact protein binding. Without this kind of permutation, the 80% maximum inhibition observed might indicate that another binding site is involved.

Studies by Harris et al.[62] and Entwistle et al.[63] have shown that the Fn-C1q interaction may involve more than one type of binding site. A low ionic strength (μ = 0.048) sodium phosphate-sodium chloride buffer with added low concentration bovine serum albumin (10 μg/ml) was used to study the binding of ^{125}I-Fn to solid phase adsorbed C1q. These studies satisfied the important criteria that (1) the Fn preparations did not contain detectable fragments, (2) essentially all of the radiolabeled Fn was capable of binding to C1q (thus the binding studies were not measuring competing Fn species), and (3) conditions were used to demonstrate near saturation of binding (and thus adding to the validity of the Scatchard analysis of the data). Nonlinear Scatchard plots of the binding data were obtained for different C1q solid phase concentrations, and a minimum of two equilibrium constants were calcuated averaging 26 *M* and 0.16 *M*. The value of 26 n*M* is similar to those of 37 n*M* and 82 n*M* found by Sorvillo et al.[58] and Bing et al.,[56] respectively, whereas the 0.16 *M* value is about 100-fold higher than any previously reported value. The latter constant could have been derived from the extended range of Fn concentrations used in the binding studies thus discerning higher affinity binding interactions. Two binding constants for the Fn-C1q interactions were also indicated by the findings of two dissociation rates of the Fn-Clq complexes of 1.5×10^{-3} min^{-1} and 3.6×10^{-5} min^{-1}. The two rates of dissociation were verified by allowing Fn to associate with solid phase C1q over longer periods of time. This resulted in a greater fraction of complexes having the slower dissociation rate. The nonlinear equilibrium data and dissociation rates, therefore, indicate more than one type of binding site interaction, cooperativity or different avidity effects in the binding between Fn and C1q. The stoichiometry of the binding was also determined to be approximately one Fn to three C1q molecules for the high affinity interaction and one to two Fn per C1q molecule for the lower affinity interaction. These data may suggest that Fn forms an initial interaction with a C1q molecule and, with time, changes conformation allowing for an additional binding site interaction with a second C1q molecule.

The question of whether Fn and C1q exhibit a binding interaction in the presence of human physiologic fluids was addressed by Carsons et al.[64] They showed that C1q present in rheumatoid-derived synovial fluids but not in normal human plasma would bind to solid phase adsorbed Fn using anti-C1q as the probe. The synovial fluid C1q binding was enhanced at low temperatures of incubation (4°C) and by the presence of EDTA. C1q binding from normal human plasma to the solid phase Fn could be demonstrated after the plasma was dialyzed against low ionic strength Tris-HCl buffer or NaCl solution. The binding from plasma was augmented by the addition of either heparin, or heat-aggregated IgG or by augmenting the Fn content of normal plasma. Synovial fluid diluted 1:20 showed maximal binding of C1q to the Fn with diminished binding occurring with further dilutions. The finding that heat-aggregated IgG (a model for immune complexes) added to plasma increased C1q binding may be interpreted to mean that C1q-aggregated IgG complexes presents the C1q molecule in the proper orientation and density to promote its interaction with Fn, similar to the results of Weiss et al.[65] Furthermore, the results of Carsons et al.[64] are particularly important in demonstrating that C1q (and C1q bound to immune complexes?) in a pathologic fluid (rheumatoid synovial fluid) binds to Fn. As discussed below, Fn in tissues could serve as a substrate for the trapping of immune complexes via interactions with C1q, or immunoglobulin, or C3 degradation products. Also, it has been shown that Fn in pathologic synovial fluids (mainly those from individuals with rheumatoid arthritis) is partially degraded.[66] Such an altered Fn may be the more important form of the molecule (either through exposure of certain binding sites or conformational change) with respect to binding C1q and/or C1q-immune complexes.

IV. BINDING INTERACTIONS BETWEEN Fn AND C1q-IMMUNE COMPLEXES, C3, AND IMMUNOGLOBULINS

A. BINDING OF Fn TO C1q BOUND TO IMMUNE COMPLEXES

The binding of Fn to immune complexes was demonstrated by Kono et al.[67] using peroxidase-labeled anti-Fn antibody as a probe. The immune complexes were formed by adsorption of BSA to the plastic of microtiter plate wells followed by the addition of anti-BSA antibodies. Fn bound to the immune complexes only if C1q was added prior to the addition of Fn. The binding of Fn could be demonstrated at physiologic ionic strength in the presence of high calcium concentrations.

Sorvillo et al.[68] also used the BSA-anti-BSA immune complex system with added C1q to show that ^{125}I-labeled Fn binding was dependent on C1q. The ^{125}I-Fn binding was saturable and was inhibitable by addition of unlabeled Fn. Scatchard analyses of the binding data yielded a single K_d of 2.8×10^{-8} *M*, using incubation conditions of ionic strength μ = 0.05. The total amount of ^{125}I-Fn bound was reduced 150-fold by increasing the ionic strength to approximately 0.15. The maximum ^{125}I-Fn bound at low ionic strength was only 26%. At the higher ionic strength, a much lower amount (2.6%) of the added ^{125}I-Fn was bound. The molar ratios of ^{125}I-Fn to C1q in their immune complex system at low ionic strength ranged between 3×10^{-3} and 18.7×10^{-3}. The ratio increased when lesser amounts of C1q bound to the immune complexes. Additional studies showed that ^{125}I-Fn binding to BSA-anti-BSA complexes in plasma, previously depleted of C1q and Fn, was also dependent on addition of C1q to the plasma. A maximum of 87.6% of the total ^{125}I-Fn was bound, and the ratio of Fn/C1q was about 2×10^{-4} at three different amounts of C1q.

Baatrup and Svehag[69] performed similar experiments in a carefully controlled manner. BSA adsorbed to wells of microplates was incubated with anti-BSA IgG or $F(ab')_2$ fragments followed by addition of serum or EDTA-serum. C1q, C3b, and Fn all bound to the solid phase complexes, but Fn was bound mainly under conditions where C1q and C3b were bound. Purified Fn also bound in a dose-dependent manner to the immune complexes after pretreatment with Fn-depleted serum. PEG-precipitated immune complexes analyzed by SDS-PAGE also showed the presence of the Fn molecule.

Weiss et al.[65] also studied the binding of radiolabeled Fn to Clq where C1q was bound to a solid phase immune complex system. The results obtained were the same for two immune complex systems: (1) anti-methotrexate antibodies bound to solid phase adsorbed methotrexate-albumin conjugates and (2) anti-di-*N*-acetylchitobiose antibodies bound to a disaccharide-carrier conjugate. Using an ionic strength buffer system with μ = 0.04 and a wide range of Fn concentrations, the nonlinear Scatchard analyses of the data yielded two binding constants of K_{d_1} = 1.0 to 2.2 n*M* and K_{d_2} = 10.4 to 18.4 n*M*. The value for the lower affinity binding constant may be interpreted to mean that Fn binds more favorably (i.e., at a lower concentration) to C1q when the C1q molecules are oriented with its collagen-like stalk regions projecting away from the immunoglobulin molecules of the solid phase adsorbed immune complexes. In additional unpublished studies by Weiss, heparin, which can bind to both Fn and C1q, was shown to allow radiolabeled Fn binding to C1q-immune complexes in one half physiologic ionic strength buffers, an ionic strength that results in minimal Fn to C1q binding in the absence of heparin. Thus, additional Fn and C1q associative molecules may promote the Fn-C1q interaction under conditions that may be considered more physiologic.

B. BINDING OF Fn TO THE C3 COMPONENT OF COMPLEMENT

Using experimental systems similiar to those that show binding of Fn to C1q, Hautenen and Keski-Oja[70] demonstrated that Fn binds to solid phase-adsorbed complement component C3. The binding occurs to the C3c and C3d fragments of C3, indicating that the complement system must be activated if this Fn interaction is to be of biological significance. Because these fragments of C3 can be bound covalently to immune complexes, C3c and C3d provide a ligand site for either direct binding of Fn to the complexes or for cooperative binding with Fn effectively cross-linked between C1q and C3 fragments, and possibly antibody and antigen molecules, as indicated in Figure 2.

Other studies have also suggested that the Fn-C3 fragment interaction may be of importance. As described above, Baatrup and Svehag[69] showed that Fn bound to solid phase immune complexes via C3b, although the binding was small in comparison to the amounts of Fn bound via C1q to the same type (BSA-anti-BSA) of complexes. Wright et al.[71] showed that Fn interaction with human monocyte plasma membranes resulted in an activation of C3bi receptors with resultant promotion of phagocytosis of complement-coated erythrocytes. Johnson et al.[72] later showed that Fn bound to agarose beads coated with C3b, C3bi, or the agarose beads pretreated with human serum but not with complement-inactivated human serum. The ^{125}I-Fn binding was inhibitable by unlabeled Fn. The Fn → C3-fragment-coated agarose beads exhibited increased association with mouse macrophages when compared to C3-fragment-coated beads. The ingestion of the coated beads apparently is mediated via the complement C3 receptors.

C. INTERACTION OF Fn WITH IMMUNOGLOBULINS

If Fn has some biologic function through its interaction with immune complexes, the possible binding of Fn to immunoglobulins must be taken into account. Such interactions have been studied, although not extensively, by several laboratories. Vuento et al.[73] demonstrated that Fn would bind to carboxyl modified human IgG. Hautenan and Keski-Oja[74] described the binding interaction of all four subclasses of human myeloma IgGs to solid phase adsorbed Fn using an indirect ELISA assay. Polyclonal human IgG demonstrated reduced binding to Fn compared to the myeloma IgGs. In contrast to the latter study, Cseh et al.[75] demonstrated that radiolabeled Fn would bind to solid phase-adsorbed human myeloma IgG1 and IgG3 but not to the IgG2 and IgG4 subclasses. Interestingly, normal human polyclonal IgG was unable to inhibit the Fn binding to the IgG1 and IgG3 myeloma proteins, nor would Fn bind to solid phase-absorbed human or rabbit IgGs. Ferraccioli et al.[76] demonstrated an interaction of Fn with rheumatoid factor that may occur at the Fc region of the immunoglobulin.

One conclusion that may be drawn from these studies is that Fn can bind to only a limited fraction of immunoglobulins and thus is only experimentally demonstrable with homogenous fractions of immunoglobulins as in the case of myeloma proteins, or certain rheumatoid factors.

D. Fn AS A COMPONENT OF IMMUNE COMPLEXES

There have been many studies on the composition of immune complexes isolated by various methods from human sera and other body fluids of individuals with a variety of diseases and from experimentally produced immune complex disease.

Examination of the data does not yield any evidence for the presence of Fn (or possible fragments of Fn) in the immune complexes. However, in almost all cases, no direct methodology was used to detect the Fn molecule. In a few studies where immune-complex-associated Fn was investigated, Fn bound *in vivo* to such complexes. Kilgallon et al.[77,78] found Fn (in addition to IgM, IgG, C3, C3c, C3d, C4, and C1q) in immune complexes isolated from Hodgkins disease, rheumatoid arthritis, and systemic lupus erythematosus sera, and in BSA-anti-BSA complexes formed in normal human sera. The presence of Fn in the immune complexes may have been due to the previously determined affinities for immunoglobulins, C1q, or C3 components of the complexes. Anderson et al.,[79] Wood et al.,[80] and Scott et al.[81] have shown that Fn was a component of cryoglobulins of both sera and synovial fluids of individuals with various arthritic and nonarthritic diseases. Cryoglobulins contain immunoglobulins as well as C1q and are thought to form at 4°C from soluble serum immune complexes. In the one study,[79] the Fn was shown not to passively adsorb to the cryoglobulins because the serum cryoglobulin fractions could be resolubilized by warming and reprecipitated with a constant amount of Fn coprecipitating. Thus, the fraction of immune-complex-associated Fn was probably an integral part of the precipitated complexes. Beaulieu et al.[82] also showed that Fn promoted cryoglobulin precipitation by removal and re-addition of purified Fn to sera followed by 4°C-dependent precipitation. The amounts of Fn-dependent precipitation was variable for different individual sera and was in part dependent on the amounts of re-added Fn to Fn-depleted sera.

V. SPECULATIONS ON THE NATURE OF Fn BINDING AND ON THE BIOLOGICAL ROLES OF Fn ASSOCIATIONS WITH COMPLEMENT PROTEINS AND IMMUNE COMPLEXES

In the immune response to a defined antigen (e.g., a protein), multiple different antibodies are elicited that can bind via one or more epitopes of the antigen molecule. Antibodies are at least bivalent (IgG, for example, having two antigen-binding sites), and if the antigen has at least two epitopes, an immune complex may be formed containing more than one antibody molecule. The antibody response is varied in that IgG, IgM, and IgA type antibodies, each of different subclasses (and with different complement-fixing abilities), may be elicited to the same antigen, or epitope of the antigen. Early in the humoral immune response, antigen levels are greater than antibody; later, as the antigen molecules are cleared via immune complexes, the antibody concentrations will be greater than antigen. Thus, for Ab_xAg_y complexes, the molar ratios of x and y for any population of immune complexes at any point in time will vary, and the x:y ratio will also vary over time. The percents of such complexes that will have attached complement proteins (as C1q and/or C3 fragments, or other complement proteins), as well as possibly bound Fn, will also vary with immune complex composition and over time. For pathogens such as bacteria or viruses, the various surface antigens would be expected to be coated with antibodies and with other subsequently bound ligands.

The geometry of the immune complexes may be defined in terms of the relative projections of the Fc portions of bound antibodies and the distances that separate any two spatially adjacent antibody Fc regions. Within a population of immune complexes, the Fc projections and spacings will be varied over a wide range of angles and distances, again dependent on concentrations of the reactants and the types of antibody and antigen molecules involved. Since complement (C1) binding and activation are dependent on the Fc orientations and spacing, the amount of activation and the subsequent molar ratios of C1, C1q, and C3 fragments (and possibly Fn) bound to the immune complexes will be different for each such complex and will change over the time of the immune response. Each immune complex then presents on its surface a different array of Fc antibody regions, the C1q aminoterminal stalk region, and orientation of C3 fragments.

The immune system phagocytes cope with the heterogeneity of immune complexes by exhibiting a redundancy of immune complex clearance recognition (binding) signals, including the Fc receptors, the C3 fragment receptors, and again possibly Fn receptors.

Smaller immune complexes are relatively inefficiently cleared by the recognition mechanisms suggesting that clearance is at least in part a function of the probability of receptor-ligand interactions.

Despite the different recognition systems for immune complexes that should remove them from the circulation, levels of complexes are elevated in a number of diseases, particularly those classified as autoimmune in nature. The higher than normal levels of immune complexes in such diseases may be due to either the relative constancy of antigen stimulation of the humoral immune response, lack of suppression of various autoantibody syntheses, or an inefficiency of the clearance mechanisms either intrinsic to the mechanisms or due to the nature of the immune complexes formed in such diseases. Regardless of the reason, immune complexes persist in the circulation and in the tissues of individuals with certain diseases initiating stimulating the inflammatory response.

What role does Fn play in immune complex metabolism? If Fn is bound to immune complexes, it is reasonable to expect that the Fn may enhance attachment and phagocytosis of the complexes through its characterized opsonic activities and cell surface binding interactions. Nearly all of the reports of Fn associations with immunoglobulins, cryoglobulins, C1q, and C3 fragments have suggested such a functional role.

However, how can one ascribe a functional role to a binding interaction, when, for example, the Fn to C1q (or C1q-immune complex binding) is either only found at low (and nonphysiologic) ionic strength conditions, or the amounts of Fn bound per C1q at higher ionic strengths are only a small fraction of the total C1q or immune complexes present in the system? Despite the large number of studies performed characterizing the Fn → C1q binding, it is clear that one or more critical factors are yet missing from the system. For example, partial cleavage or altered conformation of the Fn molecule may be necessary to expose ligand binding sites to either enhance the percent of Fn that binds to C1q molecules, or to allow attainment of a functional affinity of binding. Interaction of two Fn ligand-binding sites to two sites of the C1q molecule(s), or bridge binding to another component of the immune complex (Ag, Ab, or C3 fragments), or to a cell surface (of a phagocyte) may all be involved. This possibility was indicated in the studies of Entwistle et al.[63] and Weiss et al.[67] where a higher affinity was attained for the Fn-C1q interaction with increasing times of incubation. Also, the enhanced Fn → C1q binding at half physiologic ionic strength in the presence of heparin suggests either than another cooperative molecule may be involved in the biologically significant Fn-C1q interaction, or that heparin-Fn binding alters the conformation and binding mechanism of the Fn molecule. Furthermore, the studies by Steffes et al.[66] of a different form (partially degraded?) of Fn in synovial fluids, and by Carsons et al.[64] showing that C1q and Fn bind in the presence of rheumatoid synovial fluid, may be interpreted to suggest that an altered Fn molecule may be the important species for the Fn → C1q interaction.

In pathologic situations, the Fn interactions may be of importance in trapping immune complexes in tissues. For example, Fn is a major component of connective tissues, which may be sites of chronic inflammation in various autoimmune diseases. Also, although Fn is a minor component of the cartilage connective tissue matrix, the levels are raised when cartilage is damaged. Synovial fluid is a potential source of Fn for deposition on exposed connective tissue matrix in pathologic articular cartilage. The Fn in the tissue may trap immune complexes through one or more binding interactions including through C1q. The complexes then may be inefficiently removed because phagocyte recognition may be blocked by the Fn interactions on the tissue surfaces. The activation and persistance of the phagocytes would contribute further to the inflammatory response and to degradation of adjacent connective tissue matrix by release of various hydrolases.

In summary, the evidence is considerable that Fn has multiple binding interactions with the different components of immune complexes, and it is plausible to assume that some fraction of these Fn molecules play a functional role in promoting either attachment of complexes to phagocytes of the reticuloendothelial system, phagocytosis, or both. Because of the multiple binding sites for connective tissue components and cell surface receptors, it is also possible that Fn on immune complexes may affect immune system function via multiple mechanisms. Constructing well-defined experimental systems to study the interaction of Fn with immune complexes should elucidate its relationship to the immune system.

ACKNOWLEDGMENTS

The studies of the authors' laboratories were partially supported by grant #AI-18317 and by a Multipurpose Arthritis Center grant #AM-30692 from the National Institutes of Health. The authors appreciate the fine efforts in the preparation of the manuscript by Ms. Diantha Nicholson.

REFERENCES

1. **Reid, K. B. M. and Porter, R. R.**, The proteolytic activation systems of comcomplement, *Annu. Rev. Biochem.*, 50, 443, 1981.
2. **Fearon, D. T. and Wong, W. W.**, Complement ligand-receptor interactions that mediate biological responses, *Annu. Rev. Immunol.*, 1, 243, 1983.
3. **Reid, K. B. M.**, Proteins involved in the activation and control of the two pathways of human complement, *Biochem. Soc. Trans.*, 11, 1, 1983.
4. **Ziccardi, R. J. and Cooper, N. R.**, Active disassembly of the first complement component, C1 by C1 inactivator, *J. Immunol.*, 123, 788, 1979.
5. **Pearlstein, E., Sorvillo, J., and Gigli, I.**, The interaction of human plasma fibronectin with a subunit of the first component of complement, C1q, *J. Immunol.*, 128, 2036, 1982.
6. **Paul, J. I. and Hynes, R. O.**, Multiple fibronectin subunits and their post translational modifications, *J. Biol. Chem.*, 259, 13477, 1984.
7. **Lang-Mutschler, J.**, Acylated fibronectin: a new type of post-translational modification of cellular fibronectin, *FEBS Lett.*, 201, 210, 1986.
8. **Ruoslahti, E., Engvall, E., and Hayman, E. G.**, Fibronectin: current concepts of its structure and functions, *Collagen Relat. Res.*, 1, 95, 1981.
9. **Czop, J., Kadish, J., and Austen, K. F.**, Augmentation of human monocyte opsonin-independent phagocytosis by fragments of human plasma fibronectin, *Proc. Natl. Acad. Sci. U.S.A.*, 78, 3649, 1981.
10. **Norris, D. A., Clark, R. A. F., Swiggart, L. M., Huff, J. L., and Weston, W. L.**, Fibronectin fragment(s) are chemotactic for human peripheral blood monocytes, *J. Immunol.*, 129, 1612, 1982.
11. **Hormann, H., Richter, H., and Jelinic, V.**, Evidence for a cryptic lectin site in the cell-binding domain of plasma fibronectin, *Hoppe-Seyler's J. Physiol. Chem.*, 365, 517, 1984.
12. **Mossesson, M. W., Chen, A. B., and Huseby, R. M.**, The cold-insoluble globulin of human plasma: studies of its essential structural features, *Biochim. Biophys. Acta*, 386, 509, 1975.
13. **Rocco, M., Carson, M., Hantgan, R., McDonagh, J., and Hermans, J.**, Dependence of the shape of the plasma fibronectin molecule on solvent composition, *J. Biol. Chem.*, 258, 14545, 1983.
14. **Alexander, S. S., Jr., Colonna, G., and Edelhoch, H.**, The structure and stability of human plasma cold-insoluble globulin, *J. Biol. Chem.*, 254, 1501, 1979.
15. **Erickson, H. P. and Carrell, N. A.**, Fibronectin in extended and compact conformations, *J. Biol. Chem.*, 258, 14539, 1983.
16. **Williams, E. C., Jarney, P. A., Ferry, J. D., and Mosher, D. F.**, Conformational states of fibronectin, *J. Biol. Chem.*, 257, 14973, 1982.
17. **Tooney, N. M., Mosesson, M. W., Amrani, D. L., Hainfeld, J. F., and Wall, J. S.**, Solution and surface effects on plasma fibronectin structure, *J. Cell Biol.*, 97, 1606, 1984.
18. **Engel, J. and Furthmayr, H.**, Electron microscopy and other physical methods for the characterization of extracellular matrix components: laminin, fibronectin, collagen IV, collagen VI, and proteoglycans, *Methods Enzymol.*, 145, 3, 1987.
19. **Dessau, W., Adelmann, B. C., Timpl, R., and Martin, G. R.**, Identification of the sites in collagen α-chains that bind serum anti-gelatin factor (cold insoluble globulin), *Biochem. J.*, 169, 55, 1978.
20. **Mosher, D. F. and Johnson, R. B.**, In vitro formation of disulfide-bonded fibronectin multimers, *J. Biol. Chem.*, 258, 6595, 1983.
21. **Osterlund, E., Eronen, I., Osterlund, K., and Vuento, M.**, Secondary structure of human plasma fibronectin: conformational change induced by calf alveolar heparan sulfates, *Biochemistry*, 24, 2661, 1985.
22. **Hedman, K., Johannson, S., Vartio, T., Kjellen, L., Vaheri, A., and Hook, M.**, Structure of the pericellular matrix: association of heparan and chondroitin sulfates with fibronectin procollagen fibers, *Cell*, 28, 663, 1982.
23. **Oldberg, A. and Ruoslahti, E.**, Interactions between chondroitin sulfate proteoglycan, fibronectin, and collagen, *J. Biol. Chem.*, 257, 4859, 1982.

24. **Ruoslahti, E. and Engvall, E.,** Complexing of fibronectin, glycosaminoglycans and collagen, *Biochim. Biophys. Acta,* 631, 350, 1980.
25. **Burton, D. R., Boyd, J., Brampton, A. D., Easterbrooke-Smith, S. B., Emanuel, E. J., Novotny, J., Rademacher, T. W., van Schravendijk, M. R., Sternberg, M. J. E., and Dwek, R. A.,** The C1q receptor site on immunoglobulin G, *Nature (London),* 288, 338, 1980.
26. **Reid, K. B. M. and Porter, R. R.,** The structure and mechanism of activation of the first component of complement, in *Contemporary Topics in Molecular Immunology,* Vol. 4, Inman, F. P. and Mandey, W. J., Eds., Plenum Press, New York, 1975, 1.
27. **Cooper, N. R.,** Activation and regulation of the first complement component, *Fed. Proc.,* 42, 134, 1983.
28. **Reid, K. B. M.,** Complete amino acid sequences of the three collagen-like regions present in subcomponent C1q of the first component of human complement, *Biochem. J.,* 179, 367, 1979.
29. **Reid, K. B. M.,** Amino acid sequence of the N-terminal forty-two amino acid residues of the C chain of subcomponent C1q of the first component of human complement, *Biochem. J.,* 161, 247, 1977.
30. **Reid, K. B. M., Gagnon, J., and Frampton, J.,** Completion of the amino acid sequences of the A and B chains of subcomponent C1q of the first component of human complement, *Biochem. J.,* 203, 559, 1982.
31. **Reid, K. B. M.,** A collagen-like amino acid sequence in a polypeptide chain of human C1q (a subcomponent of the first component of complement), *Biochem. J.,* 141, 189, 1974.
32. **Reid, K. B. M.,** Amino acid sequence of the N-terminal amino acid residues of the B chain of subcomponent C1q of the first component of human complement, *Biochem. J.,* 173, 863, 1978.
33. **Shelton, E., Yonemasu, K., and Stroud, R. M.,** Ultrastructure of the human complement component, C1q, *Proc. Natl. Acad. Sci. U.S.A.,* 69, 65, 1972.
34. **Knobel, H. R., Villeger, W., and Isliker, H.,** Chemical analysis and electron microscopy studies of human C1q prepared by different methods, *Eur. J. Immunol.,* 5, 78, 1975.
35. **Chou, P. Y. and Fasman, G. D.,** Prediction of protein conformation, *Biochemistry,* 13, 222, 1974.
36. **Reid, K. B. M., Sim, R. B., and Fafers, A. B.,** Reconstitution of the haemolytic activity of the first component of human complement by a pepsin-derived fragment of subcomponent C1q, *Biochem. J.,* 161, 239, 1977.
37. **Hughes-Jones, N. C. and Gardner, B.,** Reaction between the isolated globular subunits of the complement component C1q and IgG-complexes, *Mol. Immunol.,* 16, 697, 1979.
38. **Hughes-Jones, N. C. and Gardner, B.,** The reaction between the complement subcomponent C1q, IgG complexes and polyionic molecules, *Immunology,* 34, 459, 1978.
39. **Sledge, C. R. and Bing, D. H.,** Binding properties of the human complement protein C1q, *J. Biol. Chem.,* 248, 2818, 1973.
40. **Lin, T. and Fletcher, D. S.,** Interaction of human C1q with insoluble immunoglobulin aggregates, *Immunochemistry,* 15, 107, 1978.
41. **Storrs, S. B., Kolb, W. P., and Olson, M. S.,** C1q binding and C1 activation by various isolated cellular membranes, *J. Immunol.,* 131, 416, 1983.
42. **Loos, M.,** The functions of endogenous C1q, a subcomponent of the first component of complement, as a receptor on the membrane of macrophages, *Mol. Immunol.,* 19, 1229, 1982.
43. **Bohnsack, J. F., Tenner, A. J., Laurice, G. W., Kleinman, H. R., Martin, G. R., and Browny, E. J.,** The C1q subunit of the first component of complement binds to laminin: a mechanism for the deposition and retention of immune complexes in basement membrane, *Proc. Natl. Acad. Sci. U.S.A.,* 82, 3824, 1985.
44. **Ghebrehiwet, B.,** C1q inhibitor (C1qNH): functional properties and possible relationship to a lymphocyte membrane-associated C1q precipitin, *J. Immunol.,* 126, 1837, 1981.
45. **Reid, K. B. M. and Solomon, E.,** Biosynthesis of the first component of complement by human fibroblasts, *Biochem. J.,* 167, 647, 1977.
46. **Colton, H. R., Gordon, J. M., Borsos, J., and Rapp, J. H.,** Synthesis of the first component of human complement in vitro, *J. Exp. Med.,* 128, 595, 1968.
47. **Bensa, J. C., Reboul, A., and Colomb, M. G.,** Biosynthesis in vitro of complement subcomponents C1q, C1s, C1 inhibitor by resting and stimulated human monocytes, *Biochem. J.,* 216, 355, 1983.
48. **Tenner, A. J. and Cooper, N. R.,** Analysis of receptor-mediated C1q binding to human peripheral blood mononuclear cells, *J. Immunol.,* 125, 1658, 1980.
49. **Tenner, A. J. and Cooper, N. R.,** Identification of types of cells in human peripheral blood that bind C1q, *J. Immunol.,* 126, 1174, 1981.
50. **Bordin, S., Kolb, W. P., and Page, R. C.,** C1q receptors on cultured human gingival fibroblasts: analysis of binding properties, *J. Immunol.,* 130, 1871, 1983.
51. **Wantier, J. L., Reid, K. B. M., Legrand, Y., and Caer, J. P.,** Region of the C1q molecule involved in the interaction between platelets and subcomponent C1q of the first component of complement, *Mol. Immunol.,* 17, 1799, 1980.
52. **Arviene, J., Reboul, A., Bensa, J. C., and Colomb, M. G.,** Characterization of the C1q receptor on human macrophage cell line, U937, *Biochem. J.,* 218, 547, 1984.

53. **Bordin, S., Teller, D. C., and Pope, R. C.,** Human diploid fibroblasts have two receptors for the C1q subcomponent of complement, *Fed. Proc.*, 44, 990, 1985.
54. **Ingham, K. C., Brew, S. A., and Miekka, S. I.,** Interaction of plasma fibronectin with gelatin and complement C1q, *Mol. Immunol.*, 20, 287, 1983.
55. **Isliker, H., Bing, D. H., and Hynes, R. O.,** Interaction of fibronectin with C1q, a subcomponent of the first component of complement, in *The Immune System,* Vol. 2, S. Krager, Basel, 1981, 231.
56. **Bing, D. H., Almeda, S., Isliker, H., Lahav, J., and Hynes, R. O.,** Fibronectin binds to C1q component of complement, *Proc. Natl. Acad. Sci. U.S.A.*, 79, 4198, 1982.
57. **Menzel, E.-J., Smolen, J., and Reid, K.,** Interaction of collagen with C1q via its collagen-like portion, *Biochim. Biophys. Acta,* 670, 265, 1981.
58. **Sorvillo, J., Pearlstein, E., and Gigli, I.,** Requirement for the binding of human plasma fibronectin to the C1q subunit of the first component of complement, *J. Immunol.*, 131, 1400, 1983.
59. **Reid, K. B. M., and Edmondson, J.,** Location of the binding site in subcomponent C1q for plasma fibronectin, *Acta Pathol. Microbiol. Scand. Sect. C,* 92 (Suppl. 284), 11, 1984.
60. **Sorvillo, J., Gigli, I., and Pearlstein, E.,** Fibronectin binding to complement subcomponent C1q: localization of their respective binding sites, *Biochem. J.*, 226, 207, 1985.
61. **Engvall, E. and Rouslahti, E.,** Binding of soluble forms of fibroblast surface protein, fibronectin, to collagen, *Int. J. Cancer,* 20, 1, 1977.
62. **Harris, C., Roth, S., Schmid, F. R., and Anderson, B.,** Binding of fibronectin to C1q; inhibition of binding by aggregated IgG, *Immunol. Commun.*, 10, 601, 1981.
63. **Entwistle, B., Schmid, F. R., and Anderson, B.,** Multiple binding equilibria for fibronectin and low concentration solid phase C1q, *Fed. Proc.*, 44, 1076, 1985.
64. **Carsons, S. E., Schwartzman, S., Diamond, H. S., and Berkowitz, E.,** Interaction between fibronectin and C1q in rheumatoid synovial fluid and normal human plasma, *Clin. Exp. Immunol.*, 72, 37, 1988.
65. **Weiss, J. J., Gustafson, C., Schmid, F. R., and Anderson, B.,** Fibronectin interaction with C1q bound to solid phase immune complexes, *Fed. Proc.*, 45, 263, 1986.
66. **Steffes, M. L., Iammantino, A. J., Schmid, F. R., Castor, C. W., Davis, L. E., Entwistle, R., and Anderson, B.,** Fibronectin in rheumatoid arthritic (RA) and non-RA synovial fluids and synovial fluid cryoproteins, *Ann. Clin. Lab. Sci.*, 12, 178, 1982.
67. **Kono, I., Sakuiar, Kabashima, T., Yamane, K., and Kashiwagi, H.,** Fibronectin binds to C1q: possible mechanisms for co-precipitation in cryoglobulin's from patients with systemic lupus erythematosus, *Clin. Exp. Immunol.*, 52, 305, 1983.
68. **Sorvillo, J., Gigli, I., and Pearlstein, E.,** Fibronectin binding to C1q associated with antigen-antibody complexes in EDTA-treated plasma, *Scand. J. Immunol.*, 23, 153, 1986.
69. **Baatrup, G. and Svehag, S.-E.,** Serum and plasma fibronectin bind to complement reacted immune complexes primary via C1q, *Scand. J. Immunol.*, 24, 583, 1986.
70. **Hautenan, A. and Keski-Oja, J.,** Interaction of fibronectin with complement component C3, *Scand. J. Immunol.*, 17, 225, 1983.
71. **Wright, S. D., Craigmyle, L. S., and Silverstein, S. C.,** Fibronectin and serum amyloid P component stimulate C3b- and C3bi-mediated phagocytosis in cultured human monocytes, *J. Exp. Med.*, 158, 1338, 1983.
72. **Johnson, E., Gauperaa, T., and Eskeland, T.,** Fibronectin binds to complement-coated agarose beads and increases their association to mouse macrophages, *Scand. J. Immunol.*, 22, 315, 1985.
73. **Vuento, M., Kovkolainen, M., and Stenman, U.-H.,** Association of fibronectin with carboxy-group-modified proteins *in vitro, Biochem. J.*, 205, 303, 1982.
74. **Hautenan, A. and Keski-Oja, J.,** Affinity of myeloma IgG proteins for fibronectin, *Clin. Exp. Immunol.*, 53, 233, 1983.
75. **Cseh, K., Jakab, L., Torok, J., Marticsek, J., Kalabay, L., Bendek, S. Z., and Pozsonyi, T.,** Binding of fibronectin to monoclonal immunoglobulins, complement C1q, and immune complexes, *Ann. Immunol. Hung.*, 24, 307, 1984.
76. **Ferraccioli, G., Karsh, J., and Osterland, C. K.,** Interaction between fibronectin, rheumatoid factor, and aggregated gamma globulins, *J. Rheumatol.*, 12, 680, 1985.
77. **Kilgallon, W., Amlot, P. L., and Williams, B. D.,** Anti-C1q column: ligand specific purification of immune complexes from human serum or plasma. Analysis of the interaction between C1q and immune complexes, *Clin. Exp. Immunol.*, 48, 705, 1982.
78. **Kilgallon, W., Amlot, P. L., and Williams, B. D.,** Immune complexes in Hodgkin's disease: isolation, immunochemical and physico-chemical analysis, *Clin. Exp. Immunol.*, 53, 308, 1983.
79. **Anderson, B., Rucker, M., Entwistle, R., Schmid, F. R., and Wood, G. W.,** Plasma fibronectin is a component of cryoglobulins from patients with connective tissue and other diseases, *Ann. Rheum. Dis.*, 40, 50, 1981.
80. **Wood, G. W., Rucker, M., Davis, J. W., Entwistle, R., and Anderson, B.,** Interaction of plasma fibronectin with selected cryoglobulins, *Clin. Exp. Immunol.*, 40, 358, 1980.

81. **Scott, D. L., Almond, T. J., Naqvi, S. N. H., Lea, D. J., Stone, R., and Walton, K. W.,** The significance of fibronectin in cryoprecipitation in rheumatoid arthritis and other diseases, *J. Rheumatol.*, 9, 514, 1982.
82. **Beaulieu, A. D., Valet, J. P., and Strevey, J.,** The influence of fibronectin on cryoprecipitate formation in rheumatoid arthritis and systemic lupus erythematosus, *Arthritis Rheum.*, 24, 1383, 1981.

Chapter 6

INTERACTIONS BETWEEN FIBRONECTIN AND BACTERIA

David L. Hasty, Edwin H. Beachey, Harry S. Courtney, and W. Andrew Simpson

TABLE OF CONTENTS

I. INTRODUCTION

Since Kuusela first described the interaction of fibronectin (Fn) with *Staphylococcus aureus* in 1978,[1] there have been a large number of studies on the interaction of staphylococci and other bacterial genera with this important glycoprotein. Quite a number of microorganisms have been shown to bind Fn in solution and/or to bind to substrate-adsorbed Fn with a high affinity and a high degree of specificity. In these 10 years, many interesting questions have been raised concerning the nature of this interaction and its possible functional significance in bacterial pathogenesis, but as yet only a few of the issues have been resolved to a suitable degree. Not surprisingly, the principal common feature of Fn-bacteria interactions that has emerged from these studies is shared with virtually all of the other interactions attributed to Fn: Fn functions as a ligand between cells and either other cell surfaces or extracellular matrices.

Early studies centered on the possible role of Fn as a nonimmune opsonin for phagocytic removal of bacteria. Over the years, several investigators have reported an increased attachment of Fn-coated bacteria to phagocytic cells.[2-5] One of these laboratories also found that treating bacteria with Fn promoted an increase in neutrophil chemiluminescence and in bactericidal activity, indicating an increase in actual phagocytosis.[2] However, other investigators found that Fn-depletion of serum had little if any effect upon phagocytosis of bacteria.[6-9] Although Fn does appear to effectively opsonize certain tissue debris,[10,12] the bulk of the data generated to date suggest that Fn is relatively ineffective as an opsonin for bacteria (see References 13 and 14 for reviews). Nevertheless, since there are reports of minor degrees of opsonic activity, it is possible that Fn functions in conjunction with other molecules as a co-opsonin or a cofactor for optimal opsonic activity (e.g., by binding C1q),[15] as suggested by Lanser and Saba.[7] Plasma Fn deficiency has been documented in a number of clinical situations, such as blunt trauma, burn, or surgical stress. Many patients suffering such insults become septic. Early studies suggested a relationship between depression of Fn levels and onset of sepsis and suggested that there may be a real clinical benefit of infusion of either Fn-rich cryoprecipitate or purified, opsonically active Fn in these patients. However, more recent studies question whether there is a significant or predictable effect of therapy with Fn concentrates on sepsis, organ function, or patient survival (reviewed in References 16 and 17).

Perhaps the most important functional consequence of the interaction of Fn with bacteria is the adherence of bacteria to host tissues. Quite a large body of data has been generated over the last 10 or so years, indicating that an initial and requisite event in microbial colonization and infection of host tissues is the adherence of bacteria to surface epithelial cells of the mucous membranes.[18-20] In essentially all of the thoroughly studied cases, this adherence appears to be due to specific, ligand/receptor-like interactions: reactions between adhesive molecules on the bacterial surface that are most commonly called *adhesins** and complementary molecules on the host cell or tissue surface that are called *receptors*. The adherence of bacteria to host tissues is a critical event in bacterial pathogenesis. It confers a selective advantage to the bacteria, perhaps by providing a greater access to diffusible molecules secreted by host cells or by enhancing the effects of bacterial toxins.[21] These and other complex phenomena lead eventually to outright infection.

Whereas the preponderance of the available data suggest that Fn is probably not an important opsonin, there is considerable evidence that Fn is important in bacterial pathogenesis as a receptor for the adherence of certain bacteria in the initial phase of colonization

* This terminology, specifying *adhesin* for the bacterial surface molecule(s) responsible for adherence to host tissue structures and *receptor* for the host molecules that interact with adhesins, is commonly used in the bacterial adherence literature and for this reason we will use these terms in the remainder of this chapter. Whether "receptor" is an appropriate term in the classical sense will not be dealt with here.

of host mucous membranes or wounded and damaged tissues. Interestingly, there is also evidence that in some cases Fn may function by inhibiting rather than stimulating the process of bacterial adherence. Soluble Fn or Fn bound to cells or tissues could therefore act to modulate bacterial populations by stimulating or inhibiting the number and kind of bacteria that colonize the host.

Our review of Fn-bacteria interactions will reflect the bias that the role of Fn in bacterial adherence is probably one of the most important consequences of the interaction of Fn with bacterial cells. Much has been accomplished in the last 10 years, as we hope to show, but much remains to be done to determine unequivocally how Fn-bacteria interactions function in both health and disease.

II. GRAM-POSITIVE BACTERIA

Although a large number of bacterial genera have been tested for binding soluble Fn or for binding to immobilized Fn, it is most appropriate to begin this chapter by reviewing the interactions of Fn with *Staphylococcus aureus*, for at least two reasons. The first report of Fn binding to bacteria was Kuusela's study with *S. aureus* in 1978.[1] This initial report obviously stimulated a great deal of interest in this area. Further, although the number of bacteria studied has expanded considerably, studies of the interaction of Fn with *S. aureus* still outnumber studies with other bacteria.

A. STAPHYLOCOCCI

1. Characteristics of Binding of Soluble Fn to Staphylococci

Pentii Kuusela reported in 1978[1] that radiolabeled human plasma Fn would bind to heat-killed, formalin- or TCA-fixed *S. aureus*. Fn binding was inhibited by unlabeled Fn, lysine, urea, and certain nonacetylated amino sugars. Neither divalent cations nor protein A (at 100 μg/ml) inhibited the binding of Fn to *S. aureus*. Gelatin was only slightly inhibitory, suggesting that the interaction of *S. aureus* and Fn was not via the gelatin-binding domain of Fn.

Binding of Fn by heat-killed or living *S. aureus* is saturable, specific, and, except for the earliest time intervals in the binding reaction, virtually irreversible.[1,22-25] Binding is dependent upon conditions of growth of the organism (phase of growth, medium composition, and pH) as well as on conditions of the assay (e.g., pH, temperature, etc).[22,23] Fn binding is also dependent to a great degree upon the strain of *S. aureus* tested, varying as much as 30-fold or more in some studies.[22] Even though Fn binding to *S. aureus* is essentially irreversible, several investigators have performed Scatchard analyses of data from binding experiments. Such analyses suggest an apparent dissociation constant (K_d) of approximately 5×10^{-9} *M*.[22,23] The number of binding sites per organism have also been estimated and range from less than 100 to more than 20,000.[22,23,25] Such variability in number of receptors is due in part to use of different strains of bacteria, different media, and other aspects of growth conditions.[22] Furthermore, the method used to label Fn may introduce some variability. For instance, iodination of Fn with the chloramine T method resulted in a dramatic increase in binding to C-reactive protein compared to binding of Fn radiolabeled by the periodate-NaB^3H_4 procedure.[26] Similar results were obtained in our laboratories in studies of Fn binding to *Streptococcus pneumoniae*.[135]

2. *S. aureus* Binding Domain of Fn

In 1980, Mosher and Proctor[27] found that the primary binding site for *S. aureus* was contained within the 27-kDa aminoterminal domain of Fn. This result was consistent with Kuusela's[1] finding that the gelatin-binding domain was probably not involved. The observation that intact Fn and the 27-kDa fragment could be crosslinked to *S. aureus* by coagulation factor XIII_a supports the conclusion that the amino terminus is involved in the binding of Fn to *S. aureus*.[27] Other groups, including our own,[28,30] have presented evidence for an additional, possibly minor, *S. aureus* binding site in the carboxyterminal half of the Fn monomer, but the 27-kDa fragment does appear to be the principal reactive moiety of Fn. In support of the predominance of the single reactive region within the 27-kDa aminoterminal of Fn, interaction of dimeric Fn with staphylococci leads to agglutination,[31-33] whereas reduced, carboxymethylated Fn monomers do not agglutinate staphylococci.[34] Mosher and Proctor[27] were the first to speculate that the physiological significance of the interaction between Fn and *S. aureus* might be either to enhance removal of the microorganisms by phagocytes[2] or to enhance *S. aureus* pathogenicity by providing for bacterial attachment to host tissues.[27] Although support for the former hypothesis is for the most part lacking, considerable support for the latter hypothesis has developed since 1980.

3. Staphylococcal Fn-Binding Molecules

Several groups have attempted to characterize Fn-binding molecules of staphylococci. Espersen and Clemmensen isolated a 197-kDa protein from water-soluble extracts of sonicated *S. aureus* strain E2371 by affinity chromatography of Fn-Sepharose.[35] Evidence from crossed immunoelectrophoresis experiments strongly suggested that a complex was indeed formed between the purified Fn-binding protein and Fn. However, it was not determined whether this Fn-binding protein was actually exposed at the surface of living staphylococci or whether the purified molecule could inhibit binding of Fn to the bacteria. The authors commented that a minor component of approximately 60 kDa was also present in this preparation, but they did not elaborate any further.

Rydén et al.[23] reported similar experiments designed to purify the staphylococcal Fn-binding component from lysostaphin digests of a heat-killed *S. aureus* strain Cowan I. This lysate was found to contain material that inhibited Fn binding to the staphylococci. The reactive material was postulated to be protein, since its activity was destroyed by both trypsin and pronase. An 18-kDa Fn-binding molecule was purified from the *S. aureus* lysates by affinity chromatography on a Fn-Sepharose column. The possibility that this molecule might be protein A was ruled out. In this regard, there is conflicting information concerning the possibility that protein A could be a staphylococcal Fn-binding protein, with one group reporting a close correlation between protein A content and Fn-binding activity[24] and most other groups concluding that protein A is not involved in Fn binding.[1,6,23]

Fröman et al.[36] have recently identified a high molecular weight Fn-binding molecule in lysostaphin digests of heat-killed (88°C for 20 min) *S. aureus* strain Newman. Soluble material in the lystate was capable of inhibiting Fn binding to heat-killed staphylococci. This activity was completely absorbed on affinity columns composed of 29-kDa aminoterminal fragment of Fn coupled to Sepharose. Most notable, 210- and 50-kDa molecules were eluted from the Fn column with 2 *M* guanidinium hydrochloride. Electroblots of whole lysostaphin digests probed with the radiolabeled aminoterminal Fn fragment indicated that a 210-kDa component was the most prominent of a number of Fn-binding components migrating on the gel between 66 and 210 kDa. An additional staphylococcal molecule of approximately 50 kDa was isolated from the Fn fragment-Sepharose column, but this molecule did not appear to be a prominent component of the electroblot experiment. The 210-kDa protein appeared to be multivalent, with one molecule capable of binding 6 to 9 Fn molecules. Antibodies prepared against the 210-kDa Fn-binding protein bound to heat-killed staphylococcal cells, suggesting that the molecule was exposed on the surface of the staphylococci. Whether the protein was exposed on surfaces of living staphylococci or whether this molecule plays a role as an adhesin has not been addressed. Western blots using anti-210-kDa antibodies showed that the 210-kDa molecule was degraded into smaller molecules in the absence of protease inhibitors. The most prominent peptide in this digest was approximately 50 kDa in mass and retained the capacity to bind the aminoterminal fragment of Fn. The relationship between the 210-kDa Fn-binding protein isolated earlier by this group from a different strain of *S. aureus*[23] remains unclear. The smaller molecule may represent a degradation product of the larger one since isolation was carried out in the absence of protease inhibitors.

A staphylococcal Fn-binding protein has recently been cloned in *Escherichia coli* and purified.[37,38] A gene bank of chromosomal DNA from *S. aureus* 8325-4 was screened for Fn binding protein activity (i.e., the ability to inhibit ^{125}I-Fn binding to *S. aureus* strain Cowan I) and the plasmid pFR001 from a positive clone was characterized. Gene products expressed in *E. coli* were analyzed by affinity chromatography on Fn-Sepharose and ion exchange chromatography. Two proteins with molecular weights of approximately 87 kDa and 165 kDa which inhibited binding of Fn to staphylococci were isolated. The larger protein was 30-fold more active than the 87-kDa product. Immunoprecipitation inhibition experiments using antibodies directed against whole *S. aureus* strain Newman demonstrated that the 87-kDa, 167-kDa, and native staphylococcal Fn-binding proteins shared antigenic determinants.

Subcloning and deletion mapping of pFR001 narrowed the Fn-binding region of the product to a 350-bp region of the cloned insert. This region was fused to protein A by using a 600-bp segment of the Fn-binding protein gene which spanned the 350-bp region. The fusion protein bound to Fn and totally inhibited binding of intact Fn and the 29-kDa aminoterminal Fn fragment to *S. aureus*, whereas protein A by itself had no inhibitory effect. The observation that this preparation inhibited binding of Fn to *S. aureus* suggested that this Fn receptor plays a major role in Fn binding. Because this protein appears to bind avidly to the aminoterminus of Fn, the proposed staphylococcal binding site nearer the carboxyterminal end of Fn appears to be of minor importance in the binding of soluble Fn to staphylococci. These data, however, do not rule out the possible role of a secondary binding site in binding of staphylococci to substrate-bound Fn (see below). A number of molecules have been shown to interact differently with soluble and substrate-bound Fn, and it is not unreasonable to assume that bacteria could also interact differently with the two forms of Fn.

There is no evidence as yet that the cloned proteins are adhesins or are actually involved in the binding of staphylococci to substrate-adsorbed Fn. However, there is evidence to suggest that material isolated from *S. aureus* on a Fn affinity column inhibits adherence of the organisms to substrate-adsorbed Fn. Maxe et al.[39] were able to inhibit adherence of *S. aureus* strain SA113(83A) to substrate adsorbed Fn by using material isolated from a heat-inactivated (90°C, 20 min) lysostaphin digest of *S. aureus* Newman or SA113(83A). Lysates were precipitated with ammonium sulfate, resolubilized, and then passed over an IgG-Sepharose column before isolating Fn-binding material on a Fn-Sepharose column. The affinity-purified material at a concentration of 1 mg/ml inhibited staphylococcal adherence by 75 to 85%. However, this information is difficult to interpret because of the heterogeneity of the components in the material isolated from the lysates by affinity chromatography.

Studies from other laboratories suggest that the principal Fn-binding component of staphylococci is a carbohydrate, but to date most of these data have appeared in abstract form and thus are not as fully documented as studies on staphylococcal Fn-binding proteins.[25,40,41,50] In these studies, material was isolated by gentle sonication of intact (not heat-killed) bacteria and purified by ion exchange and Fn affinity chromatography. The purified material had a molecular weight of approximately 60 kDa and was found to inhibit Fn binding to intact organisms. Treatment of the material with sodium periodate eliminated the inhibitory activity. Certain monosaccharides inhibited the adherence of staphylococci to substrate-adsorbed Fn. Lectin inhibition of Fn binding to staphylococci supported the suggestion that the Fn-binding molecule could be a carbohydrate. One report indicated that the staphylococcal Fn-binding moiety could be cell wall teichoic acid,[42] but this is in contrast to the results of others who have not found a correlation between Fn binding and cell wall teichoic acid content.[6]

Resolution of the apparently conflicting results regarding the staphylococcal Fn-binding molecule will necessitate further experimentation. Methodological differences or strain differences could be responsible. It is also possible that there are two different Fn receptors on surface of staphylococci. This latter possibility is suggested by the data referenced earlier regarding two different staphylococcal binding sites within the Fn molecule. In any future studies it will be important to determine which staphylococcal components are expressed at the surface of intact, living organisms, what growth or assay conditions affect their expression or exposure (see discussion, above), and which of the components are responsible for staphylococcal binding of soluble Fn as opposed to binding to substrate-adsorbed Fn.

4. Characteristics of Staphylococcal Binding to Substrate-Adsorbed Fn

The number of Fn receptors on clinical isolates of *S. aureus* correlated with the invasiveness of the isolate in one study,[32] but not in another.[43] The number of Fn receptors can also be altered by sub-MICs* of various antibiotics.[44] For instance, $^1/_4$ MIC penicillin increased the number of Fn-binding sites and also increased the adherence of staphylococci to substrate-bound Fn. The clinical significance of these data remains to be established.

S. aureus have also been shown by several investigators to bind to Fn and Fn fragments immobilized under various conditions. Toy et al. studying the adherence of *S. aureus* to plasma clots *in vitro*, found that depleting Fn from these plasma clots decreased staphylococcal adherence by 25 to 50%.[45] Reconstitution of Fn led to a dose-dependent return of staphylococcal adherence to previous levels. Vercellotti et al.[31] showed that several different strains of *S. aureus* bound to substrata coated with Fn, as did *Streptococcus sanguis,* whereas several genera of Gram-negative organisms adhered poorly, if at all.

Kuusela et al.[30] demonstrated a time-dependent and Fn concentration-dependent adherence of *S. aureus* to Fn-coated coverslips. Surprisingly, staphylococcal adherence was supported equally well by intact Fn and by the 120- to 140-kDa carboxyterminal fragments of Fn. In fact, although the aminoterminal fragment also supported bacterial adherence, it was not as effective as the COOH-terminal fragment. The gelatin-binding fragment did not support adherence. This study also showed that pretreating the bacteria with Fn did not inhibit the subsequent adherence of the microorganisms to Fn-coated coverslips, suggesting that binding of soluble Fn and binding to immobilized Fn occured by different mechanisms. The relationship between the two possible staphylococcal binding sites on Fn and their role in binding of soluble vs. immobilized Fn needs to be investigated further, as does the potential difference between Fn-binding proteins exhibited by viable vs. heat-killed organisms.

Waldvogel, Vaudaux, and colleagues[46-52] have studied the binding of *S. aureus* to polymethylmethacrylate (PMMA) coverslips as an *in vitro* model of so-called foreign body infections. Foreign body infections often occur after insertion of prosthetic devices such as catheters and are commonly caused by both coagulase-positive and coagulase-negative staphylococci. Although there are undoubtedly many factors that contribute to these infections, such as reduced phagocytic function by polymorphonuclear leukocytes in the presence of the foreign body,[47] attachment of bacteria to the synthetic prosthetic material and the adsorbed host molecules which coat these devices is likely to be an initial step. PMMA coverslips implanted subcutaneously into guinea pigs rapidly become coated with Fn-containing deposits. *In vitro* adherence of *S. aureus* strain Wood 46 to previously implanted coverslips was inhibited in a dose-dependent manner by anti-Fn antibody. In these studies, antibodies against fibrinogen which potentially contaminated the anti-Fn IgG preparation were removed by absorption. Inhibition of staphylococcal adherence reached 80% at 32-μg anti-Fn IgG per milliliter.[46]

* MIC = minimal inhibitory concentration.

Fn also becomes adsorbed to PMMA and polytetrafluoroethylene (PTFE) coverslips *in vitro*.[51] The number of cells of *S. aureus* strain Wood 46 binding to these coverslips increased in a dose-dependent manner; both of these processes were inhibited by the presence of other serum proteins. The presence of collagen on the coverslips led to greatly increased Fn binding and *S. aureus* adherence.[51,52] It is of interest that extracellular polymeric substances (probably similar or identical to glycocalyx or slime[53,54]) have apparently little if any effect on the ability of *S. aureus* Wood 46 to bind to Fn-coated coverslips.[49]

The possibility that Fn may serve as an adherence factor for coagulase-negative staphylococci also has been raised,[55] although Toy et al.[45] found that *S. epidermidis* strain ATCC 27626 bound to plasma clots approximately 1000-fold less well than the seven different *S. aureus* strains they tested. Sorvillo and Pearlstein[15] found that pretreating *S. epidermidis* K160 with C1q resulted in a twofold increase in the ability of these organisms to adhere to surface-bound Fn, but that level was only slightly more than 1/3 that of *S. aureus* strain 25923.

B. STREPTOCOCCI

Of the numerous genera of streptococci studied, *S. pyogenes* has been the most thoroughly characterized regarding interactions with Fn. Further, the identity of the primary adhesin of *S. pyogenes*, lipoteichoic acid (LTA), is more thoroughly documented than adhesins for other Gram-positive bacteria. Our view of this subject is admittedly biased, since most of the information on the interaction of *S. pyogenes* with Fn has emanated from our laboratories, deriving more from attempts to understand the adherence of *S. pyogenes* to cells of the oral mucosa than the interaction of Fn with the bacteria. To adequately present the studies originating from our laboratories on the interaction of *S. pyogenes* with Fn, it will first be important to consider the role of LTA as the major adhesin on the group A streptococcal cell surface. In this section, we will, of course, review the important contributions of other groups to this topic.

1. Adherence of *S. pyogenes:* Lipoteichoic acid as Adhesin

a. Inhibition of Adherence of S. pyogenes *to Epithelial Cells by LTA*

Initial reports of the adherence of *S. pyogenes* to epithelial cells suggested that M protein might be the adhesin: strains rich in M protein adhered better to epithelial cells and enzymatic treatments that removed M protein also reduced the ability of the organisms to adhere to epithelial cells.[56,57] Later studies by Beachey and Ofek[58] showed that protease treatments that removed M protein also removed LTA. To further examine the relative contribution of surface components of *S. pyogenes* on the adherence process, LTA, M protein, group A carbohydrate, and peptidoglycan were tested for their ability to block attachment of the bacteria to epithelial cells. Of course, the concept is that a solubilized adhesin would bind to receptors on the epithelial cell surface and prevent bacterial adherence. Antibodies directed against the adhesin could also inhibit adherence by binding to the bacterial surface molecules, as would receptor analogues or purified receptor molecules. Of the four streptococcal surface molecules tested, only LTA inhibited the adherence of *S. pyogenes* to epithelial cells. Likewise, when antibodies directed against various streptococcal surface molecules were tested for their ability to block the adherence of streptococci, only antibody to LTA was active.[58,59]

b. Chemical Basis for Inhibition of Streptococcal Adherence by LTA

The LTA of *S. pyogenes* is an amphipathic molecule composed of a linear polymer of glycerol phosphate linked by 1 to 3 phosphodiester bonds. The polyglycerol phosphate (PGP) backbone is substituted with ester-linked alanine and contains a glycerophosphoryl diglucosyl diglyceride moiety at one end.[60] It is the lipid moiety of the LTA molecule, rather than the PGP backbone, that enables LTA to bind to eukaryotic cells and inhibit the adherence of *S. pyogenes*. If the lipid moiety is separated from the PGP by chloroform:methanol extraction after alkaline hydrolysis, the PGP is incapable of binding to eukaryotic cells and it does not inhibit the adherence of *S. pyogenes*. By contrast, the lipid fraction retains the inhibitory activity of the original mixture. The ability to bind to cells and inhibit *S. pyogenes* adherence can be chemically restored to PGP by reacylation with palmitoyl chloride to essentially reconstitute the LTA and thereby restore the amphipathic nature of the molecule.[18,58,59,61,62]

c. *Specificity of the Interaction of LTA with Cell Receptors*

As LTA binds to host cells and inhibits streptococcal adherence, does it bind to hydrophobic pockets in specific protein components of the membrane or does it merely insert into the lipid-bilayer? Many different eukaryotic cells bind LTA in a saturable and reversible manner.[63-67] Studies of purified red cell membranes and inside-out ghosts demonstrated that the LTA-binding sites are preferentially located on the outer surface of these membranes. Additional evidence suggesting specificity of LTA-binding to membrane components comes from observations of correlated changes in binding of LTA and adherence of *S. pyogenes* to buccal epithelial cells obtained from infants during the first few days after birth. Epithelial cells from infants younger than 6 h old showed little capacity to bind either LTA or *S. pyogenes*, whereas cells obtained from infants several days old were capable of binding both.[61,62] In a similar fashion, erythrocytes obtained from infants and adults exhibited specific and saturable binding characteristics, but the cells obtained from human cord blood had an affinity for LTA that was tenfold lower than that of the red cells obtained from adults.[64] Though the differences between infant and adult buccal cells in their ability to bind LTA remain unexplained, the results of these experiments did suggest that there might be a correlation between the binding of LTA and the adherence of streptococci. Furthermore, in every case, LTA binding studies indicated that cells bind LTA in a saturable, reversible, time-dependent manner, with variable affinity constants.[62-67] Taken together these results provided circumstantial evidence for a specific receptor for LTA, but more definitive proof required isolation of specific cell-surface components.

d. *Binding of LTA to Serum Albumin, a Possible Protein Receptor Analogue*

To determine whether streptococcal LTA could interact with fatty acid binding sites of proteins and to establish characteristics that might be expected for a protein receptor for LTA, we studied the binding of LTA to bovine serum albumin. Serum albumin contains perhaps the best characterized protein-binding site for fatty acids. Specificity based on the length of the hydrocarbon chain, the degree of saturation of the fatty acid, and the head group attached to the fatty acid is exhibited by this binding site. The proposed mechanism of binding involves the insertion of the hydrocarbon chains into a hydrophobic cleft located within the albumin molecule. Based upon the available data, it was reasonable to hypothesize that a similar mechanism would be involved in LTA-receptor binding.

Serum albumin competitively blocked the ability of LTA to bind to erythrocytes, suggesting that the erythrocytes and the albumin were reacting with the same part of the LTA molecule.[68] Albumin also blocked the ability of streptococci to bind to buccal cells.[69] Subsequent studies indicated that LTA would bind to albumin in a saturable, reversible manner, with an affinity constant similar to that measured for octanoic acid.[70] These findings suggested that the mechanism of binding of LTA to albumin was similar to that of fatty acids and demonstrated that LTA could react with specific proteins in a lipid-dependent manner.

e. *Orientation of LTA on the Streptococcal Surface*

The amphiphathic nature of LTA allows it to exist in several configurations on the surface of the streptococci. The majority of the LTA found in this organism is membrane-associated with the lipid portion of the molecule probably intercalated into the lipid bilayer.[60] However, LTA can also be found in growth medium,[60] suggesting that LTA passes from the streptococcal membrane through the cell wall, surface fibrillae, and capsule, and leaks into the medium. Immunoelectron microscopic studies demonstrated that LTA can be located within the fibrillae on the streptococcal surface. *In vitro* experiments indicated that both LTA and deacylated LTA can form complexes with isolated surface proteins of streptococci, such as M protein. This interaction appears to be due to ionic attraction between the negatively charged polyglycerol phosphate backbone and positively charged residues on LTA-binding proteins of the streptococcal fibrillae. Modification of the positively charged amino acids of one of these proteins (type 24 M protein) by maleylation destroyed its ability to react with LTA. The proposed charge distribution of type 24 M protein in its coiled-coil configuration would allow alignment of M protein lysine residues with negatively charged phosphate residues of LTA. Thus, M protein or other LTA-binding proteins located in the fibrillae may anchor and properly orient the lipid moiety of LTA for its adhesin-receptor functions.[71]

2. Interaction of Streptococci with Fn

Although none of these studies conclusively proved that the lipid moiety of LTA was binding to specific receptors on the epithelial surface, the results encouraged us to develop a strategy for the identification of LTA/*S. pyogenes* adherence receptors. The following conditions should apply to streptococcal receptors on buccal epithelial cells: (1) receptors should interact with LTA in a manner consistent with observations from studies of both adherence and LTA binding; (2) receptors should be located on the surface of buccal mucosal cells; (3) purified receptor should bind to both LTA and streptococci; (4) binding of the receptor to streptococci should be inhibited by LTA but not by deacylated LTA; and (5) binding of the receptor to the surface of the streptococci should competitively inhibit the adherence of streptococci to buccal mucosal cells.

a. Location of Fn on the Oral Mucosa

Studies conducted by Zetter and colleagues[72] demonstrated Fn on the lumenal surface of the oral mucosa by immunofluorescence. Most of the Fn was localized in the subepithelial connective tissues. However, while there was no Fn found within the layers of the stratified epithelium, Fn was observed on the lumenal surface. This unusual distribution suggested that the Fn on the surface mucosal cells might be deposited from saliva rather than produced by epithelial cells. Several investigators have subsequently confirmed that Fn is present on surface epithelial cells and that Fn is contained within various salivary secretions.[73-78] Results of *in vitro* studies have been consistent with observations on the localization of Fn on buccal mucosal cells. Studies using immunofluorescence and immunoelectron microscopy showed that cells removed from the buccal surface by gentle swabbing had Fn on their surface. Many of the cells obtained in this manner exhibited intense staining for Fn, while others exhibited intermediate or negative staining patterns.[73,79] The presence of populations of epithelial cells both with and without Fn on their surface led to experiments to determine whether streptococci exhibited preferential adherence. The epithelial cells that exhibited surface Fn were the cells that bound *S. pyogenes*. Further, electron microscopic examination indicated that streptococci on individual epithelial cells were most often bound at areas stained for Fn. The co-localization of streptococci and Fn on these cells provided additional evidence that Fn was properly located to act as a receptor.[73]

The heterogeneous population of cells removed from the buccal mucosa by swabbing must come both from cells in the most superficial layer of the stratified squamous epithelium and, at least in part, from cells in slightly deeper layers not exposed to Fn in salivary secretions.[79,80] Buccal cells bind Fn specifically, saturably, and reversibly.[81] Treatment of buccal cells with saliva and exogenous Fn converts the cells to a single population with respect to the presence of immunoreactive Fn on the cell surface.[79] Thus, Fn appears to be located on cell surfaces where it can act as a streptococcal receptor in the oral cavity.

b. Interaction of Fn with LTA

Fn binds to LTA whether Fn is presented as a soluble or as an immobilized molecule.[73,82,83] The ability of soluble Fn to bind to LTA can be demonstrated by electrophoresis of mixtures of LTA and Fn. The highly charged LTA molecule is capable of changing the electrophoretic mobility of proteins with which it interacts. LTA, but not deacylated LTA, changed the mobility of Fn. Additional experiments showed that soluble Fn inhibited the ability of LTA to bind to Fn immobilized on Sepharose. When the binding of LTA to these two forms of Fn was compared at a 1:1 molar ratio, the LTA distributed equally between the two forms, suggesting that the mechanism of binding is the same for soluble and immobilized Fn.[84]

LTA bound to Fn immobilized on Sepharose could be eluted with LTA, Fn, BSA, and detergents. When substances known to bind to Fn were tested for their ability to elute LTA from Fn, only gelatin had any effect and that was limited to elution of approximately 32% of the bound LTA. Subsequent studies demonstrated that albumin competitively inhibits the binding of LTA to Fn. Furthermore, LTA binds to the fatty acid binding site of albumin with an estimated nK_a (number of binding sites × the affinity constant) of 0.63 μM^{-1}, whereas the nK_a for the binding of LTA to Fn is 250 μM^{-1}. The higher affinity of LTA for Fn may play an important role in the adherence of streptococci in the presence of complex biological fluids, such as blood or saliva.

c. Fn-Binding Molecules of Streptococci

Fn binds to all strains of *S. .pyogenes* that have been tested.[3,85] In addition to binding to group A streptococci, Fn binds to groups C and G and to certain strains of oral streptococci.[78] Representative strains of groups B, D, M, N, P, and U do not appear to bind significant amounts of Fn.[78,86] Using 15 strains representing *S. mutans* serotypes a to f, Imai et al.[86] found that all of the *S. mutans* strains bound Fn. Representative strains of *S. sanguis* bound a relatively small amount of Fn, while *S. mitior, S. milleri,* and *S. faecalis* appeared to have very little affinity for Fn. These studies demonstrate a great deal of variation in the ability of streptococcal groups, and of strains within each group, to bind Fn.

The nature of the receptor(s) for Fn on streptococci has been investigated by several groups.[78-87] The differences in techniques used by different laboratories, not to mention likely strain differences, make direct comparisons of the work difficult. Nevertheless, some of the experimental findings and the issues raised will be briefly outlined.

In the majority of the studies in which bacteria were heat-killed (usually 88°C for 20 min) before use in binding experiments, heating did not, at first glance, appear to affect Fn binding. At the least, heating does not appear to affect the amount of Fn bound. By contrast, treatment of intact streptococci with trypsin reduced their ability to bind Fn. Speziale et al.[87] found that the binding of Fn to heat-killed group A streptococci was time dependent, functionally irreversible, and specific. Treatment of *S. pyogenes* strain 1321 with either trypsin or papain led to a reduced ability to bind Fn, and the tryptic extract of the organisms contained material that blocked the ability of Fn to bind to the bacteria. LTA was present in the tryptic extract but it, or at least the PGP backbone, could be removed from the crude trypsinate on an anti-PGP immunoaffinity column without eliminating the inhibitory activity. These data suggest that the inhibitory activity remaining in this preparation was not intact LTA and also tends to support the authors' conclusion that the major receptor for Fn on *S. pyogenes* 1321 is a protein. The streptococcal Fn-binding protein has not yet been purified or completely characterized. It was not determined whether LTA-derived fatty acids remained in the trypsin extract after immunoaffinity separation.

These studies conflicted with earlier work in our laboratories which indicated that LTA was most likely the principal mediator of Fn binding to group A streptococci. A number of studies outlined above led us to postulate that LTA plays an important role in the binding of Fn to group A streptococci. The possibility that a protein, rather than LTA, was the streptococcal surface molecule that binds Fn stimulated us to attempt to clarify the role that LTA might play in Fn binding to intact streptococci. A series of experiments was performed based on the observation of Alkan and Beachey[88] that LTA could be released from streptococci with penicillin. Recent evidence showing that growth of staphylococci in the presence of clindamycin reduced their ability to bind to Fn provided an interesting control for comparison in our studies of streptococci.[44]

Pencillin caused a release of LTA from both streptococci and staphylococci.[89] Release of LTA from group A streptococci by penicillin was accompanied by a concomitant reduction in the ability of these bacteria to bind Fn. By contrast, the release of LTA from staphylococci had little or no effect on Fn binding. Clindamycin, on the other hand, had no effect on LTA release from either organism or on Fn binding to streptococci, but it led to decreased binding of Fn to staphylococci. Furthermore, while purified LTA inhibited Fn binding to group A streptococci, it had no effect on Fn binding to staphylococci.

Since Fn binds to streptococci in a functionally irreversible manner, we attempted to determine whether Fn/adhesin complexes could be released from streptococci with penicillin. Penicillin released the bound fibronectin from streptococci, but did not release Fn from staphylococci. These results provided additional evidence that the Fn-binding component of streptococci could be LTA and supported the hypothesis that the Fn-binding molecule of staphylococci was not LTA. The fibronectin released from streptococci could be specifically precipitated with antibodies to LTA, showing that, indeed, LTA and Fn were released as a complex. Attempts to precipitate the released Fn with antibodies to M protein were unsuccessful. Further experiments were performed with an antiserum raised to whole streptococci. This antiserum contained antibodies directed against multiple surface molecules and it immunoprecipitated Fn released from streptococci by penicillin as effectively as the antibody directed specifically against LTA. Adsorption of the polyclonal anti-streptococcal serum with LTA eliminated its ability to precipitate Fn. These studies again are consistent with the hypothesis that LTA plays a major role in the binding of Fn to group A streptococci, but they do not rule out the possibility that other surface components, protein or otherwise, may also participate.

Several laboratories have used heat-treated bacteria in their studies of Fn binding to streptococci. Heating the bacteria appears to have little or no effect on the quantity or kinetics of Fn binding. We attempted to determine whether heating bacteria could affect other characteristics of Fn binding. We found that LTA had no effect on Fn binding to heat-treated streptococci, suggesting that new, LTA insensitive receptors might have become exposed by heating.[90] When streptococci are treated with penicillin under conditions that reduce Fn binding to 50% of normal levels, the remaining Fn binding is still totally inhibitable by LTA. If the bacteria are then heated, their ability to bind Fn is restored to normal levels but the binding becomes insensitive to inhibition by LTA.[90]

To date, LTA is the only purified streptococcal surface molecule that inhibits binding of streptococci to Fn and cell surfaces. Whether streptococcal surface proteins may also be involved as adhesins or Fn-binding molecules remains to be demonstrated. It is possible that experimental or strain differences are responsible for the differing results obtained as it is also possible that there are two distinct mechanisms by which *S. pyogenes* can interact with Fn. It will then be important to determine how expression of the two mechanisms are regulated and under what circumstances each function.

While studying the binding of Fn to other streptococcal species, Imai et al.[86] found that proteases and NaOH decreased the binding of Fn to *S. mutans* 6715 (sometimes classified as *S. sobrinus*), whereas other substances (e.g., HCl, $NaIO_4$, and detergents) had little or no effect. Also of interest, these authors found that a monoclonal antibody raised against the polyglycerol phosphate backbone of LTA inhibited binding of Fn to *S. mutans* 6715 by 66%. Sugars and amino sugars previously reported to be important in the adherence of *S. mutans* to saliva-coated hydroxylapatite, however, did not inhibit the interaction of *S. mutans* with Fn immobilized on latex beads.[75] The ability of LTA to inhibit the binding of Fn to bacteria appears to be limited. Although extensive surveys have not been conducted, strains of *S. aureus* and *S. pneumoniae* apparently do not bind Fn in an LTA-inhibitable manner.

d. Streptococcal Binding Domain of Fn

The principal domain of the Fn molecule recognized by *S. pyogenes* appears to be within the 27-kDa aminoterminal tryptic fragment that also contains the staphylococcal binding site. Speziale et al.[87] demonstrated that *S. pyogenes* 1321 adsorbed radiolabeled tryptic peptides of Fn with apparent M_r of 27 and 23 kDa, while *S. aureus* Cowan I absorbed only the 27-kDa Fn fragment. They also found that purified 27-kDa amino terminal fragment inhibited Fn binding to *S. aureus* completely, but did not completely inhibit Fn binding to *S. pyogenes* even at the highest concentrations tested. We have obtained similar results using fragments of Fn generated by thermolysin.[29,91] *S. aureus* adsorbed three major high molecular weight polypeptides as well as a 28-kDa and a 23-kDA fragment, whereas *S. pyogenes* only adsorbed the 28-kDa and the 23-kDa fragments. The 28-kDa and 23-kDa fragments reacted with anti- aminoterminal fragment antibodies, raised by immunizing rabbits with a synthetic peptide of residues 1 to 30 of the aminoterminus of Fn. These fragments did not react with monoclonal antibodies against the collagen-binding domain or the cell-binding domain. The additional high molecular weight peptide fragments adsorbed by *S. aureus* reacted only with the anti-cell attachment domain monoclonal antibody. Interestingly, LTA completely blocked the ability of *S. pyogenes* to adsorb the 28-kDa and 23-kDa fragments. Further, two anti-Fn monoclonal antibodies that were found to block the binding of LTA to Fn reacted only with the 28- and 23-kDa fragments. These and other results from several laboratories suggest that although *S. pyogenes* and *S. aureus* appear to bind predominantly to the same domain of Fn, the binding appears to occur by different mechanisms. We would therefore expect that the two binding sites are distinct, but this remains to be demonstrated clearly.

e. Role of Fn as a Receptor for Streptococci

Results of the binding experiments and the morphologic distribution of Fn are consistent with the idea that this molecule may act as a receptor for the adherence of *S. pyogenes* to buccal mucosal cells and strongly suggest the possibility that this molecule might act as a receptor. The adherence of three different strains of group A streptococci to buccal cells was inhibited by as little as 1 μg of Fn/ml.[82] Likewise, the removal of Fn from buccal cells resulted in a proportional decrease in the adherence of group A streptococci, while the addition of Fn led to a proportional increase. Additional studies indicated that antibody to Fn can specifically block the ability of intact *S. pyogenes* to adhere to buccal cells.[83] Using a number of cultured cell lines that express various amounts of Fn, Stanislawski et al.[91] recently demonstrated that the ability of the cell lines to bind *S. pyogenes* was directly related to the amount of Fn expressed. Thus, for several types of cells, Fn appears to be a viable candidate as a receptor for *S. pyogenes*.

Several publications have demonstrated the binding of other streptococcal species to Fn. Vercellotti et al.[93] demonstrated that *S. sanguis* bound to a surface coated with Fn, but did not aggregate in the presence of soluble Fn, leading the authors to suggest that there may be a physical difference in the interaction of these bacteria with Fn when in solution or immobilized on a surface. Myhre and Kuusela[85] surveyed almost 400 different bacterial strains belonging to 35 species for Fn binding. They found positive binding in representative strains of groups A, C, and G streptococci. They found a large percentage of these strains to bind a higher percentage of Fn than did *S. aureus* strains. Most of the alpha-hemolytic strains of bovine group G streptococci, however, did not bind significant amounts of Fn. In a survey of 50 strains of group B streptococci, Butler et al.[92] found that there was uniformly poor binding of soluble Fn by these organisms. Vercellotti et al.[93] surveyed the binding of several species of bacteria to Fn-coated assay plates. Of the streptococcal strains tested, they found *S. sanguis* to bind well to Fn-coated surfaces and *S. milleri* to bind to a much lesser degree, though not as poorly as each of the strains of Gram-negative bacteria tested. Kuusela et al.[30] found two strains of group G streptococci that attached well to immobilized Fn, though at substantially different levels.

Soluble Fn inhibited the adherence of *S. pneumoniae* to human pharyngeal epithelial cells by 58% as compared to controls, whereas pretreatment of the pharyngeal cells with Fn enhanced adherence to 234% of control values.[94] In contrast to *S. pyogenes*, however, *S. pneumoniae* apparently binds to Fn via the 42-kDa thermolysin-derived collagen-binding fragment rather than the 27-kDa aminoterminal fragment, since the gelatin-binding fragment inhibited adherence of pneumococci to epithelial cells.[94]

Of the eight strains of oral streptococci tested by Babu and Dabbous,[78] four appeared to interact with Fn. The adherence of *S. mutans* strains GS-5 and OMZ-176, *S. rattus,* and *S. sobrinus* to parotid saliva-coated hydroxylapatite beads was reduced when gelatin-binding components were first removed from the saliva used to pretreat the beads. They also demonstrated that Fn is present in parotid saliva at a concentration of approximately 5 μg/ml and that it becomes incorporated into artificial tooth pellicles formed on hydroxylapatite beads. Further, when these streptococcal strains were pretreated with purified Fn, their attachment to artificial tooth pellicles was inhibited, leading these authors to conclude that Fn or a similar molecule is one of the receptors for these bacteria in tooth pellicles.

Ericson and Tynelius-Bratthall[76] studied the absorption of salivary Fn by 12 strains of oral streptococci (*S. mutans, S. salivarius,* and *S. mitior*). These strains removed between 2 and 90% of the Fn present in freshly collected submandibular/subligual gland saliva, which is, according to these authors, 1 to 25 μg Fn per milliliter as determined by ELISA.

Although most of the studies cited in this section of the chapter are primarily surveys which raise more questions than provide firm answers, they do indicate that there is a definite interaction between Fn and a large number of streptococcal species, and suggest that additional experiments to determine the significance for bacterial pathogenesis would be quite useful.

f. Differential Binding of **S. sanguis** *to Soluble vs. Substrate-Adsorbed Fn*

S. sanguis is a viridans species that commonly colonizes the oral cavity but that is usually avirulent. Intermittent bacteremia occurs with this organism, as well as with other viridans streptococci after routine dental and medical manipulations, but it appears that bacteremia alone is not sufficient to initiate infective endocarditis. Certain populations with underlying heart disease are, however, at risk of developing infective endocarditis from *S. sanguis* bacteremia. Current evidence suggests that a crucial step in the development of infective endocarditis is the initial adherence of bacteria to one or more of the molecules that would be present on damaged cardiac valvular surfaces. These surface components, which would not be exposed or deposited on healthy vascular tissues, could then act as receptors for circulating bacteria, such as *S. sanguis*. Some of the tissue components that are most likely present at sites of endocardial damage are the subepithelial collagenous matrix and the plasma components that would become bound to the matrix once it is exposed (e.g., fibrin, platelets, fibronectin, etc.). One of the factors that must be kept in mind when designing *in vitro* assays to study mechanisms of *S. sanguis* adherence to these tissue components is that the adhesin-receptor interaction takes place in the presence of plasma or blood. For instance, Hamill[95] has suggested that Fn may play a role in infective endocarditis by acting as a tissue receptor molecule for bacteria. However, *S. sanguis* has also been shown to bind soluble Fn. Because there are large concentrations of soluble Fn in plasma, soluble Fn should inhibit binding of *S. sanguis* to immobilized Fn and thus reduce the importance of tissue Fn as a receptor for *S. sanguis*.

We have undertaken a study of the mechanism of *S. sanguis* adherence using a relatively simple approach. Gelatin or collagen was deposited on assay wells. Bacteria were combined with fresh plasma or purified proteins in these assay wells and adherence was quantitated using an antiserum specific for *S. sanguis*. Data obtained from this assay indicates that Fn could indeed play a role in the adherence of *S. sanguis* to damaged endothelial tissues, since these bacteria bind to substrate-adsorbed Fn even in the presence of high concentrations of soluble Fn. This suggests that *S. sanguis* adherence may be related to conformational changes that take place when Fn is immobilized. The data also suggest that the interaction of *S. sanguis* with Fn is different from the interaction of *S. pyogenes* with Fn since soluble Fn is a competitive inhibitor of *S. pyogenes* adherence to immobilized Fn.[83] Recent results show that Fn immobilized on substrates may assume a different conformation from that in solution.[97-100] Thus, it appears possible that Fn immobilized on extracellular matrix components or in the basal lamina underling epithelial and endothelial cells could exhibit different bacterial binding sites than those presented by the soluble molecule. While this observation may have little relevance in the oral cavity, where Fn levels in saliva are low, it may be extremely important in the blood stream, where bacteria must adhere in the presence of large amounts of Fn.

III. GRAM-NEGATIVE BACTERIA

It would appear from the available literature that most Gram-negative bacteria that have been studied to date do not bind significant amounts of Fn, nor have investigators frequently found Gram-negative bacteria to bind to immobilized Fn. In fact, in many cases *E. coli* or *Pseudomonas aeruginosa* have been used as negative controls in experiments involving the interaction of Fn with Gram-positive bacteria such as *S. aureus*. In one of the first studies our laboratory reported on Fn-bacteria interactions, binding of radiolabeled Fn to *P. aeruginosa* or *K. pneumoniae* was less than 1% that of *S. aureus* Cowan I strain and averaged less than 5% of the seven strains of group A streptococci tested.[3]

A number of other investigators obtained results consistent with the idea that the capacity of Gram-negative bacteria to bind Fn is relatively poor. Myhre and Kuusela[85] found essentially all of approximately 150 Gram-negative strains (including *E. coli* and species of *Neisseria, Klebsiella, Pseudomonas, Proteus,* and *Acineobacter* among others) to be either very low or negative for binding soluble Fn. Vercellotti et al.[93] compared adherence of *E. coli, S. sanguis,* and *S. aureus* to Fn or to human umbilical vein endothelial cells and found that *E. coli* adhered very poorly to either surface while the two Gram-positive organisms bound avidly to both. Toy et al.[45] found that four strains of *E. coli* tested bound to plasma clots containing fibrin and Fn 1000-fold less well than did the seven different strains of *S. aureus* that they tested. Depletion of Fn had a relatively minor effect upon adherence of the *E. coli* strains to plasma clots. In a study comparing the binding of bacteria to Fn relative to the propensity of the organisms to cause endocarditis in rabbits, Scheld et al.[101] found *P. aeruginosa* and *E. coli* bound poorly to Fn adsorbed to tissue culture plastic. *S. aureus* and several species of *Candida* on the other hand, bound to the Fn-coated plates quite well.

Further studies in our laboratory on the relationship of Fn to adherence of *E. coli* to buccal epithelial cells were stimulated by the observation by numerous investigators that various illnesses lead to a switch in the bacteria colonizing the oropharyngeal cavity. Whereas the oropharyngeal cavities of healthy individuals are rarely colonized by Gram-negative organisms, such as *E. coli, K. pneumoniae,* or *P. aeruginosa,* the onset of acute illness or surgical stress predisposes one to colonization by these organisms. Colonization often leads to Gram-negative bacterial pneumonia by aspiration of oropharyngeal secretions into the lungs.[102-105] Such infections are frequently resistant to antimicrobial agents and result in increased morbidity and mortality. Understanding the basic mechanisms of pathogenesis underlying the switch in colonization from almost exclusively Gram-positive bacteria to high concentrations of Gram-negative bacteria with subsequent increase in lung infections might eventually lead to improved prophylaxis of populations at risk.

Studies by Johanson and Woods and their colleagues[106-110] suggested that the switch in colonization may be coupled with a loss of Fn. Oral epithelial cells of critically ill patients in intensive care units bound significantly more Gram-negative bacteria *in vitro* than did cells from control individuals, and these patients often became colonized by Gram-negative organisms.[106] Further, the ability of *P. aeruginosa* to adhere to isolated oral epithelial cells of rats increased dramatically after surgical stress, and these rats became colonized with this organism when challenged.[107] Additionally, these investigators demonstrated that a 200,000-Da protein believed to be Fn was decreased on buccal cell membranes of patients that exhibit increased *Pseudomonas* adherence.[108] Also, an increase in the adherence of *Pseudomonas* organisms to washed buccal cells obtained from acute respiratory failure patients and coronary bypass patients[109] or from cystic fibrosis patients[110] correlated with a decrease in the amount of immunoreactive Fn available on these cells compared to cells obtained from healthy controls. These investigators were not able to study the effect of Fn on adherence of *P. aeruginosa* to buccal epithelial cells directly, because they were not able to restore Fn levels on buccal cells. This was probably due, at least in part, to the poor binding of Fn to epithelial cells at pH 7.4.[81]

Results from our laboratories have consistently pointed to the correlation between increased levels of buccal epithelial cell surface-bound Fn and decreased *E. coli* adherence. Abraham et al.[73] found that *E. coli* and *P. aeruginosa* preferentially bound to buccal epithelial cells devoid of Fn, whereas group A streptococci bound to buccal cells enriched for Fn. Further, pre-treatment of buccal cells with Fn resulted in specific, saturable, and reversible binding of Fn to the cells and led to a dose-related decrease in *E. coli* adherence.[81] This inhibition occurred at concentrations of Fn that are not far in excess of recent estimates of the concentration of Fn in parotid or whole saliva or gingival crevicular fluid.[76-78] Results of those reports suggests that Fn levels in some salivary sources may range up to 10 to 26 μg/ml, whereas our data suggest that 50% maximal inhibition of *E. coli* binding to saliva-treated or untreated buccal cells can be achieved with approximately 10 μg/ml *in vitro*. Alterations of this reaction which might occur in the complex mixture of salivary components *in vivo* could affect the concentrations of Fn that are necesssary to achieve inhibition of *E. coli* adherence. In a recent report[80] we have also presented evidence suggesting that *E. coli* and Fn bind to transfer blots of salivary components, and Fn pretreatment inhibits *E. coli* binding to these salivary components, suggesting that Fn may mask *E. coli* receptors present in saliva.

By contrast, we and others have also shown that in adherence tests with certain strains of *E. coli* and certain cultured cell lines, adherence increases with increased amounts of Fn detectable on the cells. Van de Water et al.[8] found that *E. coli* bind radiolabeled Fn. In a somewhat more extensive study, Fröman et al.[111] found 4 of 17 enterotoxigenic strains of *E. coli* that had been isolated from infantile diarrhea bound radiolabeled Fn. The strain with the highest binding capacity was tested further and was found to exhibit certain characteristics of specific Fn receptors. Binding was saturable, inhibitable with unlabeled Fn but not other proteins, and was at least partially reversible. Scatchard analysis demonstrated two classes of binding sites with calculated K_ds of $3 \times 10^{-8}\,M$ and $4 \times 10^{-9}\,M$, respectively. Preliminary studies showed that this strain of *E. coli* bound to a slight degree the same aminoterminal fragment bound by *S. aureus*, but primarily bound other unidentified fragments of approximately 150 kDa. This was confirmed by inhibition experiments using the 29-kDa aminoterminal fragment. A maximum of 30% inhibition was seen with levels of the 29-kDa fragment 16-fold higher than the radiolabeled intact Fn. Further binding experiments suggested that the binding site for the 29-kDa fragment was similar to the higher affinity site seen with intact Fn, with a K_d of 8×10^{-9} *M* and approximately 7000 sites per organism. These authors also demonstrated adherence of *E. coli* to Fn-coated coverslips and Fn-containing tissue culture cells that was inhibited by soluble Fn. Such results led these authors to suggest that binding to Fn might constitute a mechanism of tissue adherence of *E. coli*.

Taken at face value, the studies cited above would appear to be in conflict, with many reports of very little, if any, binding of Fn by Gram-negative bacteria; yet there are several reports suggesting that Fn does bind to certain Gram-negative bacteria and may also serve as a tissue receptor for adherence of the organisms. Thus, the results available to date do not allow us to develop a generalized hypothesis regarding the relationship between Fn and *E. coli* adherence without a more thorough characterization of the bacterial strains and tissue culture cell lines used.

Perhaps the observations of Sorvillo and Pearlstein[15] offer another avenue to be pursued with respect to the role of Fn in *E. coli* adherence to host cells and tissues. C1q was shown to bind to *E. coli* and *S. aureus* by an antibody-independent mechanism. Further, whereas control *E. coli* bound relatively low levels of soluble Fn, preincubation of *E. coli* strain W12 with C1q led to a 20-fold increase in Fn binding to the organisms, in contrast to only a twofold increase in Fn binding when *S. aureus* strain 25923 was pretreated with C1q. The end result was a greater binding of Fn to *E. coli* than to *S. aureus*. The effects of this interaction upon phagocytosis were not investigated in this study, but pretreating *E. coli* strains W12 and 75 with C1q resulted in a significant increase in their adherence to surface-bound Fn. Thus, it is possible that in conjunction with other tissue molecules Fn may play a central role in the pathogenesis of Gram-negative bacterial infections.

IV. *TREPONEMA PALLIDUM*

T. pallidum is the spirochete responsible for syphilis. The pathogenesis of this disease is obviously a very complicated process, occurring in distinct stages. However, as for many other infectious diseases, it is thought that adherence of *T. pallidum* to host cells is an early and fundamental event in the parasitism of host cells in the initial syphilitic lesion.[112,113] *T. pallidum* adherence occurs in a distinct tip-oriented fashion.[114,115] Three distinct outer membrane proteins, P1, P2, and P3, have been implicated as adhesins on the surface of treponemes.[114,115] These three membrane proteins apparently share a common 12-kDa domain that may be responsible for their functions as adhesins.[116] In 1983, Peterson et al. found that Fn was one of the host plasma proteins bound by virulent *T. pallidum* spirochetes (avirulent *T. phagedenis* biotype Reiter did not acquire host plasma proteins) and also showed that the organisms adhered to Fn-coated, but not BSA-coated, coverslips in a typical tip-oriented fashion.[117] The adherence of treponemes to Fn-coated coverslips was inhibited by anti-Fn serum, but not by antiserum directed against albumin or transferrin. Three *T. pallidum* molecules with electrophoretic mobilties identical to outer envelope proteins P1 (89,300 Da), P2 (37,000 Da) and P3 (32,000 Da) were selectively purified from a Zwittergent extract of radiolabeled organisms on a Fn-Sepharose column. Peterson et al.[118] have recently cloned the genes for P1 and P2 into a multicopy expression plasmid, pUC19, which was used to transform *E. coli*. The *E. coli* expressed and translocated recombinant P1 and P2 into their outer membranes. Fn-binding properties of these recombinant molecules were retained. The availability of these recombinant adhesins should allow new opportunities for study of the Fn-spirochete interaction, including comparison of the structure of P1 and P2 with other Fn-binding molecules.

Thomas et al.[119] found that anti-Fn serum, but not anti-laminin or anti-type I collagen serum, inhibited adherent of *T. pallidum* to both CHO and HT1080 cells by at least 70%. These investigators also found that a monoclonal antibody directed against the gelatin-binding domain of Fn (Calbiochem-Behring Corp., San Diego, CA) had relatively little effect on adherence of *T. pallidum* to Fn-coated coverslips (approximately 22% inhibition), whereas monoclonal antibodies directed against a heparin-binding domain (Calbiochem-Behring Corp.) or the cell attachment domain[120] inhibited adherence by approximately 60% and 72% respectively, approaching the 80% inhibition obtained with polyclonal anti-Fn serum. A 2.5-fold molar excess of partially purified Fn cell-binding domain inhibited binding of saturating levels of radiolabeled intact Fn to *T. pallidum* by up to 60%, whereas preparations enriched for heparin-binding or gelatin-binding Fn fragments inhibited by only 40 or 20%, respectively. The partially purified 68-kDa cell-binding fragment was also found to bind saturably to spirochetes, and this binding was effectively inhibited by unlabeled fragment. Scatchard analysis of binding data allowed calculation of an apparent K_d of approximately 1.5×10^{-7} *M* for Fn and 2×10^{-7} *M* for the fragment. The average number of treponemal receptors for Fn and the cell-binding fragment were calculated to be approximately 4×10^6 and 1×10^6 per organism, respectively. Other investigators have confirmed the relatively low affinity binding of Fn to treponemes[121,122] but have found much lower numbers for treponemal receptors for Fn, between 3000 to 7500 per cell. Radiolabeled *T. pallidum* bound to coverslips coated with Fn or with the cell-binding fragment in approximately equal numbers. Taken together, these results suggested that the primary region of Fn that interacts with *T. pallidum* surface molecules is contained within the 68-kDa cell-binding domain.

A synthetic heptapeptide containing the RGDS sequence critical for eukaryotic cell attachment to Fn, but not control peptides, inhibited binding of the cell-binding fragment to spirochetes.[123] Importantly, the RGDS-containing peptide, but not control peptides, also inhibited spirochete binding to both HT1080 and HEp-2 cells. *T. pallidum* also adhered to bovine corneal endothelial cell extracellular matrix (ECM) that contain Fn, collagen, and laminin, and, according to Thomas et al.,[125] only anti-Fn serum inhibited attachment of the organisms.

In related experiments, however, several other investigators have presented evidence suggesting that *T. pallidum* is also capable of attaching to cultured cells by mechanisms other than those involving fibronectin (e.g., attachment to laminin, collagens, and glycosaminoglycans).[126-129] The arguments raised regarding the singular importance of Fn in treponemal attachment centers around the idea that most or all of the Fn receptors on treponemes would be saturated upon entering the host circulation. The speculation is that this would block attachment to substrate-bound Fn, but, as reviewed below, Thomas et al.[125] found a distinct enhancement of adherence of treponemes to Fn-coated coverslips when the organisms were first pre-treated with Fn. Interestingly, Baughn found that incubation of *T. pallidum* with complement C1q led to a two- to threefold increase in Fn binding and in adherence of the C1q-coated organisms to Fn-coated coverslips.[122] Whether this interaction involves the known affinity of Fn for C1q[124] remains to be determined.

To obtain *T. pallidum* for *in vitro* experiments, the organisms are grown for 10 to 14 d in rabbit testes *in vivo*. Thomas et al.[125] found that 30% of freshly extracted spirochetes exhibited surface material cross-reactive with anti-FN IgG and could be collected using magnetic protAspheres or by chromatography using glass beads coated with anti-Fn IgG. The remaining 70% did not react with the anti-Fn IgG until pretreated with soluble Fn. Baughn[122] found that the method used to extract *T. pallidum* led to different amounts of host proteins such as albumin, IgG, and Fn remaining bound to the organisms. He speculated that perhaps the increased amount of host Fn that remained bound to treponemes by the Percoll-extraction technique used by Baseman and co-workers might lead to Fn-Fn interactions and provide at least one explanation for the large number of treponemal receptors for Fn calculated by that group.

An interesting question arises as to exactly how Fn-coated organisms could use substrate-bound Fn as a receptor (see also discussion in this regard on *S. pyogenes, S. aureus,* and *S. sanguis* above). Syphilis spirochetes *in vivo* are exposed to Fn present in plasma and other body fluids. Interestingly, Thomas et al.[125] found that pretreating *T. pallidum* with 100 μg/ml Fn led to a dramatic increase in the attachment of the Fn-primed organisms to both Fn-coated coverslips and to endothelial ECM that contains Fn. Attachment of spirochetes to HT1080 or CHO cells was not affected by similar concentrations of FN. However, when the concentrations of Fn used to pretreat the spirochetes was increased from 100 to 500—1500 μg/ml, there was a resultant dose-related decrease in adherence of the organisms to both cell types, up to 50%.

Taken together, the series of studies by Alderete, Baseman, and colleagues provide important information on the molecular basis for adhesin-receptor interactions leading to the adherence of *T. pallidum* to host cells and ECM *in vitro*. Much work obviously remains to be done to extend these studies to explain the role Fn may play in the very complicated pathogenic processes involved in actual syphilitic infections. At least one other aspect of syphilis pathogenesis in which it appears Fn may be important is in the formation of immune complexes and in the induction of autoimmunity. Baughn and Musher[130] purified immune complexes from sera of secondary syphilis patients and found several host proteins, including IgG, C1q, C3, and Fn, to be consistently present. Whether Fn is present due merely to its affinity to C1q remains to be established. It is also possible that Fn is present due to its affinity for a treponemal antigen. One of the *T. pallidum* antigens present in purified immune complexes[131] measures approximately 87 kDa, similar at least in mass to the 89 kDa P1 treponemal surface Fn-binding protein described by other investigators. Western blot analysis of the Fn present in purified syphilitic immune complexes revealed that 66- and 76-kDa protein bands reacted with a monoclonal antibody directed against the cell-binding domain of Fn. Baughn and co-workers went on to demonstrate the presence of antibodies directed against human Fn in the sera from patients with secondary, but not primary, syphilis.[132] These results extended the report of Fitzgerald and Repesh[127] that antibodies directed against Fn were found in the sera of syphilitic rabbits.

It would appear evident from the overview presented above that there is good evidence for the potential role of Fn in the early events in the pathogenesis of syphilis, such as adherence of the bacteria to host cells and tissues, as well as in subsequent events, such as formation of immune complexes or autoantibodies. It should also be quite evident that the majority of our knowledge of this subject is based on *in vitro*, not *in vivo* studies and, further, that a large number of questions remain to be answered.[133] One of these questions is whether Fn could be the only host protein involved in initial attachment. Perhaps other proteins are involved. However, it is very unlikely, in our view, that investigators will discover in subsequent studies that Fn-*T. pallidum* interactions do not play a significant role in pathogenesis of syphilis.

V. CONCLUDING REMARKS/FUTURE DIRECTIONS

There has been a surge of interest in the last several years in the mechanisms by with bacteria adhere to host cells and tissues. This increased interest derives from accumulated data that support the logically attractive concept that selective adhesion of pathogenic bacteria to host cells and tissues is a critical factor in the earliest of the steps in the infectious process. It should be clear from the studies reviewed in this chapter that there is every reason to expect that Fn plays an important role in the adherence of microorganisms to host tissues. Whether Fn is involved in other aspects of the pathogenesis of bacterial infections remains to be demonstrated.

Even if we narrow our focus on the subject of bacterial pathogenesis to the role of Fn in the adherence of bacteria to host tissues, we are still left with a complex phenomenon that will pose interesting questions for years to come. One complication is the multi-domain structure of the Fn molecule itself. There is ample evidence that purified Fn exists in somewhat different conformations depending upon whether it is in solution or bound to a substrate, and upon whether it is complexed with another molecule, such as gelatin or heparin. In this regard, there is some evidence that binding of purified, soluble Fn by a bacterium does not necessarily indicate that the bacterium will bind to Fn adsorbed to a substrate, and there is evidence that the converse may also be true. Relatively little attention has been paid to the effects that changes in the conformation of Fn may have on its interactions with bacteria. Further, there are several different forms of Fn, each with a somewhat different protein and carbohydrate structure. Most experiments with Fn/bacteria interactions have been performed using plasma Fn, and possible differences in the interaction of the different forms with bacteria has not been well explored. Also, Fn interacts with quite a variety of other extracellular matrix molecules, DNA, and cell surface receptors, to give a few examples, and these interactions may affect the ability of Fn to interact with bacterial adhesins. Certain bacteria interact directly with some of these other matrix molecules, such as laminin.[134] There have been very few studies that address the adherence of bacteria to complex, "natural" matrices containing the variety of these molecules that the bacterium would be exposed to *in vivo*. It is possible, however, that using highly specific antibody probes, synthetic peptides, or well-characterized bacterial mutants which lack the ability to bind to specific host molecules, the role that adherence to Fn and other potential receptor molecules may play in the pathogenesis of bacterial infections can now be addressed more directly.

ACKNOWLEDGMENTS

The authors' own studies reported here are supported by research funds from the National Institutes of Health (DE07218 to David L. Hasty, AI-10085 and AI-13550 to Edwin H. Beachey) and by funds from the U.S. Veterans Administration (to Edwin H. Beachey and W. Andrew Simpson).

REFERENCES

1. **Kuusela, P.,** Fibronectin binds to *Staphylococcus aureus, Nature (London),* 276, 718, 1978.
2. **Proctor, R. A., Pendergrast, E., and Mosher, D. F.,** Fibronectin mediates attachment of *Staphylococcus aureus* to human neutrophils, *Blood,* 59, 681, 1982.
3. **Simpson, W. A., Hasty, D. L., Mason, J. M., and Beachey, E. H.,** Fibronectin-mediated binding of group A streptococci to human polymorphonuclear leukocytes, *Infect. Immun.,* 37, 805, 1982.
4. **Eriksen, H. O., Espersen, F., and Clemmensen, I.,** Opsonic activity of fibronectin in the phagocytosis of *Staphylococcus aureus* by polymorphonuclear leukocytes, *Eur. J. Clin. Microbiol.,* 3, 108, 1984.
5. **Jacobs, R. F., Kiel, D. P., Sanders, M. L., and Steele, R. W.,** Phagocytosis of type III group B streptococci by neonatal monocytes: enhancement by fibronectin and gammaglobulin, *J. Infect. Dis.,* 152, 695, 1985.
6. **Verbrugh, H. A., Peterson, P. K., Smith, D. F., Nguyen, B.-Y. T., Hoidal, J. R., Wilkinson, B. J., Verhoef, J., and Furcht, L. T.,** Human fibronectin binding to staphylococcal surface protein and its relative inefficiency in promoting phagocytosis by human polymorphonuclear leukocytes, monocytes and alveolar macrophages, *Infect. Immun.,* 33, 811, 1981.
7. **Lanser, M. F. and Saba, T. M.,** Fibronectin as co-factor necessary for optimal granulocyte phagocytosis of *Staphylococcus aureus, J. Reticuloendothelial Soc.,* 30, 415, 1981.

8. **Van de Water, L., Destree, A. T., and Hynes, R. O.,** Fibronectin binds to some bacteria but does not promote their uptake by phagocytic cells, *Science,* 220, 201, 1983.
9. **Oishi, K., Yamamoto, M., Yoshida, T., Ide, M., and Matsumoto, K.,** Opsonic activity of plasma fibronectin for *Staphylococcus aureus* by human alveolar macrophages: inefficiency of trypsin-sensitive staphylococcal fibronectin receptor, *Tohoku J. Exp. Med.,* 149, 95, 1986.
10. **Saba, T. M., Blumenstock, F. A., Weber, P., and Kaplan, J. E.,** Physiologic role of cold-insoluble globulin in systemic host defense: implications of its characterization as the opsonic α2-surface binding glycoprotein, *Ann. N.Y. Acad. Sci.,* 312, 43, 1978.
11. **Blumenstock, F. A., Saba, T. M., Roccario, E., Cho, E., and Kaplan, J. E.,** Opsonic fibronectin after trauma and particle injection as determined by a peritoneal macrophage assay, *J. Reticuloendothelial Soc.,* 30, 61, 1981.
12. **Dillon, B. C., Estes, J. E., Saba, T. M., Blumenstock, F. A., Cho, E., Lee, S. K., and Lewis, E. P.,** Actin-induced reticuloendothelial phagocytic depression as mediated by its interaction with fibronectin, *Exp. Mol. Pathol.,* 38, 208, 1983.
13. **Van De Water, L.,** Phagocytosis, in *Plasma Fibronectin. Structure and Function,* McDonagh, J., Ed., Marcel Dekker, New York, 1985, 175.
14. **Hormann, H.,** Fibronectin and phagocytosis, *Blut,* 51, 307, 1985.
15. **Sorvillo, J. M. and Pearlstein, E.,** C1q, a subunit of the first component of complement, enhances binding of plasma fibronectin to bacteria, *Infect. Immun.,* 49, 664, 1985.
16. **Saba, T. M., Kiener, J. L., and Holman, J. M., Jr.,** Fibronectin and the critically ill patient: current status, *Intensive Care Med.,* 12, 350, 1986.
17. **Grossman, J. E.,** Plasma fibronectin and fibronectin therapy in sepsis and critical illness, *Rev. Infect. Dis.,* 9, S420, 1987.
18. **Beachey, E. H.,** Bacterial adherence: adhesin-receptor interactions mediating the attachment of bacteria to mucosal surfaces, *J. Infect. Dis.,* 143, 325, 1981.
19. **Gibbons, R. J. and Van Houte, J.,** in *Bacterial Adherence,* Beachey, E. H., Ed., Chapman and Hall, New York, 1980, 60.
20. **Ofek, I. and Beachey, E. H.,** in: *Bacterial Adherence,* Beachey, E. H., Ed., Chapman and Hall, New York, 1980, 1.
21. **Zafriri, D., Oron, Y., Eisenstein, B. I., and Ofek, I.,** Growth advantage and enhanced toxicity of *Escherichia coli* adherent to tissue culture cells due to restricted diffusion of products secreted by the cells, *J. Clin. Invest.,* 79, 1210, 1987.
22. **Proctor, R. A., Mosher, D. F., and Olbrantz, P. J.,** Fibronectin binding to *Staphylococcus aureus,* J. Biol. Chem., 257, 14788, 1982.
23. **Rydén, C., Rubin, K., Speziale, P., Höök, M., Lindberg, M., and Wadström, T.,** Fibronectin receptors from *Staphylococcus aureus, J. Biol. Chem.,* 258, 3396, 1983.
24. **Doran, J. E. and Raynor, R. H.,** Fibronectin binding to protein A-containing staphylococci, *Infect. Immun.,* 33, 683, 1981.
25. **Proctor, R. A.,** The staphylococcal fibronectin receptor: evidence for its importance in invasive infections, *Rev. Infect. Dis.,* 9, S335, 1987.
26. **Salonen, E. M., Vartio, T., Hedman, K., and Vaheri, A.,** Binding of fibronectin by the acute phase reactant C-reactive protein, *J. Biol. Chem.,* 259, 1496, 1984.
27. **Mosher, D. F. and Proctor, R. A.,** Binding and factor XIIIa-mediated cross-linking of a 27 kilodalton fragment of fibronectin to *Staphylococcus aureus, Science,* 209, 927, 1980.
28. **Beachey, E. H. and Courtney, H. S.,** Bacterial adherence: the attachment of group A streptococci to mucosal surfaces, *Rev. Infect. Dis.,* 9, S475, 1987.
29. **Stanislawski, L., Courtney, H. S., Simpson, W. A., Hasty, D. L., Beachey, E. H., Robert, L., and Ofek, I.,** Hybridoma antibodies against the lipid binding site(s) in amino-terminal region of fibronectin inhibits binding of streptococcal lipoteichoic acid, *J. Infect. Dis.,* 156, 344, 1987.
30. **Kuusela, P., Vartio, T., Vuento, M., and Myhre, E. B.,** Attachment of staphylococci and streptococci on fibronectin, fibronectin fragments, and fibrinogen bound to a solid phase, *Infect. Immun.,* 50, 77, 1985.
31. **Vercellotti, G. M., McCarthy, J. B., Lindholm, P., Peterson, P. K., Jacob, H. S., and Furcht, L. T.,** Extracellular matrix proteins (fibronectin, laminin and type IV collagen) bind and aggregate bacteria, *Am. J. Pathol.,* 120, 13, 1985.
32. **Proctor, R. A., Christman, G., and Mosher, D. F.,** Fibronectin-induced agglutination of *Staphylococcus aureus* correlates with invasiveness, *J. Lab. Clin. Med.,* 104, 455, 1983.
33. **Espersen, F. and Clemmensen, I.,** Clumping of *Staphylococcus aureus* by human fibronectin, *Acta Path. Microbiol. Scand. Sect. B.,* 89, 317, 1981.
34. **Simpson, W. A., Courtney, H. S., and Ofek, I.,** Interaction of fibronectin with streptococci: the role of fibronectin as a receptor for *Streptococcus pyogenes, Rev. Infect. Dis.,* 9, S351, 1987.
35. **Espersen, F. and Clemmensen, I.,** Isolation of a fibronectin-binding protein from *Staphylococcus aureus, Infect. Immun.,* 37, 526, 1982.

36. **Fröman, G., Świtalski, L. M., Speziale, P., and Höök, M.,** Isolation and characterization of a fibronectin receptor from *Staphylococcus aureus, J. Biol. Chem.*, 262, 6564, 1987.
37. **Fröman, G., Świtalski, L. M., Guss, B., Lindberg, M., Höök, M., and Wadström, T.,** in *Protein-Carbohydrate Interactions in Biological Systems,* Lark, D. L., Ed., Academic Press, London, 1986, 263.
38. **Flock, J. I., Fröman, G., Jönsson, K., Guss, B., Signäs, C., Nilsson, B., Raucci, G., Höök, M., Wadström, T., and Lindberg, M.,** Cloning and expression of the gene for a fibronectin-binding protein from *Staphylococcus aureus,* EMBO J., 6, 2351, 1987.
39. **Maxe, I., Rydén, C., Wadström, T., and Rubin, K.,** Specific attachment of *Staphylococcus aureus* to immobilized fibronectin, *Infect. Immun.*, 54, 695, 1986.
40. **Vann, J. M., Kuypers, J., Proctor, R. A., and Mosher, D. F.,** Encapsulation and fibronectin binding: two independent factors enhancing, *S. aureus* infectivity, in Program and Abstracts of 22nd Interscience Conference on Antimicrobial Agents and Chemotherapy, Miami, October 4—6, 1982.
41. **Velazco, M. I. and Waldvogel, F. A.,** Molecular mechanisms of *Staphylococcus aureus* adherence and pathogenesis of catheter infections, in Program and Abstracts of 25th Interscience Conference on Antimicrobial Agents and Chemotherapy, Minneapolis, September 29 to October 2, 1985.
42. **Bibel, D. J., Aly, R., Shinefeld, H. R., and Maibach, H. I.,** The *Staphylococcus aureus* receptor for fibronectin, *J. Invest. Dermatol.*, 80, 494, 1983.
43. **Świtalski, L. M., Ljungh, Å, Rydén, C., Rubin, K., Höök, M., and Wadström, T.,** Binding of fibronectin to the surface of group A, C, and G streptococci isolated from human infections, *Eur. J. Clin. Microbiol.*, 1, 381, 1982.
44. **Proctor, R. A., Hamill, R. J., Mosher, D. F., Textor, J. A., and Olbrantz, P. J.,** Effects of subinhibitory concentrations of antibiotics on *Staphylococcus aureus* interactions with fibronectin, *J. Antimicrob. Chemother.*, 12, S85, 1983.
45. **Toy, P. T., Lai, L. W., Drake, T. A., and Sande, M. A.,** Effect of fibronectin on adherence of *Staphylococcus aureus* to fibrin thrombi *in vitro, Infect. Immun.*, 48, 83, 1985.
46. **Vaudaux, P., Suzuki, R., Waldvogel, F. A., Morgenthaler, J. J., and Nydegger, U. E.,** Foreign body infection: role of fibronectin as a ligand for the adherence of *Staphylococcus aureus, J. Infect. Dis.*, 150, 546, 1984.
47. **Vaudaux, P. E., Zulian, G., Huggler, E., and Waldvogel, F. A.,** Attachment of *Staphylococcus aureus* to polymethylmethacrylate increases its resistance to phagocytosis in foreign body infection, *Infect. Immun.*, 50, 472, 1985.
48. **Vaudaux, P., Lew, D., and Waldvogel, F. A.,** Host-dependent pathogenic factors in foreign body infection. A comparison between *Staphylococcus epidermidis* and *S. aureus, Zbl. Bakt. Suppl.*, 16, 183, 1987.
49. **Falcieri, E., Vaudaux, P., Huggler, E., Lew, D., and Waldvogel, F.,** Role of bacterial exopolymers and host factors on adherence and phagocytosis of *Staphylococcus aureus* in foreign body infection, *J. Infect. Dis.*, 155, 524, 1987.
50. **Velazco, M. I. and Waldvogel, F. A.,** Monosaccharide inhibition of *Staphylococcus aureus* adherence to human solid-phase fibronectin, *J. Infect. Dis.*, 155, 1069, 1987.
51. **Vaudaux, P., Lerch, P., Velazco, M. I., Nydegger, U. E., and Waldvogel, F. A.,** Role of fibronectin in the susceptibility of biomaterial implants to bacterial infections in *Biological and Biomechanical Performance of Biomaterials,* Christel, P., Meunier, A., and Lee, A. J. C., Eds., Elsevier Science, Amsterdam, 1986, 355.
52. **Vaudaux, P. E., Waldvogel, F. A., Morgenthaler, J. J., and Nydegger, U. E.,** Adsorption of fibronectin into polymethylmethacrylate and promotion of *Staphylococcus aureus* adherence, *Infect. Immun.*, 45, 768, 1984.
53. **Costerton, J. W., Irvin, R. T., and Cheng, K. J.,** The bacterial glycocalyx in nature and disease, *Annu. Rev. Microbiol.*, 35, 299, 1981.
54. **Christensen, G. D., Simpson, W. A., Bisno, A. L., and Beachey, E. H.,** Adherence of slime-producing strains of *Staphylococcus epidermidis* to smooth surfaces, *Infect. Immun.*, 37, 318, 1982.
55. **Wadström, T., Speziale, P., Rozgonyi, F., Ljungh, Å, Maxe, I., and Rydén, C.,** Interactions of coagulase-negative staphylococci with fibronectin and collagen as possible first step of tissue colonization in wounds and other tissue trauma, *Zbl. Bakt. Suppl.*, 16, 83, 1987.
56. **Ellen, R. P. and Gibbons, R. J.,** M protein-associated adherence of *S. pyogenes* to epithelial surfaces: prerequisite for virulence, *Infect. Immun.*, 5, 826, 1972.
57. **Ellen, R. P. and Gibbons, R. J.,** Parameters affecting the adherence and tissue tropisms of *Streptococcus pyogenes, Infect. Immun.*, 9, 85, 1973.
58. **Beachey, E. H. and Ofek, I.,** Epithelial cell binding of group A streptococci by lipoteichoic acid on fimbriae denuded of M protein, *J. Exp. Med.*, 143, 759, 1976.
59. **Ofek, I., Beachey, E. H., Jefferson, W., and Campbell, G. L.,** Cell membrane-binding properties of group A streptococcal lipoteichoic acid, *J. Exp. Med.*, 141, 990, 1975.
60. **Wicken, A. J. and Knox, K. W.,** Lipoteichoic acids: a new class of bacterial antigen, *Science,* 187, 1161, 1975.

61. **Ofek, I., Beachey, E. H., Eyal, F., and Morrison, J. C.,** Postnatal development of binding of streptococci and lipoteichoic acid by oral mucosal cells of humans, *J. Infect. Dis.*, 135, 267, 1977.
62. **Simpson, W. A., Ofek, I., Sarasohn, C., Morrison, J. C., and Beachey, E. H.,** Characteristics of the binding of streptococcal lipoteichoic acid to human oral epithelial cells, *Infect. Immun.*, 141, 457, 1980.
63. **Beachey, E. H., Chiang, T. M., Ofek, I., and Kang, A. H.,** Interaction of lipoteichoic acid of group A streptococci with human platelets, *Infect. Immun.*, 16, 649, 1977.
64. **Beachey, E. H., Dale, J. B., Simpson, W. A., Evans, J. D., Knox, K. W., Ofek, I., Wicken, A. J.,** Erythrocyte binding properties of streptococcal lipoteichoic acids, *Infect. Immun.*, 23, 618, 1979.
65. **Chiang, T. M., Alkan, M. L., and Beachey, E. H.,** Binding of lipoteichoic acid of group A streptococci to isolated human erythrocyte membranes, *Infect. Immun.*, 26, 316, 1979.
66. **Courtney, H. S., Ofek, I., Simpson, W. A., and Beachey, E. H.,** Characterization of lipoteichoic acid binding to polymorphonuclear leukocytes of human blood, *Infect. Immun.*, 32, 625, 1981.
67. **Simpson, W. A., Dale, J. B., and Beachey, E. H.,** Cytotoxicity of the glycolipid region of streptococcal lipoteichoic acid for cultures of human heart cells, *J. Lab. Clin. Med.*, 99, 118, 1982.
68. **Simpson, W. A., Ofek, I., and Beachey, E. H.,** Fatty acid binding sites of serum albumin as membrane receptor analog for streptococcal lipoteichoic acid, *Infect. Immun.*, 29, 119, 1980.
69. **Beachey, E. H., Simpson, W. A., and Ofek, I.,** Interactions of surface polymers of *Streptococcus pyogenes* with animal cells, in *Microbial Adhesion to Surfaces,* Berkeley, R. C. W., Lynch, J. M., Melling, J., Rutter, P. R., and Vincent, B., Eds., Ellis Horwood, Chichester, England, 1980, 389.
70. **Simpson, W. A., Ofek, I., and Beachey, E. H.,** Binding of streptococcal lipoteichoic acid to the fatty acid binding sites on serum albumin, *J. Biol. Chem.*, 255, 6092, 1980.
71. **Ofek, I., Simpson, W. A., and Beachey, E. H.,** The formation of molecular complexes between a structurally defined M protein and acylated or deacylated lipoteichoic acid of *Streptococcus pyogenes, J. Bacteriol.*, 149, 426, 1982.
72. **Zetter, B. R., Daniels, T. E., Quadra-White, C., and Greenspan J. S.,** LETS protein in normal and pathological human oral epithelium, *J. Dent. Res.*, 58, 484, 1979.
73. **Abraham, S. N., Beachey, E. H., and Simpson, W. A.,** Adherence of *Streptococcus pyogenes, Escherichia coli,* and *Pseudomonas aeruginosa* to fibronectin-coated and uncoated epithelial cells, *Infect. Immun.*, 41, 1261, 1983.
74. **Simpson, W. A., Courtney, H. S., and Beachey, E. H.,** Fibronectin, a modulator of the oropharyngeal bacterial flora, *Microbiology 1982,* American Society for Microbiology, Washington, D.C., 1982, 344.
75. **Babu, J. P., Simpson, W. A., Courtney, H. S., and Beachey, E. H.,** Interaction of human plasma fibronectin with cariogenic and non-cariogenic oral streptococci, *Infect. Immun.*, 41, 162, 1983.
76. **Ericson D. and Tynelius-Bratthall, G.,** Absorption of fibronectin from human saliva by strains of oral streptococci, *Scand. J. Dent. Res.*, 95, 377, 1986.
77. **Tynelius-Bratthall, G., Ericson, D., and Araujo, H. M.,** Fibronectin in saliva and gingival crevices, *J. Periodontal Res.*, 21, 563, 1986.
78. **Babu, J. P. and Dabbous, M. K.,** Interaction of salivary fibronectin with oral streptococci, *J. Dent. Res.*, 65, 1094, 1986.
79. **Simpson, W. A., Hasty, D. L., and Beachey, E. H.,** Inhibition of the adhesion of *Escherichia coli* to oral epithelial cells by FN, in *Molecular Basis of Oral Microbial Adhesion,* Mergenhagan, S. E. and Rosan, B., Eds., American Society for Microbiology, Washington, D. C., 1985, 40.
80. **Hasty, D. L. and Simpson, W. A.,** Effects of fibronectin and other salivary macromolecules on the adherence of *Escherichia coli* to buccal epithelial cells, *Infect. Immun.*, 55, 2103, 1987.
81. **Simpson, W. A., Hasty, D. L., and Beachey, E. H.,** Binding of fibronectin to human buccal epithelial cells inhibits the binding of type 1 fimbriated *Escherichia coli, Infect. Immun.*, 48, 318, 1985.
82. **Simpson, W. A. and Beachey, E. H.,** Adherence of group A streptococci to fibronectin on oral epithelial cells, *Infect. Immun.*, 39, 275, 1983.
83. **Courtney, H. S., Ofek, I., Simpson, W. A., Hasty, D. L., and Beachey, E. H.,** The binding of *Streptococcus pyogenes* to soluble and insoluble fibronectin, *Infect. Immun.*, 53, 454, 1986.
84. **Courtney, H. S., Simpson, W. A., and Beachey, E. H.,** Binding of streptococcal lipoteichoic acid to fatty acid binding sites on human plasma fibronectin, *J. Bacteriol.*, 153, 763, 1983.
85. **Myhre, E. B. and Kuusela, P.,** Binding of human fibronectin to group A, C, and G streptococci, *Infect. Immun.*, 40, 29, 1983.
86. **Imai, S., Okahashi, N., Koga, T., Nisizawa, T., and Hamada, S.,** Ability of various oral bacteria to bind human plasma fibronectin, *Microbiol. Immunol.*, 28, 863, 1984.
87. **Speziale, P., Höök, M., Świtalski, L. M., and Wadström, T.,** Fibronectin binding to a *Streptococcus pyogenes* strain, *J. Bacteriol.*, 157, 420, 1984.
88. **Alkan, M. L. and Beachey, E. H.,** Excretion of lipoteichoic acid by group A streptococci: influence of penicillin on excretion and loss of ability to adhere to human oral mucosal cells, *J. Clin. Invest.*, 61, 671, 1978.

89. **Nealon, T. J., Beachey, E. H., Courtney, H. S., and Simpson, W. A.,** Release of fibronectin-lipoteichoic acid complexes from group A streptococci with penicillin, *Infect. Immun.*, 51, 529, 1986.
90. **Simpson, W. A., Nealon, T. J., Hasty, D. L., Courtney, H. S., and Beachey, E. H.** Binding of fibronectin to *S. pyogenes:* Heat induction of fibronectin receptors, in *Mucosal Immunity and Infections at Mucosal Surfaces,* Strober, W., Lamm, M. E., McGhee, J. R., and James, S. P., Eds., Oxford University Press, New York, 1988, 349.
91. **Stanislawski, L., Simpson, W. A., Hasty, D., Sharon, N., Beachey, E. H., and Ofek, I.,** Role of fibronectin in attachment of *Streptococcus pyogenes* and *Escherichia coli* to human cell lines and isolated oral epithelial cells, *Infect. Immun.*, 48, 257, 1985.
92. **Butler, K. M., Baker, C. J., and Edwards, M. S.,** Interaction of soluble fibronectin with group B streptococci, *Infect. Immun.*, 55, 2404, 1987.
93. **Vercellotti, G. M., Lussenhop, D., Peterson, P. K., Furcht, L. T., McCarthy, J. B., Jacob, H. S., and Moldow, C. F.,** Bacterial adherence to fibronectin and endothelial cells: a possible mechanism for bacterial tissue tropism, *J. Lab. Clin. Med.*, 103, 34, 1984.
94. **Andersson, B., Dahmén, J., Frejd, T., Leffler, H., Magnusson, G., Noori, G., and Svanborg-Edén, C.,** Identification of an active disaccharide unit of a glycoconjugate receptor for pneumococci attaching to human pharyngeal epithelial cells, *J. Exp. Med.*, 158, 559, 1983.
95. **Hamill, R. J.,** Role of fibronectin in infective endocarditis, *Rev. Infect. Dis.*, 9, S360, 1987.
96. **Lowrance, J. H., Hasty, D. L., and Simpson, W. A.,** Adherence of *Streptococcus sanguis* to conformationally specific determinants in fibronectin, *Infect. Immun.*, 56, 2279, 1988.
97. **Salonen, E. M., Saksela, O., Vartio, T., Vaheri, A., Nielson, L. S., and Zeuthen, J.,** Plasminogen and tissue-type plasminogen activator bind to immobilized fibronectin, *J. Biol. Chem.*, 260, 12302, 1985.
98. **Laterra, J. and Culp, L. A.,** Differences in hyaluronate binding to plasma and cell surface fibronectins, *J. Biol. Chem.*, 257, 719, 1982.
99. **Grinnell, F. and Feld, M. K.,** Fibronectin absorption on hydrophilic and hydrophobic surfaces detected by antibody binding and analyzed during cell adhesion in serum containing medium, *J. Biol. Chem.*, 257, 4888, 1981.
100. **Akiyama, S. K. and Yamada, K. M.,** Fibronectin, *Adv. Enzymol. Relat. Areas Mol. Biol.*, 59, 1, 1987.
101. **Scheld, W. M., Strunk, R. W., Balian, G., and Calderone, R. A.,** Microbial adhesion to fibronectin in vitro correlates with production of endocarditis in rabbits, *Proc. Soc. Exp. Biol. Med.*, 180, 474, 1985.
102. **Johanson, W. G., Pierce, A. K., and Sanford, J. P.,** Changing pharyngeal bacterial flora of hospitalized patients, *N. Engl. J. Med.*, 281, 1137, 1969.
103. **Johanson, W. G., Pierce, A. K., Sanford, J. P., and Thomas, G. D.,** Nosocomial respiratory infections with Gram negative bacilli, *Ann. Intern. Med.*, 77, 701, 1972.
104. **Klastersky, J.,** Nosocomial infections due to Gram-negative bacilli in compromised hosts: considerations for prevention and therapy, *Rec. Infect. Dis.*, 7, S552, 1985.
105. **Levison, M. E. and Kaye, D.,** Pneumonia caused by Gram-negative bacilli: an overview, *Rev. Infect. Dis.*, 7, S656, 1985.
106. **Johanson, W. G., Woods, D. E., and Chaudhuri, T.,** Association of respiratory tract colonization with adherence of Gram-negative bacilli to epithelial cells, *J. Infect. Dis.*, 139, 667, 1979.
107. **Higuchi, J. H. and Johanson, W. G.,** The relationship between adherence of *Pseudomonas aeruginosa* to upper respiratory cells in vitro and susceptibility to colonization in vivo, *J. Lab. Clin. Med.*, 95, 698, 1980.
108. **Woods, D. E., Straus, D. C., Johanson, W. G., Jr., and Bass, J. A.,** Role of fibronectin in the prevention of adherence of *Pseudomonas aeruginosa* to buccal cells, *J. Infect. Dis.*, 143, 784, 1981.
109. **Woods, D. E., Straus, D. C., Johanson, W. G., Jr., and Bass, J. A.,** Role of salivary protease activity in adherence of Gram-negative bacilli to mammalian buccal epithelial cells in vivo, *J. Clin. Invest.*, 68, 1435, 1981.
110. **Woods, D. E., Bass, J. A., Johanson, W. G., and Straus, D. C.,** Role of adherence in the pathogenesis of *Pseudomonas aeruginosa* lung infections in cystic fibrosis patients, *Infect. Immun.*, 30, 694, 1980.
111. **Fröman, G., Świtalski, L., Faris, A., Wadström, T., and Höök, M.,** Binding of *Escherichia coli* to fibronectin. A mechanism of tissue adherence, *J. Biol. Chem.*, 259, 14899, 1984.
112. **Fitzgerald, T. J., Johnson, R. C., Miller, J. N., and Sykes, J. A.,** Characterization of the attachment of *Treponema pallidum* (Nichols strain) to cultured mammalian cells and the potential relationship of attachment to pathogenicity, *Infect. Immun.*, 18, 467, 1977.
113. **Hayes, N. S., Muse, K. E., Collier, A. M., and Baseman, J. B.,** Parasitism by virulent *Treponema pallidum* of host cell surfaces, *Infect. Immun.*, 17, 174, 1977.
114. **Baseman, J. B. and Hayes, E. C.,** Molecular characterization of receptor binding proteins and immunogens of virulent *Treponema pallidum, J. Exp. Med.*, 151, 573, 1980.
115. **Alderete, J. F. and Baseman, J. B.,** Surface-associated host proteins on virulent, *Treponema pallidum, Infect. Immun.*, 26, 1048, 1979.

116. **Thomas, D. D., Baseman, J. B., and Alderete, J. F.,** Putative *Treponema pallidum* cytadhesins share a common functional domain, *Infect. Immun.*, 49, 833, 1985.
117. **Peterson, K. M., Baseman, J. B., and Alderete, J. F.,** *Treponema pallidum* receptor binding proteins interact with fibronectin, *J. Exp. Med.*, 157, 1958, 1983.
118. **Peterson, K. M., Baseman, J. B., and Alderete, J. F.,** Molecular cloning of *Treponema pallidum* outer envelope fibronectin binding proteins P1 and P2, *Genitourin. Med.*, 63, 355, 1987.
119. **Thomas, D. D., Baseman, J. B., and Alderete, J. F.,** Fibronectin mediates *Treponema pallidum* cytadherence through recognition of fibronectin cell-binding domain, *J. Exp. Med.*, 161, 514, 1985.
120. **Koch, G. A., Schoen, R. C., Klebe, R. J., and Shows, T. B.,** Assignment of a fibronectin gene to human chromosome 2 using monoclonal antibodies, *Exp. Cell Res.*, 141, 293, 1982.
121. **Steiner, B. M. and Sell, S.,** Characterization of the interaction between fibronectin and *Treponema pallidum, Curr. Microbiol.*, 12, 157, 1985.
122. **Baughn, R. E.,** Antibody-independent interactions of fibronectin, C1q, and human neutrophils with *Treponema pallidum, Infect. Immun.*, 54, 456, 1986.
123. **Thomas, D. D., Baseman, J. B., and Alderete, J. F.,** Fibronectin tetrapeptide is target for syphilis spirochete cytadherence, *J. Exp. Med.*, 162, 1715, 1985.
124. **Pearlstein, E., Sorvillo, J., and Gigli, I.,** The interaction of human plasma fibronectin with a subunit of the first component of complement, C1q, *J. Immunol.*, 128, 2036, 1982.
125. **Thomas, D. D., Baseman, J. B., and Alderete, J. F.,** Enhanced levels of attachment of fibronectin-primed *Treponema pallidum* to extracellular matrix, *Infect. Immun.*, 52, 736, 1986.
126. **Fitzgerald, T. J., Repesh, L. A., Blanco, D. R., and Miller, J. N.,** Attachment of *Treponema pallidum* to fibronectin, laminin, collagen IV, and collagen I and blockage of attachment by immune rabbit IgG, *Br. J. Venereal Dis.*, 60, 357, 1984.
127. **Fitzgerald, T. J. and Repesh, L. A.,** Interactions of fibronectin with *Treponema pallidum, Genitourin. Med.*, 61, 147, 1985.
128. **Fitzgerald, T. J.,** Pathogenesis and immunology of *Treponema pallidum, Annu. Rev. Microbiol.*, 35, 29, 1981.
129. **Steiner, B. M., Sell, S., and Schell, R. F.,** *Treponema pallidum* attachment to surface and matrix proteins of cultured rabbit epithelial cells, *J. Infect. Dis.*, 155, 742, 1987.
130. **Baughn, R. E. and Musher, D. M.,** Isolation and preliminary characterization of circulating immune complexes from rabbits with experimental syphilis, *Infect. Immun.*, 42, 579, 1983.
131. **Baughn, R. E., Adams, C. B., and Musher, D. M.,** Circulating immune complexes in experimental syphilis: identification of treponemal antigens and specific antibodies to treponemal antigens in isolated complexes, *Infect. Immun.*, 42, 585, 1983.
132. **Baughn, R. E., McNeely, M. C., Jorizzo, J. L., and Musher, D. M.,** Characterization of the antigenic determinants and host components in immune complexes from patients with secondary syphilis, *J. Immunol.*, 136, 1406, 1986.
133. **Baughn, R. E.,** Role of fibronectin in the pathogenesis of syphilis, *Rev. Infect. Dis.*, 9, S372, 1987.
134. **Switalski, L. M., Murchison, H., Timpl, R., Curtis, R., and Höök, M.,** Binding of laminin to oral and endocarditis strains of viridans streptococci, *J. Bacteriol.*, 169, 1095, 1987.
135. **Courtney, H. S., et al.,** unpublished results, 1983.

Chapter 7

FIBRONECTIN AND POLYMORPHONUCLEAR NEUTROPHIL FUNCTION

André D. Beaulieu, M. La Fleur, and C. Kreis

TABLE OF CONTENTS

I. INTRODUCTION

Among the many cell types that produce fibronectin and are affected by this multifunctional protein, we must now include the polymorphonuclear neutrophil (PMN). In recent years, many observations and experimental data have been published which fully support this statement. On the one hand, the roles that PMN play in host defense mechanisms and tissue repair at sites of injury are well recognized; on the other hand, fibronectin is a glycoprotein that is also active in both of these processes, such as increasing phagocytic activity, movement, and adherence of cells in tissues. The finding that PMN and fibronectin are able to interact is therefore not unexpected. A number of studies suggest that PMN have receptors for fibronectin.[1-5] It is noteworthy that the fibronectin receptor activity only becomes functional once the cell has been activated by inflammatory molecules such as C5a.[1] Furthermore, we and others have shown that PMN, along with monocyte-macrophages, synthesize and secrete fibronectin.[6-9] An important finding made in our laboratory was that PMN significantly increase their production of fibronectin once activated at inflammatory sites. Our observations, along with the receptor findings, strongly suggest that PMN-fibronectin interactions may have important biological functions at sites of inflammation and tissue injury. This chapter reviews the current state of knowledge based on the circumstantial evidence that tends to support this concept.

II. FIBRONECTIN SYNTHESIS AND SECRETION BY PMN

Protein synthesis and secretion by PMN is an area of study which remains relatively unexplored. This is probably due to the fact that there exists a long-held belief that PMN have little capacity to synthesize proteins. This belief is likely to remain generally accepted if studies addressing this question are limited to nonactivated peripheral blood PMN. We have shown by metabolic labeling that activated PMN from the synovial fluid of patients with rheumatoid arthritis synthesize large amounts of fibronectin while nonactivated PMN from peripheral blood synthesize much less.[7-9] A number of recent studies have also clearly indicated that *in vitro* activated PMN may increase considerably their production of a number of proteins. For example, interleukin-1 production by PMN was found to be markedly increased upon lipopolysaccharide (LPS) stimulation.[10] It was also shown that PMN are inducible for heat shock protein synthesis at a transcriptional level.[11] More recently, considerable c-fos protooncogene expression in activated and nonactivated PMN has been observed.[12] By analyzing mRNA levels in these cells following exposure to a series of agents that functionally activate PMN, a significant increase in the accumulation of c-fos mRNA occurred, clearly demonstrating the capacity of these cells to synthesize RNA and proteins even in the nonactivated states. Furthermore, dexamethasone, a potent antiinflammatory agent, was shown to modulate protein synthesis in PMN by decreasing production of a certain number of proteins and increasing synthesis of at least two others.[13] It is therefore obvious that PMN have significant protein synthetic capability.

Our studies on fibronectin production by PMN were initiated in an attempt to identify the cellular source of synovial fluid fibronectin. In this chronic inflammatory joint disease, the synovial fluid levels in fibronectin far exceed the serum or plasma concentrations found in the patients afflicted with this disease.[14,15] While concentrations in plasma rarely exceed 300 μg/ml, they often reach much higher levels in synovial fluid, suggesting that there exist factors leading to local production of fibronectin in the synovial joint. At the time we initiated our studies, the known possible cellular sources of fibronectin in synovial fluid were the synovial cells themselves[16] and possibly the monocyte-macrophages.[17] However, one report suggested that PMN could also synthesize fibronectin.[18] In synovial fluid of most inflammatory joint diseases, the PMN is by far the major cell population. It is not uncommon to

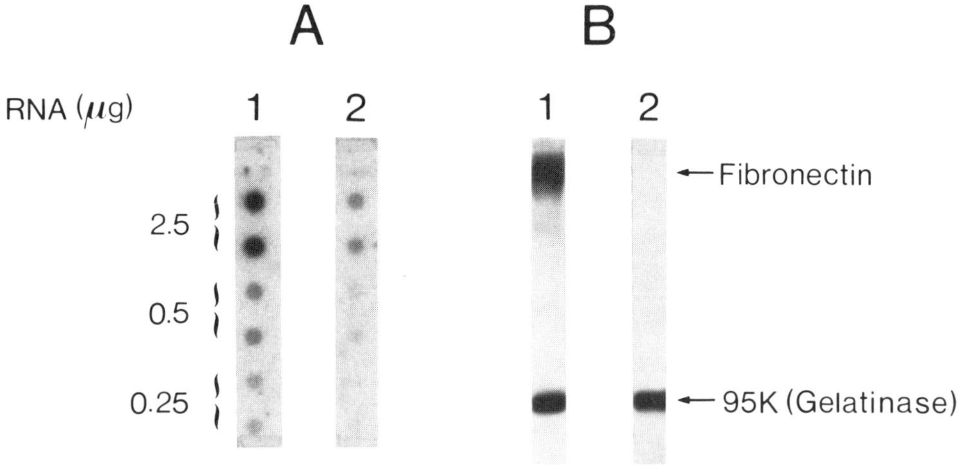

FIGURE 1. (A) Expression of fibronectin mRNA in: (1) synovial fluid PMN; (2) peripheral blood PMN. RNA was hybridized with a fibronectin cDNA probe complementary to the fibronectin cell-binding domain. (B) Fluorogram of SDS-PAGE gelatin-binding secreted products of: (1) synovial fluid PMN; (2) peripheral blood PMN.

isolate as much as 1.5- to 2 $\times$ 10^9-PMN from the synovial fluid obtained from a single knee joint of a patient with rheumatoid or psoriatic arthritis. Given this high number of cells, PMN could significantly contribute to the total protein pool in synovial fluid. Studies using immunofluorescence and flow cytometry have identified an increased cytoplasmic content of fibronectin in synovial fluid PMN when compared to peripheral blood PMN.[6] It should be pointed out that in this disease, the joint is the site of an intense inflammatory reaction, whereas in peripheral blood, the cells remain relatively inactivated. In view of the high cytoplasmic levels of fibronectin observed in synovial fluid PMN, we decided to investigate the possibility that inflammatory activated PMN increase their *de novo* synthesis of fibronectin. We first looked at fibronectin mRNA levels in inflammatory activated vs. nonactivated PMN by using cells isolated from synovial fluid of patients with inflammatory joint disease (rheumatoid and psoriatic arthritis) as a source of inflammatory activated PMN.[8] We then compared our findings with those obtained using PMN from peripheral blood of the same patient. Fibronectin mRNA levels were measured by the dot blot procedure and Northern blotting. The fibronectin cDNA probe used for these hybridizations was specific for the cell binding domain of fibronectin and is conserved in all fibronectin molecules sequenced to date. The mRNA levels were found to be markedly increased in the inflammatory activated PMN when compared to peripheral blood PMN. Results of the hybridot analysis using cells isolated from a single patient are shown in Figure 1A. When studying a greater number of patients, the increase of fibronectin mRNA in synovial fluid PMN varied from patient to patient with a 2- to 12.7-fold range.[8] Northern hybridization analysis showed that the fibronectin mRNA size in PMN was close to that of fibroblast fibronectin mRNA (8.7 to 8.8 kb). We next studied actual production and secretion of the fibronectin molecule by activated and nonactivated PMN.[8,9] We performed *in vitro* metabolic labeling of the cells with ^{35}S-methionine, collected supernatants after a 2- to 16-h labeling period in the presence of 10% fetal calf serum, and submitted these supernatants to an affinity chromatography procedure using gelatin-Sepharose in order to isolate fibronectin. We then analyzed the eluate of this column on SDS-polyacrylamide gel electrophoresis followed by fluorography. A fluorogram from a representative patient is shown in Figure 1B. Two gelatin-binding bands were consistently seen on the fluorograms when synovial fluid PMN were studied and only one band with the peripheral blood PMN. By an immunoblotting procedure, the higher molecular weight band was shown to be fibronectin, whereas the lower band was found to react with a monospecific antibody to a recently described gelatinase.[19,20] In contrast to fibronectin, the level of production and secretion of this gelatinase remained constant in the activated and nonactivated cells. Fibronectin synthesis and secretion was markedly elevated in the inflammatory activated PMN when compared to the nonactivated peripheral blood PMN. Although no fibronectin band could be seen when analyzing the peripheral blood supernatants, a faint band could be obtained when using higher amounts of ^{35}S-methionine and greater numbers of cells. This observation is in agreement with the low levels of fibronectin mRNA observed in these cells (Figure 1A). On a speculative level, the small amounts of fibronectin produced in the peripheral blood PMN may play a crucial role in the differentiation of PMN in the bone marrow, especially since some recent work indicates that myelocytes synthesize fibronectin.[21] It can therefore be concluded that inflammatory activation of PMN leads to a relatively selective increase in the synthesis and secretion of fibronectin since another gelatin-binding product was not affected by the activation process. Furthermore, the synthesis of fibronectin is a result of activation of PMN *in vivo* since our RNA studies have been done on cells that have not been cultured *in vitro*.

We next wanted to analyze another model of *in vivo* inflammatory activation of PMN in rabbit. In studying this rabbit model, we had two specific purposes in mind. First, we wanted to investigate if our findings on protein synthesis by PMN in synovial fluids were specific for this inflammatory milieu or in contrast, represented a general response of PMN to inflammatory stimuli. Second, we wanted to find an animal model which we could use on a reliable basis to study protein synthesis in PMN in relation to inflammation. In this particular model, PMN are elicited in the peritoneal cavity of rabbits by the injection of glycogen. These elicited cells have been used extensively to investigate various aspects of PMN functions and have been found to behave as activated cells in several respects. This includes their ability to spontaneously and efficiently metabolize arachidonic acid through the 5-lipoxygenase pathway,[22] as well as an increased sensitivity to chemotactic factors, as reflected by a leftward shift in dose-response curves.[23] Furthermore, nonactivated PMN can be obtained from the peripheral blood of the same animal since glycogen is injected only in the peritoneal cavity. These conditions are therefore comparable to our studies using inflammatory activated synovial fluid PMN and nonactivated peripheral blood PMN isolated from the same patient. We observed that in terms of protein synthesis and secretion, there exists a remarkable similarity of response by PMN to inflammatory stimuli in both the human and animal models. In the two cases, fibronectin synthesis (at the protein and mRNA levels) and secretion by activated PMN were increased over nonactivated peripheral blood PMN whereas the gelatinase remained stable. These findings demonstrate that increased fibronectin synthesis and secretion by PMN is an event closely associated with the activation process of these cells. At the present time it is not known what kind of fibronectin (plasma, cellular, or other) is being synthesized by PMN in the synovial fluid. This information might shed light on the function of fibronectin in the synovial fluid. Finally, it is unlikely that fibronectin synthesis in synovial fluid PMN is a nonspecific by-product of PMN activation (such as gene leakiness) since we have recently observed that thrombospondin, another major extracellular matrix protein, is synthesized by synovial fluid PMN.[37]

III. FIBRONECTIN RECEPTORS ON POLYMORPHONUCLEAR NEUTROPHILS

It has been shown that fibronectin as well as several other adhesive proteins present in the extracellular matrix and in peripheral blood contain the tripeptide Arg-Gly-Asp (RGD) in their cell-binding domain.[24] The RGD sequences of each protein can be recognized by at least one member of a family of receptors called the integrins. The so-called fibronectin receptor belongs to this family (for review see Reference 25). Several recent observations suggest that PMN have fibronectin receptors.[1-5] Using immunofluorescence and flow cytometry, we have observed the presence of surface fibronectin on PMN.[6] This finding confirmed previous observations by another group of investigators using electron microscopy and immunofluorescence.[18] However, more direct evidence of fibronectin binding to the PMN surface was obtained by studies using fibronectin-coated microspheres labeled with fluorescence and analyzed by flow cytometry.[1] In these studies, fibronectin was found to bind to both monocytes and PMN. However, differences in the pattern of binding were observed. Whereas only a subpopulation of monocytes bound relatively high amounts of fibronectin, all PMN bound fibronectin but in considerably lower amounts. In independent studies by another group of investigators, the IIb-IIIa glycoprotein complex was identified on both monocytes and PMN.[5] This is a receptor which belongs to the integrin family of receptors.[25] It was originally isolated from platelets, and it can bind fibrinogen, fibronectin, von Willebrand factor, and vitronectin (for review see Reference 26). The IIb-IIIa complex is not identical but is homologous to the fibronectin receptor.[25,27] Using a monoclonal antibody to an epitope present on the IIb-IIIa receptor, this protein complex was immunoprecipitated from the surface of both monocytes and PMN.[5] An interesting observation made in this study was that only 12 to 33% of monocytes bound the fluorescein-labeled antireceptor antibody with an intensity varying from weak to strong, whereas nearly all PMN stained for the antibody but in a uniform weak pattern. The parallel between the IIIb-IIIa receptor findings and the previous observations using the fibronectin-coupled microspheres is striking and certainly suggests that the fibronectin binding activity on monocytes and PMN is at least in part mediated by the IIb-IIIa receptor complex.

Additional evidence for the presence of specific fibronectin binding sites on PMN came from studies investigating the effect of this protein on the phagocytic activity of PMN and monocytes. It was clearly shown that fibronectin significantly enhances the phagocytosis of erythrocytes opsonized with either IgG or C3b.[1] In these experiments, it was interesting to note that monocytes and PMN again respond differently to the effect of fibronectin. In the case of monocytes, fibronectin enhanced phagocytosis through an interaction with the cell surface without any prior activation of the cells. This is in contrast to PMN, which needed to be activated by an inflammatory molecule such as C5a before any phagocytic-enhancing effect of fibronectin could be seen. In spite of this, both PMN and monocytes were shown to bind to the same cell binding domain of the fibronectin molecule since monoclonal antifibronectin antibodies which inhibited the binding of fibronectin to monocytes also inhibited binding to PMN. It was subsequently shown that the RGD sequence of the fibronectin molecule is responsible for the binding activity of fibronectin to monocytes.[28] In view of the fact that the same monoclonal antibodies could inhibit binding to both monocytes and PMN, it was presumed that the RGD sequence is also responsible for fibronectin binding to PMN. This assumption is now strengthened by the observation that the IIb-IIIa receptor complex, an RGD-recognizing receptor, is present on both monocytes and PMN.[5]

IV. THE EFFECTS OF FIBRONECTIN ON PMN FUNCTIONS

A. PHAGOCYTOSIS

There can now remain little doubt that fibronectin enhances, at least *in vitro,* the phagocytic activity of PMN. Much of the evidence supporting the opsonic role of fibronectin for PMN has been summarized above. The mechanism by which fibronectin might play this role has not been elucidated. Based on the studies performed with monocytes, it has been suggested that fibronectin might activate complement receptors on monocytes.[1] In view of the similarities noted between the monocyte and PMN fibronectin receptors, it was suggested that fibronectin might exert its opsonic activity on PMN through a closely related mechanism. In the experimental system whereby monocytes were studied, it had been observed that erythrocytes opsonized by IgG are phagocytosed by monocytes in the absence of fibronectin, although fibronectin significantly enhanced phagocytosis.[1] C3b opsonized erythrocytes only bound to monocytes without any phagocytosis occurring. In the presence of fibronectin, efficient phagocytosis took place. Since it had been clearly shown that fibronectin binding to monocytes was essential for the phagocytosis enhancing effect of fibronectin to occur, it was suggested that the most likely mechanism of action of fibronectin involved some form of interaction with complement receptors. As previously stated, however, no enhancement of phagocytosis occurs with PMN in the presence of fibronectin with either IgG or C3b opsonized erythrocytes. However, when PMN are activated by chemotactic agents such as C5a and FMLP prior to incubation with the erythrocytes, fibronectin enhances significantly the phagocytosis of both IgG and C3b opsonized erythrocytes. Whatever the mechanisms responsible for this phenomenon, it does not involve an increase in the number of fibronectin receptors on PMN as determined by the binding of radiolabeled fibronectin before and after PMN activation. However, both C5a and FMLP are known to enhance the expression of complement receptors on PMN.[29] If fibronectin indeed exerts its opsonic activity through a complement receptor mechanism, making these receptors available in greater numbers for fibronectin interactions could certainly constitute a valid explanation for the effect of PMN activation on fibronectin-induced phagocytic activity.

A number of studies have also addressed the question of the opsonic activity of fibronectin on PMN for bacteria. Considerable controversy exists in this field of study.[30] Chapter 6 of this book is devoted to interactions between fibronectin and bacteria.

B. ADHERENCE

The role of fibronectin in promoting the adherence of PMN to substrata has been studied by several investigators. In most studies, evidence was obtained strongly suggesting that fibronectin plays an active role in mediating the adherence of PMN to endothelial cells, to gelatin, and collagen-coated surfaces, as well as to uncoated glass surfaces.[3,4,31] However, in one study, plasma fibronectin inhibited the adherence of PMN to glass and protein-coated surfaces.[32] Although the explanation for this conflicting finding was not elucidated, it should be noted that other investigators have provided evidence demonstrating that PMN-derived fibronectin has adherence-promoting effects while exogenous plasma fibronectin is devoid of this activity.[3] This would make physiological sense in view of the high concentration of fibronectin present in blood and the obvious necessity for nonactivated circulating PMN to be in a nonadherent state. In this same study, it was shown that PMN needed to be activated before the adherence-promoting effect of fibronectin could be observed. It was again found that PMN-derived fibronectin was likely to be the active molecule since activation of the PMN was accompanied by an increased release of fibronectin in the incubation media. It was finally shown by others that release products of activated PMN are able to alter fibronectin, rendering this protein a more adherent substrate for PMN.[31] The alteration of fibronectin by the PMN release products involved proteolytic cleavage of the fibronectin molecule by a cathepsin-D-like enzyme. A striking finding in these studies was that activated PMN release products were also able to increase the expression of fibronectin on the surface of endothelial cells. The mechanism by which this occurred was not elucidated, but it was proposed that altered fibronectin and PMN interactions are likely to be important in PMN adherence and injury to endothelial cells during inflammation. Furthermore, the fibronectin synthesized by PMN may activate specific gene expression in endothelial cells by receiving and integrating signals from PMN and possibly other cells. For example, macrophages are induced by fibronectin to synthesize GM-CSF and c-sis mRNA when cultured *in vitro*.[33]

V. CONCLUSION

Inflammatory stimuli lead to a dramatic increase in the number of PMN at inflammatory sites such as the synovial fluid of patients with synovitis, along with an increase in the synthesis and secretion of fibronectin. Another extracellular matrix protein, thrombospondin, is also synthesized and secreted in large amounts by inflammatory activated PMN. The extracellular matrix proteins, as well as their receptors, play a role in cell-cell and cell-matrix interactions. Fibronectin may also have an influence on gene expression in differentiated cells. Based on the studies reviewed here, it is tempting to conclude that fibronectin-PMN and fibronectin-PMN-endothelial interactions occur *in vivo* at sites of inflammation and lead to enhancement of important biological functions such as phagocytosis, migration, adherence, and stimulation of other cells. Since fibronectin receptor and various other integrins may influence the biological functions of fibronectin, it is important to characterize these receptors in activated and nonactivated PMN. Questions remain as to the type of fibronectin produced and secreted by activated PMN as well as the type of fibronectin receptors present on these activated cells. Does PMN secrete fibronectin in an inflammatory milieu directly or indirectly induce the over-stimulation of the host defense mechanisms? Alternatively, is the fibronectin secreted by PMN contributing to immunopathological reactions such as seen in rheumatoid arthritis where abnormally high amounts of fibronectin are present both in synovial tissue and fluid? The possibility that these events might occur is certainly suggested by studies showing that fibronectin secreted by PMN has adherence-promoting activity for PMN while plasma fibronectin does not. The possibility now exists that there is actually a series of fibronectin molecules produced by differences in splicing by various cell types, or even within one cell.[34-36] This mechanism could generate a whole range of closely related adhesive proteins with differing properties, each of which could be better suited to a specific normal or pathological event. Future studies will have to focus on the chemistry and biochemistry of PMN-secreted fibronectin on the one hand, and a detailed analysis of its biological properties on the other.

ACKNOWLEDGMENT

This work was supported in part by The Medical Research Council of Canada Grant MA-7604 and The Arthritis Society Grant 3-226 to A. D. Beaulieu.

REFERENCES

1. **Pommier, C. G., O'Shea, J., Chused, T., Yancey, K., Frank, M. M., Takahashi, T., and Brown, E. J.,** Studies on the fibronectin receptors of human peripheral blood leukocytes, *J. Exp. Med.*, 159, 137, 1984.
2. **Hosein, B. and Bianco, C.,** Monocyte receptors for fibronectin characterized by a monoclonal antibody that interferes with receptor activity, *J. Exp. Med.*, 162, 157, 1985.
3. **Marino, J. A., Pensky, J., Culp, L. A., and Spagnuolo, P. J.,** Fibronectin mediates chemotactic factor-stimulated neutrophil substrate adhesion, *J. Lab. Clin. Med.*, 105, 725, 1985.
4. **Harris, M. C., Levitt, J., Douglas, S. D., Gerdes, J. S., and Polin, R. A.,** Effect of fibronectin on adherence of neutrophils from newborn infants, *J. Clin. Microbiol.*, 21, 243, 1985.
5. **Burns, G. F., Cosgrove, L., Triglia, T., Beall, J. A., López, A. F., Werkmeister, J. A., Begley, C. G., Haddad, A. P., d'Apice, A. J. F., Vadas, M. A., and Cawley, J. C.,** The IIb-IIIa glycoprotein complex that mediates platelet aggregation is directly implicated in leukocyte adhesion, *Cell*, 45, 269, 1986.
6. **Beaulieu, A. D., Audette, M., Ménard, C., Parent, C., and Duval, M.,** Studies on fibronectin in inflammatory vs. non-inflammatory polymorphonuclear leukocytes of patients with rheumatoid arthritis. I. Immunofluorescent and flow cytometric analysis, *Clin. Exp. Immunol.*, 60, 339, 1985.
7. **Ménard, C., Beaulieu, A. D., Audette, M., Corbeil, J., and Latulippe, L.,** Studies on fibronectin in inflammatory vs. non-inflammatory polymorphonuclear leukocytes of patients with rheumatoid arthritis. II. Synthesis and release of fibronectin in vitro, *Clin. Exp. Immunol.*, 60, 347, 1985.

8. **La Fleur, M., Beaulieu, A. D., Kreis, C., and Poubelle, B.,** Fibronectin gene expression in polymorphonuclear leukocytes, *J. Biol. Chem.*, 262, 2111, 1987.
9. **Beaulieu, A. D., Lang, F., Belles-Isles, M., and Poubelle, P.,** Protein biosynthetic activity of polymorphonuclear leukocytes in inflammatory arthropathies. Increased synthesis and release of fibronectin, *J. Rheum.*, 14, 656, 1987.
10. **Tiku, K., Tiku, M. L., and Skosey, J. L.,** Interleukin 1 production by human polymorphonuclear neutrophils, *J. Immunol.*, 136, 3677, 1986.
11. **Eid, N. S., Kravath, R. E., and Lanks, K. W.,** Heat-shock protein synthesis by human polymorphonuclear cells, *J. Exp. Med.*, 165, 1448, 1987.
12. **Colotta, F., Wang, J. M., Polentarutti, N., and Mantovani, A.,** Expression of c-fos protooncogene in normal human peripheral blood granulocytes, *J. Exp. Med.*, 165, 1224, 1987.
13. **Blowers, L. E., Jayson, M. I., and Jasani, M. K.,** Effect of dexamethasone on polypeptides in polymorphonuclear leukocytes, *FEBS Lett.*, 181, 362, 1985.
14. **Vartio, T., Vaheri, A., Von Essen, R., Isomäki, H., and Stenman, S.,** Fibronectin in synovial fluid and tissue in rheumatoid arthritis, *Eur. J. Clin. Invest.*, 11, 207, 1981.
15. **Clemmensen, I. and Anderson, R. B.,** Different molecular forms of fibronectin in rheumatoid synovial fluid, *Arthritis Rheum.*, 25, 25, 1982.
16. **Mapp, P. I. and Revell, P. A.,** Fibronectin production by synovial intimal cells, *Rheumatol. Int.*, 5, 229, 1985.
17. **Alitalo, K., Hovi, T., and Vaheri, A.,** Fibronectin is produced by human macrophages, *J. Exp. Med.*, 151, 602, 1980.
18. **Hoffstein, S. T., Weissman, G., and Pearlstein E.,** Fibronectin is a component of the surface coat of human neutrophils, *J. Cell Sci.*, 50, 315, 1981.
19. **Vartio, T.,** Gelatin-degrading activity secreted by cultured macrophages from human blood, *Eur. J. Biochem.*, 152, 323, 1985.
20. **Vartio, T., Hedman, K., Jansson, S. E., and Hovi, T.,** The M_r 95,000 gelatin-binding protein in differentiated human macrophages and granulocytes, *Blood*, 65, 1175, 1985.
21. **Marino, J. A., Davis, A. H., and Spagnulo, P. J.,** Fibronectin is stored but not synthesized in mature human peripheral blood granulocytes, *Biochem. Biophys. Res. Commun.*, 146, 1132, 1987.
22. **Borgeat, P. and Samuelsson, B.,** Arachidonic acid metabolism in polymorphonuclear leukocytes: unstable intermediate in formation of dihydroxy acids, *Proc. Natl. Acad. Sci. U.S.A.*, 76, 3213, 1979.
23. **Becker, E. L., Showell, H. J., Naccache, P. H., and Sha'afi, R.,** The role and interrelationships of Ca^{++} and arachidonic acid metabolism in stimulated neutrophil functions, in *Biochemistry of the Acute Allergic Reactions*, Becker, E. L., Simon, A. S., and Austin, K. F., Eds., Alan R. Liss, New York, 1981, 257.
24. **Pierschbacher, M. D. and Ruoslahti, E.,** Cell attachment activity of fibronectin can be duplicated by small synthetic fragments of the molecule, *Nature (London)*, 309, 30, 1984.
25. **Ruoslahti, E. and Pierschbacher, M. D.,** New perspectives in cell adhesion: RGD and integrins, *Science*, 238, 491, 1987.
26. **Lawler, J.,** The structural and functional properties of thrombospondin, *Blood*, 67, 1197, 1986.
27. **Hynes, R. O.,** Integrins: a family of cell surface receptors, *Cell*, 48, 549, 1987.
28. **Wright, S. D. and Meyer, C.,** Fibronectin receptor of human macrophages recognizes the sequence Arg-Gly-Asp-Ser, *J. Exp. Med.*, 162, 762, 1985.
29. **Fearson, D. T. and Collins, L. A.,** Increased expression of C3b receptors on polymorphonuclear leukocytes induced by chemotactic factors and by purification procedures, *J. Immunol.*, 130, 370, 1983.
30. **Doran, J. E.,** A critical assessment of fibronectin's opsonic role for bacteria and microaggregates, *Vox. Sang.*, 45, 337, 1983.
31. **Vercellotti, G. M., McCarthy, J., Furcht, L. T., Jacob, H. S., and Moldow, C. F.,** Inflammed fibronectin: an altered fibronectin enhances neutrophil adhesion, *Blood*, 62, 1063, 1983.
32. **Brown, A. F.,** Neutrophil granulocytes: adhesion and locomotion on collagen substrata and in collagen matrices, *J. Cell Sci.*, 58, 455, 1983.
33. **Thorens, B., Mermod, J. J., and Vassalli, P.,** Phagocytosis and inflammatory stimuli induce GM-CSF mRNA in macrophages through posttranscriptional regulation, *Cell*, 48, 671, 1987.
34. **Schwarzbauer, J. E., Tamkun, J. W., Lemischka, I. R., and Hynes, R. O.,** Three different fibronectin mRNAs arise by alternative splicing within the coding region, *Cell*, 35, 421, 1983.
35. **Kornblihtt, A. R., Vibe-Pedersen, K., and Baralle, F. E.,** Human fibronectin: cell specific alternative mRNA splicing generates polypeptide chains differing in the number of internal repeats, *Nucleic Acids Res.*, 12, 5853, 1984.
36. **Paul, J. I., Schwarzbauer, J. E., Tamkum, J. W., and Hynes, R. O.,** Cell-type-specific fibronectin subunits generated by alternative splicing, *J. Biol. Chem.*, 261, 12258, 1986.
37. **Kreis, C., La Fleur, M., Menard, C., Paduin, R., and Beaulieu, A. D.,** Thrombospondin and fibronectin are synthesized by neutrophils in human inflammatory joint disease and in a rabbit model of *in vivo* neutrophil activation, *J. Immunol.*, in press.

Chapter 8

FIBRONECTIN AND MONOCYTE RECEPTOR FUNCTION

Celso Bianco

TABLE OF CONTENTS

I. INTRODUCTION

Functions and properties of cells of the mononuclear phagocyte system have been described in increasing detail during the past 20 years.[1] These cells are present both in the circulation (e.g., monocytes) and immobilized in tissue (e.g., Kupffer cells of the liver) and are responsible for: (1) the clearance of debris and dead cells,[2] bacteria,[3] and potentially infectious agents;[4] (2) resolution of inflammation; (3) tissue remodeling;[5] and (4) antigen presentation and production of modulators of the immune response,[1] growth regulators,[6-8]

It has been clearly demonstrated that several of these activities are modulated by specific, receptor-mediated interactions between mononuclear phagocytes and components of the extracellular matrix, notably fibronectin.[9-12] This matrix-associated protein appears to play a fundamental role in localizing mononuclear phagocytes at sites of inflammation, thereby potentiating the resolution of infections and promoting wound healing.

The structure of fibronectin has been reviewed in accompanying chapters. It is a 450,000-Da protein present both in the circulation and deposited in the extracellular matrix. Two types of fibronectin have been identified, namely, plasma and cellular.[13] Plasma fibronectin is synthesized in the liver[14] and is present in plasma at concentrations of about 300 μg/ml.[15] Cellular fibronectin is produced by a variety of cell types including fibroblasts and megakaryocytes.[16] Complete amino acid and DNA sequences of both forms of the protein have been reported.[17,18] The molecule is composed of two nonidentical polypeptide chains of 230 kDa and 220 kDa, respectively, which are disulfide linked near the carboxy-termini.[17]

For our purpose, the most relevant features of fibronectin are the several structural domains with binding specificities for macromolecules and for cells. Binding sites for fibrin, gelatin/collagen, glycosaminoglycans (e.g., heparin), actin, and DNA[19] mediate the incorporation of fibronectin into extracellular matrices[20] and into fibrin clots.[21] A specific cell binding domain containing the sequence Arg-Gly-Asp (RGD) is found in Fns of all species,[22] as well as in a variety of other molecules[23-24] and appears to be required for the specific binding of fibronectin to specific cell surface fibronectin receptors.[25] This sequence is located at residues 1493 to 1495 of human fibronectin.[26] Recent studies have described other putative cell binding domains in the C-terminus of the molecule.[27]

II. CELL SURFACE RECEPTORS FOR FIBRONECTIN

Cell surface receptors for fibronectin have been identified and/or isolated from a number of cell types including fibroblasts,[28] hepatocytes,[29] monocytes,[30,31] and monocytoid cell lines,[30-34] erythroblasts,[35] and other hematopoietic cells.[27,36-38] Specificity of fibroblast receptors for fibronectin has been demonstrated through inhibition of attachment of fibroblasts to fibronectin-containing surfaces[39-41] by proteolytic fragments containing the cell binding domain. Similarly, we have demonstrated the inhibition of binding of human peripheral blood monocytes and monocytoid cells to gelatin-fibronectin surfaces, both by receptor-specific monoclonal antibodies and by specific fibronectin fragments.[30,42] Fibronectin fragments and antireceptor monoclonals have also been used in receptor purification.[43]

Cell surface receptors for fibronectin are members of the integrin family of proteins.[44] The most extensively studied are the fibroblast Fn receptor and the IIb-IIIa platelet surface glycoproteins[45,46] for which the complete primary sequences are known. cDNA and genomic clones have been isolated from fibroblasts, and a complete amino acid sequence has been reported.[47] SDS-PAGE has revealed that these receptors are composed of two nonidentical, disulfide-linked polypeptides of molecular weights ~150 kDa, (α-chain) and 110 to 120 kDa (β-chain) respectively. The β-chains of members of the integrin family are quite similar, whereas there are significant differences in the molecular weights of the α-chains. Binding of fibronectin to these receptors requires divalent ions.[44] Structural studies of the fibronectin receptors of hematopoietic cells have been limited.

III. FIBRONECTIN RECEPTORS OF MONONUCLEAR PHAGOCYTES

The monocyte fibronectin receptor has been studied by us[30,42] and by others.[48] It is clearly related to other members of the integrin family, with respect to polypeptide chain composition and molecular weight, specificity for the RGD sequence, and ion dependency of binding.

Monoclonal antibodies which recognize the human monocyte receptor have been used for their characterization and isolation. They were raised by immunization of Balb/c mice with human peripheral blood monocytes and screened for the ability to inhibit monocyte attachment to gelatin-fibronectin surfaces. One antibody, A6F10, inhibited attachment of monocytes to fibronectin-gelatin surfaces in the dose-dependent manner and was used for further studies.[42] In immunofluorescence assays, this antibody interacted with human monocytes, very weakly with neutrophils, and not at all with lymphocytes, platelets, or erythrocytes. A6F10 also bound to guinea-pig-tissue macrophages. The antibody did not bind to primary human foreskin fibroblasts or to the human fibroblastic cell line WI38. Western blots of detergent lysates of monocytes developed with A6F10 revealed a ~110-kDa polypeptide which was susceptible to trypsin. Subsequent analysis supported the view that the polypeptide identified by A6F10 was a component of the fibronectin receptor complex. In retrospect, it is likely that this antibody recognized the β-chain of the receptor.

Other interactions of A6F10 with monocytoid cells pointed to its specificity for the fibronectin receptor:

1. The antibody blocked the binding of fibronectin-coated microspheres and fibronectin-induced EC3b ingestion by monocytes, activated neutrophils, and HL-60 cells, but did not block these effects in induced HL-60 cells;[48]
2. Antibody binding sites increased three-to-fivefold on monocytes and U937 cells exposed to gamma-interferon;[115]
3. A6F10 showed great specificity for binding to tissue macrophage receptors.[49]

Unfortunately, A6F10 was produced by a very unstable hybridoma cell line which has ceased to secrete the antibody.

The relatively high affinity of the fibronectin receptor of monocytes and U937 cells for an 80-kDa fragment of fibronectin containing the cell-binding site has permitted isolation and characterization of fibronectin-binding proteins from these cells by affinity chromatography. When octylglucoside lysates of surface-iodinated U937 cells were applied to columns containing the 80-kDa fragment covalently linked to Sepharose, a protein complex could be specifically eluted with the peptide GRGDSPC but not with the analog GRGESP. Nonreducing SDS-PAGE analysis revealed two proteins, of apparent M_r 152 and 125 kDa, in the eluate. Under reducing conditions, this complex appeared as a single band of 144 kDa. The 152/125 complex could also be eluted with EDTA. It also bound to WGA-Sepharose columns and could be eluted with *N*-acetylglucosamine. These results supported the contention that the fibronectin-binding proteins of U937 cells are typical of receptors of the integrin family in molecular weight, polypeptide chain composition, ionic dependence of binding, and glycoprotein nature.

Analyses of octylglucoside lysates of PMA-treated U937 cells and of monocytes revealed similar patterns by SDS-PAGE, although subtle differences in the molecular weights could be observed.[116]

IV. FIBRONECTIN BINDING TO MONONUCLEAR PHAGOCYTE RECEPTORS

Interactions between monocytes and intact fibronectin occur preferentially on a solid phase (e.g., extracellular matrix, clots, opsonized particles) rather than in solution. It is assumed that effective receptor-ligand interaction requires conformational changes of the fibronectin and/or multiple receptor-ligand interactions. This is especially relevant in light of the high concentrations of fibronectin normally found in plasma[15] and of the relatively low affinity and specificity for the binding of intact, soluble fibronectin to monocytes, fibroblasts, and other cells.[29,50] In contrast, substrate-bound fibronectin and some proteolytic fragments of fibronectin bind to receptor-bearing cells with relatively high affinity and specificity ($K_d < 10^{-6}$ *M*).

Fibronectin is highly susceptible to proteolysis by physiologically relevant enzymes (plasmin, thrombin, elastase, cathepsins) as well as by other enzymes (e.g., trypsin).[17] These fragments typically contain isolated macromolecule- or cell-binding domains. It has, therefore, been possible to test the effects of these isolated regions of fibronectin on cell function.[51-54] Fibronectin fragments are produced *in vivo* under certain pathological circumstances (e.g., chronic corneal ulcers,[55] disseminated intravascular coagulation (DIC),[56] rheumatoid synovial fluid,[57] lupus erythematosus.[58] One school of thought states that, in order for biological activities to be elicited, fibronectin must be modified (e.g., degraded to specific fragments *in situ*).[59,60] While still controversial, this view supports the contention that fibronectin fragments have potent biological activities.[61-66] Some of the activities promoted by large fibronectin fragments are similar to those elicited by the intact molecule, as for instance opsonization,[63] phagocytosis of gelatin-coated particles and bacteria,[9] enhancement of the activity of receptors for complement,[64] and binding to fibrin polymers.[65,66] However, chemotaxis of monocytes[61] and fibroblasts[67] mediated by fibronectin fragments seems to be enhanced relative to the intact molecule.

We have studied quantitative aspects of the binding of an 80-kDa tryptic fragment of fibronectin to the fibronectin receptor. The 80-kDa fragment contains the RGD sequence in the cell binding domain, but does not contain the gelatin/collagen domain. The position of this polypeptide in the fibronectin molecule was confirmed by N-terminal amino acid analysis. The sequence obtained was SD( )VPSPR( )LQF, indicating that the it begins at position 874.[26,117] Equilibrium binding of this fragment was studied in the monocytoid cell line U937. It was both time and concentration dependent. Scatchard analysis revealed 3 to 4 $\times$ 10^5 sites per cell with a K_d of 2×10^{-7} *M*.

Circulating levels of fibronectin are reduced in some patients suffering from severe sepsis following surgery, traumatic injury, burns, and coagulopathies.[68-71] In these instances, depletion of fibronectin may be due to enhanced utilization, for example by incorporation into clots,[21] matrices,[20] or by binding to debris.[72] According to some authors, depletion of fibronectin results in a general reduction of phagocytic activity.[73] These authors have suggested that repletion of fibronectin, given in the form of cryoprecipitate or purified protein, may normalize the activity of mononuclear phagocytes.[74] While animal studies on fibronectin replacement therapy point to the potential value of the procedure,[75] human clinical trials have reached contradictory conclusions.[76,77] Generally, it has been noted that the levels of fibronectin are not diagnostic of coagulopathies[78,79] and that replacement therapy has marginal effects on patient survival[80] or restoration of mononuclear phagocyte function,[76,77,81] but it is efficacious in topical applications to wounds.[82]

Limited information is available about the concentrations and/or nature of fibronectin fragments found *in vivo*. This is due in part to technical difficulties in measurement of degradation products in the presence of intact fibronectin by available immunoassays.[83] However, in the course of various pathological processes, proteases produced by endothelial cells and leukocytes at a site of injury could generate fibronectin degradation products. These fragments, if present in sufficient concentration, could compete for binding to fibronectin receptors of monocytes, prevent their interaction with intact fibronectin in the matrix or with specific chemotactic peptides, and reduce their ability to migrate to, and to localize at the inflammatory site, thereby delaying the resolution of the injury. This could be extended to opsonic activity, such that circulating fragments could interfere with binding of intact fibronectin to particles, and interfere with phagocytic recognition.[84]

V. ADDITIONAL FIBRONECTIN-BINDING SITES

The region of fibronectin primarily involved in the binding to cells is located toward the center of the molecule and contains the RGD sequence.[23] Recent work from several laboratories has indicated the existence of other potential cell-binding sites in the fibronectin molecule, located within the central cell binding domain[85] and within the C-terminal heparin-binding domain,[27,86] respectively. This latter site appears to mediate binding of fibronectin to melanoma cells and to erythroleukemia cells only at certain stages of differentiation.[87,88] Because monocytes are one of the few cell types which differentiate in the presence of fibronectin, it is possible that recognition of specific fibronectin domains is regulated during monocyte differentiation.

VI. MONOCYTES AND INFLAMMATION

The cascade of events which ensues following tissue injury involves monocytes and macrophages at several critical steps. Three phases can be clearly distinguished, namely, migration to the site of injury, attachment of monocytes to the extracellular matrix, and monocyte differentiation to macrophage. Circulating monocytes and neutrophils are recruited to the site of injury by the chemotactic activities of products of local proteolytic degradation of extracellular matrix components,[89] including fibrin[90] and fibronectin,[61] and by cleaved proteins of the complement system. Migrating cells first encounter a gell containing fibrin or fibrinogen, plasma and cellular fibronectin, clotting factors, platelets, and red cells.[91] This so called ''provisional matrix'' serves to localize mononuclear phagocytes at the site of injury[92] and provides support for subsequent monocyte differentiation and fibroblast migration as the wound resolves.[93] Matrix degradation products, generated by degradative enzymes of matrix-associated macrophages[94] and neutrophils,[95] recruit additional phagocytic cells from the circulation.[96]

Platelets can also serve as a ''substrate'' for monocyte attachment. Platelets contain substantial amounts of fibronectin in their alpha-granules. During activation, some of the fibronectin is released and some remains associated with the platelet surface.[97]

Highly purified resting platelets prepared by gel-filtration on Sepharose CL-2B columns[98] do not bind to human monocytes. Activation of the platelets by exposure to 0.05 IU thrombin per 10^8 platelets, an amount insufficient to produce visible aggregation, induces binding and rosette formation. When cells from different individuals were tested, between 400 and 1200 activated platelets bound per 100 monocytes. Adhesion of activated platelets could be inhibited by 1 μg/ml of A6F10 or by 12 μg/ml of Fab fragments of A6F10.

Platelet adhesion to monocytes could also be inhibited by a 70-kDa tryptic fragment of fibronectin which contained the cell-binding domain, but not the collagen-binding domain. Monoclonal antibodies to the cell-binding domain of fibronectin were quite effective, while antibodies to the N-terminus, to the elastase cleavage site, and to the C-terminus produced only minimal inhibition. Maximal inhibition could be reached with 20 μg/ml of antibodies to the midmolecule and to the cell-binding site of fibronectin.

These experiments strongly suggest that monocytes recognize and bind to activated platelets through the fibronectin receptor: the interaction is specifically inhibited by a monoclonal antibody to the receptor, by fibronectin fragments containing the RGD sequence, and by monoclonals to the cell binding domain of fibronectin. They also support the hypothesis that fibronectin fragments may interfere with monocyte function. In this case, fragments could prevent the adhesion of monocytes to a clot and delay resolution of the thrombus.[118]

VII. MONOCYTE DIFFERENTIATION

Following association with the matrix, monocytes differentiate into macrophages.[99] *In vitro* exposure of peripheral blood monocytes to fibronectin-coated surfaces, for example, results in increased phagocytic activity,[100,101] the production of neutral serine proteases,[100] the release of lysosomal[102] and other degradative enzymes,[94,100,103] increased activity of Fc and CR1 and CR3 receptors,[104] and production of fibroblast growth-promoting factors.[7] Each of these activities is associated with specific phases of the resolution of the inflammation or healing of the wound. While the role of fibronectin in differentiation is not unique to monocytes (e.g., embryonic neural crest cells differentiate upon exposure to ECM[105]) it is of interest to note that: (1) erythroblasts[35] and other hemopoietic cells apparently lose their capacity to interact with the matrix as differentiation proceeds; and (2) in most other circumstances (e.g., cell migration, haptotaxis, dissemination of metastatic cells), interaction of cells with extracellular matrices does not result in cell differentiation. Some authors have indicated that fibronectin is an effective opsonin for bacteria and other infectious agents.[3,4] Fibronectin binds to bacteria, notably *Staphylococcus aureus,*[106] through specific sites at the N-terminus of the molecule.[106] Fibronectin, immobilized on the surface of the bacterium, can then interact with cell surface receptors for fibronectin on monocytes or macrophage, thereby potentiating phagocytosis of the particle. A similar mechanism might be envisioned for the clearance of other materials at the site of inflammation. Further, it has been suggested that the ability of the organism to combat bacterial infection might be related to levels of circulating fibronectin.[107] While this evidence suggests that fibronectin is an effective opsonin,[76] recent studies indicate that fibronectin promotes attachment, but does not promote uptake of opsonized particles such as bacteria or coated beads.[108]

Monocytes cultivated for 24 h on fibronectin-collagen surfaces acquire enhanced (>8-fold) ability to secrete plasminogen activator and elastase. This does not occur when monocytes are cultivated in the absence of fibronectin. Secretion requires a phagocytic stimulus (e.g., latex beads).[109]

VIII. CONCLUSIONS

The previous pages describe some of the evidence for the importance of fibronectin in the modulation of monocyte and macrophage function. Fibronectin is a substrate for monocyte attachment at sites of injury and inflammation. In addition, fibronectin provides a signal for monocyte differentiation.

It is also becoming clear that proteolytic fragments of fibronectin may play a role in mononuclear phagocyte function. For instance, proteolytic fragments of fibronectin containing the cell-binding domain, but lacking the collagen-binding domain, bind to the receptor with higher affinity than native fibronectin and prevent attachment of monocytes to fibronectin-containing substrates. It is possible that proteolysis of fibronectin occurring in some pathological conditions leads to the release of fragments with only some of the binding domains of the molecule. These incomplete molecules may bind to circulating monocytes and, as a result, prevent their localization at sites of injury. These fragments could also modify or interfere with monocyte differentiation or other monocyte functions. Large amounts of circulating fibronectin fragments could explain the infections, the impaired recanalization of vessels obstructed by thrombi, and the delayed wound repair sometimes observed in patients with major trauma and certain clotting disorders.

The resolution of inflammation and wound repair is associated with the remodeling of the extracellular matrix.[110] Neutral proteases (plasminogen activator, elastase, cathepsin D, collagenases) are in part responsible for this remodeling. The role of serine proteases is supported by various observations: mononuclear phagocytes isolated from inflammatory exudates produce and secrete plasminogen activator,[100,111] inflammatory agents stimulate[112] and antiinflammatory glucocorticoids depress production of this enzyme.[113] We have previously demonstrated that monocytes plated on fibronectin secrete neutral serine proteases, such as plasminogen activator (PA) and elastase[100] upon phagocytic stimulation. It has also been shown that plasmin-derived fragments of fibronectin reduce erythrophagocytosis by macrophages.[114]

Based on the indications above, we anticipate that fibronectin fragments which do not contain the gelatin domain might be the most effective in preventing the localization of monocytes at inflammatory sites by competing efficiently for access to the monocyte fibronectin receptor. Thus, large fragments (e.g., the ~200-kDa fragment generated by limited proteolysis by plasmin) may promote attachment because all of the relevant binding domains are present. The effectiveness of fragments to inhibit monocyte attachment *in vivo,* and, therefore, their ability to modulate the effects of mononuclear phagocytes on the inflammatory response may be dictated by the extent of proteolysis. Indeed, both the presence of fibronectin fragments[78] and some potential biological activities[52] have been documented.

One might speculate that confirmation of the hypothesis suggesting a deleterious role for fibronectin fragments in certain pathological studies would result in better understanding of the impaired wound healing and enhanced infection observed in trauma and burn patients. Diagnostic assays able to distinguish desirable and undesirable forms of circulating fibronectin could be developed. The therapeutic approach to these patients would be changed from attempts to replace circulating fibronectin to procedures for removal of circulating fibronectin fragments that interfere with mononuclear phagocyte function.

REFERENCES

1. **van Furth, R.,** *Mononuclear Phagocytes: Characteristics, Physiology and Function,* M. Nijhuis Publishers, Dordrecht, 1985.
2. **Saba, T. M., Blumenstock, F. A., Weber, F. A., and Kaplan, J. E.,** *Ann. N.Y. Acad. Sci.,* 312, 43, 1978.
3. **Wright, S. D., Craigmyle, L. S., and Silverstein, S. D.,** *J. Exp. Med.,* 158, 1338, 1983.
4. **Kuusela, P.,** *Nature (London),* 276, 718, 1978.
5. **Peterson, K. M., Baseman, J. B., and Alderete, J. F.,** *J. Exp. Med.,* 157, 1958, 1983.
6. **Werb, Z., Banda, M. J., and Jones, P. A.,** *J. Exp. Med.,* 1980.
7. **Schmidt, J. A., Mizel, S. B., Cohen, D., and Green, I.,** *J. Immunol.,* 128, 2177, 1982.
8. **Martin, B. M., Gimbrone, M. A., Jr., Unanue, E. R., and Cotron, R. S.,** *J. Immunol.,* 126, 1510, 1981.
9. **Wright, S. D., Licht, M. R., Craigmyle, L. S., and Silverstein, S. C.,** *J. Cell Biol.,* 99, 336, 1984.
10. **Norris, D. A., Clark, R. A. F., Swigart, L. M., Huff, J. C., Weston, W. L., and Howell, S. E.,** *J. Immunol.,* 129, 1612, 1982.
11. **Remold, H. G., Shaw, J. E., and David, J. R.,** *Cell Immunol.,* 58, 175, 1981.
12. **Pommier, C. G., Inada, S., Fries, L. F., Takahashi, T., Frank, M. M., and Brown, E. J.,** *J. Exp. Med.,* 157, 1844, 1983.
13. **Brown, E. J.,** *J. Leuk. Biol.,* 39, 579, 1986.
14. **Owens, M. R. and Cimino, C. D.,** *Blood,* 59, 1305, 1982.
15. **Mosesson, M. W. and Umfleet, R. A.,** *J. Biol. Chem.,* 245, 5728, 1970.
16. **Hynes, R. O.,** *Proc. Natl. Acad. Sci. U.S.A.,* 70, 3170, 1973.
17. **Petersen, T. E. and Skorstengaard, K.,** in *Plasma Fibronectin,* McDonagh, J., Ed., Marcel Dekker, New York, 1985, 7.
18. **Hirano, H., Yamada, Y., Sullivan, M., B de Crombrugghe, A., Pastan, I., and Yamada, K. M.,** *Proc. Natl. Acad. Sci. U.S.A.,* 80, 46, 1983.

19. **Sekiguchi, K. and Hakamori, S.,** *J. Biol. Chem.*, 258, 3967, 1983.
20. **Hedman, K., Johansson, S., Vartio, T., Kjellan, L., Vaheri, A., and Hook, M.,** *Cell*, 28, 663, 1982.
21. **McDonagh, J., Hada, M., and Kaminski, M.,** in *Plasma Fibronectin*, McDonagh, J., Ed., Marcel Dekker, New York, 1985, 121.
22. **Pierschbacher, M. D., Ruoslahti, E., Sundelin, J., Lind, P., and Peterson, P. A.,** *J. Biol. Chem.*, 257, 9593, 1982.
23. **Ruoslahti, E. and Pierschbacher, M. D.,** *Cell*, 44, 517, 1986.
24. **Haverstick, D. M., Cowan, J. F., Yamada, K. M., and Santoro, S. A.,** *Blood*, 66, 946, 1985.
25. **Yamada, K. M.,** *Ann. Rev. Biochem.*, 52, 761, 1983.
26. **Kornblihtt, A. R., Umezawa, K., Vibe-Petersen, K., and Barelle, F.,** *EMBO J.*, 4, 1755, 1985.
27. **Bernardi, P., Patel, V. P., and Lodish, H. F.,** *J. Cell. Biol.*, 105, 489, 1987.
28. **Akiyama, S. K., Yamada, S. S., and Yamada, K. M.,** *J. Cell Biol.*, 102, 442, 1986.
29. **Johansson, S., Foresberg, E., and Lundgren, B.,** *J. Biol. Chem.*, 262, 7879, 1987.
30. **Bevilaqua, M. P., Amrani, D., Mosesson, M. W., and Bianco, C.,** *J. Exp. Med.*, 153, 42, 1981.
31. **Wright, S. D. and Meyer, B. C.,** *J. Exp. Med.*, 162, 762, 1985.
32. **Pommier, C., O'Shea, J., Chused, T., Takahashi, T., Ochoa, M., Brown, E., and Bianco, C.,** *Clin. Res.*, 32, 2, 1984.
33. **Law, S. K., Gagnon, J., Hildreth, J. E., Wells, C. E., Willis, A. C., and Wong, A. J.,** *EMBO J.*, 6, 915, 1987.
34. **Giancotti, F. G., Comoglio, P. M., and Tarone, G.,** *J. Cell Biol.*, 103, 429, 1986.
35. **Patel, V. P. and Lodish, H. F.,** *J. Cell Biol.*, 102, 449, 1986.
36. **Gardner, J. M. and Hynes, R. O.,** *Cell*, 42, 439, 1985.
37. **Liao, N. S., St. John, J., Du, Z. J., and Cheung, H. T.,** *Exp. Cell. Res.*, 171, 306, 1987.
38. **Cardarelli, P. M. and Pierschbacher, M. D.,** *J. Cell Biol.*, 105, 499, 1987.
39. **Gensberg, M. H., Pierschbacher, M. D., Ruoslahti, E., Marguerie, G., and Plow, E. J.,** *Biol. Chem.*, 260, 3931, 1985.
40. **Yamada, K. M. and Kennedy, D. W.,** *J. Cell Biol.*, 99, 29, 1984.
41. **Pierschbacher, M. D. and Ruoslahti, E.,** *Proc. Natl. Acad. Sci. U.S.A.*, 81, 5985, 1984.
42. **Hosein, B. and Bianco, C.,** *J. Exp. Med.*, 162, 157, 1985.

42a. **Pommier, C., O'Shea, J., Chused, T., et al.,** *Blood*, 64, 858, 1984.

43. **Pytela, R., Pierschbacher, M. D., and Ruoslahti, E.,** *Cell*, 40, 191, 1985.
44. **Hynes, R. O.,** *Cell*, 48, 549, 1987.
45. **Poncz, M., Eisman, R., Heidenreich, R., et al.,** *J. Biol. Chem.*, 262, 8467, 1987.
46. **Fitzgerald, L. A., Steiner, B., Rall, S. C., Lo, S.-S., and Phillips, D. R.,** *J. Biol. Chem.*, 262, 3936, 1987.
47. **Argraves, W. S., Suzuki, S., Arai, H., Thompson, K., Pierschbacher, M. D., and Ruoslahti, E.,** *J. Cell Biol.*, 105, 1183, 1987.
48. **Brown, E. J. and Goodwin, J. L.,** *J. Exp. Med.*, 167, 777, 1988.
49. **Kradin, R. L., Zhu, Y., Hales, C. A., Bianco, C., and Colvin, R. B.,** Response of pulmonary macrophages to hyperoxic pulmonary injury. Acquisition of surface fibronectin and fibrinogen and enhanced expression of a fibronectin receptor, *Am. J. Pathol.*, 125, 349, 1986.
50. **Molnar, J., Hoekstra, S., Ku, C. S. L., and Van Alten, P.,** *J. Cell. Physiol.*, 131, 374, 1987.
51. **De Petro, G., Barlati, S., Vartio, T., and Vaheri, A.,** *Proc. Natl. Acad. Sci. U.S.A.*, 78, 4965, 1981.
52. **Ruoslahti, E.,** *Can. Metas. Rev.*, 3, 43, 1988.
53. **Humphries, M. J. and Ayad, S. R.,** *Nature (London)*, 305, 811, 1983.
54. **Ruoslahti, E., Hayman, E. G., Engvall, E., Cothran, W. C., and Butler, W. T.,** *J. Biol. Chem.*, 256, 7277, 1981.
55. **Berman, M., Manseau, E., and Aiken, D.,** *Invest. Opthal. Vis. Sci.*, 24, 1358, 1983.
56. **Mosher, D. F.,** *Annu. Rev. Med.*, 35, 561, 1984.
57. **Carsons, S., Lavietes, B. B., Diamond, H. S., and Kinney, S. G.,** *Arthritis Rheum.*, 28, 601, 1985.
58. **Carsons, S.,** *J. Rheum.*, 14, 5, 1987.
59. **Vartio, T., Vaheri, A., De Petro, G., and Barlati, S.,** *Inv. Metas.*, 3, 125, 1983.
60. **MacDonald, J. A., Senior, R. M., Griffin, G. L., Broekelmann, T. J., and Prevedel, P.,** *J. Cell Biol.*, 95, 123a, 1982.
61. **Postlethwaite, A. E., Keski-Oja, J., Balian G., and Kang, A. H.,** *J. Exp. Med.*, 153, 494, 1981.
62. **Colvin, R. B. and Kradin, R. L.,** *Surv. Synth. Pathol. Res.*, 2, 10, 1983.
63. **Czop, J. K. and Austen, H. F.,** *J. Immunol.*, 129, 2678, 1982.
64. **Czop, J. K., Kadish, J. L., and Austen, H. F.,** *Proc. Natl. Acad. Sci. U.S.A.*, 78, 3649, 1981.
65. **Hormann, H., Richter, H., and Jelinic, V.,** *Thromb. Res.*, 46, 39, 1987.
66. **Hormann, H., Richter, H., and Jelinic, V.,** *Thromb. Res.*, 39, 183, 1985.
67. **Seppa, H. E. J., Yamada, K. M., Seppa, S. T., Silver, M. H., Leinman, H. K., and Schiffman, E.,** *Cell. Biol. Int. Rep.*, 5, 813, 1981.

68. **Cembrowski, G. S., Griffin, J. H., and Mosher, D. F.,** *Arch. Int. Med.*, 146, 1997, 1986.
69. **Hesselvik, F., Brodin, B., Blomback, M., Cedergren, B., Lieden, G., and Maller, R.,** *Scand. J. Clin. Lab. Invest.*, 178 (Suppl.), 67, 1985.
70. **Clemmensen, I.,** *Haematologica*, 17, 101, 1984.
71. **Kaplan, J. E. and Saba, T. M.,** *Am. J. Physiol.*, 234, 323, 1978.
72. **Grinnel, F., Head, J. R., and Hoffpauier, J.,** *J. Cell Biol.*, 94, 597, 1982.
73. **Saba, T. M. and Jaffe, E.,** *Am. J. Med.*, 68, 577, 1980.
74. **Rubli, E., Brussard, S., Frei, E., Lundsgaard-Hansen, P., and Pappova, E.,** *Ann. Surg.*, 197, 310, 1983.
75. **Coliter, L. F., Saba, T. M., Lewis, E., and Vincent, P. A.,** *J. Appl. Physiol.*, 63, 623, 1987.
76. **van de Water, L., III,** in McDonagh, J., New York: Marcel Dekker, New York, 1985, 175.
77. **Saba, T. M., Blumenstock, F. A., Scovill, W. A., and Bernard, H.,** *Science*, 201, 622, 1978.
78. **Cembrowski, G. S. and Mosher, D. F.,** *Thromb. Res.*, 36, 437, 1984.
79. **Boughton, B. J.,** *Cell Biochem. Funct.*, 3, 79, 1985.
80. **Hesselvik, F., Brodin, B., Carlsson, C., Cedergren, B., Jorfeldt, L., and Lieden, G.,** *Crit. Care Med.*, 15, 475, 1987.
81. **Seghieri, G., Bartolomei, G., and De Giorgio, L. A.,** *Diabet. Metabol.*, 12, 186, 1986.
82. **Nishida, T., Nakagawa, S., and Manabe, R.,** *Opthalmologica*, 92, 213, 1985.
83. **Bykowska, K., Wegzynowicz, Z., Lopacuik, S., and Kopek, M.,** *Thromb. Hemostasis*, 53, 377, 1985.
84. **Brown, R. A.,** *Lancet*, 2, 1058, 1983.
85. **Humphries, M. J., Obara, M., Kang, M. S., Komoriya, A., Olden, K., and Yamada, K. M.,** *J. Cell Biol.*, 105, 90a, 1987.
86. **Humphries, M. J., Komoriya, A., Akiyama, S. K., Olden, K., and Yamada, K. M.,** *J. Biol. Chem.*, 262, 6886, 1987.
87. **Patel, V. P. and Lodish, H. F.,** *J. Cell Biol.*, 102, 449, 1986.
88. **Patel, V. P. and Lodish, H. F.,** *Science*, 224, 996, 1984.
89. **Postlethwaite, A. E., Seyergaard, J. M., and Kang, A. H.,** *Proc. Natl. Acad. Sci. U.S.A.*, 75, 871, 1978.
90. **Kay, A. B., Pepper, D. S., and McKenzie, R.,** *Br. J. Hematol.*, 27, 669, 1988.
91. **Clark, R. A. F., Lanigan, J. M., Monceau, E., DellaPelle, P., Dvorak, H. F., and Colvin, R. B.,** *J. Invest. Dermatol.*, 1988.
92. **Clark, R. A. F., Dvorak, H. F., and Colvin, R. B.,** *J. Immunol.*, 126, 787, 1981.
93. **Hynes, R. O. and Yamada, K. M.,** *J. Cell Biol.*, 95, 369, 1982.
94. **Werb, Z. and Gordon, S.,** *J. Exp. Med.*, 139, 834, 1975.
95. **Ciano, P. S., Colvin, R. B., Suenram, C. A., and Schwartz, C. J.,** *Exp. Mol. Pathol.*, 46, 266, 1987.
96. **Clark, R. A. F. and Colvin, R. B.,** in *Plasma Fibronectin*, McDonagh, J., Ed., Marcel Dekker, New York, 1985, 197.
97. **Packham, M. A. and Mustard, J. F.,** *Prog. Hemostasis Thromb.*, 7, 211,
98. **Groscurth, P., Cheng, S., Vollenweider, I., and von Felter, A.,** *Acta. Hematol. (Basel)*, 77, 150, 1987.
99. **Kelley, J. L., Rozek, M. M., Suenram, C. A., and Schwartz, C. J.,** *Exp. Mol. Pathol.*, 46, 266, 1987.
100. **Bianco, C.,** *Ann. N.Y. Acad. Sci.* 408, 602, 1983.
101. **Hoffman, H.,** *Blut*, 51, 307, 1985.
102. **Schorlemmer, M. U., Davies, P., Hylton, W., Gusig, M., and Allison, A. C.,** *Br. J. Exp. Pathol.*, 58, 315, 1977.
103. **Mainardi, C. L.,** *Biochim. Biophys. Acta*, 805, 137, 1984.
104. **Wright, S. D., Craigmyle, L. S., and Silverstein, S. C.,** *J. Exp. Med.*, 1, 1338, 1983.
105. **Thiery, J.-P, Duband, J. L., Rocher, S., and Yamada, K. M.,** *Prog. Clin. Biol. Res.*, 217, 155, 1986.
106. **Mosher, D. F. and Proctor, R. A.,** *Science*, 209, 927, 1980.
107. **Blumenstock, F., Saba, T. M., Weber, P., and Laffin, R.,** *J. Biol. Chem.*, 4287, 1978.
108. **Dorna, J. E., Mansbogen, A. K., and Reese, A. C.,** *J. Reticuloendotheliol. Soc.*, 27, 471, 1980.
109. **Bianco, C.,** Fibrin, fibronectin and macrophages, in *Molecular Biology of Fibrinogen and Fibrin, Ann. N.Y. Acad. Sci.*, 408, 602, 1983.
110. **Jones, P. A. and Werb, Z.,** *J. Exp. Med.*, 152, 1527, 1980.
111. **Unkeless, J. C., Gordon, S., and Reich, E.,** *J. Exp. Med.*, 139, 834, 1974.
112. **Vassalli, J. D. and Reich, E.,** *J. Exp. Med.*, 145, 429, 1977.
113. **Vassalli, J. D., Hamilton, J., and Reich, E.,** *Cell*, 8, 271, 1976.
114. **Ehrlich, M. I., Kriskell, J. S., Blumenstock, F. A., and Kaplan, J. E.,** *J. Lab. Clin. Med.*, 98, 263, 1981.
115. **Hosein, B., Brock, Valinsky, J., and Biano, C.,** unpublished.
116. **Garcia-Pardo, A., Ferreira, O., Valinsky, J., and Bianco, C.,** submitted.
117. **Garcia-Pardo, A., Valinsky, J., and Bianco, C. J.,** manuscript in preparation.
118. **Strauss, D. and Bianco, C.,** in preparation.

Chapter 9

FIBRONECTIN AND PLATELET FUNCTION

Mark H. Ginsberg and Edward F. Plow

TABLE OF CONTENTS

I. INTRODUCTION

A. ADHESIVE REACTIONS IN PLATELET FUNCTION

Platelets are anucleate 3-μm diameter cell fragments which circulate at a concentration of 3×10^8 cells per milliliter in blood. The primary function of platelets is in the hemostatic response, i.e., they play an essential role in the arrest of bleeding. Platelets perform this function by adhering to and spreading on subendothelium and connective tissue exposed as a consequence of vessel injury and by adhering to each other for form aggregates which may occlude large openings in injured blood vessels. These basic reactions — aggregation, adhesion, and spreading along surfaces — appear to be mediated by at least four distinct large glycoproteins: fibronectin (Fn), fibrinogen (Fg), thrombospondin (TSP), and von Wilebrand factor (vWF). Although this chapter will focus on the role of Fn in platelet function and the possible value of the platelet system as a model for understanding the binding and processing of Fn by other cells, we will also highlight similarities and differences between Fn and the other three members of the "Big Four".

B. CHEMICAL AND FUNCTIONAL PROPERTIES OF THE "BIG FOUR"

These four large glycoproteins share a variety of chemical and functional similarities. In addition to being large (M_r = 340,000 to several million) and glycosylated, they each possess intramolecular symmetry. Such symmetry permits a single molecule to exhibit bridging functions between other molecules, cells, or between cells and substrata. Further, all are present in platelets; at least three (Fn, TSP, vWF) are made by endothelial cells[1-3] and are incorporated into the matrix, and three (Fn, vWF, Fg) circulate at substantial concentrations in plasma (Table 1).

To participate in adhesive reactions, a protein must bind to the cell surface. Each of these four proteins binds to platelets[4-7] (among others). To bridge cells to substrata, the protein must also interact with the substratum. In blood vessels, relevant substrata may be connective tissue or subendothelium containing collagen, proteoglycans, and noncollagenous proteins or fibrin clots. Certain members of this quartet bind to each of these components. For example, Fg, Fn,[8] TSP,[9] and vWF[10] may bind to fibrin or Fg. It is therefore clear that each of these proteins has chemical and functional properties to serve a role in platelet adhesive function. In this review, we shall consider the following issues concerning Fn and platelets: (1) the evidence for a role for Fn in platelet function; (2) the localization of pools of Fn involved in platelet function; (3) the processing of the intraplatelet Fn pool; (4) the identification of and properties of the platelet-binding site(s) for Fn and how occupancy of such sites may influence cellular functions.

II. ROLE OF Fn IN PLATELET FUNCTION

A. ROLE IN ADHESION TO SUBSTRATA

Inasmuch as Fn serves as an attachment protein for several cell types, binds to platelets, and binds to constituents of the subendothelium and connective tissues, it is a likely candidate to be a platelet attachment protein. Nevertheless, as discussed above, platelets have many potential attachment proteins to choose among. In addition, because platelets may adhere to certain substrata in the absence of any exogenous adhesive protein,[11] the role of endogenous stores of these proteins must be carefully considered. Finally, adhesion as measured by the deposition of cells per unit area of substratum is dependent on shear rate[12] and because blood vessel diameters and blood flow rates vary, no single shear rate can be selected as being most representative. Thus, measurement of effects of Fn on platelet adhesion is a difficult task. Variable results have been reported on the effect of plasma Fn on platelet adhesion to collagenous substrata. Two groups found little effect of plasma Fn on the number

TABLE 1
Large Adhesive Platelet Glycoproteins

Protein	M.W. ($\times 10^{-3}$)	Platelet content (per 10^9 cells)	Plasma content (per ml)
Fibrinogen	340	50 μg[a]	300 μg[a]
Fibronectin	≥ 450	3 μg[a]	300 μg[a]
von Willebrand factor	> 10^3	0.25 U[b]	1 U[b]
Thrombospondin	450	20 μg[a]	20 ng[a]

[a] Based on our measurements in radioimmunoassay and those of others.
[b] Based on Nachman, R. L. and Jaffe, E. A., *J. Exp. Med.*, 141, 1101, 1975.

of platelets adherent to collagenous substrata,[13,14] while a third reported substantial enhancement.[15] A fourth investigator reported that anti-Fn slightly inhibited platelet adhesion to fibrillar collagen in a filtration assay.[16] These potential controversies obviously may relate to the previous discussion and to variable contributions of endogenous secreted adhesive proteins and aggregation in the different assays used. Nevertheless, there is clear agreement[13-15] that plasma Fn promotes platelet spreading on collagenous surfaces. Inasmuch as such spreading probably causes an increase in the area of cell-substratum contact, Fn is likely to promote increased strength of cell attachment. This point has been directly established by the work of Houdijk and co-workers, who found that plasma or matrix Fn play a crucial role in platelet adhesion in flowing systems.[17] The same group of investigators have also directly demonstrated a similar role could be played by matrix Fn in the subendothelium.[18]

B. ROLE OF Fn IN PLATELET AGGREGATION

Platelets require a plasma cofactor in order to aggregate in response to ADP. Fn does not serve this function.[19] In addition, platelet aggregation in response to ADP or thrombin proceeds normally in Fn-depleted plasma.[20] Thus, plasma Fn does not appear to play an important role in aggregation responses to ADP or thrombin. Hammerschmidt et al.[21] reported a patient with Ehlers-Danlos Syndrome whose collagen-induced platelet aggregation was abnormal: the abnormality could be corrected by addition of purified plasma Fn back to the patient's plasma. This observation is difficult to reconcile with data indicating that collagen-induced platelet aggregation is normal[20] or increased[22] in Fn-depleted plasma. Thus, present experimental evidence does not make a compelling case for a role for plasma Fn in aggregation. The possibility that secreted platelet Fn plays a contributory role in aggregation among secreting platelets cannot be excluded, as suggested by the finding that a monoclonal anti-Fn antibody inhibits platelet aggregation.[23]

III. LOCALIZATION OF Fn

As is noted elsewhere in this volume, Fn is a constituent of connective tissue matrices, is a plasma protein (concentration ~ 600 nM), and is synthesized by cultured endothelial cells.[1] TSP is made by endothelial cells[2] and fibroblasts.[24] vWF appears to be made by megakaryocytes[25] and endothelial cells[3] and is a plasma protein. Each of these three proteins is also found in association with cell matrices.[24,26] In contrast, Fg is made primarily in the liver and is a plasma protein which does not appear to be a normal matrix constituent. In this section we will focus on the evidence for platelet pools of each of these proteins and their localization in resting platelets.

A. PLATELET Fn

The presence of Fn antigen in platelets was first suggested by the observation of a precipitin line when a platelet extract was reacted with an antiserum to "cold-insoluble globulin".[27] Subsequently, platelet Fn antigen was detected by radioimmunoassay[28] and electroimmunoassay,[19] and levels of ~3 μg per 10^9 cells were found by both methods. Since platelet Fn antigen was immunochemically identical to the plasma form, and approximately 300 μg/ml are present in normal human plasma, it was important to exclude plasma contamination as the source of the platelet antigen. This was done by use of exogenous radiolabeled plasma Fn as a control for plasma contamination of suspensions of washed platelets and direct immunofluorescent demonstration of the platelet Fn antigen. The latter experiment also served to exclude the possibility that the measured Fn antigen was associated with a contaminating cell type.[28] It should be noted that platelet Fn antigen also shares the property of binding to gelatin[28] and the work of Paul et al.[29] suggests that in the rat, platelet Fn is enriched in extra domain (ED) containing (cellular) Fn.

Although initial workers proposed that platelet Fn might be a cell surface molecule,[30] later studies bring this concept into question. Cell surface straining for Fn has not been observed consistently in resting platelets by immunofluorescence.[13,31] Second, lactoperoxidase catalyzed iodination of intact platelets does not result in labeling of a protein of the size of Fn.[6] Third, on subcellular fractionation, the bulk of platelet Fn antigen associates with alpha-granule rather than membrane fractions.[19] Thus, it appears that the bulk of platelet Fn antigen is not on the cell surface. Moreover (see below), since at least two mechanisms exist for the appearance of Fn on the surface of activated platelets, the finding of Fn on the surface of resting cells may be due to low level activation. Indeed, we have found that the binding of radiolabeled affinity-purified f(ab)′2 anti-Fn to "resting" platelets could be completely inhibited when the cells were washed in the presence of an inhibitor of their activation.[32] This caveat also pertains to the detection of other members of the "Big Four" on the surface of resting cells.

In contrast to the lack of convincing data concerning the presence of Fn on the surface of resting cells, there is compelling evidence for storage of platelet Fn in organelles termed alpha granules. First, on subcellular fractionation, platelet Fn cosediments with authentic alpha-granule markers.[19] Second, platelets congenitally deficient in alpha granule have reduced Fn content.[33,34] Third, platelet Fn is actively secreted along with other alpha-granule constituents.[19,31] Finally, by immunofluorescence, platelet Fn has a similar intracellular localization to other alpha-granule constituents,[31,35,36] and immunoelectron microscopy of frozen thin sections of platelets has directly established its presence in this organelle.[37]

B. SURFACE EXPRESSION

If the four large platelet alpha-granule glycoproteins are to serve an adhesive function, the proteins must reach the cell surface. There is compelling evidence that each of these proteins is expressed on the surface of activated platelets.[6,38-40] In the case of Fn, this has been established by immunofluorescence and by binding of affinity-purified radiolabeled antibody.[38] Two mechanisms could account for these observations. First, secreted proteins could rebind to the cell surface or, second, proteins may be prebound to granule membranes and expressed on the cell surface as a "sticky patch" as a direct consequence of the secretory granule-plasma membrane fusion event. In the case of Fn, the initial surface expression is probably too rapid to be accounted for by rebinding,[41] suggesting that at early timepoints the second mechanism predominates. Such a rapid localized expression of adhesive proteins on the cell surface has interesting implications. First, it may partially account for the close kinetic relationship between platelet secretion and secondary aggregation.[42] Second, if as occurs in other cell types,[43] secretion along an activating surface may be polarized towards that surface. Polarized expression of adhesive proteins in platelets adherent to a vessel wall would result in these proteins being present only on the vessel side and not the luminal side. Thus, the luminal surface of such adherent platelets might be relatively less thrombogenic than the adherent side so that aggregation would not be a necessary concomitant of the adhesive event.

In order for such a mechanism of polarized rapid secretion to occur, several requirements should be met. First, the proteins should be prebound to receptors on the inner surface of the alpha-granule membrane. Gogstad[44] reported that a molecule of similar size to TSP co-isolated with a particulate fraction of ruptured alpha-granules. Second, the same group[45] presented evidence to suggest that membrane glycoprotein IIb-IIIa (a receptor for at least three of these proteins)[46] was present in alpha granules. Immunofluorescence and immunoelectron microscopy has recently been employed to document the presence of GP IIb-IIIa in alpha granules,[47] thus providing potential receptors for these proteins. A third requirement of this hypothesis is that the alpha granules fuse with the plasma membrane. Although secretion of alpha-granule constituents is nonlytic, fusion of these granules with the platelet plasma membrane has not been documented, suggesting that they may not secrete by a simple exocytotic mechanism. Behnke[48] and White[49] provided evidence that platelets contain an open canalicular system in free communication with the extracellular space, leading White[49] to propose that secretion is into this system. White and Clawson[50] have found that the canaliculi are an interconnected meshwork which is unlikely to be evaginated during secretion. Thus, the hypothesized insertion of adhesive proteins would occur in the depths of the canalicular system, a site unlikely to be accessible to substrata or other cells. In our studies of release of alpha-granule constituents using platelet factor 4 as a marker, we found that the constituents were consolidated into larger pools prior to secretion. These pools appeared as vacuolar structures in transmission electron microscopy and were closed to probes of the extracellular space.[51] These data are compatible with alpha-granule secretion into an intermittently open canalicular system as suggested by Zucker-Franklin,[52] but are also compatible with a compound exocytotic mechanism. In the latter case, we propose that alpha-granule constituent release occurs by initial fusion of alpha granules with each other and/or another intracellular compartment to form a morphologically distinct compound granule followed by formation of a compound granule with the plasma membrane. Possible instances of the latter have been observed in transmission electron microscopy.[53] Finally, we have recently found that the ''Big Four'' appear to undergo a similar sequence of redistributions as platelet factor 4 in thrombin stimulated platelets.[54] Thus, we hypothesize that these adhesive proteins may be transported to the platelet surface by a compound excocytotic mechanism resulting in their rapid expression at the cell surface initially in localized, potentially polarized patches.

Once on the platelet surface, Fn undergoes a time-dependent redistribution to a more compact distribution. In immunofluorescence, this redistribution bears a passing resemblance to ''capping'' as occurs in other cell types.[38] Since capping may involve transmembrane interactions between cell surface and cytoskeletal proteins,[55] the possibility that cell-surface Fn or other adhesive proteins enters into such transmembrane associations exists. Phillips et al.[56] reported that a portion of platelet membrane glycoprotein IIb and IIIa associated with the triton-insoluble residue of thrombin-aggregated platelets; as noted above, these proteins may be the receptor for Fg. In addition, we[57] rigorously established that membrane glycoproteins, IIb and IIIa, were physically associated with the cytoskeleton of concanavalin A-activated platelets. First, these glycoproteins cosedimented with the triton-insoluble ''cytoskeleton'' in concanavalin A activated platelets; and, second the glycoproteins could be solubilized by depolymerization of F-actin by DNAaseI. Third, the ''cytoskeleton''-associated concanavalin A was demonstrated to be bound to the periphery of the filamentous cytoskeleton in frozen thin sections of these triton residues. Finally, surface-bound concanavalin A could be induced to ''cap'' by a thrombin-stimulated redistribution of the underlying cytoskeleton, providing evidence of a transmembrane interaction in intact cells. Since each of the four adhesive proteins has intramolecular symmetry, the possibility that they, like concanavalin A, act as multivalent ligand seems likely, and it is not surprising that evidence suggesting their association with cytoskeleton has been obtained.[58,59]

IV. Fn BINDING TO PLATELET(S)

There is considerable interest in the definition of cellular Fn-binding sites. In this regard the platelet has proven valuable in that the cell saturably binds plasma Fn when stimulated with appropriate agonists.

A. REQUIREMENTS FOR EXPRESSION OF PLATELET Fn BINDING

Incubation of ^{125}I-plasma Fn with thrombin-stimulated suspensions of washed human platelets results in time-dependent binding of Fn to the cells.[4] Binding requires ~ 20 min to reach a maximum and does not occur in resting cells. The specificity of this reaction was documented by the ability of Fn but not unrelated proteins or Fn-depleted serum to inhibit binding. The ability of Fn to inhibit binding indicates a saturable process, and by Scatchard analysis, 120,000 molecules per cell are bound at saturation. Half saturation of cellular Fn-binding sites is achieved at 300-nM Fn. Binding occurs in the presence of 5-m*M* MgEGTA, suggesting a lack of an absolute calcium requirement, but was inhibited in the presence of 5-m*M* EDTA indicating a requirement for divalent cations.

B. IDENTIFICATION OF PLATELETS CONGENITALLY DEFICIENT IN Fn BINDING

Glanzmann's thrombasthenia is an autosomally inherited recessive severe bleeding disorder that results from an intrinsic platelet defect. *In vitro* abnormalities of individuals with this disorder included reduced or absent platelet aggregation in response to collagen, thrombin, and ADP, reduced clot retraction, and abnormal platelet spreading on glass surfaces.[60] These platelets are not always[61,62] deficient in a specific platelet membrane glycoprotein, GPIIb-IIIa.[63] Platelets from patients with thrombasthenia bind very little plasma Fn when stimulated with thrombin.[34] This failure to bind plasma Fn is not due to a lack of platelet stimulation by thrombin, nor to a soluble inhibitor of Fn binding, but rather to a reduced number of Fn-binding sites.[34] Thus, this autosomal recessive disorder is associated with reduced platelet Fn and may also be associated with reduced fibroblast retraction of fibrin clots[64] and collagen gels.[65]

C. ROLE OF SECRETED PROTEINS IN Fn BINDING

As noted above, thrombin induces Fn sites on platelets. Immunofluorescence suggests that collagen may do so as well.[13] Both of these stimuli induce secretion of platelet alpha-granule constituents and surface expression of the adhesive proteins (see above). In contrast, in our assay conditions neither epinephrine nor ADP are potent inducers of Fn-binding sites although they do induce Fg receptors.[4] This raises the possibility that a newly surface-expressed alpha-granule constituent mediates Fn binding to platelets.

One candidate alpha-granule protein is platelet Fg, which is present at a level of approximately 100,000 molecules per cell. Thrombasthenic platelets often have a reduced Fg content and fail to express Fg receptors in response to ADP,[66] raising the possibility that a Fg-related defect may indeed account for the defect in Fn binding. Further, fibrin interacts with Fn, although the noncovalent interaction is of low affinity at 37°, the incubation temperature of our binding assays.[8] Nevertheless, platelet Fg is not absolutely required for Fn binding since afibrinogenemic platelets, containing less than 2% the normal Fg, bind Fn.[34] There may, however, be a Fg-dependent pathway of Fn binding to platelets, since a portion of platelet-bound Fn appears to be covalently crosslinked in a manner compatible with factor XIIIa-mediated cross-linking to fibrin.[4] Moreover, this cross-linking has an absolute calcium dependence, in contrast to the noncovalent association of Fn with platelets[67] and can be reconstituted *in vitro* with exogenous fibrin bound to platelet Fg receptors.

A second candidate alpha-granule protein is TSP, which is present at approximately 40,000 molecules per cell. TSP is a trimer,[68] which means that TSP could accommodate Fn binding at a one to one stoichiometry. TSP binds to Fn,[69] is synthesized by fibroblasts[24] and endothelial cells,[2,70] appears to be a matrix protein,[24] and may therefore support Fn interaction in other cell types as well. Lahav et al.[40] found that when platelets react with surface-bound Fn derivatized with *N*-succinimidyl-3[2-nitro-4-azidophenyl)-2-aminoethyl-dithio] proprionate, photoactivation results in cross-linking to TSP. Nevertheless, TSP is unlikely to serve as a major univalent Fn receptor on thrombin-activated platelets because alpha-granule-deficient platelets, containing less than 5% of normal TSP, bind plasma Fn normally in response to thrombin.[71] Moreover, since these platelets are deficient in other alpha-granule constituents as well, such constituents are unlikely to be required for thrombin-induced Fn binding.

D. GPIIb-IIIa FUNCTIONS AS A COMMON ADHESIVE PROTEIN RECEPTOR ON PLATELETS AND IS A PROTOTYPE OF THE CYTOADHESIN FAMILY

As noted above, Fg, Fn, and vWF bind to thrombin-activated platelets in a saturable fashion consistent with a discrete receptor site.[4,5,7,66] These stimulated interactions share the requirement for divalent cations and ADP dependence (reviewed in Reference 46). Moreover, there is a reduction in this interaction in Glanzmann's thrombasthenic platelets.[34,66,72] Such platelets are selectively deficient in membrane glycoproteins GPIIb-IIIa.[73,74] Moreover, certain monoclonal antibodies reactive with GPIIb-IIIa which inhibit Fg binding, also inhibit the binding of Fn and vWF.[75] All three of these adhesive proteins contain the Arg-Gly-Asp (RGD) sequence[76-78] originally identified by Pierschbacher and Ruoslahti as a potential common cell adhesion signal.[79] Peptides containing RGD inhibit the binding of Fg, Fn, and vWF to activated platelets.[80-83] Affinity chromatography of whole platelet extracts on immobilized RGD peptide matrix results in the purification of GPIIb-IIIa,[84] and purified GPIIb-IIIa incorporated into liposomes interacts with Fg, Fn, and vWF.[84-87] Thus, it appears quite likely that platelet membrane glycoprotein IIb-IIIa is a constituent of a common membrane receptor for Fn, Fg, and vWF. Moreover, GPIIb-IIIa binds to RGD-containing peptides and such peptides inhibit the binding of these three proteins which contain the sequence. Thus, it is attractive to hypothesize that this shared sequence amongst these otherwise distinct proteins forms the common site recognized by GPIIb-IIIa.

Peptides derived from the gamma chain of Fg also inhibit the binding of Fg[88] to platelets. They also inhibit the binding of Fn and vWF.[89,90] Surprisingly, there is no obvious homology to the gamma chain sequence in Fn and vWF. Recent data has provided an explanation for this apparent paradox. Specifically, we found gamma chain peptides could inhibit the binding of an RGD peptide to platelets with parameters consistent with those of competitive antagonism. Second, gamma chain peptides could elute GPIIb-IIIa bound to insolubilized RGD peptides, and conversely, the RGD peptides can elute GPIIb-IIIa from insolubilized gamma chain peptides.[91] These data indicate that the two peptide sets either bind to the same site, or bind in a mutually exclusive manner to a GPIIb-IIIa-containing receptor. Recently reported cross-linking studies[92] seem to favor the latter hypothesis. These findings may also explain the capacity of RGD peptides to inhibit binding to platelets of Fg deficient in the RGDS sequence of the alpha chain.[93]

As noted above, the RGD sequence has been implicated in a variety of cell adhesion events, and GPIIb-IIIa bears striking gross structural relationship to a variety of cell adhesion receptors.[84] Epitopes shared with platelet membrane GPIIb-IIIa have been found on a wide variety of cell types[94-98] including endothelial cells. These related proteins are biosynthetic products of these cells[94-96,98] and have similar subunit structure to GPIIb-IIIa. This antigenic, structural, and functional relationship has led to the proposal that there is a family of GPIIb-IIIa-related proteins termed "cytoadhesins" present in a variety of cells.[96] Moreover, recent studies have directly demonstrated that GPIIIa is similar or identical to the beta subunit of another cytoadhesin, the vitronectin receptor. Furthermore, GPIIb is distinct from the α subunit of the vitronectin receptor; however, it appears to be homologous to this protein and to the leukocyte adhesion receptor α subunits LFA-1 and MAC-1.[95,99] This GPIIb-IIIa-related family of heterodimers thus appears to share a common β subunit and distinct α subunits. These cytoadhesins appear to differ in ligand recognition specificity;[84] thus it is tempting to speculate that ligand specificity is regulated by the α subunit. The cytoadhesins also resemble the family of leukocyte adhesion molecules[100] as well as the very late lymphocyte activation antigen (VLA) family described by Hemler et al.,[101] in having distinct α and common β subunits. Finally, the sequences of β subunits of these three families have been found to show remarkable similarity, providing definitive proof of their relatedness.[102-104] Hynes has proposed that the entire group of adhesion receptors be termed "integrins", since some of them integrate the extracellular matrix with the cytoskeleton.[105]

Members of the VLA family have been found on platelets.[106,107] Interestingly, Pischel et al. have identified VLA-2 as an alpha beta heterodimer of platelet GPIa-IIa.[107] Since GPIa has been reported to bind to collagen[108] and be deficient on the platelets of a patient who does not respond to collagen,[109] VLA-2 may function as a collagen receptor on platelets. Indeed, direct immunochemical evidence for this has recently been obtained.[110] The β subunit of the other VLA member on platelets has the mobility of GPIc[107] and may therefore be VLA-3, or VLA-5 α subunit or an undescribed VLA protein. Recently, Sonnenberg and colleagues have reported a monoclonal antibody which reacts with GPIc, immunoprecipitates a heterodimer of GPIc-IIa, and whose epitope is also present in a variety of epithelial and tumor cells.[111] In collaboration with Hemler's group,[112] they have found that it is distinct from both VLAs 3 and 5. The nature of the ligand for this VLA-6 is presently unknown. In addition, since VLA-5 is also on platelets,[112] it seems likely that this species accounts for some or all of the activation-independent Fn receptors[113] on platelets.

V. CONFORMATIONAL CHANGES WITHIN ADHESION RECEPTORS

Occupancy of adhesion receptors leads to a host of cellular responses.[114] The occurrence of such postoccupancy events implies that signal transduction mechanisms are engaged as a consequence of the binding of adhesive proteins to these receptors. One phenomenon which has been implicated in transduction of cellular responses which follow hormone binding is lateral aggregation of receptors in the plane of the membrane. Indeed, such aggregation of GPIIb-IIIa has been observed following fibrinogen binding to platelets[115] or addition of Arg-Gly-Asp peptides to platelets.[116] Clustering of these receptors may be a prelude to their association with the cytoskeleton[57] and the formation of focal contacts in adherent cells.[117] Such lateral aggregation might be mediated by simple cross-bridging of receptors by multivalent ligands such as adhesive proteins. Alternatively, the work of Isenberg et al.[116] suggests that simple occupancy of these receptors is sufficient to trigger clustering. It is thus probable that a conformational change in the receptor might expose sites which could then mediate such lateral aggregation or other postoccupancy events. Experimental evidence for such conformational changes in soluble GPIIb-IIIa following the binding of Arg-Gly-Asp peptides was obtained by Parise et al.,[118] however those changes were relatively slow and were not demonstrated in intact cells following the binding of the physiologic ligands. An additional consequence of such conformational changes might be the expression of neoantigens on occupied adhesion receptors, i.e., that antibodies could be produced which would recognize occupied but not vacant receptors. Such antibodies might also possess the property of inhibiting postoccupancy events and thus providing experimental validation of the hypothesis as well as reagents with therapeutic and diagnostic potential. An antibody with these properties has been identified.

The PMI-1 antibody binds to the heavy chain of GPIIb[119] and defines a region of GPIIb that is associated with four distinct functional activities. First, PMI-1 inhibits platelet adhesion to collagen.[119] Second, the surface orientation of this region is regulated by divalent cations as millimolar concentrations of calcium or magnesium suppress expression of the PMI-1 epitope.[61] Third, abnormal divalent cation regulation of the conformation of this site is associated with a functional thrombasthenic state.[61] Finally, the interaction of RGD-containing ligands, including fibrinogen, with GPIIb-IIIa alters the conformation of this region resulting in increased exposure of the PMI-1 epitope.[120]

Based on the reactivity of this antibody in Western blotting of reduced GPIIb,[119] it seemed that expression cloning might offer a means to define its antigenic site. In our initial screening of 200,000 λ gt11 recombinant from a human erythroleukemia cell cDNA library with rabbit polyclonal GPIIb-IIIa antiserum, we identified six immunoreactive clones. One of these clones, HEL41, contained a 1.1-kDa insert which directed the synthesis of a bacterial fusion protein of $M_r \sim 140$ kDa with characteristics consistent with it possessing GPIIb epitopes. These characteristics included: (1) reactivity of the polyclonal serum with the fusion protein encoded by this clone was blocked by prior absorption with purified GPIIb-IIIa; (2) antibodies affinity-purified on the fusion protein reacted specifically with GPIIb in immunoblots of platelet lysates and purified GPIIb-IIIa; and (3) GPIIb-IIIa was immunoprecipitated from surface-labeled platelets by the affinity-purified antiserum. The affinity-purified antibody failed to react with the α subunit of the vitronectin receptor by Western blotting and failed to immunoprecipitate the related endothelial cell cytoadhesin.[121] The fusion protein encoded by HEL41 was immunoreactive with the monoclonal antibody PMI-1, and the size of the fusion protein suggested an insert-coded polypeptide size of 23 kDa, indicating a long open reading frame in the clone.

The cDNA sequence had an open reading frame of 682 bases with a TAG termination codon at position 683 followed by 419 bases of 3′ untranslated sequence. The first 120 nucleotides at the 5′ end of the clone possess the characteristics of the consensus Alu repetitive sequence,[122] indicating that this clone may have been derived from an incompletely processed mRNA. The cDNA sequence of clone HEL41 encoded an open reading frame of 227 amino acids. The determined N-terminal sequence for the light chain of GPIIb, as reported by Charo et al.,[95] was identified within the clone as the sequence between L(157) and T(171). The consensus Alu repetitive sequence translated in frame and accounted for the first 40 amino acids encoded by HEL41. To investigate whether this sequence was present in mature GPIIb, the reactivity of antibodies to peptides from within this region with GPIIb was determined by immunoblotting. Two peptides, W(19)-S(34) which is encoded by the repetitive sequence and R(53)-K(64) which lies outside this region, were synthesized and antipeptide antibodies were produced. Antibodies to W(19)-S(34) failed to blot to GPIIb, whereas antibodies to R(53)-K(64) were reactive with GPIIb heavy chain in immunoblot.

At this point, the next logical step would have been to proceed to construction of an epitope library derived from subclones of this clone. Nevertheless, we had performed preliminary experiments which indicated that a 9-kDa staphylococcal V8 protease fragment of GPIIb contained the PMI-1 epitope. Inspection of the deduced amino acid sequence of clone HEL41 identified a single predicted V8 fragment of appropriate size between L(71) and E(150), and this fragment lay within the predicted heavy chain sequence. We next reasoned that if this antibody recognized a segmental epitope, then it should react with the appropriate small synthetic peptide derived from this V8 protease fragment of the GPIIb heavy chain.

In recent years there has been a lively effort to establish rules for identification of protein sequences likely to function as continuous epitopes, i.e., for synthetic peptides which will elicit antibodies which react with the native protein. Indeed, several highly informative studies and reviews of this subject have been published.[123-126] Initial efforts focused on simple algorithms to predict hydrophilic segments of proteins based on the logic that such sequences are more likely to be surface exposed.[126] Subsequent studies have suggested that additional considerations such as secondary structure, local protein mobility,[127] charge, and protein shape[128] may also enter into the equation. Accordingly, we utilized the algorithm of Jameson and Wolf[129] to analyze potential continuous epitope within this predicted V8 fragment. This algorithm estimates antigenicity based on predictions of secondary structure, flexibility, and surface accessibility as well as hydrophilicity. The results of such an analysis identified a single continuous region with a high predicted value for an antigenic site. Two peptides, one flanking and containing this region (V-43) and one immediately carboxyterminal to it (V-41), were synthesized, and their ability to inhibit the binding of PMI-1 to platelets was examined. Peptide V-43 (P_{129}-Q_{145}) inhibited the binding of PMI-1 in a dose-dependent manner. Since 50% inhibition was achieved at 1.6-μM peptide when the antibody was present at 1 μM, the approximate K_d of the peptide-PMI-1 interaction is 1.2 μM. Conversely, peptide V-41 (R_{144}-R_{156}) had no effect on the binding of PMI-1. In control experiments, neither peptide had any effect on the binding of a monoclonal anti-GPIIIa antibody, 22C4,[96] to platelets.

Direct binding of PMI-1 to peptide V-43 was demonstrated by ELISA assay. PMI-1 bound to purified GPIIb-IIIa and peptide V-43, but failed to react with peptide V-41. The monoclonal antibody, Tab,[130] bound to purified GPIIb-IIIa but failed to bind to either peptide. In a competitive assay, peptide V-43 (20 μM) completely inhibited the binding of PMI-1 to purified GPIIb-IIIa but had no effect on Tab binding to GPIIb-IIIa. Peptide V-41 had no effect on the binding of either antibody.

These data have permitted the assignment of this epitope to a 17-residue peptide sequence near the carboxyterminus of GPIIb heavy chain, but what are the potential implications of these results?

1. We know that the PMI-1 antibody inhibits platelet spreading[119] on collagenous surfaces. Is this because the region of GPIIb recognized by the antibody is involved in events which lead to spreading and stabilization of attachment? Obviously, synthetic peptides based on this sequence should prove most useful in experimental evaluation of this possibility.
2. This region of GPIIb undergoes a change in its microenvironment or structure following the binding of fibrinogen or Arg-Gly-Asp-containing peptides to platelets.[120] These binding events are also associated with self-association of GPIIb-IIIa within the plane of the plasma membrane.[115,116] Again, these synthetic peptides should prove valuable in testing the hypothesis that the change in the carboxyterminus of the GPIIb heavy chain mediates this self-association or other postoccupancy events involving GPIIb-IIIa.
3. There has been considerable interest in the identification of synthetic peptides which inhibit platelet function as inhibitors of adhesive protein binding.[80,88,131] Peptides based on the V-43 sequence may offer a novel class of such inhibitors which block selected postoccupancy events. As such they may prove more cell-type- or response-specific than the previous peptides.
4. Antibodies which are selective for activated platelets may prove useful in detection of such cells in the circulation[132-134] or in imaging thrombi. The PMI-1 antibody represents a novel group of such reagents which recognize occupied fibrinogen receptors. Thus antibody or antibodies directed against its synthetic epitope may thus prove useful for detection of platelet activation *in vitro* and *in vivo*.

5. Patients with pseudothrombocytopenia often have antibodies which react with GPIIb-IIIa in the presence of EDTA, but not at physiologic divalent cation concentrations.[135] The PMI-1 epitope is an obvious target[61] for such antibodies and the V-43 peptide would seem to offer a suitable reagent for assaying for PMI-1-like antibodies in these patients.

VI. CONCLUDING REMARKS

The discovery, characterization, chemistry, and biology of adhesive proteins has provided fruitful insights into a broad group of problems in cell biology. Platelets are anucleate cell fragments which are programmed to express adhesive activities in response to vessel injury. As indicated in this review, there are at least four candidate well-characterized adhesive proteins which may subserve these activities and considerable data concerning the identity of and regulation of receptors for these proteins. This "cell" thus offers a useful model for the study of cell adhesion and cell cohesion. Moreover, due to the existence of deficiency diseases we know that Fg (M_r = 340,000) and vWF (M_r $>10^6$) function in hemostasis. In the case of TSP (M_r = 450,000) and Fn (M_r = 450,000), although there are good reasons to believe they function in hemostasis, the lack of selective deficiency diseases makes their function less certain. Nevertheless, the *in vitro* data is sufficiently compelling to suggest that Fn plays an important role in hemostasis.

ACKNOWLEDGMENTS

The authors are deeply indebted to Jane Forsyth for the technical conduct of the majority of our experiments described here. We also acknowledge the expert secretarial assistance of Lynn LaCivita. This work was supported by grants HL-16411, HL-28235, and AR-27214 from the National Institutes of Health. This is publication number 5405-IMM from the Research Institute of Scripps Clinic.

REFERENCES

1. **Jaffe, E. A. and Mosher, D. F.,** Synthesis of fibronectin by cultured human endothelial cells, *J. Exp. Med.*, 147, 1779, 1978.
2. **Mosher, D. F., Doyle, M. J., and Jaffe, E. A.,** Synthesis and secretion of thrombospondin by cultured human endothelial cells, *J. Cell Biol.*, 93, 343, 1982.
3. **Jaffe, E. A., Hoyer, L. W., and Nachman, R. L.,** Synthesis of antihemophilic factor antigen by cultured human endothelial cells, *J. Clin. Invest.*, 52, 2757, 1973.
4. **Plow, E. F. and Ginsberg, M. H.,** Specific and saturable binding of plasma fibronectin to thrombin-stimulated human platelets, *J. Biol. Chem.*, 256, 9477, 1981.
5. **Marguerie, G. A., Plow, E. F., and Edgington, T. S.,** Human platelets possess an inducible and saturable receptor specific for fibrinogen, *J. Biol. Chem.*, 254, 5357, 1979.
6. **Phillips, D. R., Jennings, L. K., and Prasanna, H. R.,** Ca^{2+}-mediated association of glycoprotein G (thrombin sensitive protein, thrombospondin) with human platelets, *J. Biol. Chem.*, 255, 11629, 1980.
7. **Fujimoto, T., Ohara, S., and Hawiger, J.,** Thrombin-induced exposure and prostacyclin inhibition of the receptor for factor VIII/von Willebrand factor on human platelets, *J. Clin. Invest.*, 69, 1212, 1982.
8. **Ruoslahti, E. and Vaheri, A.,** Interaction of soluble fibroblast surface antigen with fibrinogen and fibrin identity with cold insoluble globulin of human plasma, *J. Exp. Med.*, 141, 497, 1975.
9. **Leung, L. L. and Nachman, R. L.,** Complex formation of platelet thrombospondin with fibrinogen, *J. Clin. Invest.*, 70, 542, 1982.
10. **Amrani, D. L., Mosesson, M. W., and Hoyer, L. W.,** Distribution of plasma fibronectin (cold-insoluble globulin) and components of the factor VIII complex after heparin-induced precipitation of plasma, *Blood*, 59, 657, 1982.

11. **Shadle, P. J. and Barondes, S. H.,** Adhesion of human platelets to immobilized trimeric collagen, *J. Cell Biol.*, 95, 361, 1982.
12. **Weiss, H. J., Turitto, V. T., and Baumgartner, H. R.,** Effect of shear rate on platelet interaction with subendothelium in citrated and native blood. I. Shear rate-dependent decrease of adhesion in von Willebrand's disease and the Bernard-Soulier syndrome, *J. Lab. Clin. Med.*, 92, 750, 1978.
13. **Hynes, R. O., Ali, I. U., Destree, A. T., Mautner, V., Perkins, M. E., Senger, D. R., Wagner, D. D., and Smith, K. K.,** A large glycoprotein lost from the surfaces of transformed cells, *Ann. N.Y. Acad. Sci.*, 312, 317, 1978.
14. **Grinnell, F. and Hays, D. G.,** Cell adhesion and spreading factor. Similarity to cold insoluble globulin in human serum, *Exp. Cell Res.*, 115, 221, 1978.
15. **Chazov, E. I., Alexeev, A. V., Antonov, A. S., Koteliansky, V. E., Leytin, V. L., Lyubimova, E. V., Repin, V. S., Sviridov, D. D., Torchilin, V. P., and Smirnov, V. N.,** Endothelial cell culture on fibrillar collagen: model to study platelet adhesion and liposome targeting to intercellular collagen matrix, *Proc. Natl. Acad. Sci. U.S.A.*, 78, 5603, 1981.
16. **Santoro, S. A. and Cunningham, L. W.,** Fibronectin and the multiple interaction model for platelet-collagen adhesion, *Proc. Natl. Acad. Sci. U.S.A.*, 76, 2644, 1979.
17. **Houdijk, W. P. M., Sakariassen, K. S., and Sixma, J. J.,** Role of factor VIII-von Willebrand factor and fibronectin in the interaction of platelets in flowing blood with human collagen types I and III, *J. Clin. Invest.*, 75, 531, 1983.
18. **Houdijk, W. P. M., deGroot, P. G., Nievelstein, P. F., and Sakariassen, K. S.,** Subendothelial proteins and platelet adhesion. von Willebrand factor and fibronectin, not thrombospondin, are involved in platelet adhesion to extracellular matrix of human vascular endothelial cells, *Arteriosclerosis*, 6, 24, 1986.
19. **Zucker, M. B., Mosesson, M. W., Broekman, M. J., and Kaplan, K. L.,** Release of platelet fibronectin (cold-insoluble globulin) from alpha granules induced by thrombin or collagen; lack of requirement for plasma fibronectin in ADP-induced platelet aggregation, *Blood*, 54, 8, 1979.
20. **Cohen, I., Potter, E. V., Glaser, T., Entwistle, R., Davis, L., Chediak, J., and Anderson, B.,** Fibronectin in von Willebrand's disease and thrombasthenia: role in platelet aggregation, *J. Lab. Clin. Med.*, 97, 134, 1981.
21. **Hammerschmidt, D. E., Arneson, M. A., Larson, S. L., Van Tassel, R. A., and McKenna, J. L.,** Maternal Ehlers-Danlos syndrome type X. Successful management of pregnancy and parturition, *JAMA*, 248, 2487, 1982.
22. **Moon, D. G., Kaplan, J. E., and Mazurkewicz, J. E.,** The inhibitory effect of plasma fibronectin on collagen-induced platelet aggregation, *Blood*, 67, 450, 1986.
23. **Dixit, V. M., Haverstick, D. M., O'Rourke, K., Hennessy, S. W., Broekelmann, T. J., McDonald, J. A., Grant, G. A., Santoro, S. A., and Frazier, W. A.,** Inhibition of platelet aggregation by a monoclonal antibody against human fibronectin, *Proc. Natl. Acad. Sci. U.S.A.*, 82, 3844, 1985.
24. **Raugi, G. J., Mumby, S. M., Abbott-Brown, D., and Bornstein, P.,** Thrombospondin: synthesis and secretion by cells in culture, *J. Cell Biol.*, 95, 351, 1982.
25. **Nachman, R., Levine, R., and Jaffe, E. A.,** Synthesis of factor VIII antigen by cultured guinea pig megakaryocytes, *J. Clin. Invest.*, 60, 914, 1977.
26. **Wagner, D. D., Olmstead, J. B., and Marder, V. J.,** Immunolocalization of von Willebrand protein in Weibel-Palade bodies of human endothelial cells, *J. Cell Biol.*, 95, 355, 1982.
27. **Mosesson, M. W. and Umfleet, R. A.,** The cold-insoluble globulin of human plasma. I. Purification, primary characterization, and relationship to fibrinogen and other cold-insoluble fraction components, *J. Biol. Chem.*, 245, 5728, 1970.
28. **Plow, E. F., Birdwell, C., and Ginsberg, M. H.,** The identification and quantitation of platelet fibronectin antigen, *J. Clin. Invest.*, 63, 540, 1979.
29. **Paul, J. I., Schwarzbauer, J. E., Tamkun, J. W., and Hynes, R. O.,** Cell-type-specific fibronectin subunits generated by alternative splicing, *J. Biol. Chem.*, 261, 12258, 1986.
30. **Bensusan, H. B., Koh, T. L., Henry, K. G., Murray, B. A., and Culp, L. A.,** Evidence that fibronectin is the collagen receptor on platelet membranes, *Proc. Natl. Acad. Sci. U.S.A.*, 75, 5864, 1978.
31. **Ginsberg, M. H., Painter, R. G., Birdwell, C., and Plow, E. F.,** The detection, immunofluorescent localization and thrombin induced release of platelet fibronectin antigen, *J. Supramol. Struct.*, 11, 167, 1979.
32. **Ginsberg, M. H., Plow, E. F., and Forsyth, J.,** Fibronectin expression on the platelet surface occurs in concert with secretion, *J. Supramol. Struct.*, 17, 91, 1981.
33. **Nurden, A. T., Kunicki, T. J., Dupuis, D., Soria, C., and Caen, J. P.,** Specific protein and glycoprotein deficiencies in platelets isolated from two patients with the gray platelet syndrome, *Blood*, 59, 709, 1982.
34. **Ginsberg, M. H., Forsyth, J., Lightsey, A., Chediak, J., and Plow, E. F.,** Reduced surface expression and binding of fibronectin by thrombin-stimulated thrombasthenic platelets, *J. Clin. Invest.*, 71, 619, 1983.
35. **Giddings, J. C., Brookes, L. R., Piovella, F., and Bloom, A. L.,** Immunohistological comparison of platelet factor 4 (PF4), fibronectin (Fn) and factor VIII related antigen (VIIIR:Ag) in human platelet granules, *Br. J. Haematol.*, 52, 79, 1982.

36. **Wencel-Drake, J. D., Plow, E. F., Zimmerman, T. S., and Ginsberg, M. H.,** Adhesive glycoproteins reach the platelet surface by a common mechanism, *Circulation,* 66, 700a, 1982.
37. **Wencel-Drake, J. D., Painter, R. G., Zimmerman, T. S., and Ginsberg, M. H.,** Ultrastructural localization of human platelet thrombospondin, fibrinogen, fibronectin, and von Willebrand factor in frozen thin section, *Blood,* 65, 929, 1985.
38. **Ginsberg, M. H., Painter, R., Forsyth, J., Birdwell, C., and Plow, E.,** Thrombin increases expression of fibronectin antigen on the platelet surface, *Proc. Natl. Acad. Sci. U.S.A.,* 77, 1049, 1980.
39. **George, J. N., Lyons, R. M., and Morgan, R. K.,** Membrane changes associated with platelet activation. Exposure of actin on the platelet surface after thrombin-induced secretion, *J. Clin. Invest.,* 66, 1, 1980.
40. **Lahav, J., Schwartz, M. A., and Hynes, R. O.,** Analysis of platelet adhesion with a radioactive chemical crosslinking reagent: interaction of thrombospondin with fibronectin and collagen, *Cell,* 31, 253, 1982.
41. **George, J. N. and Onofre, A. R.,** Human platelet surface binding of endogenous secreted factor VIII-von Willebrand factor and platelet factor 4, *Blood,* 59, 194, 1982.
42. **Charo, I. F., Feinman, R. D., and Detwiler, T. C.,** Interrelations of platelet aggregation and secretion, *J. Clin. Invest.,* 60, 866, 1977.
43. **Morrison, D. C., Roser, J. F., Cochrane, C. G., and Henson, P. M.,** The initiation of mast cell degranulation: activation at the cell membrane, *J. Immunol.,* 114, 966, 1975.
44. **Gogstad, G. O., Hagen, I., Korsmo, R., and Solum, N. O.,** Evidence for release of soluble, but not of membrane-integrated, proteins from human platelet a-granules, *Biochim. Biophys. Acta,* 702, 81, 1982.
45. **Gogstad, G. O., Hagen, I., Korsmo, R., and Solum, N. O.,** Characterization of the proteins of isolated human platelet alpha-granules. Evidence for a separate alpha-granule-pool of the glycoproteins IIb and IIIa, *Biochim. Biophys. Acta,* 670, 150, 1981.
46. **Plow, E. F., Marguerie, G. A., and Ginsberg, M. H.,** Expression and function of platelet adhesive proteins at the platelet surface, in *Biochemistry of Platelets,* Shuman, M. and Phillips, D. R., Eds., Harcourt Brace Jovanovich, Orlando, FL, 1986, 226.
47. **Wencel-Drake, J. D., Plow, E. F., Kunicki, T. J., Woods, V. L., Keller, D. M., and Ginsberg, M. H.,** Localization of internal pools of membrane glycoproteins involved in platelet adhesive responses, *Am. J. Pathol.,* 124, 324, 1986.
48. **Behnke, O.,** Electron microscopic observations on the membrane systems of the rat blood platelet, *Anat. Rec.,* 158, 121, 1967.
49. **White, J. G.,** Exocytosis of secretory organelles from blood platelets incubated with cationic polypeptides, *Am. J. Pathol.,* 69, 41, 1972.
50. **White, J. G. and Clawson, C. C.,** The surface-connected canalicular system of blood platelets — a fenestrated membrane system, *Am. J. Pathol.,* 101, 353, 1980.
51. **Ginsberg, M. H., Taylor, L., and Painter, R. G.,** The mechanism of thrombin-induced platelet factor 4 secretion, *Blood,* 55, 661, 1980.
52. **Zucker-Franklin, D.,** Endocytosis by human platelets: metabolic and freeze-fracture studies, *J. Cell Biol.,* 91, 706, 1981.
53. **Painter, R. G. and Ginsberg, M. H.,** Centripetal myosin redistribution in thrombin-stimulated platelets. Relationship to platelet Factor 4 secretion, *Exp. Cell Res.,* 155, 198, 1984.
54. **Wencel-Drake, J. D., Plow, E. F., Zimmerman, T. S., Painter, R. G., and Ginsberg, M. H.,** Immunofluorescent localization of adhesive glycoproteins in resting and thrombin-stimulated platelets, *Am. J. Pathol.,* 115, 156, 1984.
55. **Flanagan, J. and Koch, G. L. E.,** Cross-linked surface Ig attaches to actin, *Nature (London),* 273, 278, 1978.
56. **Phillips, D. R., Jennings, L. K., and Edwards, H. H.,** Identification of membrane proteins mediating the interaction of human platelets, *J. Cell Biol.,* 86, 77, 1980.
57. **Painter, R. G. and Ginsberg, M. H.,** Concanavalin A induces interactions between surface glycoproteins and the platelet cytoskeleton, *J. Cell Biol.,* 92, 565, 1982.
58. **Tuszynski, G. P., Kornecki, E., Cierniewski, C., Knight, L. C., Koshy, A., Srivastava, S., Niewiarowski, S., and Walsh, P. N.,** Association of fibrin with the platelet cytoskeleton, *J. Biol. Chem.,* 259, 5247, 1984.
59. **Niewiarowska, J., Cierniewski, C. S., and Tuszynski, G. P.,** Association of fibronectin with the platelet cytoskeleton, *J. Biol. Chem.,* 259, 6181, 1984.
60. **Nurden, A. T. and Caen, J. P.,** The different glycoprotein abnormalities in thrombasthenic and Bernard-Soulier platelets, *Semin. Hematol.,* 16, 234, 1979.
61. **Ginsberg, M. H., Lightsey, A., Kunicki, T. J., Kaufman, A., Marguerie, G., and Plow, E. F.,** Divalent cation regulation of the surface orientation of platelet membrane glycoprotein IIb: correlation with fibrinogen binding function and definition of a novel variant of Glanzmann's thrombasthenia, *J. Clin. Invest.,* 78, 1103, 1986.
62. **Nurden, A. T., Rosa, J.-P., Fournier, D., Legrand, C., Didry, D., Parquet, A., and Pidard, D.,** A variant of Glanzmann's thrombasthenia with abnormal glycoprotein IIb-IIIa complexes in the platelet membrane, *J. Clin. Invest.,* 79, 962, 1987.

63. **Jennings, L. K. and Phillips, D. R.,** Purification of glycoproteins IIb and III from human platelet plasma membranes and characterization of a calcium-dependent glycoprotein IIb-III complex, *J. Biol. Chem.*, 257, 10458, 1982.
64. **Donati, M. B., Balconi, G., Remuzzi, G., Borgia, R., Morasca, L., and de Gaetano, G.,** Skin fibroblasts from a patient with Glanzmann's thrombasthenia do not induce fibrin clot retraction [letter], *Thromb. Res.*, 10, 173, 1977.
65. **Steinberg, B. M., Smith, K., Colozzo, M., and Pollack, R.,** Establishment and transformation diminish the ability of fibroblasts to contract a native collagen gel, *J. Cell Biol.*, 87, 304, 1980.
66. **Bennett, J. S. and Vilaire, G.,** Exposure of platelet fibrinogen receptors by ADP and epinephrine, *J. Clin. Invest.*, 64, 1393, 1979.
67. **Plow, E. F., Marguerie, G. A., and Ginsberg, M. H.,** Fibronectin binding to thrombin-stimulated platelets: evidence to fibrin(ogen) independent and dependent pathways, *Blood*, 66, 26, 1985.
68. **Lawler, J. W., Slayter, H. S., and Coligan, J. E.,** Isolation and characterization of a high molecular weight glycoprotein from human blood platelets, *J. Biol. Chem.*, 253, 8609, 1978.
69. **Lahav, J., Lawler, J., and Gimbrone, M. A.,** Thrombospondin interactions with fibronectin and fibrinogen. Mutual inhibition in binding, *Eur. J. Biochem.*, 145, 151, 1984.
70. **McPherson, J., Sage, H., and Bornstein, P.,** Isolation and characterization of a glycoprotein secreted by aortic endothelial cells in culture. Apparent identity with platelet thrombospondin, *J. Biol. Chem.*, 256, 11330, 1981.
71. **Ginsberg, M. H., Wencel, J. D., White, J. G., and Plow, E. F.,** Binding of fibronectin to alpha-granule-deficient platelets, *J. Cell Biol.*, 97, 571, 1983.
72. **Ruggeri, Z. M., Bader, R., and DeMarco, L.,** Glanzmann thrombasthenia: deficient binding of von Willebrand factor to thrombin-stimulated platelets, *Proc. Natl. Acad. Sci. U.S.A.*, 79, 6038, 1982.
73. **Nurden, A. T. and Caen, J. P.,** An abnormal platelet glycoprotein pattern in three cases of Glanzmann's thrombasthenia, *Br. J. Haematol.*, 28, 253, 1974.
74. **Phillips, D. R. and Agin, P. P.,** Platelet membrane defects in Glanzmann's thrombasthenia, *J. Clin. Invest.*, 60, 535, 1977.
75. **Plow, E. F., McEver, R. P., Coller, B. S., Woods, V. L., Jr., Marguerie, G. A., and Ginsberg, M. H.,** Related binding mechanisms for fibrinogen, fibronectin, von Willebrand factor, and thrombospondin on thrombin-stimulated human platelets, *Blood*, 66, 724, 1985.
76. **Doolittle, R. F.,** Fibrinogen and fibrin, *Sci. Am.*, 106, 126, 1981.
77. **Pierschbacher, M. D., Ruoslahti, E., Sundelin, J., Lind, P., and Peterson, P. A.,** The cell attachment domain of fibronectin. Determination of the primary structure, *J. Biol. Chem.*, 257, 9593, 1982.
78. **Sadler, J. E., Shelton-Inloes, B. B., Sorace, J. M., Harlan, J. M., Titani, K., and Davie, E. W.,** Cloning and characterization of two cDNAs coding for human von Willebrand factor, *Proc. Natl. Acad. Sci. U.S.A.*, 82, 6394, 1985.
79. **Pierschbacher, M. D. and Ruoslahti, E.,** Variants of the cell recognition site of fibronectin that retain attachment-promoting activity, *Proc. Natl. Acad. Sci. U.S.A.*, 81, 5985, 1984.
80. **Ginsberg, M. H., Pierschbacher, M. D., Ruoslahti, E., Marguerie, G., and Plow, E.,** Inhibition of fibronectin binding to platelets by proteolytic fragments and synthetic peptides which support fibroblast adhesion, *J. Biol. Chem.*, 260, 3931, 1985.
81. **Plow, E. F., Pierschbacher, M. D., Ruoslahti, E., Marguerie, G. A., and Ginsberg, M. H.,** The effect of Arg-Gly-Asp-containing peptides on fibrinogen and von Willebrand factor binding to platelets, *Proc. Natl. Acad. Sci. U.S.A.*, 82, 8057, 1985.
82. **Gartner, T. K. and Bennett, J. S.,** The tetrapeptide analogue of the cell attachment site of fibronectin inhibits platelet aggregation and fibrinogen binding to activated platelets, *J. Biol. Chem.*, 260, 11891, 1985.
83. **Haverstick, D. M., Cowan, J. F., Yamada, K. M., and Santoro, S. A.,** Inhibition of platelet adhesion to fibronectin, fibrinogen, and von Willebrand factor substrates by a synthetic tetrapeptide derived from the cell-binding domain of fibronectin, *Blood*, 66, 946, 1985.
84. **Pytela, R. P., Pierschbacher, M. D., Ginsberg, M. H., Plow, E. F., and Ruoslahti, E.,** Platelet membrane glycoprotein IIb/IIIa: member of a family of Arg-Gly-Asp-specific adhesion receptors, *Science*, 231, 1559, 1986.
85. **Baldassare, J. J., Kahn, R. A., Knipp, M. A., and Newman, P. J.,** Reconstruction of platelet proteins into phospholipid vesicles. Functional proteoliposomes, *J. Clin. Invest.*, 75, 35, 1985.
86. **Parise, L. V. and Phillips, D. R.,** Reconstitution of the purified platelet fibrinogen receptor fibrinogen binding properties of the glycoprotein IIb-IIIa complex, *J. Biol. Chem.*, 260, 10698, 1985.
87. **Parise, L. V. and Phillips, D. R.,** Fibronectin-binding properties of the purified platelet glycoprotein IIb-IIIa complex, *J. Biol. Chem.*, 261, 14011, 1986.
88. **Kloczewiak, M., Timmons, S., and Hawiger, J.,** Localization of a site interacting with human platelet receptor on carboxy-terminal segment of human fibrinogen gamma chain, *Biochem. Biophys. Res. Commun.*, 107, 181, 1982.
89. **Timmons, S., Kloczewiak, M., and Hawiger, J.,** ADP-dependent common receptor mechanism for binding of von Willebrand factor and fibrinogen to human platelets, *Proc. Natl. Acad. Sci. U.S.A.*, 81, 4935, 1984.

90. **Plow, E. F., Srouji, A. H., Meyer, D., Marguerie, G., and Ginsberg, M. H.,** Evidence that three adhesive proteins interact with a common recognition site on activated platelets, *J. Biol. Chem.,* 259, 5388, 1984.
91. **Lam, S.C.-T., Plow, E. F., Smith, M. A., Andrieux, A., Ryckwaert, J.-J., Marguerie, G., and Ginsberg, M. H.,** Evidence that arginyl-glycyl-aspartate peptides and fibrinogen gamma chain peptides share a common binding site on platelets, *J. Biol. Chem.,* 262, 947, 1987.
92. **Santoro, S. A. and Lawing, W. J., Jr.,** Competition for related but nonidentical binding sites on the glycoprotein IIb-IIIa complex by peptides derived from platelet adhesive proteins, *Cell,* 48, 867, 1987.
93. **Plow, E. F., Pierschbacher, M. D., Ruoslahti, E., Marguerie, G., and Ginsberg, M. H.,** Arginyl-glycyl-aspartic acid sequences and fibrinogen binding to platelets, *Blood,* 70, 110, 1986.
94. **Fitzgerald, L. A., Charo, I. F., and Phillips, D. R.,** Human and bovine endothelial cells synthesize membrane proteins similar to human platelet glycoproteins IIb and IIIa, *J. Biol. Chem.,* 260, 10893, 1985.
95. **Charo, I. F., Fitzgerald, L. A., Steiner, B., Rall, S. C., Jr., Bekeart, L. S., and Phillips, D. R.,** Platelet glycoproteins IIb and IIIa: evidence for a family of immunologically and structurally related glycoproteins in mammalian cells, *Proc. Natl. Acad. Sci. U.S.A.,* 83, 8351, 1986.
96. **Plow, E. F., Loftus, J., Levin, E., Fair, D., Dixon, D., Forsyth, J., and Ginsberg, M. H.,** Immunologic relationship between platelet membrane glycoprotein GPIIb/IIIa and cell surface molecules expressed by a variety of cells, *Proc. Natl. Acad. Sci. U.S.A.,* 83, 6002, 1986.
97. **Burns, G. F., Cosgrove, L., Triglia, T., Beall, J. A., Lopez, A. F., Werkmeister, J. A., Begley, C. G., Haddad, A. P., d'Apice, A. J. F., Vadas, M. A., and Cawley, J. C.,** The IIb-IIIa glycoprotein complex which mediates platelet aggregation is directly implicated in leukocyte adhesion, *Cell,* 45, 269, 1986.
98. **Thiagarajan, P., Shapiro, S. S., Levine, E., DeMarco, L., and Yalcin, A.,** A monoclonal antibody to human platelet glycoprotein IIIa detects a related protein in cultured human endothelial cells, *J. Clin. Invest.,* 75, 896, 1985.
99. **Ginsberg, M. H., Loftus, J. C., Ryckwaert, J.-J., Pierschbacher, M., Pytela, R., Ruoslahti, E., and Plow, E. F.,** Immunochemical and amino-terminal sequence comparison of two cytoadhesins indicates they contain similar or identical beta subunits and distinct alpha subunits, *J. Biol. Chem.,* 262, 5437, 1987.
100. **Sanchez-Madrid, F., Nagy, J. A., Robbins, E., Simon, P., and Springer, T. A.,** A human leukocyte differentiation antigen family with distinct a-subunits and a common G-subunit: the lymphocyte function-associated antigen (LFA-1), the C3bi complement receptor (OKMI/Mac-1), and the p150,95 molecule, *J. Exp. Med.,* 158, 1785, 1983.
101. **Hemler, M. E., Huang, C., and Schwarz, L.,** The VLA protein family. Characterization of five distinct cell surface heterodimers each with a common 130,000 molecular weight beta subunit, *J. Biol. Chem.,* 262, 3300, 1987.
102. **Kishimoto, T. K., O'Connor, K., Lee, A., Roberts, T. M., and Springer, T. A.,** Cloning of the B subunit of the leukocyte adhesion proteins: homology to an extracellular matrix receptor defines a novel supergene family, *Cell,* 48, 681, 1987.
103. **Tamkun, J. W., DeSimone, D. W., Fonda, D., Patel, R. S., Buck, C., Horwitz, A. F., and Hynes, R. O.,** Structure of integrin, a glycoprotein involved in transmembrane linkage between fibronectin and actin, *Cell,* 46, 271, 1986.
104. **Fitzgerald, L. A., Steiner, B., Rall, S. C., Jr., Lo, S.-S., and Phillips, D. R.,** Protein sequence of endothelial glycoprotein IIIa derived from a cDNA clone. Identity with platelet glycoprotein IIIa and similarity to "integrin", *J. Biol. Chem.,* 262, 3936, 1987.
105. **Hynes, R. O.,** Integrins: a family of cell surface receptors, *Cell,* 48, 549, 1987.
106. **Giancotti, F. G., Languino, L. R., Zanetti, A., Peri, G., Tarone, G., and Dejana, E.,** Platelets express a membrane protein complex immunologically related to the fibroblast fibronectin receptor and distinct from GPIIb-IIIa, *Blood,* 69, 1535, 1987.
107. **Pischel, K. D., Bluestein, H. G., and Woods, V. L.,** Platelet glycoproteins Ia, Ic, and IIIa are physiochemically indistinguishable from the very late activation antigens adhesion/related proteins of lymphocytes and other cell types, *J. Clin. Invest.,* 81, 505, 1988.
108. **Santoro, S. A.,** Identification of a 160,000 dalton platelet membrane protein that mediates the initial divalent cation-dependent adhesion of platelets to collagen, *Cell,* 46, 913, 1986.
109. **Nieuwenhuis, H. K., Akkerman, J. W. N., Houdijk, W. P. M., and Sixma, J. J.,** Human blood platelets showing no response to collagen fail to express surface glycoprotein Ia, *Nature (London),* 318, 470, 1985.
110. **Kunicki, T. J., Nugent, D. J., Staats, S. J., Orchekowski, R. P., Wayner, E. A., and Carter, W. G.,** The human fibroblast class II extracellular matrix receptor mediates platelet adhesion to collagen and is identical to the platelet glycoprotein Ia-IIa complex, *J. Biol. Chem.,* 263, 4516, 1988.
111. **Sonnenberg, A., Janssen, H., Hogervorst, F., Calafat, J., and Hilgers, J.,** A complex of platelet glycoproteins Ic and IIa identified by a rat monoclonal antibody, *J. Biol. Chem.,* 262, 10376, 1987.

112. **Hemler, M. E., Crouse, C., Takada, Y., and Sonnenberg, A.,** Multiple very late antigen (VLA) heterodimers on platelets: evidence for distinct VLA-2, VLA-5 (fibronectin receptor) and VLA-6 structures, *J. Biol. Chem.,* 263, 7660, 1988.
113. **Piotrowicz, R. S., Orchekowski, R. P., Nugent, D. J., Yamada, K. Y., and Kunicki, T. J.,** Glycoprotein Ic-IIa functions as an activation-independent fibronectin receptor on human platelets, *J. Cell Biol.,* 106, 1359, 1988.
114. **Ruoslahti, E. and Pierschbacher, M. D.,** New prespectives in cell adhesion: RGD and integrins, *Science,* 238, 491, 1987.
115. **Loftus, J. C. and Albrecht, R. M.,** Redistribution of the fibrinogen receptor of human platelets after surface activation, *J. Cell Biol.,* 99, 822, 1984.
116. **Isenberg, W. M., McEver, R. P., Phillips, D. R., Shuman, M. A., and Bainton, D. F.,** The platelet fibrinogen receptor: an immunogold-surface replica study of agonist-induced ligand binding and receptor clustering, *J. Cell Biol.,* 104, 1655, 1987.
117. **Singer, I. I., Kawka, D. W., Scott, S., Mumford, R. A., and Lark, M. W.,** The fibronectin cell attachment sequence Arg-Gly-Asp-Ser promotes focal contact formation during early fibroblast attachment and spreading, *J. Cell Biol.,* 104, 573, 1987.
118. **Parise, L. V., Helgerson, S. L., Steiner, B., Nannizzi, L., and Phillips, D. R.,** Synthetic peptides derived from fibrinogen and fibronectin change the conformation of purified platelet glycoprotein IIb-IIIa, *J. Biol. Chem.,* 262, 12597, 1987.
119. **Shadle, P. J., Ginsberg, M. H., Plow, E. F., and Barondes, S. H.,** Platelet-collagen adhesion: inhibition by a monoclonal antibody that binds glycoprotein IIb, *J. Cell Biol.,* 99, 2056, 1984.
120. **Frelinger, A. L., III, Lam, S.C.-T., Smith, M. A., Plow, E. F., and Ginsberg, M. H.,** Arg-Gly-Asp peptides induce changes in platelet membrane glycoproteins IIb-IIIa associated with loss of adhesive protein binding function, *Clin. Res.,* 35, 598A, 1987.
121. **Loftus, J., Sorge, J., Plow, E. F., Marguerie, G. A., and Ginsberg, M. H.,** Molecular cloning of cDNA coding for a divalent cation regulated epitope in platelet membrane GPIIb, *Clin. Res.,* 34, 660A, 1986.
122. **Schmid, C. W. and Jelinek, W. R.,** The Alu family of dispersed repetitive sequences, *Science,* 216, 1065, 1982.
123. **Lerner, R. A., Green, N., Alexander, H., Liu, F.-T, Sutcliffe, J. G., and Shinnick, T. M.,** Chemically synthesized peptides predicted from the nucleotide sequence of the hepatitis B virus genome elicit antibodies reactive with the native envelope protein of Dane particles, *Proc. Natl. Acad. Sci. U.S.A.,* 78, 3403, 1981.
124. **Tanaka, T.,** Efficient generation of antibodies to oncoproteins by using synthetic peptide antigens (oncogenes/transformed cells), *Proc. Natl. Acad. Sci. U.S.A.,* 82, 3400, 1985.
125. **Doolittle, R. F.,** in *Of Urfs and Orfs, A Primer on How to Analyze Derived Amino Acid Sequences,* Eds. University Science Books, Mill Valley, Calif., 4, 187, 1986.
126. **Hopp, T. P. and Woods, K. R.,** Prediction of protein antigenic determinants from amino acid sequences, *Proc. Natl. Acad. Sci. U.S.A.,* 78, 3824, 1981.
127. **Tainer, J. A., Getzoff, E. D., Alexander, H., Houghten, R. A., Olson, A. J., Lerner, R. A., and Hendrickson, W. A.,** The reactivity of anti-peptide antibodies is a function of the atomic mobility of sites in a protein, *Nature (London),* 312, 127, 1984.
128. **Geysen, H. M., Tainer, J. A., Rodda, S. J., Mason, T. J., Alexander, H., Getzoff, E. D., and Lerner, R. A.,** Chemistry of antibody binding to a protein, *Science,* 235, 1184, 1987.
129. **Wolf, H., Modrow, S., Motz, M., Jameson, B., Hermann, G., and Fortsch, B.,** An integrated family of amino acid sequence analysis programs, *CABIOS,* 1987.
130. **McEver, R. P., Baenziger, N. L., and Majerus, P. W.,** Isolation and quantitation of the platelet membrane glycoprotein deficient in thrombasthenia using a monoclonal hybridoma antibody, *J. Clin. Invest.,* 66, 1311, 1980.
131. **Plow, E. F. and Marguerie, G.,** Inhibition of fibrinogen binding to human platelets by the tetrapeptide glycyl-L-prolyl-L-arginyl-L-proline, *Proc. Natl. Acad. Sci. U.S.A.,* 79, 3711, 1982.
132. **George, J. N., Pickett, E. B., Saucerman, S., McEver, R. P., Kunicki, T. J., Kieffer, N., and Newman, P. J.,** Platelet surface glycoproteins. Studies on resting and activated platelets and platelet membrane microparticles in normal subjects, and observations in patients during adult respiratory distress syndrome and cardiac surgery, *J. Clin. Invest.,* 78, 340, 1986.
133. **Hsu-Lin, S.-C., Berman, C. L., Furie, B. C., August, D., and Furie, B.,** A platelet membrane protein expressed during platelet activation and secretion. Studies using a monoclonal antibody specific for thrombin-activated platelets, *J. Biol. Chem.,* 259, 9121, 1984.
134. **Shattil, S. J., Hoxie, J. A., Cunningham, M., and Brass, L. F.,** Changes in the platelet membrane glycoprotein IIb-IIIa complex during platelet activation, *J. Biol. Chem.,* 260, 11107, 1985.
135. **Pegels, J. G., Bruynes, E. C. E., Engelfriet, C. P., and von dem Borne, A. E. G. Kr.,** Pseudothrombocytopenia: an immunologic study on platelet antibodies dependent on ethylene diamine tetra-acetate, *Blood,* 59, 157, 1982.

Chapter 10

FIBRONECTIN AND BLOOD COAGULATION

Dudley G. Moon and John E. Kaplan

TABLE OF CONTENTS

I. INTRODUCTION

Blood coagulation is the process by which the soluble plasma protein fibrinogen is converted into an insoluable matrix (clot) of polymerized fibrin. This coagulum promotes hemostasis while serving as a provisional matrix for cell migration and wound healing. Coagulation is a highly regulated process requiring the participation of a series of proteolytic enzymes and cofactors. The final events which result directly in clot formation are, first, the cleavage of fibrinogen by the serine protease thrombin resulting in fibrin monomer formation; second, the spontaneous noncovalent polymerization of fibrin monomers, and finally, the covalent stabilization of fibrin-fibrin interactions via the action of activated fibrin stabilizing factor (factor XIIIa). The initiation and process of blood coagulation is influenced by a number of plasma, cellular, and matrix components including fibronectin. This chapter will review the evidence suggesting that fibronectin modulates coagulation by virtue of its interactions with fibrin(ogen), incorporation into fibrin clots, and possible involvement in the interactions between fibrin and cells and other proteins. Finally, the potential physiological significance of such interactions will be discussed. Before proceeding, we will begin with a brief review of the actual mechanism of fibrin clot formation. For the reader not familiar with the coagulation cascade leading to thrombin generation, the reaction sequence is well reviewed by Colman et al.[1]

II. REVIEW OF FIBRIN CLOT FORMATION

Fibrinogen is a hexameric glycoprotein of 340,000 Da normally found in the plasma at a concentration of 2 to 3 mg/ml (see Reference 2 for review). The two pairs of the three polypeptide chains (A-alpha, B-beta, gamma) within each fibrinogen molecule are covalently linked by 29 intra- and interchain disulfide bonds. Early electron microscopic studies of fibrinogen by Hall and Slayter[3] showed a trinodular structure with a central globular region now known to correspond to the E domain of the molecule containing the aminotermini of all six polypeptide chains. The two terminal globular regions in the Hall and Slayter model correspond to the D domains of the molecule. While the Hall and Slayter model has been shown to be generally correct, recent studies at higher resolution have revealed more of the morphological details of subdomain structure.[4] Normal fibrinogen is soluble at physiological pH and ionic strength. However, following activation of the coagulation cascade, fibrinogen is converted to the far less soluble fibrin form by the action of the proteolytic enzyme thrombin (Factor IIa). Thrombin cleaves the aminoterminal 16 amino acids (fibrinopeptide A) from the alpha chains and the aminoterminal 14 residues (fibrinopeptide B) from the beta chain.

Once fibrinopeptide A is released, the fibrin ''monomers'' begin to self-associate. The contribution of fibrinopeptide B release to fibrin formation is not fully understood, but it does affect clot structure.[5,6] Fibrin monomers polymerize as a half-staggered overlap of molecules with the E domain of one fibrin interacting with the D domains of two other fibrins. Initial polymerization appears to occur via chain elongation of a two-molecule-thick protofibril followed by subsequent lateral polymer growth and fibril branching.[2,7] This process can be monitored *in vitro* by turbidity measurements at 600 nm.[7]

Coincident with fibrin gel formation, factor XIII (plasma transglutaminase, fibrinoligase, or fibrin stabilizing factor) is converted to the active form (XIIIa) by the action of thrombin in the presence of calcium (see Reference 8 for review). Factor XIIIa catalyzes a transamidation reaction between a glutamine residue and lysine residue at several positions in the gamma and alpha chains of adjacent fibrin monomers. The gamma chain sites appear to react first with Glu-397 and Lys-405 cross-linking these residues to another gamma chain in an antiparallel manner.[9] Formation of such gamma dimers is very rapid and may be complete by the end of the protofibril phase of gelation. Cross-linking of the alpha chains is slower and contributes to stabilization and lateral strength of the forming fibrin clot.

III. INTERACTION OF FIBRONECTIN WITH FIBRIN(OGEN)

The original association between fibrinogen and fibronectin was made by Morrison et al.[10] who reported a "cold-insoluble globulin" as a component of crude fibrinogen preparations purified from Cohn fraction I. A preparation enriched in cold-insoluble globulin could be prepared by cold precipitation. Edsall et al.[11] suggested an interaction between the cold-insoluble globulin and fibrinogen in the enriched fraction based upon electrophoretic and sedimentary behavior. These initial observations have been reinforced and extended by additional studies of cold-induced and heparin-induced precipitation of normal and pathological plasmas. In this regard, a series of studies in the 1950s and 1960s indicated that cryoprecipitates readily occur in heparinized plasma and were increased in plasma under certain pathological conditions.[12-15] A component identical to the previously identified cold-insoluble globulin as well as fibrinogen was considered essential to this process.[13]

The purification and characterization of the cold-insoluble globulin, now known to be plasma fibronectin, by Mosesson and Umfleet[16] allowed more complete evaluation of the process of cryoprecipitation, in the presence[17] and absence[18] of heparin. It was found that in the presence of heparin, fibronectin and fibrinogen form complexes at 22°C, but these complexes do not precipitate until the temperature is lowered to 2°C. Fibrinogen was not essential for precipitation and did not form a cryoprecipitate with heparin. However, fibrinogen could complex with the fibronectin-heparin precipitate during formation as a result of its affinity for fibronectin at low temperatures. Fractions of fibrinogen lacking the carboxyterminus of the alpha chain did not precipitate, suggesting that the component of fibrinogen which interacts with fibronectin is located in that region. In the absence of heparin, the reaction of fibronectin and fibrin(ogen) was necessary and sufficient for cryoprecipitation if the plasma was first treated with small amounts of thrombin. Although fibronectin bound to fibrinogen at 4°C, mixtures of fibronectin and fibrinogen did not cryoprecipitate. Fibronectin also bound to fibrin at both 4°C and 22°C (37°C was not evaluated). Therefore, the authors suggested that fibronectin acted as a nucleus with multiple binding sites for fibrin-fibrinogen complexes and those complexes formed the cryoprecipitate. As with heparin-induced cryoprecipitation, fibrin-fibrinogen complexes lacking the carboxyterminal region of the alpha chain failed to precipitate. Ruoslahti and Vaheri[19] produced evidence, based primarily on loss from soluble phase during clotting, that fibronectin bound to fibrin.

IV. FIBRONECTIN INCORPORATION INTO FIBRIN CLOTS

Fibronectin is incorporated into the clot during fibrin polymerization. Indeed, the concentration of fibronectin in serum is 20 to 50% less than that in plasma.[16] Much of the incorporation of fibronectin into the clot appears to be the result of factor XIIIa covalently crosslinking fibronectin to the alpha chain of fibrin. Fibronectin is one of three plasma proteins that bind to the fibrin clot; the other two are alpha$_2$-plasmin inhibitor[20] and thrombospondin.[21] Fibronectin is the major nonfibrin protein component of the clot and can constitute 4 to 5% of the mass of a plasma-derived clot.[23]

Noncovalent binding of fibronectin to fibrin appears to be stronger than its interaction with fibrinogen. There are at least two fibrin binding sites on each subunit of fibronectin. The first (and probably most significant) binding site is the aminoterminal region of the molecule in an area of repeats of type I homology.[23] Localization of the site was determined by affinity chromatography of proteolytically fragmented fibronectin. Such studies have shown that the 27- to 29-kDa plasmin-generated fragment consisting of the aminoterminal portion of the fibronectin molecule binds to fibrin-Sepharose. Binding of this fragment to fibrin-Sepharose occurs at physiological pH and ionic strength. A second fibrin binding site has been described by Sekiguchi et al.[24] This site occurs in the extreme carboxyterminal region of fibronectin and does not bind well to fibrin at physiological salt concentrations. A third fibrin-binding site on fibronectin has been proposed by Seidl and Hormann.[25] This site overlaps the gelatin-binding site.

The existence of multiple binding sites for fibrin on fibronectin suggests noncovalent binding might be an important interaction between these two molecules. However, Mosher demontrated that fibronectin incorporation into plasma-derived clots requires calcium and factor XIIIa at room temperature or 37°C.[26] Only at 2°C was there significant fibronectin incorporation into clots formed from factor XIII-deficient plasma. These data suggest that at physiological temperatures, noncovalent binding of fibronectin to fibrin contributes little to the effects of fibronectin on clot formation.

The importance of factor XIIIa in covalently coupling fibronectin to polymerizing fibrin is supported by a number of observations. Fibronectin has been demonstrated to be a substrate for factor XIIIa[26] and fibronectin can be crosslinked to the alpha chain of fibrin.[27] Tamaki and Aoki[28] have examined the relative incorporation of $alpha_2$-plasmin inhibitor and fibronectin by factor XIIIa into clotting fibrin and reported that the two proteins are cross-linked to the alpha chains of fibrin independently. Okada et al.[29] observed that batroxobin-activated factor XIIIa, which catalyzes gamma-gamma cross-links but not alpha-alpha chain cross-links in fibrin, did not covalently incorporate fibronectin into the clot. Many of the aspects of factor XIIIa cross-linking of fibronectin to fibrin have been reviewed in depth by McDonagh et al.[30]

V. INFLUENCE OF FIBRONECTIN ON FIBRIN POLYMERIZATION AND STRUCTURE

Whereas the observation that fibronectin can be incorporated into the forming fibrin clot is well accepted, its influence on initiation of clot formation is less clear. Kaplan and Snedeker[31] reported that fibronectin could inhibit urea-solubilized fibrin monomer precipitation. Holm and Brosstad[32] reported no effect of fibronectin upon polymerization of fibrin formed with reptilase. Interestingly, Sherman and Lee[33] reported data suggesting that unlike fibrin formed from the action of thrombin on fibrinogen, fibrin formed by the action on reptilase on fibrinogen did not interact with fibronectin. Bang et al.[34] demonstrated the presence of fibronectin in soluble fibrin complexes. Thus, one "physiological" function of plasma fibronectin may be to maintain fibrin in a soluble complex and help prevent inappropriate systemic fibrin deposition. On the other hand, fibrinogen is a much better "solubilizing" agent when complexed with fibrin[35] and normally circulates at a tenfold greater molar concentration than fibronectin. The observation that fibronectin enhances macrophage binding of fibrin (see Section VII.A. Mononuclear Phagocytes) may relate directly to this point. This would then be expected to accelerate clearance of soluble fibrin:fibrinogen:fibronectin complexes from the vasculature. Thus, the primary role of fibrinogen in the complex may be maintenance of solubility, while fibronectin may contribute to removal of complexes from the circulation.

The effect of fibronectin on clot polymerization and structure is also complex. Consistent with the observation by Kaplan and Snedeker[31] is the report by Niewiarowska and Cierniewski[36] that fibronectin increased the lag time of thrombin-induced plasma derived fibrin polymerization and decreased clot turbidity at 600 nm. Lag time correlates with the end of formation of linear polymers of protofibrils and final turbidity changes correlate with the extent of lateral polymerization and branching. The fibronectin effects observed by Niewiarowska and Cierniewski[36] were dependent upon the ratio of fibrinogen to fibronectin with complete inhibition occurring at a ratio of 1:2. Bale et al.[37] observed that fibronectin alone (i.e., without XIIIa or thrombospondin) slightly increased lag time (time to increase in turbidity at 600 nm) and had no effect on final turbidity at a fibrinogen/fibronectin ratio of 3.5:1 and 1:1. In the presence of XIIIa, fibronectin decreased lag time by 7 to 10% and decreased final turbidity by 15%. Fibronectin did not interfere with the effect of thrombospondin on clot formation. The major differences between these two studies were the thrombin concentrations (1.0 U/ml vs. 0.008 U/ml, respectively) and fibrinogen concentrations (7.0 mg/ml vs. 0.270 mg/ml, respectively). Since these differences resulted in control lag times of less than 1 min vs. 18.9 min (7.8 min with XIIIa), it appears that the polymerization rates were quite different in these two studies and therefore the mechanism of polymerization may have been altered. The issue remains to be resolved, since Okada et al.[29] reported that, in the presence of factor XIIIa, fibronectin did not affect clotting time and that fibronectin increased turbidity of polymerizing fibrin. These authors used a purified protein system with a fibrinogen concentration of 0.54 mg/ml and fibronectin concentration of 1.8 mg/ml.

Fibronectin also has been reported to affect mechanical strength of the clot. Kamykowski et al.[38] reported that under conditions promoting fine clots, incorporation of factor XIIIa-cross-linked to fibronectin decreased shear modulus (a measure of clot stiffness) by a factor of 0.48, whereas under conditions promoting coarse clots, shear modulus increased by a factor of 2.0. It should be noted that these authors used a purified system with 7.0 mg/ml fibrinogen and 1.2 mg/ml fibronectin or a fibrinogen:fibronectin ratio of 6:1 (normal = 10 to 20:1). Using plasma-derived clots, Chow et al.[39] observed that fibronectin slightly decreased clot rigidity (2 to 3% decrease in dynamic rigidity modulus per 0.1 mg/ml fibronectin added).

It is difficult to reconcile all of the diverse results which have been reported. However, an overall synthesis of these findings leads to the conclusion that fibronectin interacts with polymerizing fibrin primarily via factor XIIIa cross-linking of fibronectin to the COOH-terminal end of the fibrin alpha chain. Under some experimental conditions *(ex plasma)* fibronectin can modulate fibrin polymerization. However, in plasma (where the molar ratio of fibronectin:fibrinogen is 1:10) fibronectin does not significantly inhibit clot formation despite incorporation of 40 to 50% of the plasma fibronectin into the clot. Thus, the direct effects of fibronectin on clot polymerization would appear to be minor. On the other hand, incorporation of fibronectin into the clot may be very important in the interaction of cells with fibrin. Additionally, slight differences in clot structure, due to the presence of fibronectin, may have little effect on the overall mechanical stability of the clot but allow more rapid penetration of fibrinolytic factors leading to more rapid resolution of the clot.

VI. INFLUENCE OF FIBRONECTIN ON FIBRINOLYSIS

Because of its incorporation into fibrin clots, several investigators have evaluated the influence of fibronectin on fibrinolysis. Iwanaga et al.[27] reported that fibronectin enhanced urokinase activation on plasminogen in a concentration-dependent manner in purified systems (as assessed by esterolysis, caseinolysis, and fibrinolysis). Gilboa and Kaplan[40] reported effects of fibronectin on the fibrinolytic system at several levels *in vitro*. Fibronectin caused a time- and concentration-dependent increase in the amidolytic activity of tissue plasminogen activator (TPA) but not urokinase. The effect was attributable to a decrease in K_m with no change in V_{max}. Fibronectin produced a concentration-dependent increase in the activation of lys-plasminogen by both TPA and urokinase. These effects were also the result of a decrease in K_m. A weaker enhancing effect of fibronectin was noted on the amidolytic activity of plasma. Fibronectin accelerated clot lysis in isolated systems with both TPA activation of plasminogen and direct addition of plasmin. This effect was independent of both Ca^{2+} and lysine-binding sites. Clot lysis was delayed in clots prepared from fibronectin-depleted plasma. These findings are consistent with those of Niewiarowska and Cierniewski[41] who reported that fibronectin incorporation into clots rendered them more susceptible to lysis.

The mechanism of these effects is unclear. It is possible that fragments as opposed to intact fibronectin exert the enhancement of fibrinolysis. Plasmin degrades fibronectin extensively[23,42] and limited cleavages by TPA have been recently demonstrated.[43] In addition, Banyai et al.[44] have demonstrated homology between the fibrin binding domain of fibronectin and a 43-amino-acid residue from TPA. Increased fibrinolysis may potentially result from the augmentation of enzyme activity or physical changes in the clot which render it more susceptible to lysis as mentioned in the previous section.

Fibronectin may influence both plasmin and nonplasmin mediated fibrinolysis initiated by phagocytic cells. As discussed in the following section, fibronectin promotes attachment of soluble and particle-bound fibrin to macrophages and the attachment of monocytes to fibrin surfaces. These cells release plasminogen activators as well as enzymes which are fibrinolytic in their own right, such as elastase and cathepsins. Bianco[45] reported that monocytes bound to a surface coated with fibronectin released significantly more plasminogen activator and elastolytic activity than did cells adhering to plastic in the absence of fibronectin.

VII. FIBRONECTIN AND THE INTERACTION OF CELLS AND PROTEINS WITH FIBRIN

Perhaps the most important consequence of the participation of fibronectin in coagulation may relate to the interaction of cells with clots and soluble fibrin. A role for fibronectin has been implicated in the interaction of mononuclear phagocytes[45-50] with fibrin. The demonstration of fibronectin binding to other cell types including platelets, granulocytes, fibroblasts, and endothelial cells[51-55] suggests that fibronectin may influence the interaction of these cells with fibrin(ogen) as well.

A. MONONUCLEAR PHAGOCYTES

The interaction of fibrin and soluble fibrin complexes with mononuclear phagocytic cells has been studied in the presence and absence of plasma fibronectin. Bang et al.[56] and Sherman and co-workers[57-59] have reported soluble fibrin complexes adhere to alveolar and peritoneal macrophages. Most adherent fibrin remains on the cell surface with 10 to 20% internalized,[56] closely corresponding to observations made in our laboratory. The phagocytes examined possessed fibrinolytic activity which resulted in proteolysis of fibrin and the liberation of small molecular weight fragments.[56,58] The fibrinolytic activity was attributed to lysosomal enzymes, and it was determined that the major proteases involved were cathepsin B, an unidentified sulfhydryl protease,[56] and elastase.[57] In a series of studies, Sherman and co-workers[57-59] reported that labelled fibrin-fibrinogen complexes were specifically bound by elicited peritoneal macrophages in a saturable manner. Lysis of this fibrin was enhanced by the presence of plasminogen suggesting a role for macrophage plasminogen activator,[54] although the lysis observed was predominantly plasminogen-independent. Although neither the studies of Bang or Sherman directly addressed the role of fibronectin, they both utilized serum in their media, thus including fibronectin in their experimental systems. Therefore, final interpretation of the results of these studies should take this into account.

Bianco[45] reported tenfold enhancement in the number of human monocytes adherent to fibrin films when these films were preexposed to plasma fibronectin. Cells bound to a surface coated with fibronectin released significantly more plasminogen activator and elastolytic activity than did cells adhering to plastic in the absence of fibronectin. The finding by Bianco[45] that fibronectin augments plasminogen activator release is especially interesting in the light of our findings that fibronectin augments plasminogen activation described in the previous section. Moreover, these studies suggest that release of fibrinolytic enzymes occurs in response to cell surface binding and does not require fibrin internalization. Falcone[48] reported that fibronectin enhanced binding but not internalization of fibrin-coated particles to macrophages. Horsburgh et al.[60] reported that monocyte adherence was greater on films of fibrinogen and fibronectin than on films of either protein alone. Preliminary studies in our laboratory have addressed the effect of incorporation of fibronectin into clots on mononuclear phagocyte accumulation *in vivo*. Fibronectin was incorporated into clots made of fibronectin-deficient plasma and the clots were implanted into the peritoneal cavity of rats. After 4 h, there were significantly greater numbers of mononuclear cells in clots containing fibronectin.

Gonda and Shainoff[61] observed that purified des-A-fibrin monomer incubated with macrophages in the presence of fibrinogen and albumin was specifically bound to the cell surface. The surface-bound fibrin was then internalized by vesicular uptake with much of it being incorporated into lysosomal bodies. This uptake, which took place in the absence of plasma fibronectin, was specifically blocked by a peptide corresponding to the aminoterminus of the fibrin alpha chain. These data indicate that the alpha-chain aminoterminus may have been recognized by a fibronectin-independent macrophage receptor for fibrin.

Hormann and co-workers have published a series[47,49,50] of studies indicating a role for fibronectin and/or an aminoterminal fragment in macrophage uptake of soluble fibrin. Their studies were performed with trypsinized peritoneal exudate cells rather than purified macrophages. Studies in our laboratory and others[42,59,62] have shown that trypsin removes both fibronectin and fibrin receptors from mononuclear cells. Additionally, incubations were carried out overnight at room temperature during which time significant fibronectin is synthesized by phagocytes and fibrin and fibronectin degradation may have occurred.

Colvin[46] has found that inflammatory peritoneal macrophages stained with immunofluorescent reagents for both fibronectin and fibrin(ogen). There was less fibronectin on the surface of resident peritoneal macrophages and alveolar macrophages. Alveolar macrophages lacked surface fibrinogen. Fibrin(ogen) showed a net-like pattern on the cell surface, while fibronectin displayed a granular pattern. When macrophages were incubated with fluorescein-conjugated $F(ab)'_2$ antifibronectin, surface fibronectin aggregated or "capped" in specific regions in 10 to 20 min followed by ingestion. Surface fibrin(ogen) related antigen co-capped or co-aggregated with the fibronectin. Colvin postulated that fibronectin on macrophages might serve as a receptor for fibrin.

We studied the ability of plasma fibronectin to augment macrophage uptake of fibrin as well as the ability of fibrin to influence other fibronectin-mediated macrophage responses. Fibrin, but not fibrinogen, inhibited fibronectin-mediated uptake of gelatinized fixed sheep erythrocytes by monolayers of peritoneal macrophages. Fibronectin increased uptake of soluble fibrin, but not fibrinogen, by peritoneal macrophages and Kupffer cells. Macrophage-associated fibrin in the absence of fibronectin was primarily internalized, whereas the increment in fibrin uptake in the presence of fibronectin remained primarily surface bound as indicated by susceptibility to removal of trypsin. A 25-kDA aminoterminal fibrin-binding fragment also augmented fibrin uptake. More fibrin was associated with the macrophages in the presence of inhibitors of elastase activity, suggesting that an elastase-like enzyme degrades surface-bound fibrin. Trypsinization of cells inhibited uptake of both fibrin and fibronectin. Competitive binding studies indicated the presence of approximately 50,000 high affinity (25 n*M*) fibronectin receptors per cell on peritoneal macrophages.

Based on the collective findings of our laboratory, Gonda and Shainoff,[61] Sherman,[59] Colvin,[46] and Bianco,[45] it appears plausible that fibrin may bind to mononuclear phagocytes in both a direct fibronectin-independent manner and in a fibronectin-dependent manner. If this is the case, then fibrin may well be handled differently by the cells depending upon the amount of fibronectin present.

B. PLATELETS

The interaction of fibronectin with platelets is discussed in detail in Chapter 9 and thus will not be reviewed in detail here. We will, however, consider possible implications of this interaction within the context of the overall hemostatic and thrombotic processes. Platelets subserve several major functions in hemostasis. The first is to aggregate at sites of vascular injury and thereby limit initial blood loss. Second, in the process of forming the "platelet plug", platelets express procoagulant activity, bind factor Va and significantly enhancing prothrombinase activity. Third, platelets mediate clot retraction which is an important aspect of wound closure. Finally, upon activation, platelets secrete a wide variety of substances (including fibronectin) which play a role in initial hemostasis and ultimately wound healing.

Fibronectin influences a number of platelet functions *in vitro*. Fibronectin inhibits platelet aggregation as induced by thrombin or A23187[63] and collagen.[64] On the other hand, attachment and spreading of platelets on collagen substrates[65,66] is enhanced by fibronectin. In work in progress in our laboratory, we have found that clot retraction is accelerated in fibronectin-depleted plasma and retarded in a dose-dependent manner when fibronectin is added back. Thus, a role for fibronectin has been implicated in platelet adhesive and aggregatory processes. Fibronectin binds to the same receptor on activated platelets (the glycoprotein IIbIIIa complex) as does fibrin(ogen) and von Willebrand factor.[67] Therefore, the potential clearly exists for these proteins to mutually influence their interactions with platelets. This, in turn, could modify platelet interaction with platelets and other cells, matrix components, and the fibrin clot.

The net effect *in vivo* of the potential interactions outlined above is poorly understood. Based on studies from our laboratory (described in Section VIII, Disseminated Intravascular Coagulation and Thrombosis), our current working hypothesis is that the presence of fibronectin *in vivo* enhances platelet attachment and spreading on exposed vascular/interstitial matrix components and simultaneously limits platelet aggregation. The inhibition of platelet aggregation may result from fibronectin occupancy of the GPIIbIIIa receptor during the process of forming platelet-matrix attachments. Ultimately, the GPIIbIIIa complexes bound to fibronectin would be sequestered between the basal surface of the platelet and the matrix. Fibronectin binding to GPIIbIIIa would thus preclude fibrinogen binding to this receptor complex. Since fibrinogen binding to GPIIbIIIa is a key step in platelet aggregation,[68] platelet aggregation would be attentuated or blocked.

C. OTHER CELLS AND PROTEINS

The interaction of several other cell types with clots is, at least in part, dependent on or modified by fibronectin. Grinnell et al. demonstrated that fibronectin was necessary for fibroblast adhesion to films of fibrin or fibrinogen.[69] He found dependence on cross-linking and fibronectin concentration and suggested a critical role for fibronectin orientation. Knox et al.,[70] using plasma clots, found that fibronectin was an absolute requirement for fibroblast migration into clots, and migration is dependent upon fibronectin incorporation into the clots. Unlike Grinnell's study, Knox et al.[70] reported that fibronectin was not essential for adhesion and spreading, but accelerated the rate. These differences may be due to orientation of the proteins in these different systems based upon Grinnell's suggestion of the importance of orientation. Histological studies[71] of fibroblast migration and adherence to peritoneally implanted sponges suggest a role for fibronectin-fibrin adhesion. Colvin et al.[72] have reported that fibrin(ogen) and fibronectin codistribute on the surface of fibroblasts. If fibronectin binding was reduced as a result of viral transformation, adherence of fibrinogen was correspondingly reduced.

Histological studies have suggested a role for fibrin-fibronectin binding in vascular endothelial cell migration,[73] and in skin epidermal cell,[74] and corneal epithelial cell[75,76] adherence and migration during wound repair. Recent studies[77] have demonstrated that hepatocytes elaborate a matrix consisting primarily of fibrin and fibronectin.

The potential role of fibronectin in mediating interaction among its protein ligands has not been addressed in detail. However, data suggest a role for fibronectin in the binding of collagen[30] and gelsolin[78] to fibrin clots.

VIII. FIBRONECTIN IN DISSEMINATED INTRAVASCULAR COAGULATION AND THROMBOSIS

Decreased circulating concentration of plasma fibronectin has been associated with disseminated intravascular coagulation in a variety of clinical syndromes and experimental animal models.[79-103] It has been suggested that fibronectin is incorporated into clots and soluble fibrin-fibrinogen complexes. Circulating fibrin-containing complexes may be cleared from the circulation in a fibronectin-dependent manner, and this process may contribute to fibronectin depletion.

Plasma fibronectin deficiency has been reported to be decreased in patients with both disseminated intravascular coagulation[79] and consumptive coagulopathy.[80] Such deficiency has been reported in patients following burn, injury, sepsis, and shock,[81-93] each of which can be associated with coagulopathy. Similarly, fibronectin levels are decreased in animal models of Rocky Mountain spotted fever,[94] some models of sepsis,[95-97] and soft tissue and burn injury.[98,99] Other animal models of sepsis and disseminated intravascular coagulation are not associated with fibronectin deficiency.[100-103]

Isotopic, biochemical, and immunological techniques have indicated that fibronectin may circulate in association with fibrin(ogen) after initiation of coagulation.[33,34,104,105] Sherman and Lee[33] reported that there was a parallel disappearance of radiolabeled pools of fibronectin and fibrinogen after administration of thrombin.

A large component of circulating fibrin is removed from the circulation by the hepatic Kupffer cells, a sessile sinus-lining macrophage. Studies in our laboratory[106,107] (and unpublished observations) have indicated that clearance of radiolabeled fibrin from the blood and accumulation in the liver is decreased following administration of antibody to fibronectin and accelerated by passive administration of exogenous fibronectin.

Fibronectin may also serve as a determinant of fibrin deposition on artificial surfaces. Ihlenfeld et al.[108] demonstrated that fibronectin absorbed to polymeric materials resulted in augmented fibrin deposition.

In vivo studies performed in our laboratory[109] have led us to the working hypothesis that *in vivo* plasma fibronectin plays a largely inhibitory role in regulation of thrombosis. Using an anesthetized rat model, we observed that systemic depletion of fibronectin (by monospecific polyclonal antibody) resulted in an enhanced susceptibility to thrombotic stimuli (intra-arterial infusion of thrombin or ADP) as compared to vehicle controls and antibody controls (pretreated with anti-rat albumin). In this study, "susceptibility" was measured in terms of fibrinogen consumption, fibrin(ogen) degradation product generation, consumption of platelets, and mortality. Mean arterial blood pressures were also measured as an index of the hemodynamic effects of fibronectin depletion. While these observations clearly indicate the importance of fibronectin in response to thrombin or ADP infusion, they do not address mechanism(s) of such action. In view of the known opsonic activity of fibronectin with regard to the reticuloendothelial phagocyte system, one mechanism could be to enhance phagocytic clearance of fibrin and activated platelets from the circulation. As discussed in other sections of this chapter, phagocytosis/lysis of fibrin is, in part, fibronectin enhanced. However, *in vitro* studies[110] have not shown fibronectin to be an opsonin for activated platelets. Alternatively, the direct action of fibronectin on platelets (i.e., binding to GPIIbIIIa, inhibition of aggregation, and enhanced spreading of collagen-attached platelets) may account for the observed effects *in vivo*.

In subsequent studies,[111] we used a porcine vascular injury model to investigate the direct effects of fibronectin on fibrin and platelet deposition. The carotid arteries of six pigs were isolated and clamped at the sternal notch of the common carotid and the external and internal branches. The segment was entered through the external branch and rinsed with phosphate-buffered saline. The distal half of each vessel segment was stripped of its endothelium using an embolectomy catheter. The vessels were then incubated for 5 min with either a fibronectin or control solution.[111] Indium-labeled platelets and ^{125}I-labeled fibrinogen were injected through the femoral vein. Flow was reestablished through the common carotid and the internal branch for 20 min. The animal was sacrificed and 2 cm of normal and stripped segments of artery which had been incubated with and without fibronectin was removed. Platelet and fibrin localization were quantitated relative to a 2-cm femoral artery segment. Stripping increased platelet and fibrin deposition 12- and 36-fold, respectively. The stripped fibronectin-treated segment showed 57% less platelet deposition and 51% less fibrin deposition compared to the stripped control segment. Thus, prior exposure of damaged vessel segments to plasma fibronectin reduced fibrin and platelet deposition.

IX. CONCLUSION

The studies reviewed in this chapter, in aggregate, support a role for fibronectin in the process of blood coagulation. These data indicate that fibronectin can interact with fibrin, that this interaction is enhanced by factor XIIIa, and that subtle changes in clot formation and mechanical properties result. Collectively, however, the available data suggest that the most important role for the interaction of fibronectin with fibrin relates to interactions with cellular elements, promotion of wound healing, the vascular clearance of soluble fibrin from the circulation, and clot susceptibility to lysis. *In vivo* studies suggest that the net effect of fibronectin is antithrombotic.

ACKNOWLEDGMENTS

The authors wish to acknowledge the assistance of Maureen Davis in the preparation of this manuscript. Studies were supported by grants GM-25946 and HL-32418 from the National Institutes of Health.

REFERENCES

1. **Colman, R. W., Marder, V. J., Salzman, E. W., and Hirsh, J.,** Overview of hemostasis, in *Hemostasis and Thrombosis,* 2nd ed., Colman, R. W., Hirsh, J., Marder, V. J., and Salzman, E. W., Eds., J. B. Lippincott, Philadelphia, 1987, chap. 1.
2. **Doolittle, R. F.,** Fibrinogen and fibrin, *Annu. Rev. Biochem.,* 53, 195, 1984.
3. **Hall, C. E. and Slayter, H. S.,** The fibrinogen molecule: its size, shape and model of polymerization, *J. Biophys. Biochem. Cytol.,* 5, 11, 1959.
4. **Weisle, J. W., Stauffacher, C. V., Bullitt, E., and Cohen, C. A.,** A model for fibrinogen: domains and sequence, *Science,* 230, 1386, 1985.
5. **Shen, L. L., Hermans, J., McDonagh, J., and McDonagh, R. P.,** Role of fibrinopeptide B release: comparison of fibrins produced by thrombin and ancrod, *Am. J. Physiol.,* 232, 629, 1977.
6. **Mosesson, M. W., Diorio, J. P., Muller, M. F., Shainoff, J. R., Siebenlist, K. R., Amrani, D. L., Homandberg, G. A., Soria, J., Soria, C., and Samana, M.,** Studies on the ultrastructure of fibrin lacking fibrinopeptide B, *Blood,* 69, 1073, 1987.
7. **Carr, M. E. and Hermans, J.,** Size and density of fibrin fibers from turbidity, *Macromolecules,* 11, 46, 1978.
8. **Lorand, L. and Gotoh, T.,** Fibrinoligase: the fibrin stabilizing factor system, *Meth. Enzymol.,* 19, 770, 1970.

9. **Olexa, S. and Budzynski, A. S.,** Localization of a fibrin polymerization site, *J. Biol. Chem.*, 256, 3544, 1981.
10. **Morrison, P. R., Edsall, J. T., and Miller, S. G.,** Preparation and properties of serum and plasma proteins. XCIII. The separation of purified fibrinogen from fraction I of human plasma, *J. Am. Chem. Soc.*, 70, 3103, 1948.
11. **Edsall, J. T., Gilbert, G. F., and Scheraga, H. A.,** The nonclotting component of the human plasma fraction I-1 ("cold-insoluble globulin"), *J. Am. Chem. Soc.*, 77, 157, 1955.
12. **Thomas, L., Smith, R. T., and von Korff, R.,** Cold-precipitation by heparin of a protein in rabbit and human plasma, *Proc. Soc. Exp. Biol. Med.*, 86, 813, 1954.
13. **Smith, R. T. and von Korff, R. W.,** A heparin-precipitable fraction of human plasma. I. Isolation and characterization of the fraction, *J. Clin. Invest.*, 36, 596, 1957.
14. **Smith, R. T.,** A heparin-precipitable fraction of human plasma. II. Occurrence and significance of the fraction in normal individuals and in various disease states, *J. Clin. Invest.*, 36, 605, 1957.
15. **Heinrich, R. A., von der Heide, E. C., and Climie, A. R. W.,** Cryofrinogen: formation and inhibition in heparinized plasma, *Am. J. Physiol.*, 204, 419, 1963.
16. **Mosesson, M. W. and Umfleet, R. A.,** The cold-insoluble globulin of human plasma. I. Purification, primary characterization, and relationship to fibrinogen and other cold insoluble fraction components, *J. Biol. Chem.*, 245, 5728, 1970.
17. **Stathakis, N. E. and Mosesson, M. W.,** Interactions among heparin, cold-insoluble globulin, and fibrinogen in formation of the heparin-precipitable fraction of plasma, *J. Clin. Invest.*, 60, 855, 1977.
18. **Engvall, E. and Ruoslahti, E.,** Binding of soluble form of fibroblast surface protein, fibronectin, to collagen, *Int. J. Cancer*, 20, 1, 1977.
19. **Ruoslahti, E. and Vaheri, A.,** Interaction of soluble fibroblast surface antigen with fibrinogen and fibrin. Identify with cold insoluble globulin of human plasma, *J. Exp. Med.*, 141, 497, 1975.
20. **Sakata, Y. and Aoki, N.,** Crosslinking of alpha$_2$-plasmin inhibitor to fibrin by fibrin stabilizing factor, *J. Clin. Invest.*, 66, 1374, 1980.
21. **Bale, M. D., Westrick, L. G., and Mosher, D. F.,** Incorporation of thrombospondin into fibrin clots, *J. Biol. Chem.*, 260, 7502, 1985.
22. **Mosher, D. F. and Johnson, R. B.,** Specificity of fibronectin-fibrin crosslinking, *Ann. N. Y. Acad. Sci.*, 408, 583, 1983.
23. **Jilek, F. and Hormann, H.,** Cold-insoluble globulin. III. Cyanogen bromide and plasminolysis fragments containing a label introduced by transamidation, *Hoppe-Seyler's Z. Physiol. Chem.*, 358, 1165, 1977.
24. **Sekiguchi, K. and Hakomori, S.,** Identification of two fibrin binding domains in plasma fibronectin and unequal distribution of these two domains in two different subunits, *Biochem. Biophys. Res. Commun.*, 97, 709, 1980.
25. **Seidl, M. and Hormann, H.,** Affinity chromatography on immobilized fibrin monomer. IV. Two fibrin-binding peptides of a chymotryptic digest of human plasma fibronectin, *Hoppe-Seyler's Z. Physiol. Chem.*, 364, 83, 1983.
26. **Mosher, D. F.,** Cross-linking of cold-insoluble globulin by fibrin-stabilizing factor, *J. Biol. Chem.*, 250, 6614, 1975.
27. **Iwanaga, S., Suzuki, K., and Hashimoto, S.,** Bovine plasma cold-insoluble globulin: gross structure and function, *Ann. N.Y. Acad. Sci.*, 312, 56, 1978.
28. **Tamaki, T. and Aoki, N.,** Cross-linking of alpha$_2$-plasmin inhibitor and fibronectin to fibrin by fibrin-stabilizing factor, *Biochem. Biophys. Acta*, 661, 280, 1981.
29. **Okada, M., Blomback, B., Chang, M.-D., and Horowitz, B.,** Fibronectin and fibrin gel structure, *J. Biol. Chem.*, 260, 1811, 1985.
30. **McDonagh, J., Hada, M., and Kaminski, M.,** Plasma fibronectin and fibrin formation, in *Plasma Fibronectin: Structure and Function*, Vol. 5, McDonagh, J., Ed., Marcel Dekker, New York, 1985, chap. 7.
31. **Kaplan, J. E. and Snedeker, P. W.,** Maintenance of fibrin solubility by plasma fibronectin, *J. Lab. Clin. Med.*, 96, 1054, 1980.
32. **Holm, B. and Brosstad, F.,** Interactions between fibrin and fibronectin. No effect of fibronectin on the plasma solubility of fibrin monomers, in *Fibrinogen — Structure, Functional Aspects, Metabolism*, Vol. 2, Haverkate, F., Henschen, A., Nieuwenhuizen, W., and Straub, P. W., Eds., Walter de Gruyter, New York, 1983, 323.
33. **Sherman, L. A. and Lee, J.,** Fibronectin: blood turnover in normal animals and during intravascular coagulation, *Blood*, 60, 558, 1982.
34. **Bang, N. V., Hansen, M. S., Smith, G. F., and Mosesson, M. W.,** Properties of soluble fibrin polymers encountered in thrombotic states, *Thromb. Diath. Haemorrh. (Suppl.)*, 56, 75, 1973.
35. **Sasaki, T., Page, I. H., and Shainoff, J. R.,** Stable complexes of fibrinogen and fibrin, *Science*, 152, 1069, 1966.
36. **Niewiarowska, J. and Cierniewski, C. S.,** Inhibitory effect of fibronectin on the fibrin formation, *Thromb. Res.*, 27, 611, 1982.

37. **Bale, M. and Mosher, D. F.**, Effects of thrombospondin on fibrin polymerization and structure, *J. Biol. Chem.*, 261, 862, 1986.
38. **Kamykowski, G. W., Mosher, D. F., Lorand, L., and Ferry, J. D.**, Modification of shear modulus and creep compliance of fibrin clots by fibronectin, *Biophys. Chem.*, 13, 25, 1981.
39. **Chow, Y. W., McIntire, L. V., and Petersen, D. M.**, Importance of plasma fibronectin in determining PFP and PRP clot mechanical properties, *Thromb. Res.*, 29, 243, 1983.
40. **Gilboa, N. and Kaplan, J. E.**, Plasma fibronectin enhances fibrinolytic system *in vitro*, *Thromb. Haemostas.*, 54, 639, 1985.
41. **Niewiarowska, J. and Cierniewski, C. S.**, Quantitation of the fibronectin interaction with fibrin monomers, *Thromb. Haemostas.*, 50, 28, 1983.
42. **Rourke, F. J., Blumenstock, F. A., and Kaplan, J. E.**, Effect of fibronectin fragments on macrophage phagocytosis of gelatinized particles, *J. Immunol.*, 132, 1931, 1984.
43. **Quigley, J. P., Gold, L. I., Schwimmer, R., and Sullivan, L. M.**, Limited cleavage of cellular fibronectin by plasminogen activator purified from transformed cells, *Proc. Natl. Acad. Sci. U.S.A.*, 84, 2776, 1987.
44. **Banyai, L., Varaddi, A., and Patthy, L.**, Common evolutionary origin of the fibrin-binding structures of fibronectin and tissue-type plasminogen activator, *FEBS Lett.*, 163, 37, 1983.
45. **Bianco, C.**, Fibrin, fibronectin and macrophages, *Ann. N.Y. Acad. Sci.*, 408, 603, 1983.
46. **Colvin, R. B.**, Fibrinogen-fibrin interactions with fibroblasts and macrophages, *Ann. N.Y. Acad. Sci.*, 408, 621, 1983.
47. **Jilek, F., and Hormann, H.**, Fibronectin (cold insoluble globulin). VI. Mediation of fibrin monomer binding to macrophages, *Hoppe Seyler's Physiol. Chem.*, 359, 1603, 1978.
48. **Falcone, D. J.**, Fluorescent opsonization assay: binding of plasma fibronectin to fibrin-derivatized fluorescent particles does not enhance their uptake by macrophages, *J. Leuk. Biol.*, 39, 1, 1986.
49. **Hormann, H., Richter, H., and Jelinic, V.**, Fibrin monomer binding to macrophages mediated by fibrin-binding fibronectin fragments, *Thromb. Res.*, 39, 183, 1985.
50. **Hormann, H., Richter, H., and Jelinic, V.**, The role of fibronectin fragments and cell-attached transamidase on the binding of soluble fibrin to macrophages, *Thromb. Res.*, 46, 39, 1987.
51. **Plow, E. F. and Ginsberg, M. H.**, Specific and saturable binding of plasma fibronectin to thrombin-stimulated human platelets, *J. Biol. Chem.*, 256, 9477, 1981.
52. **Pommier, C. G., O'Shea, J., Chused, T., Yancey, K., Frank, M. M., Takahashi, T., and Brown, E. J.**, Studies in the fibronectin receptors of human peripheral blood leukocytes, *J. Exp. Med.*, 159, 137, 1984.
53. **Akiyama, S. K., Yamada, S. S., and Yamada, K. M.**, Characterization of a 140-kd avian cell surface antigen as a fibronectin-binding molecule, *J. Cell Biol.*, 102, 442, 1986.
54. **McKeown-Longo, P. J. and Mosher, D. F.**, Interaction of the 70,000 mol-wt amino-terminal fragment of fibronectin with the matrix assembly receptor of fibroblasts, *J. Cell Biol.*, 100, 364, 1985.
55. **Grinnell, F.**, Fibronectin and wound healing, *J. Cell Biochem.*, 26, 107, 1984.
56. **Bang, N. U., Chang, M. L., Mattler, L. E., Burck, P. J., Van Frank, R. M., Zuimmerman, R. E., Marks, C. A., and Boxer, L. J.**, Monocyte/macrophage-mediated catabolism of fibrinogen and fibrin, *Ann. N.Y. Acad. Sci.*, 370, 568, 1981.
57. **Sherman, L. A., Lee, J., and Stewart, C. C.**, Release of fibrinolytic enzymes by macrophages in response to soluble fibrin, *J. Reticuloendothel. Soc.*, 30, 317, 1981.
58. **Sherman, L. A.**, Binding of soluble fibrin to macrophages, *Ann. N.Y. Acad. Sci.*, 408, 610, 1983.
59. **Sherman, L. A. and Lee, J.**, Specific binding of soluble fibrin to macrophages, *J. Exp. Med.*, 145, 76, 1977.
60. **Horsburgh, C. R., Clark, R. A. F., and Kirkpatrick, C. H.**, Lymphokines and platelets promote human monocyte adherence to fibrinogen and fibronectin *in vitro*, *J. Leuk. Biol.*, 41, 14, 1987.
61. **Gonda, S. R. and Shainoff, J. R.**, Adsorptive endocytosis of fibrin monomer by macrophages: evidence of a receptor for the amino terminus of the fibrin chain, *Proc. Natl. Acad. Sci. U.S.A.*, 79, 4565, 1982.
62. **Bevilacqua, M. P., Amrani, D., Mosesson, M. W., and Bianco, C.**, Receptors for cold-insoluble globulin (plasma fibronectin) on human monocytes, *J. Exp. Med.*, 153, 42, 1981.
63. **Santoro, S.**, Inhibition of platelet aggregation by fibronectin, *Biochem. Biophys. Res. Commun.*, 116, 135, 1983.
64. **Moon, D. G., Kaplan, J. E., and Mazurkiewicz, J. E.**, The inhibitory effect of plasma fibronectin on collagen-induced platelet aggregation, *Blood*, 67, 50, 1986.
65. **Koteliansky, V. E., Leytin, V. L., Sviridov, D. D., Repin, V. S., and Smirnov, V. N.**, Human plasma fibronectin promotes the adhesion and spreading of platelets on surfaces coated with fibrillar collagen, *FEBS Lett.*, 123, 59, 1981.
66. **Ill, C. R., Engvall, E., and Ruoslahti, E.**, Adhesion of platelets to laminin in the absence of activation, *J. Cell Biol.*, 99, 2140, 1984.
67. **Plow, E. F., Srouki, A., Meyer, D., Marguerie, G., and Ginsberg, M. H.**, Evidence that three adhesive proteins interact with a common recognition site on activated platelets, *J. Biol. Chem.*, 259, 5388, 1984.

68. **Peerschke, E. I., Zucker, M. B., Grant, R. A., Egan, J. J., and Johnson, M. M.,** Correlation between fibrinogen binding to human platelets and platelet aggregability, *Blood,* 55, 841, 1980.
69. **Grinnell, F., Feld, M., and Minter, D.,** Fibroblast adhesion to fibrinogen and fibrin substrata: requirement for cold-insoluble globulin (plasma fibronectin), *Cell,* 19, 517, 1980.
70. **Knox, P., Crooks, S., and Rimmer, C. S.,** Role of fibronectin in the migration of fibroblasts into plasma clots, *J. Cell Biol.,* 102, 2318, 1986.
71. **Kurkinen, M., Vaheri, A., Roberts, P. J., and Steinman, S.,** Sequential appearance of fibronectin and collagen in experimental granulation tissue, *Lab. Invest.,* 43, 47, 1980.
72. **Colvin, R. B., Gardner, P. I., Roblin, R. O., Verderber, E. L., Lanigan, J. M., and Mosesson, M. W.,** Cell surface fibrinogen-fibrin receptors on cultured human fibroblasts, *Lab. Invest.,* 41, 464, 1979.
73. **Clark, R. A. F., DellePella, P., Manseau, E., Lanigan, J. M., Dvorak, H. F., and Colvin, R. B.,** Blood vessel fibronectin increases in conjunction with endothelial cell proliferation and capillary ingrowth during wound healing, *J. Invest. Dermatol.,* 79, 269, 1982.
74. **Clark, R. A. F., Lanigan, J. M., DellePella, P., Manseau, E., Dvorak, H. F., and Colvin, R. B.,** Fibronectin and fibrin provide a provisional matrix for epidermal cell migration during wound reepithelialization, *J. Invest. Dermatol.,* 79, 264, 1982.
75. **Fujikawa, L. S., Foster, C. S., Harrist, T. J., Lanigan, J. M., and Colvin, R. B.,** Fibronectin in the healing rabbit corneal wound, *Lab. Invest.,* 45, 120, 1981.
76. **Fujikawa, L. S., Foster, C. S., Gipson, I. K., and Colvin, R. B.,** Basement membrane components in healing rabbit corneal epithelial wounds: immunofluorescence and ultrastructural studies, *J. Cell Biol.,* 98, 128, 1984.
77. **Stamatoglou, S. C., Hughes, R. C., and Lindahl, U.,** Rat hepatocytes in serum-free primary culture elaborate an extensive extracellular matrix containing fibrin and fibronectin, *J. Cell Biol.,* 105, 2417, 1987.
78. **Smith, D. B., Janmey, P. A., Herbert, T.-J., and Lind, S. E.,** Quantitative measurement of plasma gelsolin and its incorporation into fibrin cots, *J. Lab. Clin. Med.,* 110, 189, 1987.
79. **Mosher, D. F. and Williams, E. M.,** Fibronectin concentration is decreased in plasma of severely ill patients with disseminated intravascular coagulation, *J. Lab. Clin. Med.,* 91, 729, 1978.
80. **Cembrowski, G. S. and Mosher, D. F.,** Plasma fibronectin concentration in patients with acquired consumptive coagulopathies, *Thromb. Res.,* 36, 437, 1984.
81. **Scovill, W. A., Saba, T. M., Kaplan, J. E., Bernard, H., Powers, S. R., Jr.,** Deficits in reticuloendothelial humoral control mechanisms in patients after trauma, *J. Trauma,* 16, 898, 1976.
82. **Pott, G., Voss, B., Lohmann, J., and Zundorf, P.,** Loss of fibronectin in plasma of patients with shock and septicaemia and after haemoperfusion in patients with severe poisoning, *J. Clin. Chem. Biochem.,* 20, 333, 1982.
83. **Couland, J. M., LaBrousse, J., Salmona, J. P., Tanaillon, A., Lissac, J., Jacqueson, A., Beyne, P., Rapin, J., Allard, D., and Jerome, H.,** Plasma fibronectin concentrations in critically ill patients, *Ric. Clin. Lab.,* 12, 137, 1982.
84. **O'Connell, M. T., Becker, D. M., Steele, B. W., Peterson, G. S., and Hellman, R. L.,** Plasma fibronectin in medical ICU patients, *Crit. Care Med.,* 12, 479, 1984.
85. **Maunder, R. J., Harlan, J. M., Pepe, P. E., Paskell, S., Carrico, C. J., and Hudson, L. D.,** Measurement of plasma fibronectin in patients who develop the adult respiratory distress syndrome, *J. Lab. Clin. Med.,* 104, 583, 1984.
86. **Gerdes, J. S., Yoder, M. C., Douglas, S. D., and Polin, R. A.,** Decreased plasma fibronectin in neonatal sepsis, *Pediatrics,* 72, 877, 1983.
87. **Richards, W. A., Scovill, W. A., and Shin, B.,** Opsonic fibronectin deficiency in patients with intra-abdominal infection, *Surgery,* 94, 210, 1983.
88. **Rubli, E., Bussard, S., Frei, E., Lundsgaard-Hansen, P., and Pappova, E.,** Plasma fibronectin and associated variables in surgical intensive care patients, *Ann. Surg.,* 197, 310, 1983.
89. **Lundsgaard-Hansen, P., Doran, J. E., Rubli, E., Papp, E., Morgenthaler, J., and Spath, P.,** Purified fibronectin administration to patients with severe abdominal infections: a controlled clinical trial, *Ann. Surg.,* 202, 745, 1985.
90. **Goldman, A. S., Rudolff, H. B., McNamee, R., Loose, L. D., and Di Luzio, N. R.,** Deficiency of plasma humoral recognition factor activity following burn injury, *J. Rediculoendothel. Soc.,* 15, 193, 1974.
91. **Lanser, M. E., Saba, T. M., and Scovill, W. A.,** Opsonic glycoprotein (plasma fibronectin) levels after burn injury: relationship to extent of burn and development of sepsis, *Ann. Surg.,* 192, 776, 1980.
92. **Ekindjian, O. G., Murien, M., Wasserman, D., Bruxelle, J., Cazalet, C., Konter, E., and Yonger, J.,** Plasma fibronectin time course in burn patients: influence of sepsis, *J. Trauma,* 24, 214, 1984.
93. **Brodin, B., von Schench, H., Schildt, B., and Liljedahl, S-O.,** Low plasma fibronectin indicates septicaemia in major burns, *Acta Chir. Scand.,* 150, 5, 1984.
94. **Mosher, D. F.,** Changes in plasma cold-insoluble globulin concentration during experimental Rocky Mountain spotted fever infection in rhesus monkeys, *Thromb. Res.,* 9, 37, 1976.

95. **Niehaus, G. D., Schumacker, P. T., and Saba, T. M.,** Influence of opsonic fibronectin deficiency on lung fluid balance during bacterial sepsis, *J. Appl. Physiol.*, 49, 693, 1980.
96. **Saba, T. M., Niehaus, G. D., Scovill, W. A., Blumenstock, F. A., Newell, J. C., Holman, J., Jr., and Powers, S. R., Jr.,** Lung vascular permeability after reversal of fibronectin deficiency in septic sheep: correlation with patient studies, *Ann. Surg.*, 198, 654, 1983.
97. **Dillon, B. C. and Saba, T. M.,** Fibronectin deficiency and intestinal transvascular fluid balance during bacteremia, *Am. J. Physiol.*, 242, H557, 1982.
100. **Loegering, D. J. and Schneidkraut, M. J.,** Effect of endotoxin on a_2-SB opsonic protein activity and reticuloendothelial system phagocytic function, *J. Reticuloendothel. Soc.*, 26, 197, 1976.
101. **Kaplan, J. E., Scovill, W. A., Bernard, H., Saba, T. M., Gray, V.,** Reticuloendothelial phagocytic response to bacterial challenge after traumatic shock, *Circ. Shock,* 4, 1, 1977.
102. **Richards, P. S. and Saba, T. M.,** Effect of endotoxin on fibronectin and Kupffer cell activity, *Hepatology,* 5, 32, 1985.
103. **Aukburg, S. J. and Kaplan, J. E.,** The influence of elevated fibronectin levels during sepsis on the distribution of blood-borne particles, *Adv. Shock Res.*, 6, 37, 1981.
104. **Klingemann, H. G., Kosikavak, M., Hofeler, H., and Havemann, K.,** Fibronectin and factor VIII related antigen in acute leukaemia, *Hoppe Seyler's Z. Physiol. Chem.*, 364, 269, 1983.
105. **Reilly, J. T., Mackie, M. J., and McVerry, B. A.,** Observations on the heterogeneity of fibronectin C, *Thromb. Res.*, 33, 289, 1984.
106. **Snedeker, P. W., Kaplan, J. E., and Saba, T. M.** Effect of traumatic shock and alteration of reticuloendothelial function on the vascular clearance of soluble fibrin, *Physiologist,* 21, 113, 1978.
107. **Snedeker, P. W. and Kaplan, J. E.,** Reticuloendothelial clearance and vascular localization of soluble fibrin monomers, *Circ. Shock,* 7, 207, 1980.
108. **Ihlenfeld, J. V., Mathis, T. R., Barber, T. A., et al.,** Transient *in vivo* thrombus deposition onto polymeric biomaterials: role of plasma proteins, *Trans. Am. Soc. Artif. Intern. Organs,* 24, 727, 1978.
109. **Kaplan, J. E., Snedeker, P. W., Baum, S. H., Moon, D. G., and Minnear, F. L.,** Influence of plasma fibronectin on the response to infusion of thrombin and adenosine diphosphate, *Thromb. Haemostas.*, 49, 217, 1983.
110. **Moon, D. G. and Kaplan, J. E.,** Plasma fibronectin inhibits phagocytosis of platelets *in vitro, Fed. Proc.*, 43, 979, 1984.
111. **Matayoshi, B. M., Moon, D. G., and Kaplan, J. E.,** Reduction of the thrombotic response to vascular injury by plasma fibronectin, *Thromb. Haemostas.*, 54, 218, 1985.

Chapter 11

FIBRONECTIN AND CANCER: IMPLICATIONS OF CELL ADHESION TO FIBRONECTIN FOR TUMOR METASTASIS

Martin J. Humphries, Kenneth Olden, and Kenneth M. Yamada

TABLE OF CONTENTS

I. TUMOR METASTASIS AND INVASION

A. CLINICAL RELEVANCE

The propensity of tumor cells to spread from their site of origin to another part of the body, i.e., to invade and metastasize, is one of the most distinctive features of neoplastic disease. It is increasingly apparent that the high rate of cancer mortality is due principally to the biological consequences of tumor cell dissemination. Currently, about half of all diagnosed cancer patients survive at least 5 years when treated with conventional regimens based on surgery, chemotherapy, and radiotherapy; the treatment failure group are generally victims of the incomplete eradication of their metastases.[1] Usually, even at the time of initial presentation, patients with a detectable malignant tumor already have occult metastases, and the simple surgical removal of the primary tumor does little to prolong survival. If oncologists were able to successfully treat, or even prevent, the formation of metastases, then the incidence of death from cancer would be dramatically reduced.

In order to significantly improve survival statistics for cancer patients, recent research has been concentrated on developing new approaches directed at inhibiting metastasis-specific events that will be more effective than current modalities for treatment of patients with malignant disease.[2-7] In this chapter, we will review the role of fibronectin in cancer, with particular emphasis on its roles in adhesion and the metastatic process. We then describe how elucidation of the molecular mechanisms of cell interaction with fibronectin has led to the development of a novel class of antimetastatic agent which may have future applications in the treatment of malignant disease.

B. THE METASTATIC CASCADE

1. Spatiotemporal Events

The dissemination of malignant cells from a primary to a secondary site involves a complex series of biological events that are linked both temporally (in that they take place in a defined sequence) and spatially (in that they occur at particular sites within the body).[8-12] An inherent feature of tumor metastasis is that the translocation of malignant cells can take place over great distances within the body. Consequently, in order for a cell to successfully colonize its target organ, it must be able to perform a multitude of functions and overcome a number of obstacles. A diagrammatic outline of the events comprising the metastatic process is presented in Figure 1 (right).

Initially, malignant cells detach from the primary tumor mass and enter the circulatory system. It is likely that certain aspects of this intravasation require a combination of degradative and active migratory and invasive events;[13-16] however, at present, the biochemical details of this process are not well understood. Tumor cells may come into close proximity to microcapillaries by direct contact, as in certain highly malignant sarcomas, or may frequently infiltrate capillaries and/or lymphatics after penetrating surrounding interstitial tissues.[17,18] Direct entry into the venous system favors metastasis to organs such as the lung and liver, while invasion of peritumoral connective tissue usually results in lymph node colonization.[17,18] Where tumor cells are shed directly into the circulation, they have no natural barriers to penetrate; however, active intravasation from vessels into tissues requires malignant cells to pass through both the endothelial cell layer and the subendothelial matrix, in a process which probably requires controlled secretion of specific proteases[13-16] as well as various adhesive and migratory steps.

The vast majority of cells that are initially released from tumors into the circulatory environment are killed as a result of the high hemodynamic forces that they encounter, e.g., during the passage of blood through a narrow capillary bed.[19] Others are destroyed by host effector cell populations that are able to recognize and kill their targets without prior immune stimulation. One of the most prominent cell types thought to be responsible for this activity

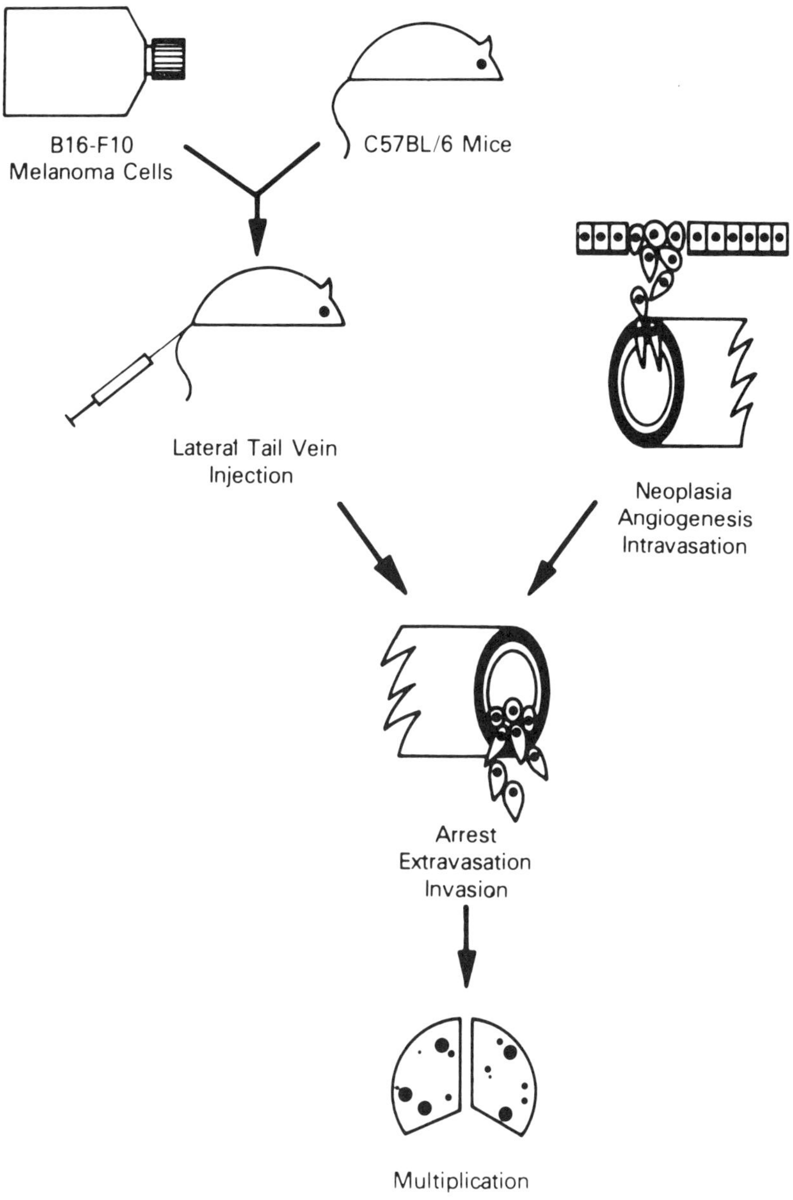

FIGURE 1. Authentic and experimental tumor metastasis. In the authentic hematogenous metastatic cascade (right pathway), neoplastic cells first intravasate into the circulatory system. Cells may detach from the primary tumor mass by a combination of enzymatic degradation of interstitial tissues and directed migration, or by simple mechanical sloughing. Once in the circulation, tumor cells form emboli with other tumor cells or with blood cells, a process which aids their retention in the target organ. Specific arrest can potentially occur by several mechanisms including adhesion to endothelial cell surfaces, exposed basement membrane, or subendothelial connective tissue. Tumor cells then extravasate out of the circulation, invade underlying tissues, and multiply to form metastatic colonies (shown diagrammatically as melanotic lung lesions). In the experimental metastasis assay (left pathway; Reference 354), cultured B16-F10 murine melanoma cells, selected *in vivo* for their lung-colonizing potential, are detached with EDTA and injected into the lateral tail vein of syngeneic C57BL/6 mice. Two to three weeks later, the number of melanotic nodules in the lungs is determined after sacrifice. The experimental assay therefore bypasses the early intravasation phases of metastasis, but is well suited for application to studies of the role of cell adhesion in arrest of metastatic cells.

is the natural killer, or NK, cell;[20,21] animals containing defective NK cell populations are much more susceptible to experimental induction of metastasis.[22,23]

While in the circulation, tumor cells frequently adhere both to other tumor cells,[24,25] and to blood cells, including platelets.[26,27] These homo- and heterotypic adhesions produce enlarged emboli and, as a result, directly affect the ability of the tumor cells to implant in the target organ. In general, the larger the embolus, the more likely it is to remain lodged in a narrow capillary bed.[24,25,28] Tumor cell retention in the target organ is also prolonged by the formation of a fibrin meshwork around the arrested cells.[27,29] This tumor cell-induced procoagulation is a consistent feature of metastatic lesions. In addition to aiding extravasation, the formation of emboli may also serve to protect the tumor cells from immune surveillance. Alternative, more specific mechanisms of tumor cell arrest include adhesion to protein receptors on the luminal surface of the endothelium,[9,26,30] or adhesion to exposed extracellular matrix molecules in the subendothelial basement membrane and connective tissue.[10,11,30-32]

Subsequent to their arrest in the target organ, malignant cells pass through the endothelium and invade through its associated subendothelial layer in a process which, at least on a superficial level, may be the reverse of the events that take place during intravasation. However, it is likely that the molecular architecture of the subendothelium varies in different organs and that the mechanism of colonization will also vary.[9,30,33,34] The interaction of tumor cells with the individual components of the endothelium or subendothelium may, therefore, play a role in dictating the organ specificity of metastasis of a particular tumor type.

2. Role of Extracellular Matrix Components

One recurring theme during the progression of tumor cells through the metastatic cascade is their ability to successfully invade and penetrate extracellular matrices (Figure 1). During the initial stages of intravasation, malignant cells often must invade into and migrate through the interstitial matrix that surrounds the tumor and then, for metastasis via the hematogenous route, they may have to penetrate at least one basement membrane, the anatomical barrier that underlies capillary endothelial cells. Furthermore, although the actual arrest of the tumor cells in the target organ may be facilitated by nonspecific trapping of emboli, it is likely that metastasizing cells recognize specific adhesive proteins in the endothelial layer or especially in the subendothelial extracellular matrix. These, or related, molecules may also support the subsequent migration of the tumor cells during their invasion of the subendothelial connective tissue of the target organ. Adhesive molecules that have *in vivo* distributions consistent with a possible role in these key events include fibronectin, laminin, vitronectin, collagens, and heparan sulfate proteoglycans.[11,35-37]

When the complexity of the metastatic process is considered, coupled with the fact that the malignant cell must successfully accomplish each component step, it is no surprise that the process is so inefficient. It has been estimated that 10^5 to 10^7 cells are shed from a primary mouse tumor per day,[25,38] yet only a small percentage of these go on to form metastases.[19,25,39] A further benefit of this complexity is that it potentially provides many different ways to intervene therapeutically in the metastatic cascade; it is conceivable that if only a single critical step could be blocked, e.g., by specifically and completely blocking the interaction of tumor cells with a crucial extracellular matrix molecule, then the whole metastatic process would fail.

II. FIBRONECTIN STRUCTURE AND FUNCTION

Fibronectin is currently the best characterized extracellular matrix molecule (for recent reviews, see References 40 to 44 and the reviews cited therein and Chapter 1). Its structure is known in great detail, and it has been implicated in an impressive array of normal biological functions, including cell migration, embryonic development, extracellular matrix assembly, and certain facets of the wound healing process (see References 42, 44 to 46, and this volume). Almost all of the functions of fibronectin, however, can be directly attributed to its adhesive activity, and it is the interaction of fibronectin with the cell surface that has attracted the most experimental attention; recent insights into this complex process will be reviewed later.

A. *IN VIVO* DISTRIBUTION

In vivo, fibronectin has an ubiquitous distribution in tissues and body fluids and, in the past, it has been classified into two main forms based on location: *plasma* and *cellular* fibronectin. In normal plasma, fibronectin is present at fairly high concentration (approximately 300 μg/ml), but as discussed below, this value changes in certain disease states. Plasma fibronectin is synthesized predominantly in the liver.[47] Fibronectin is also found in other body fluids, such as synovial and amniotic fluids, which are essentially filtrates of plasma.[48-50] Fibronectin is also present in most tissues where it is found juxtaposed to basement membranes and as an integral part of interstitial matrices. Since there is only a single fibronectin gene,[51,52] the designation plasma and cellular is somewhat arbitrary since both molecules differ only slightly in structure, and each type is probably a mixture of different variants, some of which may be shared. Nevertheless, these differences may be relevant to determining whether fibronectin becomes incorporated into an extracellular matrix (cellular type) or whether only a small proportion does (plasma type) (see Chapter 1). *In vitro,* fibronectin is synthesized and secreted by a wide variety of cells including fibroblasts, endothelial cells, chondrocytes, melanoma cells, myoblasts, macrophages, neutrophils, hepatocytes, glial cells, Schwann cells, and intestinal, mammary, liver, kidney, prostate, and thyroid epithelial cells.

B. STRUCTURAL DETERMINANTS OF CELL ADHESION

1. Gene Structure and Processing

Fibronectin is a modular glycoprotein composed of a number of independent structural domains (see References 53, 54, and Chapter 1). Through the use of cDNA and gene cloning techniques, it is now clear that the overall organization of these binding domains within fibronectin is the same for all of the members of the fibronectin family. Except for post-translational modifications, the only differences in structure are due to alternative splicing of precursor mRNA molecules.[40,55-61] Primary sequence analysis of fibronectin from a number of species, including human, has shown that almost all of the molecule is composed of a series of homologous repeating units.[62-64] This intriguing finding suggests that in evolution fibronectin may have arisen through duplication of a small number of primitive ligand-binding structures. There are three types of repeat with characteristic disulfide-bonding patterns; these are distributed asymmetrically along the subunit, but are localized to particular ligand-binding domains. For example, type II repeats are found only in the collagen-binding domain, and these are thought to be important in the interaction of fibronectin with collagen.[65] At the genomic level, the concept that gene duplication played an important role in determining the structure of fibronectin is supported, since the exon structure of the fibronectin gene coincides precisely with the repeating unit structure of the protein.[66-69]

Currently, three sites of alternative splicing are known: the spliced regions are termed ED-A, ED-B (extra domain A and B; also termed ED or EIIIA and EDII or EIIIB, respectively), and IIICS (type III connecting segment, also known as V). ED-A[56] and ED-B[58-61] represent whole type III homology units that can be included or spliced out of the final fibronectin product. In general, the content of these ED domains is high in cellular fibronectin, possibly higher in fibronectin from transformed cells, and is undetectable in plasma fibronectin,[56,58-61,70,71] suggesting that these regions may contribute to the unique properties

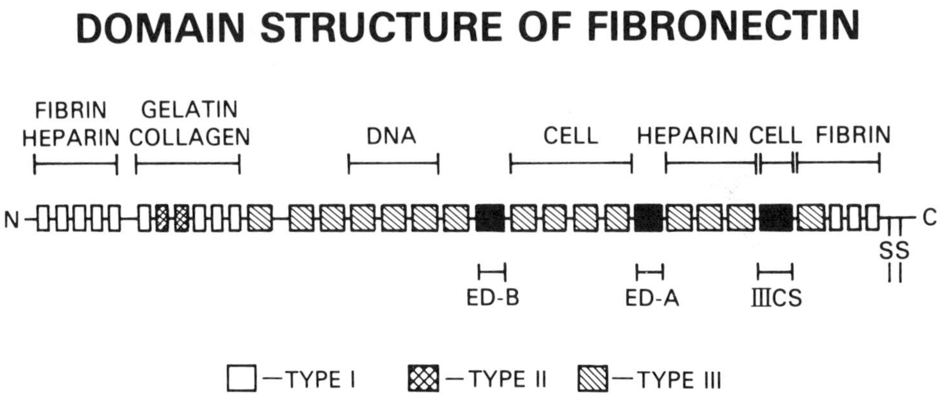

FIGURE 2. Modular structure of fibronectin. The fibronectin molecule comprises two similar subunits, joined near their COOH-termini, and differing in their content of the IIICS region. ED-A, ED-B, and IIICS (solid boxes) are regions that can be selectively removed from fibronectin by alternative splicing of precursor mRNA. The ligand-binding properties of the structural domains of fibronectin are shown at the top. Each domain is composed of a series of polypeptide units which fall into three different groups based on their homology to each other. Type I and II units (open and cross-hatched boxes, respectively) have characteristic patterns of intradomain disulfide-bond formation, while type III repeats are approximately twice as large and have no disulfide bonds.

of tissue fibronectin. The IIICS region is spliced in a complex manner; from a single exon, five potential variants are possible in human fibronectin, three in rat, and two in chicken fibronectin.[55,57,59] In human fibronectin, therefore, when the ED regions are taken into account, there are 20 possible variants. The pattern of splicing of the IIICS is broadly similar to that of the ED regions; it is present in greater amounts in cellular and transformed cell fibronectins than in plasma fibronectin, although it is not completely excluded from the latter.[55,59,71-75] Rather than distinguishing between the different tissue forms of fibronectin, splicing of the IIICS region appears to account for size differences in the subunits of either plasma or cellular fibronectins.[71,72] As discussed below, the identification of the IIICS as a cell interaction domain has heightened interest in the role of alternative splicing in the function of fibronectin.

2. Cell Interaction Sites

In order to obtain probes for the adhesive function of fibronectin, there has been considerable recent interest in identifying the active sites on the fibronectin molecule that are recognized by cells. At present, two distinct cell-binding domains are known, one located near the center of the molecule and the other located in the IIICS region. A variety of cell types, including fibroblasts and certain tumor cells, adhere to the central cell-binding domain,[46] while current evidence suggests that the IIICS is recognized in a cell type-specific manner by derivatives of the neural crest (peripheral neurons, melanoma cells, corneal fibroblasts; for review, see Reference 44). Localization of each domain was originally performed using defined proteolytic fragments of intact fibronectin; for the central cell-binding domain, a 75-kDa tryptic fragment contains all of the adhesive activity of intact human plasma fibronectin for fibroblasts,[76] and for the IIICS, a 113-kDa tryptic fragment is maximally active for melanoma cells.[77]

a. Central Cell-Binding Domain

Several complementary approaches have been used to narrow down the active sites within the central cell-binding domain (see Figure 3). The results of each of these studies is consistent with the hypothesis that the pentapeptide sequence Gly-Arg-Gly-Asp-Ser (GRGDS) is critical for the functioning of this domain.

APPROACHES FOR IDENTIFICATION OF CELL INTERACTION SITES

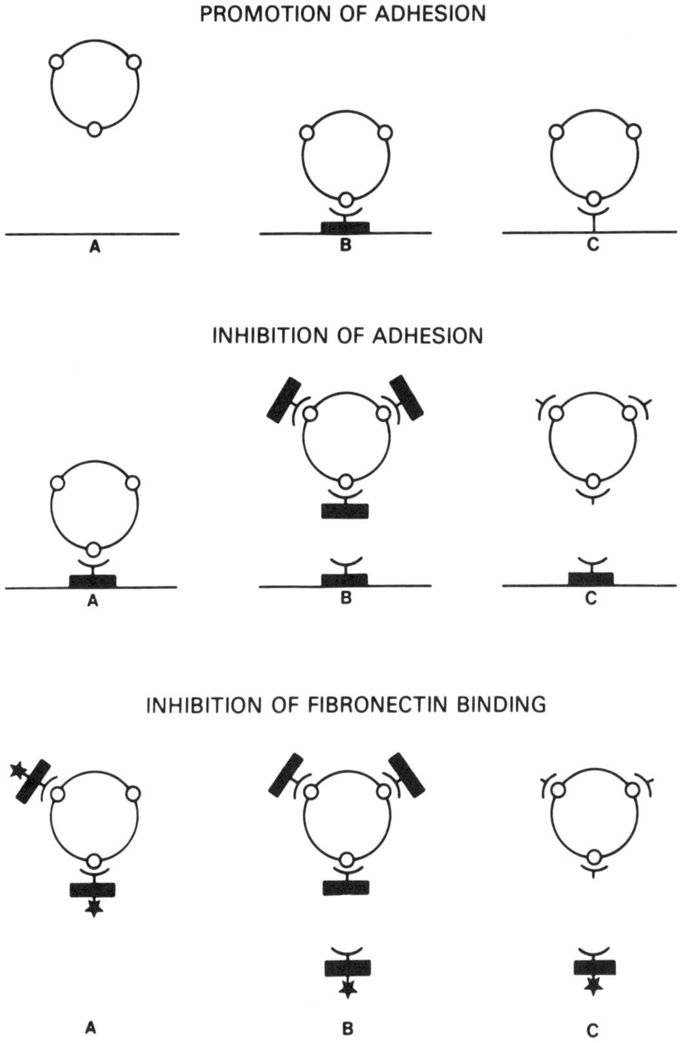

FIGURE 3. Diagrammatic representation of the general approaches used for identification of cell interaction sites. Cell-binding sequences within fibronectin have been identified by a combination of experimental techniques based on promotion or inhibition of adhesion and inhibition of fibronectin binding. Top panel: intact fibronectin (B) or its proteolytic fragments and synthetic peptides (C) are able to promote adhesion directly when absorbed onto a suitable substrate, while the uncoated substrate is inactive (A). Middle panel: the adhesive activity of fibronectin (A) is sensitive to autoinhibition either by co-incubation of cells with an excess of intact fibronectin itself (B) or with cell-binding fragments and peptides (C). Bottom panel: the direct binding of radiolabeled fibronectin to cells in suspension (A) is also sensitive to inhibition by fibronectin (B) or its cell-binding fragments and peptides (C).

By analyzing the adhesive activity of progressively smaller fragments of the central cell-binding domain, activity was localized to an 11.5-kDa peptic fragment.[78,79] Overlapping synthetic peptides were then designed to span the whole sequence of this fragment, and this ultimately led to the assignment of adhesive activity to the GRGDS sequence.[80-83] In a related approach, based on the autoinhibitory activity of fibronectin and its fragments, progressively smaller molecules were found to block the adhesion of fibroblastic cells to intact fibronectin[82] (Figure 3). Peptides containing the GRGDS sequence were then selected from an evolutionary conserved, hydrophilic region of the domain, and were found to be effective inhibitors of fibronectin-mediated adhesion.[82] RGDS is the minimal active sequence of this site, but its activity is reproducibly lower than that of GRGDS.[83,84]

Perhaps the most direct evidence for the role of the GRGDS sequence in cellular interactions with fibronectin has come from studies of the binding of labeled fibronectin to fibroblasts in suspension. In these assays, under conditions in which it is possible to study the simple interaction of fibronectin with its cell-surface receptor, cell-binding fragments and synthetic GRGDS inhibited fibronectin binding in a reversible, competitive manner[85-87] (Figure 3). Following identification of the GRGDS pentapeptide as a critical cell-interaction sequence within the central cell-binding domain, structure-function analyses of the specificity of the GRGDS sequence for fibroblast adhesion have been performed. The results are consistent with an absolute requirement for the central RGD tripeptide; peptides with core residues such as KGD, RAD, and RGE have minimal activity[81,83,84] (reviewed in References 41, 42, 46).

Recently, cDNA cloning and expression approaches have pinpointed a second adhesion site within the central cell-binding domain.[88,89] In these studies, mini-gene constructions containing the central cell-binding domain were expressed as fusion proteins linked to bacterial beta-galactosidase. After purification, these fusion proteins were examined for their adhesive activity in cell spreading or cell attachment assays with BHK fibroblasts. Site-directed mutation of the GRGDS sequence to GRGES or even deletion of RGDS caused a substantial drop in activity consistent with the importance of this site, but activity was not totally lost, indicating the possible presence of further adhesion site(s). Stepwise exonuclease digestion of the region corresponding to the NH_2-terminus of the central cell-binding domain subsequently narrowed down this "second site" to a region 20 to 30 kDa NH_2-terminal to the GRGDS sequence (Reference 89; Figure 4). It appears that this site and the GRGDS determinant function cooperatively within the central cell-binding domain and that the presence of both sites is required for full activity. Deletion of either RGDS or the region containing the second site produced >95% losses of adhesive activity in the resulting fusion proteins, but co-coating with both mutants produced a synergistic stimulation of cell spreading and cytoskeletal organization.[89] Thus, two low-affinity sites interact synergistically to mediate adhesion by the central cell-binding domain. The sequence of amino acids constituting the second site is currently unknown, but it is quite possible that synthetic probes from this region will also have antiadhesive activity when used as competitive inhibitors of fibronectin function.

b. Type III Connecting Segment

The IIICS is a 120-amino-acid region located between the COOH-terminal heparin- and fibrin-binding domains. As for the central cell-binding domain, active sites within the IIICS have also been narrowed down by analyzing the adhesive and antiadhesive activities of synthetic peptides. Originally, six molecules were synthesized that spanned the entire region. Two peptides, termed CS1 and CS5, were found to promote melanoma cell adhesion directly, and to inhibit the adhesion of these cells to intact fibronectin.[77,90] Subsequently, the active site within the CS5 peptide has been narrowed down to the tetrapeptide Arg-Glu-Asp-Val (REDV; Figure 4), a sequence that resembles RGDS. Synthetic REDV was inhibitory in

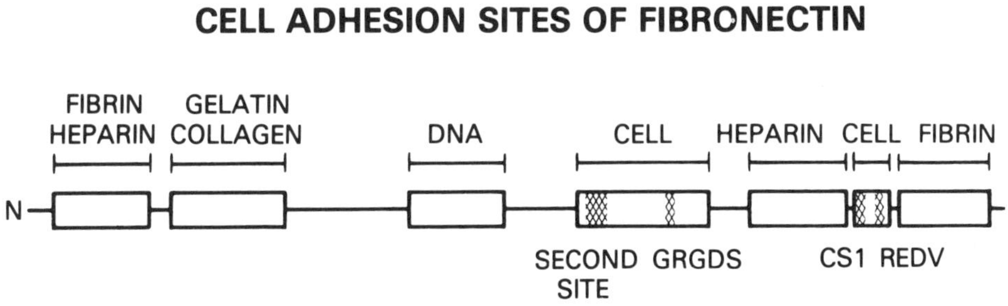

FIGURE 4. Fibronectin contains two cell-binding domains, one located in the center of the molecule and the second located in the IIICS region between the COOH-terminal heparin- and fibrin-binding domains. Each adhesive domain possesses at least two active sites. In the central cell-binding domain, critical amino acid sequences are the pentapeptide GRGDS and a second site located 20- to 30-kDa NH_2-terminal to GRGDS identified by recombinant DNA techniques. In the IIICS, the two active amino acid sequences are represented by the synthetic peptides CSI and REDV.

assays measuring melanoma cell adhesion to fibronectin, but, consistent with the results of structure-function studies of the RGDS sequence, it had no effect on fibroblast adhesion.[77] The principal activity of the IIICS region appears to reside in the CS1 peptide, however, since in *in vitro* assays, CS1 was approximately two orders of magnitude more active than CS5[90] (Figure 4). The CS1 sequence is also recognized by peripheral neurons, corneal fibroblasts, and neural crest precursor cells.[91,377,378]

C. THE FIBRONECTIN RECEPTOR

1. Identification, Isolation, and Structure

Recently, the receptor molecule recognizing the central cell-binding domain of fibronectin has been identified, purified, and sequenced. The receptor was first detected as an avian cell surface antigen recognized by antiadhesive monoclonal antibodies (CSAT and JG22).[92-95] The antigen (recently termed integrin) has been isolated by immunoaffinity chromatography of detergent-extracted cells as a mixture of three distinct polypeptides of molecular weight 155 to 160 kDa (band 1), 135 kDa (band 2), and 110 to 120 kDa (band 3).[96-98] It is now thought likely that this mixture is actually a pair of heterodimeric molecules with a distinct large (or α) subunit (either band 1 or band 2) and a shared small (or β) subunit (band 3). The ability of this complex to interact with fibronectin has been examined in several model systems. First, in equilibrium gel filtration experiments, the receptor complex was found to bind to fibronectin in a CSAT- and GRGDS-sensitive manner.[99] Second, JG22E, a subclone of the original JG22 antibody, blocked the direct binding of the 75-kDa cell-binding fragment of fibronectin to chicken embryo fibroblasts.[100] Finally, an affinity matrix of 75-kDa Sepharose was found to retard the passage of the receptor complex;[100] this retardation was blocked by GRGDS, but not by the control GRGES peptide.[100] The receptor could not be purified by this technique, however, because of the relatively low affinity of its interaction with fibronectin.

Fibronectin receptors have also been identified on mammalian cells.[101-106] It appears that the interaction of this group of molecules with fibronectin is of significantly higher affinity than the chicken receptor, since they are retained in a GRGDS-sensitive manner on affinity columns containing the central cell-binding domain of fibronectin as the ligand.[102,103,105,106] All mammalian receptors appear to be complexes of two dissimilar, noncovalently linked subunits of molecular weight 140 to 160 kDa and 115 to 135 kDa.[101-106] The receptor recognizing the IIICS region of fibronectin has not yet been identified, although since the antiadhesive monoclonal antibody JG22E is able to disrupt adhesion on a substrate of CS1 peptide,[91,377] it appears likely that the receptor is related to that for the central cell-binding domain.

2. Supergene Family

Receptors of similar size and subunit composition to the fibronectin receptor have also been identified for other adhesion proteins, including laminin, vitronectin, collagen, fibrinogen, and von Willebrand factor. As yet, there is no evidence for two receptors sharing an α subunit; however, in some instances, receptors have been found to contain a common β subunit; in fact, there now appear to be at least three groups of adhesion receptor, which can be classified according to their particular β subunit (for recent reviews, see References 43, 46, and 107 to 109). The first group contains the avian and mammalian fibronectin receptors together with a collection of five dimeric molecules isolated from T lymphocytes termed "very late antigens" (VLA).[110] The second comprises the vitronectin receptor[111] and the platelet surface protein complex IIb/IIIa.[112-114] The final group of receptors are the molecules LFA-1,[115,116] Mac-1,[117,118] and p150,95.[119] These are found on lymphoid and myeloid cells and appear to have important roles in immune cell recognition processes.

Recently, primary sequence information has been obtained for several of the adhesion protein receptors,[120-131] as well as for the position-specific antigens of *Drosophila* which may serve related functions in insects.[132] All the α subunits for which sequence information is available share evolutionarily conserved NH_2-termini and have a similar overall structure, including a short cytoplasmic domain, a predicted transmembrane domain, and a large extracellular domain containing predicted calcium-binding sites. Similarly, the β subunits of the three receptor groups are highly homologous, again with a short cytoplasmic domain, a predicted transmembrane domain, and a large, cysteine-rich extracellular domain. There is no apparent homology between the α and β subunits of any individual receptor, but, at the amino acid level, the degree of homology between the different α or β subunits is 40 to 50%.

3. Role in Cell Migration

In general, the level of the cell surface receptor for fibronectin is elevated on cell populations undergoing morphogenesis, but is low in stationary, mature tissues.[133-135] The role of fibronectin receptor-fibronectin interactions in embryonic cell migration is indicated by studies showing that antireceptor agents block certain major morphogenetic events. For example, microinjecting the GRGDS peptide into embryos in order to competitively inhibit receptor function results in blockage of amphibian and *Drosophila* gastrulation and of neural crest cell migration.[136,137] Furthermore, antibodies to the fibronectin receptor block gastrulation,[138] the migration of hemopoietic precursor cells to the embryonic thymus gland,[139] the movement of neural crest cells away from the neural tube,[140] and newt epidermal cell migration in healing skin wounds.[141]

Examination of the subcellular distribution of fibronectin receptors has revealed similar, striking differences between motile and nonmotile cell populations. On highly migratory neural crest cells and somitic fibroblasts, which contain low numbers of cell-to-substrate focal contacts, the receptor is diffusely distributed on the cell surface.[133] In contrast, on stationary cells, such as long-term cultures of somitic fibroblasts and certain ectodermal cell types, the receptor forms cell surface aggregates and is found in prominent linear arrays that codistribute with extracellular fibronectin and intracellular alpha-actinin.[133] Receptors in each pattern of localization have been compared for their ability to move laterally in the plane of the membrane. Membrane protein mobility can be quantitated using fluorescence recovery after photobleaching (FRAP) methods. Fibronectin receptors in the diffuse distribution are freely and rapidly mobile in the plane of the membrane, whereas those in streaks are immobile.[142] As cells mature and become less migratory, they display an intermediate stage in which the receptors are not necessarily in streaks, yet are partially immobilized in a temperature-dependent manner. Maturation to a nonmigratory state is accompanied by immobilization of receptor.[142] These results indicate that the receptor can exist in different mobility states, and that partial or full immobilization correlates with the loss of capacity for cell migration. The diffuse, laterally mobile arrangement of receptor on migratory cells, coupled with their lack of stable cell-substrate contact sites, suggests that rapidly motile cells exhibit only weak interactions with fibronectin. This would be important for cells that need to transiently attach and rapidly detach from an adhesive substrate during active migration, for example during the invasion of subendothelial connective tissue by metastatic cells.

III. FIBRONECTIN LEVELS IN NEOPLASTIC CELLS

A. CELLS IN TISSUE CULTURE

1. Transformation-Associated Changes

Early observations of a quantitative decrease in the level of surface fibronectin on virally transformed fibroblasts[143-152] provided the first indication that this protein may have an important role in influencing tumorigenicity or malignancy. The reported decrease was five-fold in chick fibroblasts[153] and tenfold in human fibroblasts.[154-155] In kinetic studies performed with temperature-sensitive mutant viruses, fibronectin reappeared at the cell surface 1 to 4 h after shift-up to nonpermissive temperatures,[146,149-153,156] a similar time course to its appearance after the seeding of trypsinized, normal cells in culture.[152] In shift-down experiments, the kinetics of decrease of surface fibronectin were complicated by the slow turnover of molecules already incorporated into the pericellular matrix; however, after subculture, most surface fibronectin was lost.[146,149-153,156]

In addition to fibroblasts, decreases in the level of cell-associated fibronectin are a consistent feature of the transformation of a variety of cell types (including glial cells,[157] kidney cells,[143,146,148,150,158] and myoblasts[158,159]) from a number of different species (including human,[154,160] chicken,[149-152] hamster,[143-146,148,158] rat,[150,157-161] mouse,[144,147] and quail.[162]) In certain cases, these decreases parallel the level of *in situ* expression of fibronectin in tumors of the same tissue origin (see below). In addition to viral infection, chemical[163-166] and spontaneous transformation[158,167,168] are also accompanied by a decrease in fibronectin. Although the overwhelming majority of studies have reported a loss of cell-associated fibronectin upon transformation, there is a small proportion of cells in which fibronectin levels do not change (viral transformation of rat[164,169-171] and mouse[167,170,172,173] fibroblasts, mouse epithelial cells,[174] and hamster kidney cells,[170] chemical transformation of rat epithelial cells[163,175-178] and mouse salivary gland and bladder epithelial cells,[179] and spontaneous transformation of rat myoblasts and hamster kidney cells[170]).

Consistent with its sensitivity to oncogenic transformation, the loss of surface fibronectin also appears to be an early event in the promotion stage of the initiation-promotion, two-step model of chemical carcinogenesis. Tumor promoters, such as the phorbol esters[180-187] and teleocidin,[188] have been shown to induce the rapid release of cell surface fibronectin in cells susceptible to promotion, but not in nonpromotable cells. In immunophotoelectron microscopy studies, where cell membrane- and pericellular matrix-associated fibronectins were distinguishable, the phorbol ester TPA induced the selective release of membrane-bound fibronectin 10 to 30 min after addition.[186] This rapid effect, coupled with the reported ability of TPA to induce fibronectin release in enucleated cells,[183] implies a direct effect of tumor promoters on pericellular matrix assembly at the membrane level. Although its exact mechanism of action is currently unknown, TPA has no effect on the binding or release of exogenously added fibronectin,[187] suggesting that its effects may not be mediated by direct action on the fibronectin receptor.

2. Mechanisms of Fibronectin Loss

Many mechanisms have been proposed to account for the decrease in fibronectin after transformation. These include a reduction in biosynthesis, increased degradation, defective retention at the cell surface (i.e., an altered fibronectin receptor system), and failure to incorporate the molecule into the pericellular matrix (due to structural alterations in the fibronectin molecule). There is now evidence for the involvement of each of these, but perhaps the most consistent defects are reduced biosynthesis and increased degradation.

Fibronectin is frequently still synthesized by transformed cells,[153,154] but both the intracellular and extracellular pools[153] and the fibronectin messenger RNA copy number are reduced approximately fivefold.[189-193] The factors which mediate the suppression of the fibronectin gene are not known. In metabolic labeling experiments in normal and transformed chick fibroblasts, the transit of newly synthesized fibronectin to the cell surface was found to be normal,[153] but the turnover rate was elevated.[149,153,156,194] Surface fibronectin is extremely sensitive to proteolytic digestion,[143,195,196] and it is likely that tumor cell-associated proteases can account for at least part of the decrease on transformation. Nevertheless, protease inhibitors have shown no ability to restore fibronectin to transformed cells[186,189] or to prevent the "trans" removal of fibronectin from normal cells cocultured with transformed cells.[194,197,198]

It has been reported that fibronectin binding to the surface of transformed cells is defective.[199-201] However, since recent studies have found similar levels of functional receptor on normal and transformed cells, the defect does not appear to be at the receptor level[59,202] (but see Reference 203). Similarly, there is no evidence that structural alterations in the fibronectin molecule itself cause defective binding, since normal and transformed cell fibronectins possess equal activity in adhesion assays.[201] Fibronectin isolated from transformed cells is of slightly higher molecular weight than that isolated from corresponding normal cells, but the difference appears to be the result of increased branching and sialylation of N-linked oligosaccharide subunits;[201,204] current evidence suggests that this alteration would have an insignificant effect on biological activity.[205] Proteolytic cleavage patterns of normal and transformed cell fibronectins are very similar,[201,206] although there may be subtle alterations in the alternative splicing profile of the ED and/or IIICS regions between normal and transformed cells.[59-61,70,73,74] Fibronectin is phosphorylated on serine and possibly threonine, but not on tyrosine,[206,207] and the degree of phosphorylation has been reported to be increased after transformation[206] (but see Reference 207). However, the functional consequences of phosphorylation of fibronectin, if any, are still not clear.

3. Fibronectin and Tumorigenicity

The finding that exogenous addition of purified fibronectin transiently restores a more normal morphology,[208-211] but not nutrient transport rate or growth rate,[208,211,212] to transformed cells, originally stimulated research on the protein. Reversion of the transformed phenotype, concomitant with restoration of surface fibronectin or enhancement of fibronectin matrix assembly, is also induced by treatment with bromodeoxyuridine,[213] interferon,[214,215] butyrate,[216] glycocorticoids,[217,218] and cAMP.[219] Early reports that attempted to correlate fibronectin levels with either the tumorigenicity or metastatic potential of transformed cell lines demonstrated a close inverse correlation between the amount of cell-associated fibronectin and the ability of the cells to produce a tumor *in vivo*.[158,160] In addition, serial passage of rat liver epithelial cells *in vitro* caused a progressive loss of fibronectin concomitant with the acquisition of a tumorigenic phenotype.[220] However, more recently, many studies have reported discordant results in which tumorigenic cell lines have been generated that still express normal, or even elevated, levels of surface fibronectin, together with lines which have lost their ability to form a fibronectin-containing extracellular matrix, yet are unable to form a tumor *in vivo* (for example, see References 169, 221). Furthermore, in a series of chemically transformed guinea pig lines, no correlation was found between the number of cells required to form a tumor and the amount of cell surface fibronectin.[222] Similarly, no link was found between fibronectin expression and the ability of a series of rat mammary adenocarcinoma lines to colonize the lungs in an experimental metastasis model system.[223]

Studies examining the properties of hybrid clones produced by somatic cell fusion of tumorigenic (fibronectin-negative) and nontumorigenic (fibronectin-positive) cell lines have been the most definitive in demonstrating the lack of a link between fibronectin expression and tumorigenicity. In almost every case, no cosegregation was detected between the degree of fibronectin expression and tumorigenicity either in first generation or back-selected hybrids; tumorigenic clones possessing surface fibronectin and nontumorigenic clones lacking it were found.[221,224-232] Nevertheless, it has also been suggested that the subcellular distribution of fibronectin should be considered rather than the presence or absence of the molecule per se. In this respect, a strong link between tumorigenicity and the expression of fibronectin as short, unbranched surface fibrils as opposed to longer, branched fibrils has been reported.[233,234]

B. Tumors and Tumor Metastases

Fibronectin has now been the subject of a multitude of studies examining its distribution in tumors and their metastases. A comprehensive listing is presented in Table 1. These studies have generally been performed with fixed, embedded surgical specimens, and fibronectin has been detected by indirect immunofluorescence, immunoperoxidase, or immunoelectron microscopy techniques. Based on the results of these investigations, a number of generalizations can be made:

1. Fibronectin is greatly decreased or absent from human carcinoma cells and their metastases (see References 236, 262, 264, and 272 for rare exceptions). In benign or well-differentiated neoplasms, fibronectin is localized to intact basement membranes bordering the tumor cells, but in invasive tumors, the characteristic fragmentation of basement membranes is accompanied by reduced fibronectin staining. Fibronectin loss appears to be progressive with the stage of the disease and with increasing anaplasia. Despite its loss from tumor cells, fibronectin is consistently retained in the tumoral stroma and in the basement membranes underlying blood vessels. In some instances, the amount of stromal fibronectin is actually elevated,[235,242,254,273] presumably as part of the reactive connective tissue response of tumor desmoplasia.
2. Fibronectin is present at normal levels in human sarcoma and soft tissue tumors (see Reference 235 for exceptions). In tumors which may originate from more than one cell type, such as rhabdoid renal tumors and nephroblastomas/Wilm's tumors, the presence of fibronectin is variable.[236,258-260]
3. Fibronectin is absent from brain tumors of glial cell origin, but is present in nonglial tumors. Melanomas and lymphomas are also fibronectin-free. In contrast, fibronectin is abundant in basal cell carcinomas, as well as in benign tumors of epithelial origin such as breast fibroadenoma and dysplastic lesions of the uterus.

Investigations of the expression of fibronectin in tumor-bearing animals are generally in accord with the results of examining human tumor biopsies. Human basal cell carcinoma cells implanted into nude mice were found to be fibronectin positive, and in addition were stained with antisera specific for human fibronectin, thereby demonstrating that carcinoma cells were the biosynthetic source of the tumoral fibronectin.[280] Similarly, a rat chondrosarcoma was found to retain its fibronectin,[281] while a chemically induced hamster pancreatic carcinoma[282] and metastases from a rat mammary adenocarcinoma were fibronectin negative.[223]

A number of studies have examined the level of cell surface fibronectin in tumor cells explanted from patients and propogated in tissue culture. Although in most cases the amount of cell-associated fibronectin paralleled its *in situ* expression, there have been many exceptions reported. For example, human gliomas,[238-287] melanomas,[288,289] and hepatomas,[290-292] which usually lack fibronectin, consistently express it in culture. Similarly, explants of

TABLE 1
Distribution of Fibronectin in Tumors[a]

Tumor	Fibronectin present	Fibronectin absent
Bladder tumors		
Carcinoma	235[b], 236[b]	235[s]—237[s]
Myosarcoma	235	
Bone tumors		
Giant cell sarcoma	237	
Osteosarcoma	236	
Brain tumors		
Astrocytoma		238—240
Glioblastoma		238, 240
Gliosarcoma	241	
Meningioma	238, 239, 241	
Oligodendroglioma		240
Breast tumors		
Carcinomas (cystic/ductal/lobular)		235[s], 236[s], 237[s], 242—244, 245[s], 246[s], 247
Fibroadenoma	248	237[s]
Gynecomastia		237[s]
Colorectal carcinoma	249[d]	237[s], 242, 245[s], 249, 250[s]—255[s]
Endothelial cell tumors		
Hemangioblastoma	239	
Hemangiosarcoma	256, 257	235
Fat cell tumors		
Angiolipoma	235	
Liposarcoma	236, 237	235
Lipoma	235, 236	
Fibrosarcoma	236, 237, 256	235
Gastric carcinoma	249[d]	249, 254
Kaposi's sarcoma	237	
Kidney tumors		
Bone-metastasizing renal tumor of childhood		259
Clear cell carcinoma		258
Hypernephroma		237[s]
Mesoblastic nephroma	258, 259	
Rhabdoid renal tumor	258	258
Nephroblastoma/Wilm's tumor	259, 260	236, 258
Liver tumors		
Cholangiocarcinoma		261[s]
Hepatocellular carcinoma	261[d], 262[d], 263	261
Lung tumors		
Adenocarcinoma		236[s], 237[s], 264
Epidermoid carcinoma/squamous cell carcinoma		235[s], 236[s], 237[s], 264
Large cell carcinoma	264	236[s]
Laryngeal carcinoma		265[s]
Small cell carcinoma	264	237[s]
Lymphomas (follicular center cell/Hodgkin's/plasmacytoma)		266—268
Mediastinal endodermal tumor		269
Melanoma		237[s]
Myxoma	270	
Oropharyngeal tumors		
Esophageal granular cell tumor		271
Oral mucosal carcinoma	247[d], 272	247
Salivary gland tumors	247[d]	247, 254[s]

TABLE 1 (continued)
Distribution of Fibronectin in Tumors[a]

Tumor	Fibronectin present	Fibronectin absent
Pancreatic carcinoma		273[s]
Parathyroid carcinoma		237[s]
Prostate carcinoma		235, 237[s]
Reticulum cell sarcoma	237	
Schwann cell tumors		
Neurilemoma	237	
Neurofibroma	236, 237, 256, 274, 275	
Schwannoma	236, 256, 257	
Seminoma		236
Skin tumors		
Basal cell carcinoma	276—278	
Dermatofibroma/histiocytoma	236, 256, 257, 279	235
Desmoid tumor	237	
Squamous cell carcinoma		237[s], 278
Smooth muscle tumors		
Hemangioma	237, 256	
Hemangiopericytoma	236, 237, 256	235
Leiomyofibroma	236	
Leiomyoma	237,256	
Leiomyosarcoma	236, 237, 256, 257	
Synoviocytoma	256	
Teratocarcinoma		236
Thyroid tumors		
Follicular adenoma	235, 236	
Follicular carcinoma	235	236
Papillary carcinoma	236	235, 237[s]
Uterine tumors		
Carcinoma	235[d], 236	235, 237[s]
Dysplastic lesions	236	

[a] A comprehensive list of studies that have examined the presence or absence of fibronectin in human tumors. Presence or absence refers to immunostaining of tumor cell surfaces, not tumor stroma. Superscript ''s'' denotes the absence of fibronectin on tumor cell surfaces with concomitant staining of tumor stroma. Superscript ''b'' denotes fibronectin presence in benign tumors, but not in malignant tumors. Superscript ''d'' denotes fibronectin presence on well-differentiated tumor cells, but not on poorly differentiated cells.

primary breast carcinomas and normal mammary epithelial cells have been reported to synthesize similar amounts of fibronectin.[293-297] There have also been isolated reports of human squamous cell,[298] transitional cell,[299] and colon carcinomas[294] expressing fibronectin *in vitro*. One major concern in this type of study is the potential for tissue culture artifacts; in fact, it has been reported that the acquisition of a fibronectin-positive phenotype in glioma cell cultures progresses with increasing passage number and is concomitant with the loss of the glial fibrillary acidic protein marker characteristic of these tumors *in vivo*.[283-287] It seems unlikely, therefore, that expression of fibronectin by cultured tumor cells is a valid diagnostic marker for either the neoplastic or malignant state.

C. FLUID FIBRONECTINS AS TUMOR MARKERS

1. Plasma Fibronectin

In addition to alterations in the level of tumor-associated fibronectin, there are now a number of reports documenting changes in the concentration of plasma fibronectin in cancer patients. Normally, based on immunological quantitation, plasma fibronectin is present at a concentration of approximately 300 μg/ml (mean values varying from 238 to 650 μg/ml).[300-324] Many different techniques have been used for measuring plasma fibronectin levels, including immunodiffusion, immunoturbidimetry, electroimmunoassay, radioimmunoassay, and enzyme-linked immunosorbent assay. In addition to the normal variability within the population, measurement of fibronectin is subject to errors arising from the use of these different quantifying methods, as well as different sampling methods, handling procedures, and storage conditions. Furthermore, it should be noted that immunological methods of measuring fibronectin concentration are unable to distinguish between the intact molecule and functionally altered proteolytic fragments, because both molecules possess similar antigenic determinants. It is likely, therefore, that proteolytically degraded fibronectin, arising either from cleavage *in vivo* or even as a result of mishandling in the clinic, contributes to the final value. In this regard, it has been reported that a specific fragment of fibronectin called MAD-2 is elevated in the plasma of patients with malignant lung, breast, testicular, and ovarian cancer.[325,326] Although the biological relevance of this finding is unclear, fragmentation of fibronectin has been reported to uncover novel activities such as enhancement of morphological cell transformation[327] and stimulation of DNA synthesis.[328] Similar cleavage of fibronectin has been observed in studies of synovial fluid in rheumatoid arthritis patients; electrophoretic analysis of fibronectin purified from synovial fluid samples revealed a significant level of immunoreactive, but fragmented, fibronectin.[329]

With potentially important exceptions, plasma fibronectin is consistently present at a significantly higher level in patients with neoplastic disease compared to age- and sex-matched controls[301,315,316,322-324,330-332] (but see References 301, 305, 319, 330). This finding is true for a wide variety of tumor types from a number of organs, including breast, lung, colon, brain, and liver. The one prominent exception is in leukemia patients, where a slight depression in chronic lymphocytic leukemia,[330] and more substantial decreases in acute myeloblastic leukemia,[321] non-Hodgkin's malignant lymphoma,[321] and other leukemias[307,318,330] were observed. It should be noted, however, that in two other studies,[315,317] no change in fibronectin level was found in either acute myelocytic leukemia, acute myeloblastic leukemia, acute lymphocytic leukemia, or chronic myelocytic leukemia.

Despite these results, determination of the concentration of a blood protein like fibronectin is necessarily subject to a great many internal variables (as well as the external variables discussed above), including the stage of the disease, prior therapy, and complications such as accompanying infections and organ failure. Without strict standardization and greater sampling numbers, the general conclusions discussed above must remain equivocal. A promising approach to examining fibronectin levels under more defined conditions is the use of animal model systems (see Chapter 17). In an experimental setting it is possible to use large numbers of animals in order to obtain significance, and to examine in detail the effects of external agents as well as the kinetics of the change. To date, without exception, administration of tumorigenic cells (fibrosarcoma, colon carcinoma, melanoma, malignant breast carcinoma, Ehrlich ascites, Walker 256 carcinoma, and sarcoma 180 cells) into either mice or rats has resulted in an elevation of plasma fibronectin,[333-337] ostensibly confirming the studies in human patients documented above.

Despite the apparent correlation between fibronectin levels and cancer, fibronectin currently appears to have only limited usefulness as either a diagnostic or prognostic parameter in tumor patients. The principal reason for this is the sensitivity of plasma fibronectin concentration to a number of other factors, resulting in its lack of specificity as a cancer marker. Fibronectin levels are locally elevated in rheumatoid arthritis,[338] and the concentration in plasma is raised in various dermatological disorders,[339] and in some autoimmune diseases,[340] and is decreased during traumatic shock,[303,304,341] surgery,[303,304,341] sepsis,[342] starvation,[311] and in response to burn injury[343] and liver dysfunction.[309,310] In addition to these disease conditions, almost nothing is known about the factors that normally control the concentration of fibronectin in plasma. Without an improved understanding of which parameters to measure before assessing whether a particular fibronectin level is normal, there would appear to be too many unknown factors that may contribute to an altered plasma concentration.

Our current lack of understanding of the normal regulation of plasma fibronectin levels also makes it impossible to judge why elevated or lowered levels are found in cancer patients. The mechanisms of these alterations are clearly important to study, but at present can only be the subject of speculation. Increases in fibronectin could be due to shedding of cellular fibronectin from the tumor cells, enhancement of liver-derived plasma fibronectin synthesis, decreased clearance, or synthetic contributions from other normal cell populations such as lymphocytes. Since cancer cells usually have reduced rates of fibronectin synthesis, it is unlikely that they might be able to produce enough protein to cause an increase in the substantial plasma fibronectin pool, even though many transformed cell populations are defective in their ability to retain the fibronectin that they make. In support of this, there is no evidence for a link between the increase in fibronectin levels observed in patients with cancer and the mass of their tumors.[315] A more likely explanation for raised fibronectin would be a contribution from host cell populations, such as in desmoplastic tissue, but there is currently no information on this possibility. On the other hand, a decrease in fibronectin concentration could be attributed to the proposed opsonic function of the molecule in mediating the removal of circulating debris and foreign agents, including tumor cells.[344,345] In fact, a decrease in the resistance to experimental tumor challenge with lowered fibronectin levels has been reported for Walker 256 carcinoma cells injected into rats.[333]

2. Other Body Fluids

In addition to alterations in the level of plasma fibronectin in tumor patients, there are several reports documenting fluctuations in the concentration of fibronectin in other body fluids. Fibronectin was elevated in (1) the pleural fluid of patients with mesothelioma compared to their plasma or to the pleural fluid of patients with other pleurisies,[319,346] (2) the urine of patients with prostatic carcinoma,[347] and (3) the cerebrospinal fluid of patients with brain tumors.[348] It remains to be determined whether these changes are reproducible and whether the fluid level of fibronectin might have diagnostic or prognostic value in a specific setting.

Perhaps the most promising diagnostic use of plasma fibronectin concentration is in distinguishing between malignant and cirrhotic ascites, arising as a result of either malignant neoplastic disease or chronic liver disease, respectively. Although the mechanisms involved have not been evaluated, in all reported instances the level of ascitic fibronectin was substantially raised in patients with ascites developing as a consequence of a malignant neoplasm (in hepatocellular carcinoma, biliary duct carcinoma, and in peritoneal metastases from ovarian cancer and liver metastases), compared to the level in cirrhotic ascites.[349-353]

IV. DISSECTING THE ROLE OF FIBRONECTIN IN TUMOR METASTASIS

The presence of fibronectin in and around tumors, together with its ability to support the adhesion of a variety of normal and transformed cells, suggests that fibronectin may play a role in certain adhesive aspects of target organ colonization by metastatic cells. Until recently, this kind of indirect correlation was the only basis for hypothesizing its involvement. However, with the development of specific peptide inhibitors of the adhesive activity of fibronectin, it is now possible to probe the role of these recognition signals *in vivo* and to obtain more direct information regarding the biological functions of the molecule.

A. FIBRONECTIN IN THE ARREST OF METASTATIC CELLS

1. Experimental Metastasis Assay

An experimental metastasis model system can be used to examine the role of fibronectin in the spread of tumor cells.[354] In one version of this widely used assay, metastatic B16-F10 murine melanoma cells, propagated in tissue culture, are detached and injected into the lateral tail vein of syngeneic mice (Figure 1). After 2 to 3 weeks, the animals are examined for evidence of metastasis at necropsy. This assay has certain advantages for testing anti-metastatic therapies. First, the assay can be quantitated by determining the number of melanotic lesions in the target organ at a specific time after injection, and second, by selection of an appropriate tumor cell inoculum, it can be adapted into a method for examining animal survival and for determining the potential clinical efficacy of a particular treatment. Although intravenous administration of tumor cells has the disadvantage of bypassing the initial intravasation phases of the metastatic cascade, it appears to faithfully reproduce the events of hematogenous metastasis subsequent to the release of tumor cells into the circulation. Furthermore, since the arrest of tumor cells in the target organ occurs after injection, this model is also ideal for studying the role of cell adhesion in colonization.

The B16-F10 melanoma line was originally developed by Fidler,[355] who selected for enhanced metastatic activity by repeated passaging of successive, lung-colonizing, parental tumor lines through mice (cell were injected intravenously, and resulting pulmonary metastases were excised, cultured *in vitro*, and reinjected into mice). After ten passages (B16-F10), the cells were found to possess greatly enhanced lung-colonizing activity compared to the parental line.

2. Effects of Synthetic Peptides

As a first test of the system, GRGDS, representing the principal recognition site in the central cell-binding domain of fibronectin, was used as a prototype antiadhesive synthetic peptide. Premixing and coinjection of GRGDS with B16-F10 cells elicited a dose-dependent inhibition of pulmonary metastasis.[356] At a dose of 3 mg GRGDS per mouse, $\geq$90% inhibition of the formation of visible tumor nodules was consistently achieved with inocula of 50,000 to 100,000 cells. When fewer cells were injected into peptide-treated mice, complete inhibition of colonization was obtained. This reduction translated into a substantial prolongation of the survival of recipient mice, demonstrating the potential clinical efficacy of GRGDS-like molecules. Histological sectioning and staining confirmed the decreased level of tumor colonization in peptide-treated mouse lungs, thereby ruling out the possibility that GRGDS only blocked expression of the melanotic phenotype. The effectiveness of GRGDS treatment indicates an important role for this signal in tumor metastasis, and implies that adhesive recognition of the GRGDS sequence by metastasizing tumor cells occurs at least once during colonization and is crucial for successful completion of the metastasic process. Inhibition of experimental metastasis by GRGDS was not an artifact caused by cytotoxicity or suppression of tumorigenicity, and the peptide possessed almost equal inhibitory activity when cells and peptide were injected into different tail veins without premixing.[356]

The specificity of GRGDS inhibition has been tested by comparing the antiadhesive and antimetastatic activities of a number of homologous peptides. These included GRGES, the prototype control peptide in which the aspartic acid residue in position 4 is replaced with a glutamic acid residue, GRDGS, in which the central glycine and aspartic acid residues are transposed; RGDS, which lacks the NH_2-terminal glycine residue; GRGD, which lacks the COOH-terminal serine residue; and SDGR, the reverse of RGDS. The relative inhibitory activities of these molecules in the experimental metastasis assay was found to closely match their established capacities to disrupt cell adhesion *in vitro*; for example, RGDS and SDGR were active, but less so than GRGDS, and GRGES, GRDGS, and GRGD were either minimally active or completely inactive.[356,357] The inhibition by GRGDS is therefore highly specific, since minor sequence alterations result in a dramatic loss of activity. Since it was not possible to dissociate antiadhesive and antimetastatic activity through testing homologous peptides, these results strongly suggest, as expected, that GRGDS blocks experimental metastasis through its ability to disrupt cell adhesion.

To examine the potential role of the other defined cell interaction domains of fibronectin in metastatic colonization, antiadhesive synthetic peptides taken from the active sites of these regions were tested for their effects in the experimental metastasis assay. Both CS1, the 25-mer peptide representing the dominant active site from the IIICS region, and REDV, the active site contained within CS5, the second active region of the IIICS, were found to have no detectable inhibitory activity.[377] This interesting result suggests that the IIICS cell-binding domain plays an insignificant role in the metastasis of B16-F10 cells and that it is the central cell-binding domain of fibronectin that is the principal region recognized by malignant melanoma cells during their arrest in the target organ. The differential activity of the synthetic peptides from the two adhesive domains is also evidence that the different cell-binding regions of fibronectin may support different functions *in vivo*. At present, the active site within CS1 has not been narrowed down further than 25 amino acids, and because it is difficult to inject an amount of CS1 equivalent in molar terms to 3-mg GRGDS, confirmation of the lack of activity of this domain will need to await the testing of smaller peptides from within CS1.

3. Sites of Action

Although it is now apparent that GRGDS blocks experimental metastasis through its inhibition of cell adhesive processes, identification of exactly which processes are crucial for metastatic colonization is not simple. There are potentially many key sites at which GRGDS-dependent adhesion could be involved — for example, during the arrest of tumor cells, during their adhesion and migration through basement membranes, and during their invasion of subendothelial connective tissue. Furthermore, an effect of GRGDS on tumor cell arrest could itself be mediated by a number of different mechanisms, including direct effects on tumor cell adhesion to fibronectin, effects on heterotypic and homotypic cell-cell adhesion, and less well-defined effects on blood cell function.

As one approach to more closely define the particular site of action of GRGDS, kinetic analyses of the retention of radiolabeled melanoma cells in the target organ have been undertaken. In these studies, GRGDS rapidly promoted the loss of B16-F10 cells from the lung;[356] the rate of loss was approximately twice as fast in mice administered the peptide. Almost all of the decrease in colony number caused by GRGDS could be accounted for by the early removal of cells from the lungs (e.g., 7 h after injection of B16-F10 cells, the level of pulmonary radioactivity in peptide-treated mice was fivefold lower than in control mice[356]). This time course of activity is consistent with both the proposed antiadhesive mechanism of action of GRGDS (since malignant cells adhere and extravasate rapidly following their arrest in the target organ), and with the time frame over which the peptide is present at a high enough concentration to interfere with tumor cell adhesion. Radiolabeled GRGDS has a circulatory half-life of only 8 min,[357] and assuming, therefore, that the mouse has an extracellular fluid volume of 6 ml, the concentration of the peptide would fall below an active level within 1 to 2 h.

In addition to a simple effect on adhesion, it is also conceivable that GRGDS might interfere with certain metastasis-related blood cell functions restricted to the early stages of colonization, including inhibition of platelet aggregation and/or enhancement of NK-cell reactivity. To examine the contribution of either platelets or NK cells to GRGDS-mediated inhibition of experimental metastasis, peptide activity was compared in normal mice or in mice defective in either platelet or NK-cell function. Platelet deficiency was induced by injection of rabbit anti-mouse platelet antiserum to make the animals thrombocytopenic,

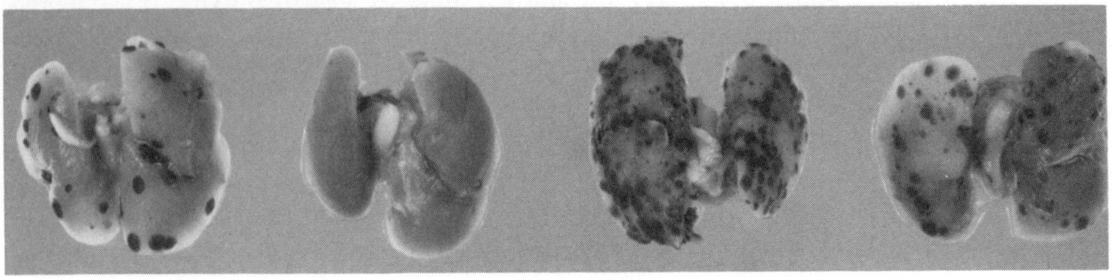

FIGURE 5. Antimetastatic activity of GRGDS peptide in normal and NK-deficient beige mice. Representative lungs from groups of normal C57BL/6 (a, b) or C57BL/6$^{bg/bg}$ beige mice (c, d) injected with either 7 × 10^4 B16-F10 melanoma cells in the presence (b, d) or absence (a, c) of 3 mg GRGDS. Mice were sacrificed 14 d after injection and the lungs were excised and fixed in 10% formaldehyde. Consistent with the genotypic defect that renders them NK-deficient, the number of visible colonies in the nonpeptide-treated beige mice was 11-fold higher than in nonpeptide-treated regular mice. However, the degree of inhibition of colonization by GRGDS was similar in both types of animal.

while mutant beige mice, which lack factors required for cytolysis by NK cells, were used as a model for NK deficiency. GRGDS was found to retain all of its antimetastatic activity in both types of animals[357] (Figure 5), indicating that neither platelets nor NK cells play a significant role in GRGDS-mediated inhibition of experimental metastasis.

At present, the various potential connective tissue sites at which GRGDS may act await identification. Since coinjection of GRGDS with radiolabeled melanoma cells causes a significant decrease in the number of cells initially arresting in the lung,[356] this implies an effect of the peptide on attachment to the endothelium or exposed basement membrane. However, while this decrease may be one mechanism of action, it does not account for all of the antimetastatic activity of the peptide, since the rate of tumor cell loss from the lungs of peptide-treated mice is still greater than that in mice bearing lower numbers of initially arrested cells from smaller tumor cell inocula.[357] Alternative mechanisms of action, which will undoubtedly be the subject of future investigations, including promotion of cell detachment, blockage of tumor cell binding to the subendothelial matrix, and interference with cell migratory events subsequent to extravasation.

B. FIBRONECTIN AND ITS RECEPTOR IN TUMOR CELL MIGRATION AND INVASION

Since the migration of cells on fibronectin requires the function of cell surface fibronectin receptors, both the number of such receptors and their distribution may be important for determining the ability of a cell to adhere, migrate, and invade. As indicated in previous sections, fibronectin and its receptor are involved in a variety of transient, normal cell migratory events during embryonic development and wound healing. Cells can also apparently use fibronectin for invasion. Tumor cells, like normal cells, can use fibronectin as a migratory substrate *in vitro*.[32,358,359] The frequently observed decrease of cell-associated fibronectin on transformed cells may be important in this regard, since rapid cell migration may require that levels of fibrillar fibronectin on the cell surface be low.[45,360] For example, the accumulation of fibronectin fibrils on normal embryonic heart fibroblasts is accompanied by a loss of migratory activity, and the addition of exogenous fibronectin to migrating cells at a susceptible stage promotes the stationary phenotype.[360,361] It appears likely that low endogenous production of fibronectin makes a cell more responsive to exogenously available fibronectin substrates.[45]

The role of fibronectin in invasion remains to be examined in detail, but preliminary studies indicate that it is required for invasion across a basement membrane. Lymphoid precursor cells transiently become highly invasive in order to enter the thymus;[139] *in vitro*, such cells readily invade across a thick amniotic basement membrane used to test invasiveness of tumor cells. This invasion is inhibited by antifibronectin antibodies.[139] Since such antibodies only modestly inhibit melanoma cell adhesion to subendothelial extracellular matrix,[362] they may block invasion at an additional step besides initial cell attachment to the basement membrane.

Further insight can be obtained using inhibitors of fibronectin receptor function. The GRGDS peptide blocks lymphoid cell invasion across basement membranes.[139] Similarly, passage of tumor cells across basement membranes has also recently been shown to be inhibited by such peptides, and the site of action is reportedly at the level of migration and invasion rather than on initial cell attachment.[363, 379] These synthetic peptide studies implicate an RGD recognition signal in tumor cell invasion, but they do not indicate whether the target molecule is fibronectin or some other molecule using a related sequence; this uncertainty will require resolution using more specific inhibitors of adhesive protein function. Nevertheless, integrin receptors are most likely involved in the invasive process, since antireceptor antibodies can block lymphoid precursor invasion of basement membranes.[139]

Since recent studies have indicated that levels of fibronectin receptors are substantially decreased after the completion of morphogenetic events during embryogenesis,[133-135] it appears logical that tumor cells might need to reexpress fibronectin receptors in order to successfully colonize and invade target organs. A recent study has addressed this question in the classical Rous sarcoma virus tumor system of chickens, where the injection of virus produces rapid, predictable tumor formation with minimal complication by a host desmoplastic reaction. Such tumors reproducibly show fivefold increases in levels of fibronectin receptor (Figure 6[380]). The tumor cell receptor complexes contain α and β subunits similar to those found in normal cells. This overall elevation in fibronectin receptor levels may at least partially account for the ability of these cells to invade. It is therefore possible that the GRGDS peptide may block experimental metastasis by inhibiting the interaction of such newly expressed receptors with fibronectin or other related proteins.

As discussed above, fibronectin is often retained in the tumoral stroma of human carcinomas, as well as in both the cellular and stromal compartments of sarcomas. These locations are consistent with fibronectin mediating the migration of invasive tumor cells as well as being a target of tumor-derived proteases involved in the controlled degradation of connective tissue. There is now strong *in vitro* evidence that transformed cells express surface proteases at points of membrane-substrate contact where they can contact and degrade immobilized fibronectin.[364-368]

The overall organization or distribution of fibronectin receptors is also altered in transformed cells. The fibrillar, streak-like pattern characteristic of normal fibroblasts is lost on transformed avian cells, which display a diffuse distribution of receptor both *in vitro*[202] and *in vivo*.[380] A similar diffuse organization of receptor is also present on a variety of malignant human cell lines.[381] Reconstitution of fibronectin on the cell surface by treatment with purified exogenous cellular fibronectin partially, but not completely, restores a more normal distribution of receptors to such cells.[202] This result suggests that receptor distribution can be regulated by exogenous fibronectin, but also that other factors can interfere, including possible alterations in the receptor itself.

A difference between integrin-type receptors of normal and transformed cells has been found in the degree of phosphorylation of tyrosine residues.[369] Although it was necessary to use long-term vanadate treatment to document this difference, which may artificially change observed levels of phosphorylation due to inhibition of tyrosine phosphatases, it represents the first evidence for structural differences between normal and transformed cell fibronectin receptors. Phosphorylation of tyrosine residues may inhibit the binding of these receptors to fibronectin and the intracellular molecule talin.[382] A current hypothesis is thus that such phosphorylation may regulate fibronectin receptor function.

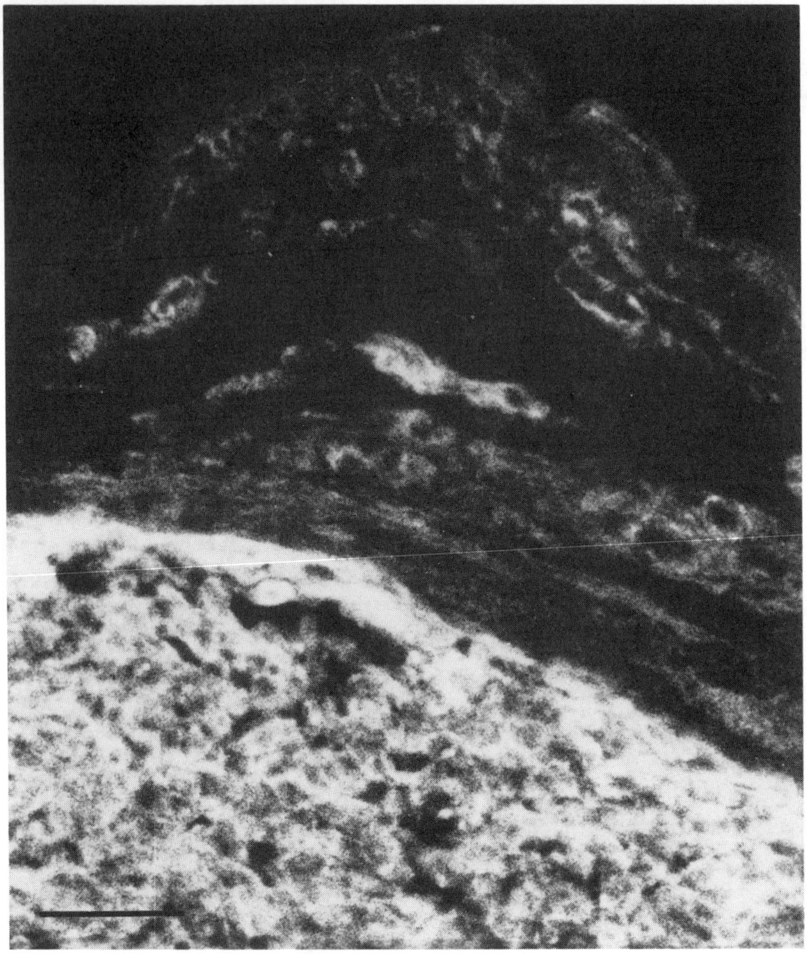

FIGURE 6. Localization of the chicken fibronectin receptor in a Rous sarcoma virus-induced tumor. Rous sarcoma virus was injected subcutaneously into the wing web skin of 1-week-old chickens and 7-d later tumors were fixed for 16 h at 4°C with 3% paraformaldehyde in PBS containing 60-m*M* sucrose. Frozen thin sections (5 μm) were cut in OCT compound and stained by indirect immunofluorescence with polyclonal anti-chicken fibronectin receptor IgG. The entire tumor mass (bottom of figure) stained heavily for the fibronectin receptor, while normal tissue (top of figure) exhibited low staining intensity. The normal cells that were stained with antifibronectin receptor antibody were fibrocytes, capillary endothelial cells, and the basal layer of the epidermis. Bar = 50 μm. We are grateful to Dr. Shinsuke Saga for providing this micrograph.

There are strong parallels between the disruption of receptor distribution in transformed cells and developmental changes in receptor distribution. The diffuse organization of receptor on embryonic neural crest cells and somitic cells is transient, however, and changes to a fibrillar organization as the cells mature to a more stationary phenotype.[133] Unexpectedly, the diffuse organization of receptor in rapidly migrating embryonic cells does not appear to be regulated by tyrosine phosphorylation, since the receptor on neural crest cells is not detectably phosphorylated, while stationary somitic fibroblasts with a normal receptor distribution paradoxically display phosphorylated receptors;[370] thus, the role of phosphorylation in receptor function remains uncertain. Although it remains unclear what regulates receptor distribution, the correlation between a diffuse distribution, rapid lateral mobility of the receptor in the plane of the membrane, and cell migratory activity appears to be relevant to embryonic and tumor cell behavior. It will be important in the future to determine whether any subtle changes in ratios of various α subunits are associated with these changes in overall integrin receptor distribution and function.

Whether fibronectin receptors are induced and altered in distribution in human tumors *in situ* remains to be examined. However, in a recently developed model system to quantitate human tumor cell migration, the outward migration of human colon carcinoma cells from an artificial aggregate was found to be dependent on the presence of fibronectin on the substrate.[32] This effect was completely abolished by incubation of the cells with purified anti-human fibronectin receptor IgG but not by a variety of control antibodies.[32] Thus, a similar role for fibronectin-fibronectin receptor interactions in cell migration and invasion probably exists for human tumor cells as for the animal model systems studied previously. The ability to inhibit migration by specific synthetic peptides and antireceptor antibodies provides hope for developing novel approaches to understand and possibly inhibit human tumor cell invasion and metastasis.

V. FUTURE DIRECTIONS

Although GRGDS appears to block metastasis through its ability to interfere with tumor cell adhesion, the actual site(s) of action of the peptide await identification. Kinetic studies suggest that GRGDS may, in fact, have several activities, including inhibiting tumor cell attachment to the target organ, enhancing cell detachment, and preventing tumor cell invasion through connective tissues. In the future, examination of the biological effects of GRGDS-like peptides in a variety of *in vivo* and *in vitro* model systems, together with microscopic analyses of the process of tumor cell colonization in peptide-treated animals, should lead to a more exact definition of its sites of action.

The relative antimetastatic activity of a series of peptides modeled on the GRGDS signal most closely matches their ability to interfere with fibronectin-mediated adhesion, as opposed to adhesion supported by other extracellular matrix molecules containing an RGDX sequence whose adhesion is disrupted by peptides based on this motif (collagens, fibrinogen, vitronectin, thrombospondin, and von Willebrand factor; for reviews see References 41 and 43). It is highly likely, although not absolutely proven, therefore, that fibronectin can be used by malignant melanoma cells during target organ colonization. The recent report that YIGSR-amide, a synthetic peptide derivative representing one cell adhesion site in laminin, also possesses antimetastatic activity for melanoma cells[371] supports the conclusions of previous studies that this basement membrane glycoprotein plays an important role in tumor cell arrest.[372-375] It is currently unknown whether other connective tissue components are able to mediate the arrest of malignant cells *in vivo*, although this would be interesting to study. While the use of peptide probes such as GRGDS and YIGSR-amide is a novel approach to studying adhesive function, complementary to more traditional immunological approaches, it is now becoming apparent that, due to their small size, peptides may not always possess complete specificity for the particular adhesive ligand from which they are derived. Studies are therefore in progress to modify this approach and to develop peptide derivatives which possess greater specificity and activity. Once available, these agents should permit more definitive assignment of function to particular extracellular matrix molecules.

A further aim from optimizing the biological activity of antiadhesive peptides will of course be the development of clinically useful agents. Although most cancer patients that present with a detectable primary tumor already have occult micrometastases, thereby rendering prophylaxis meaningless, there are situations where prevention of seeding of malignant cells is a potential clinical reality. For example, during autologous bone marrow transplantation protocols, malignant cells contaminating the bone marrow are often incompletely purged by chemotherapy and are unfortunately transferred back into the patient. Reinfusion with an antiadhesive peptide may be able to prevent this inopportune seeding of metastatic cells. The use of peptides in this type of setting may in fact be one of their earliest applications, because the threat of metastasis is limited to the time of infusion and therefore the short half-life of the peptide would not be a problem. Indeed, it may in fact be an advantage because it is unlikely that acute administration would produce any deleterious side-effects. In a similar situation, the release of tumor cells during surgical manipulation of a tumor could be combatted by antiadhesive agents, and to extrapolate further, it is conceivable that further dissemination of tumor cells in patients who may already have a limited number of metastases may be prevented by continuous administration of antiadhesive drugs. Application of this technology to such high risk populations, in contrast to that in bone marrow transplantation protocols, would require substantial prolongation of the activity of the peptide. Consequently, it is likely that, in the near future, studies examining the effects of modification of the structure of GRGDS-like peptides or of conjugation to inert carriers that confer an extended vascular half-life will be performed. As a necessary prelude to the use of GRGDS-like peptides clinically, it will be obligatory to examine their effects on normal body functions. It might be anticipated, because the peptide has no reported selectivity for tumor cells, that interference in normal cell adhesive processes would be the greatest concern; for example, disruption of hemostasis through prevention of platelet aggregation or breakdown of the integrity of microcapillaries through interference with endothelial cell adhesion. It may, however, be possible to develop peptide derivatives that can block tumor cell adhesion without affecting normal cell functions. In this regard, the reversed peptide SDGR is reported to possess antimetastatic, but not antiplatelet, activity.[357,376]

The involvement of fibronectin in the adhesion of neoplastic cells *in vivo* has been hypothesized for some time, but the most direct evidence comes from recent use of the GRGDS peptide. The advent of this new class of biological inhibitor has been instrumental in identifying this sequence as a crucial recognition signal in the metastatic cascade. We therefore speculate that significant, future advances in our understanding of the biological functions of fibronectin, including its role in neoplastic disease, will rely heavily on the development of new probes and model systems through which the participation of fibronectin can be studied *in vivo*.

One area which would benefit greatly from the availability of new probes is the biochemical analysis of pericellular matrix assembly. Now that the fibronectin receptor has been cloned and sequenced, identification of the regions of the molecule mediating its biological activity will be a top priority, and this will hopefully lead to the development of new agents which could be used to dissect its role in cytoskeleton-plasma membrane-extracellular matrix interactions. A second area of investigation, which is highly relevant to studies of the role of fibronectin in cancer, is the search for factors that control the normal expression and turnover of cellular fibronectin. By analysis of the structure of the fibronectin gene, and in particular, its promoter region, it may be possible to shed light on these questions and to understand (1) why fibronectin synthesis is altered in transformed cells, and (2) what ramifications these changes have for tumor cells *in situ*. In addition to answering important biological questions, an improved understanding of fibronectin-mediated adhesion may also lead to the development of clinically useful agents for suppressing the biosynthesis of various components of the adhesive machinery, or of antiadhesive agents which interfere with the function of the fibronectin molecule itself, its receptor, or its cytoskeletal-binding proteins.

The role of fibronectin and its receptor in tumor cell migration and invasion remain to be clarified. While it seems intuitively likely that up-regulation of receptor levels and maintenance of a diffuse, freely mobile receptor distribution is important for the capacity of tumor cells to migrate and invade, this hypothesis should be tested rigorously by using specific antibodies to inhibit receptor functions. One obvious future experiment is to use monoclonal antibodies or Fab fragments to attempt to inhibit tumor cell invasion *in vitro*.

The complementary *in vivo* experiment would be to attempt to inhibit experimental metastasis of human tumor cells in animal models with antireceptor antibodies. Although the increase in receptor levels in tumor cells may not be large enough, it is conceivable that elevated receptor levels might eventually prove to be a useful marker for tumor cell detection or for targeting of chemotherapeutic agents to tumor tissues. For example, one could imagine a type of combination therapy in which synergistic tumoricidal agents are present on antireceptor antibodies and on antibodies directed against an entirely distinct antigen characteristic of tumor cells; although either might not be sufficiently specific, only tumor cells might simultaneously display both target molecules and be subjected to cytotoxic doses of agent.

ACKNOWLEDGMENTS

These studies were supported in part by the Howard University Faculty Support Grant and by Grants CA-14718 and CA-45290 from the National Institutes of Health and PDT-312 from the American Cancer Society to Kenneth Olden. Martin J. Humphries was supported by a Cancer Research Institute Fellowship. We thank Dr. Shinsuke Saga for providing the micrograph shown as Figure 6.

REFERENCES

1. **Sugarbaker, E. V., Weingard, D. N., and Roseman, J. M.,** in *Cancer Invasion and Metastasis,* Liotta, L. A. and Hart, I. R., Eds., Martinus Nijhoff, The Hague, 1982, 427.
2. **Liotta, L. A. and Hart, I. R., Eds.,** *Cancer Invasion and Metastasis,* Martinus Nijhoff, The Hague, 1982.
3. **Nicolson, G. L. and Milas, L., Eds.,** *Cancer Invasion and Metastasis: Biologic and Therapeutic Aspects,* Raven Press, New York, 1984.
4. **Schirrmacher, V.,** Cancer metastasis: experimental approaches, theoretical concepts and impact for treatment strategies, *Adv. Cancer Res.,* 43, 1, 1985.
5. **Weiss, L., Ed.,** *Principles of Metastasis,* Academic Press, Orlando, FL 1985.
6. **Honn, K. V., Powers, W. E., and Sloane, B. F., Eds.,** *Mechanisms of Cancer Metastasis: Potential Therapeutic Implications,* Martinus Nijhoff, Boston, 1986.
7. **Poste, G.,** Pathogenesis of metastatic disease: implications for current therapy and for the development of new therapeutic strategies, *Cancer Treat. Rep.,* 70, 183, 1986.
8. **Fidler, I. J., Gersten, D. M., and Hart, I. R.,** The biology of cancer invasion and metastasis, *Adv. Cancer Res.,* 28, 149, 1978.
9. **Nicolson, G. L.,** Cancer metastasis. Organ colonization and the cell-surface properties of malignant cells, *Biochim. Biophys. Acta,* 695, 113, 1982.
10. **McCarthy, J. B., Basara, M. L., Palm, S. L., Sas, D. F., and Furcht, L. T.,** The role of cell adhesion proteins — laminin and fibronectin — in the movement of malignant and metastatic cells, *Cancer Metastasis Rev.,* 4, 125, 1985.
11. **Liotta, L. A., Rao, C. N., and Wewer, U. M.,** Biochemical interactions of tumor cells with the basement membrane, *Annu. Rev. Biochem.,* 55, 1037, 1986.
12. **Nicolson, G. L.,** Tumor cell instability, diversification, and progression to the metastatic phenotype: from oncogene to oncofetal expression, *Cancer Res.,* 47, 1473, 1987.
13. **Reich, E.,** Activation of plasminogen: a general mechanism for producing localized extracellular proteolysis, in *Molecular Basis of Biological Degradative Processes,* Berlin, R. D., Herrmann, H., Lepow, I. H., and Tanzer, J. M., Eds., Academic Press, New York, 1978, 155.
14. **Liotta, L. A., Thorgeirsson, U. P., and Garbisa, S.,** Role of collagenases in tumor cell invasion, *Cancer Metastasis Rev.,* 1, 277, 1982.
15. **Goldfarb, R. H. and Liotta, L. A.,** Proteolytic enzymes in cancer invasion and metastasis, *Semin. Thromb. Hemost.,* 12, 294, 1986.
16. **Nakajima, M., Welch, D. R., Irimura, T., and Nicolson, G. L.,** Basement membrane degradative enzymes as possible markers of tumor metastasis, *Prog. Clin. Biol. Res.,* 212, 113, 1986.
17. **Willis, R. A.,** *The Spread of Tumors in the Human Body,* Butterworth, London, 1952.

18. **Weiss, L. and Ward, P.M.,** Cell detachment and metastasis, *Cancer Metastasis Rev.*, 2, 111, 1983.
19. **Fidler, I. J.,** Metastasis: quantitative analysis of distribution and fate of tumor emboli labeled with ^{125}I-5-iodo-2′-deoxyuridine, *J. Natl. Cancer Inst.*, 45, 773, 1970.
20. **Hanna, N.,** The role of natural killer cells in control of tumor growth and metastasis, *Biochim, Biophys. Acta,* 780, 213, 1985.
21. **Herberman, R. B.,** Natural killer cells, *Annu. Rev. Med.*, 37, 347, 1986.
22. **Roder, J. and Duwe, A.,** The beige mutation in the mouse selectively impairs natural killer cell function, *Nature (London),* 278, 451, 1979.
23. **Hanna, N. and Fidler, I. J.,** Role of natural killer cells in the destruction of circulating tumor emboli, *J. Natl. Cancer Inst.*, 65, 801, 1980.
24. **Fidler, I. J.,** The relationship of embolic homogeneity, number, size, and viability to the incidence of experimental metastasis, *Eur. J. Cancer,* 9, 233, 1973.
25. **Liotta, L. A., Kleinerman, J., and Saidel, G. M.,** The significance of hematogenous tumor cell clumps in the metastatic process, *Cancer Res.*, 36, 889, 1976.
26. **Gasic, G. J.,** Role of plasma, platelets, and endothelial cells in tumor metastasis, *Cancer Metastasis Rev.*, 3, 99, 1984.
27. **Gasic, G. J., Tuszynski, G. P., and Gorelik, E.,** Interaction of the hemostatic and immune systems in the metastatic spread of tumor cells, *Int. Rev. Exp. Pathol.*, 29, 173, 1986.
28. **Knisely, W. H. and Mahaley, M. JS.,** Relationship between size and distribution of ''spontaneous'' metastases and three sizes of intravenously injected VX2 carcinoma, *Cancer Res.*, 18, 900, 1958.
29. **Dvorak, H. F., Senger, D. R., and Dvorak, A. M.,** Fibrin as a component of the tissue stroma: origins and biological significance, *Cancer Metastasis Rev.*, 2, 41, 1983.
30. **Nicolson, G. L.,** Organ specificity of tumor metastasis: role of preferential adhesion, invasion and growth of malignant cells at specific secondary sites, *Cancer Metastasis Rev.*, in press.
31. **Terranova, V. P., Hujanen, E. S., and Martin, G. R.,** Basement membrane and the invasive activity of metastatic tumor cells, *J. Natl. Cancer Inst.*, 77, 311, 1986.
32. **Humphries, M. J., Obara, M., Olden, K., and Yamada, K. M.,** Role of fibronectin in adhesion, migration, and metastasis, *Cancer Invest.*, in press.
33. **Netland, P. A. and Zetter, B. R.,** Organ-specific adhesion of metastatic tumor cells in vitro, *Science,* 224, 1113, 1984.
34. **Kawaguchi, T. and Nakamura, K.,** Analysis of the lodgement and extravasation of tumor cells in experimental models of hematogenous metastasis, *Cancer Metastasis Rev.*, 5, 77, 1986.
35. **Linder, E., Vaheri, A., Ruoslahti, E., and Wartiovaara, J.,** Distribution of fibroblast surface antigen in the developing chick embryo, *J. Exp. Med.*, 142, 41, 1975.
36. **Stenman, S. and Vaheri, A.,** Distribution of a major connective tissue protein, fibronectin, in normal human tissue, *J. Exp. Med.*, 147, 1054, 1978.
37. **Hayman, E. G., Pierschbacher, M. D., Ohgren, Y., and Ruoslahti, E.,** Serum spreading factor (vitronectin) is present at the cell surface and in tissues, *Proc. Natl. Acad. Sci. U.S.A.*, 80, 4003, 1983.
38. **Butler, T. and Gullino, P.,** Quantitation of cell-shedding into efferent blood of mammary adenocarcinoma, *Cancer Res.*, 35, 512, 1975.
39. **Weiss, L., Mayhew, E., Glaves Rapp, D., and Holmes, J. C.,** Metastatic inefficiency in mice bearing B16 melanomas, *Br. J. Cancer,* 45, 44, 1982.
40. **Hynes, R.,** Molecular biology of fibronectin, *Annu. Rev. Cell Biol.*, 1, 67, 1985.
41. **Ruoslahti, E. and Pierschbacher, M. D.,** Arg-Gly-Asp: a versatile cell recognition signal, *Cell,* 44, 517, 1986.
42. **Akiyama, S. K. and Yamada, K. M.,** Fibronectin, in *Advances in Enzymology and Related Areas of Molecular Biology,* Meister, A., Ed., John Wiley & Sons, New York, 1987, 1.
43. **Ruoslahti, E. and Pierschbacher, M. D.,** New perspectives in cell adhesion: RGD and integrins, *Science,* 238, 491, 1987.
44. **Humphries, M. J. and Yamada, K. M.,** Cell interaction sites of fibronectin in adhesion and metastasis, in *The Cell in Contact: Adhesions and Junctions as Morphogenetic Determinants,* Vol. 2, Edelman, G. M. and Thiery, J. P., Eds., John Wiley & Sons, New York, in press.
45. **Thiery, J. P., Duband, J. L., and Tucker, G. C.,** Cell migration in the vertebrate embryo: role of cell adhesion and tissue environment in pattern formation, *Annu. Rev. Cell Biol.*, 1, 91, 1985.
46. **Yamada, K. M.,** Fibronectin domains and receptors, in *Fibronectin,* Mosher, D. F., Ed., Academic Press, New York, in press.
47. **Tamkun, J. W. and Hynes, R. O.,** Plasma fibronectin is synthesized and secreted by hepatocytes, *J. Biol. Chem.*, 258, 4641, 1983.
48. **Crouch, E., Balian, G., Holbrook, K., Duksin, D., and Bornstein, P.,** Amniotic fluid fibronectin. Characterization and synthesis by cells in culture, *J. Cell Biol.*, 78, 701, 1978.
49. **Carsons, S., Mosesson, M. W., and Diamond, H. S.,** Detection and quantitation of fibronectin in synovial fluid from patients with rheumatic disease, *Arthritis Rheum.*, 24, 1261, 1981.

50. **Vartio, T., Vaheri, A., Von Essen, R., Isomaki, H., and Stenman, S.,** Fibronectin in synovial fluid and tissue in rheumatoid arthritis, *Eur. J. Clin. Invest.*, 11, 207, 1981.
51. **Kornblihtt, A. R., Vibe-Pedersen, K., and Baralle, F. E.,** Isolation and characterization of cDNA clones for human and bovine fibronectins, *Proc. Natl. Acad. Sci. U.S.A.*, 80, 3218, 1983.
52. **Tamkun, J. W., Schwarzbauer, J. E., and Hynes, R. O.,** A single rat fibronectin gene generates three different mRNAs by alternative splicing of a complex exon, *Proc. Natl. Acad. Sci. U.S.A.*, 81, 5140, 1984.
53. **Alexander, S. S., Colonna, G., Yamada, K. M., Pastan, I., and Edelhoch, H.,** Molecular properties of a major cell surface protein from chick embryo fibroblasts, *J. Biol. Chem.*, 253, 5820, 1978.
54. **Colonna, G. Alexander, S. S., Yamada, K. M. Pastan, I., and Edelhoch, H.,** The stability of cell surface protein to surfactants and denaturants, *J. Biol. Chem.*, 253, 7787, 1978.
55. **Schwarzbauer, J. E., Tamkun, J. W., Lemischka, I. R., and Hynes, R. O.,** Three different fibronectin mRNAs arise by alternative splicing within the coding region, *Cell*, 35, 421, 1983.
56. **Kornblihtt, A. R., Vibe-Pedersen, K., and Baralle, F. E.,** Human fibronectin: molecular cloning evidence for two mRNA species differing by an internal segment coding for a structural domain, *EMBO J.*, 3, 221, 1984.
57. **Kornblihtt, A. R., Vibe-Pedersen, K., and Baralle, F. E.,** Human fibronectin: cell specific alternative mRNA splicing generates polypeptide chains differing in the number of internal repeats, *Nucleic Acids Res.*, 12, 5853, 1984.
58. **Gutman, A. and Kornblihtt, A. R.,** Identification of a third region of cell-specific alternative splicing in human fibronectin mRNA, *Proc. Natl. Acad. Sci. U.S.A.*, 84, 7179, 1987.
59. **Norton, P. A. and Hynes, R. O.,** Alternative splicing of chicken fibronectin in embryos and in normal and transformed cells, *Mol. Cell. Biol.*, 7, 4297, 1987.
60. **Schwarzbauer, J. E., Patel, R. S., Fonda, D., and Hynes, R. O.,** Multiple sites of alternative splicing of the rat fibronectin gene transcript, *EMBO J.*, 6, 2573, 1987.
61. **Zardi, L., Carnemolla, B., Siri, A., Petersen, T. E., Paolella, G., Sebastio, G., and Baralle, F. E.,** Transformed human cells produce a new fibronectin isoform by preferential alternative splicing of a previously unobserved exon, *EMBO J.*, 6, 2337, 1987.
62. **Petersen, T. E., Thogerson, H. C., Skorstengaard, K., Vibe-Pedersen, K., Sottrup-Jensen, L., and Magnusson, S.,** Partial primary structure of bovine plasma fibronectin: three types of internal homology, *Proc. Natl. Acad. Sci. U.S.A.*, 80, 137, 1983.
63. **Kornblihtt, A. R., Umezawa, K., Vibe-Pedersen, K., and Baralle, F. E.,** Primary structure of human fibronectin: differential splicing may generate at least 10 polypeptides from a single gene, *EMBO J.*, 4, 1755, 1985.
64. **Skorstengaard, K., Jensen, M. S., Sahl, P., Petersen, T. E., and Magnusson, S.,** Complete primary structure of bovine plasma fibronectin, *Eur. J. Biochem.*, 161, 441, 1986.
65. **Owens, R. J. and Baralle, F. E.,** Mapping the collagen-binding site of human fibronectin by expression in *Escherichia coli*, *EMBO J.*, 5, 2825, 1986.
66. **Hirano, H., Yamada, Y., Sullivan, M., de Crombrugghe, B., Pastan, I., and Yamada, K. M.,** Isolation of genomic DNA clones spanning the entire fibronectin gene, *Proc. Natl. Acad. Sci. U.S.A.*, 80, 46, 1983.
67. **Vibe-Pedersen, K., Kornblihtt, A. R., and Baralle, F. E.,** Expression of a human alpha-globin/fibronectin
68. **Odermatt, E. Tamkun, J. W., and Hynes, R. O.,** The repeating modular structure of the fibronectin gene: relationship to protein structure and subunit variation, *Proc. Natl. Acad. Sci. U.S.A.*, 82, 6571, 1985.
69. **Patel, R. S., Odermatt, E., Schwarzbauer, J. E., and Hynes, R. O.,** Organization of the fibronectin gene provides evidence for exon shuffling during evolution, *EMBO J.*, 6, 2565, 1987.
70. **Colombi, M., Barlati, S., Kornblihtt, A., Baralle, F. E., and Vaheri, A.,** A family of fibronectin mRNAs in human normal and transformed cells, *Biochim. Biophys. Acta.*, 868, 207, 1986.
71. **Paul, J. I., Schwarzbauer, J. E., Tamkun, J. W., and Hynes, R. O.,** Cell-type-specific fibronectin subunits generated by alternative splicing, *J. Biol. Chem.*, 261, 12258, 1986.
72. **Schwarzbauer, J. E., Paul, J. I., and Hynes, R. O.,** On the origin of species of fibronectin, *Proc. Natl. Acad. Sci. U.S.A.*, 82, 1424, 1985.
73. **Castellani, P., Siri, A., Rosellini, C., Infusini, E., Borsi, L., and Zardi, L.,** Transformed human cells release different fibronectin variants than do normal cells, *J. Cell Biol.*, 103, 1671, 1986.
74. **Borsi, L., Carnemolla, B., Castellani, P., Rosellini, C., Vecchio, D., Allemanni, G., Chang, S. E., Taylor-Papadimitriou, J., Pande, H., and Zardi, L.,** Monoclonal antibodies in the analysis of fibronectin isoforms generated by alternative splicing of mRNA precursors in normal and transformed human cells, *J. Cell Biol.*, 104, 595, 1987.
75. **Garcia-Pardo, A., Rostagno, A., and Frangione, B.,** Primary structure of human plasma fibronectin. Characterization of a 38kDa domain containing the C-terminal heparin-binding site (Hep III site) and a region of molecular heterogeneity, *Biochem. J.*, 241, 923, 1987.
76. **Hayashi, M. and Yamada, K. M.,** Domain structure of the carboxyl-terminal half of human plasma fibronectin, *J. Biol. Chem.*, 258, 3332, 1983.

77. **Humphries, M. J., Akiyama, S. K., Komoriya, A., Olden, K., and Yamada, K. M.,** Identification of an alternatively spliced site in human plasma fibronectin that mediates cell type-specific adhesion, *J. Cell Biol.*, 103, 2637, 1986.
78. **Pierschbacher, M. D., Hayman, E. G., and Ruoslahti, E.,** Location of the cell attachment site in fibronectin with monoclonal antibodies and proteolytic fragments of the molecule, *Cell,* 26, 259, 1981.
79. **Pierschbacher, M. D., Ruoslahti, E., Sundelin, J., Lind, P., and Peterson, P. A.,** The cell attachment domain of fibronectin. Determination of the primary structure, *J. Biol. Chem.*, 257, 9593, 1982.
80. **Pierschbacher, M. D., Hayman, E. G., and Ruoslahti, E.,** Synthetic peptide with cell attachment activity of fibronectin, *Proc. Natl. Acad. Sci. U.S.A.*, 80, 1224, 1983.
81. **Pierschbacher, M. D. and Ruoslahti, E.,** Cell attachment activity of fibronectin can be duplicated by small synthetic fragments of the molecule, *Nature (London),* 309, 30, 1984.
82. **Yamada, K. M. and Kennedy, D. W.,** Dualistic nature of adhesive protein function: fibronectin and its biologically active peptide fragments can autoinhibit fibronectin function, *J. Cell Biol.*, 99, 29, 1984.
83. **Yamada, K. M. and Kennedy, D. W.,** Amino acid sequence specificities of an adhesive recognition signal, *J. Cell. Biochem.*, 28, 99, 1985.
84. **Pierschbacher, M. D. and Ruoslahti, E.,** Variants of the cell recognition site of fibronectin that retain attachment-promoting activity, *Proc. Natl. Acad. Sci. U.S.A.*, 81, 5985, 1984.
85. **Akiyama, S. K. and Yamada, K. M.,** The interaction of plasma fibronectin with fibroblastic cells in suspension, *J. Biol. Chem.* 260, 4492, 1985.
86. **Akiyama, S. K. and Yamada, K. M.,** Synthetic peptides competitively inhibit both direct binding to fibroblasts and functional biological assays for the purified cell-binding domain of fibronectin, *J. Biol. Chem.*, 260, 10402, 1985.
87. **Akiyama, S. K., Hasegawa, E., Hasegawa, T., and Yamada, K. M.,** The interaction of fibronectin fragments with fibroblastic cells, *J. Biol. Chem.*, 260, 13256, 1985.
88. **Obara, M., Kang, M. S., Rocher-Dufour, S., Kornblihtt, A., Thiery, J. P., and Yamada, K. M.,** Expression of the cell-binding domain of human fibronectin in E. coli, *FEBS Lett.*, 213, 261, 1987.
89. **Obara, M., Kang, M. S., and Yamada, K. M.,** Site-directed mutagenesis of the cell-binding domain of human fibronectin: separable, synergistic sites mediate adhesive function, *Cell,* in press.
90. **Humphries, M. J., Komoriya, A., Akiyama, S. K., Olden, K., and Yamada, K. M.,** Identification of two distinct regions of the type III connecting segment of human plasma fibronectin that promote cell type-specific adhesion, *J. Biol. Chem.*, 262, 6886, 1987.
91. **Humphries, M. J., Akiyama, S. K., Komoriya, A., Olden, K., and Yamada, K. M.,** Neurite extension of chicken peripheral nervous system neurons on fibronectin: relative importance of specific adhesion sites in the central cell-binding domain and the alternatively-spliced type III connecting segment, *J. Cell Biol.*, in press.
92. **Greve, J. M. and Gottlieb, D. I.,** Monoclonal antibodies which alter the morphology of cultured chick myogenic cells, *J. Cell. Biochem.*, 18, 221, 1982.
93. **Neff, N. T., Lowrey, C., Decker, C., Tovar, A., Damsky, C., Buck, C., and Horwitz, A. F.,** A monoclonal antibody detaches embryonic skeletal muscle from extracellular matrices, *J. Cell Biol.*, 95, 654, 1982.
94. **Chapman, A. E.,** Characterization of a 140 kD cell surface glycoprotein involved in myoblast adhesion, *J. Cell. Biochem.*, 25, 109, 1984.
95. **Decker, C., Greggs, R., Duggan, K., Stubbs, J., and Horwitz, A.,** Adhesive multiplicity in the interaction of embryonic fibroblasts and myoblasts with extracellular matrices, *J. Cell Biol.*, 99, 1398, 1984.
96. **Chen, W. T., Hasegawa, E., Hasegawa, T., Weinstock, C., and Yamada, K. M.,** Development of cell surface linkage complexes in cultured fibroblasts, *J. Cell Biol.*, 100, 1103, 1985.
97. **Hasegawa, T., Hasegawa, E., Chen, W. T., and Yamada, K. M.,** Characterization of a membrane-associated glycoprotein complex implicated in cell adhesion to fibronectin, *J. Cell. Biochem.*, 28, 307, 1985.
98. **Knudsen, K. A., Horwitz, A., and Buck, C. A.,** A monoclonal antibody identifies a glycoprotein complex involved in cell-substratum adhesion, *Exp. Cell Res.*, 157, 218, 1985.
99. **Horwitz, A., Duggan, K., Greggs, R., Decker, C., and Buck, C. A.,** Cell substrate attachment (CSAT) antigen has properties of a receptor for laminin and fibronectin, *J. Cell Biol.*, 103, 2134, 1985.
100. **Akiyama, S. K., Yamada, S. S., and Yamada, K. M.,** Characterization of a 140 kD avian cell surface antigen as a fibronectin-binding molecule, *J. Cell Biol.*, 102, 442, 1986.
101. **Brown, P. and Juliano, R. L.,** Selective inhibition of fibronectin-mediated cell adhesion by monoclonal antibodies to a cell surface glycoprotein, *Science,* 228, 1448, 1985.
102. **Pytela, R., Pierschbacher, M. D., and Ruoslahti, E.,** Identification and isolation of a 140 kd cell surface glycoprotein with properties expected of a fibronectin receptor, *Cell,* 40, 191, 1985.
103. **Patel, V. P. and Lodish, H. F.,** The fibronectin receptor on mammalian erythroid precursor cells: characterization and developmental regulation, *J. Cell Biol.* 102, 449, 1986.
104. **Giancotti, F. G., Comoglio, P. M., and Tarone, G.,** Fibronectin-plasma membrane interaction in the adhesion of hemopoietic cells, *J. Cell Biol.*, 103, 429, 1986.

105. **Giancotti, F. G., Comoglio, P. M., and Tarone, G.,** A 135,000 molecular weight plasma membrane glycoprotein involved in fibronectin-mediated cell adhesion, *Exp. Cell Res.*, 163, 47, 1986.
106. **Cardarelli, P. M. and Pierschbacher, M. D.,** Identification of fibronectin receptors on T lymphocytes, *J. Cell Biol.*, 105, 499, 1987.
107. **Buck, C. A. and Horwitz, A. F.,** Cell surface receptors for extracellular matrix molecules, *Annu. Rev. Cell Biol.*, 3, 179, 1987.
108. **Hynes, R. O.,** Integrins: a family of cell surface receptors, *Cell*, 48, 549, 1987.
109. **Springer, T. A., Dustin, M. L., and Kishimoto, T. K.,** The lymphocyte function-associated LFA-1, CD2, and LFA-3 molecules: cell adhesion receptors of the immune system, *Annu. Rev. Immunol.*, 5, 223, 1987.
110. **Hemler, M. E., Huang, C., and Schwarz, L.,** The VLA protein family. Characterization of five distinct cell surface heterodimers each with a common 130,000 molecular weight beta subunit, *J. Biol. Chem.*, 262, 3300, 1987.
111. **Pytela, R., Pierschbacher, M. D., and Ruoslahti, E.,** A 125/115 kDa cell surface receptor specific for vitronectin interacts with the arginine-glycine-aspartic acid adhesion sequence derived from fibronectin, *Proc. Natl. Acad. Sci. U.S.A.*, 82, 5766, 1985.
112. **Gardner, J. M. and Hynes, R. O.,** Interaction of fibronectin with its receptor on platelets, *Cell*, 42, 439, 1985.
113. **Ginsberg, M. H., Pierschbacher, M. D., Ruoslahti, E., Marguerie, G., and Plow, E.,** Inhibition of fibronectin binding to platelets by proteolytic fragments and synthetic peptides which support fibroblast adhesion, *J. Biol. Chem.*, 260, 3931, 1985.
114. **Pytela, R., Pierschbacher, M. D., Ginsberg, M. H., Plow, E. F., and Ruoslahti, E.,** Platelet membrane glycoprotein IIb/IIIa: member of a family of Arg-Gly-Asp-specific adhesion receptors, *Science*, 231, 1539, 1986.
115. **Hildreth, J. E. K., Gotch, F. M., Hildreth, P. D. K., and McMichael, A. J.,** A human lymphocyte-associated antigen involved in cell-mediated lympholysis, *Eur. J. Immunol.*, 13, 302, 1982.
116. **Sanchez-Madrid, F., Krensky, A. M., Ware, C. F., Robbins, E., Strominger, J. L., Burakoff, S. J., and Springer, T. A.,** Three distinct antigens associated with human T lymphocyte-mediated cytolysis: LFA-1, LFA-2 and LFA-3, *Proc. Natl. Acad. Sci. U.S.A.*, 79, 7489, 1982.
117. **Springer, T. A., Galfre, G., Secher, D. S., and Milstein, C.,** Mac-1: a macrophage differentiation antigen identified by monoclonal antibody, *Eur. J. Immunol.*, 9, 301, 1979.
118. **Wright, S. D., Rao, P. E., van Voorhis, W. C., Craigmyle, L. S., Iida, K., Talle, M. A., Westberg, E. F., Goldstein, G., and Silverstein, S. C.,** Identification of the C3bi receptor of human monocytes and macrophages by using monoclonal antibodies, *Proc. Natl. Acad. Sci. U.S.A.*, 80, 5699, 1983.
119. **Sanchez-Madrid, F., Nagy, J. A., Robbins, E., Simon, P., and Springer, T. A.,** A human leukocyte differentiation antigen family with distinct alpha subunits and a common beta subunit: the lymphocyte function-associated antigen (LFA-1), the C3bi complement receptor (OKM1/Mac-1), and the p150,95 molecule, *J. Exp. Med.*, 158, 1785, 1983.
120. **Sastre, L., Roman, J. M., Teplow, D. B., Dreyer, W. J., Gee, C. E., Larson, R. S., Roberts, T. M., and Springer, T. A.,** A partial genomic DNA clone for the alpha subunit of the mouse complement receptor type 3 and cellular adhesion molecule Mac-1, *Proc. Natl. Acad. Sci. U.S.A.*, 83, 5644, 1986.
121. **Tamkun, J. W., DeSimone, D. W., Fonda, D., Patel, R. S., Buck, C., Horwitz, A. F., and Hynes, R. O.,** Structure of integrin, a glycoprotein involved in the transmembrane linkage between fibronectin and actin, *Cell*, 46, 271, 1986.
122. **Argraves, W. S., Suzuki, S., Arai, H., Thompson, K., Pierschbacher, M. D., and Ruoslahti, E.,** Amino acid sequence of the human fibronectin receptor, *J. Cell Biol.*, 105, 1183, 1987.
123. **Fitzgerald, L. A., Poncz, M., Steiner, B., Rall, S. C., Bennett, J. S., and Phillips, D. R.,** Comparison of cDNA-derived protein sequences of the human fibronectin and vitronectin receptor alpha-subunits and platelet glycoprotein IIb, *Biochemistry*, 26, 8158, 1987.
124. **Fitzgerald, L. A., Steiner, B., Rall, S. C., Lo, S. S., and Phillips, D. R.,** Protein sequence of endothelial glycoprotein IIIa derived from a cDNA clone. Identity with platelet glycoprotein IIIa and similarity to "integrin", *J. Biol. Chem.*, 262, 3936, 1987.
125. **Kishimoto, T. K., O'Connor, K., Lee, A., Roberts, T. M., and Springer, T. A.,** Cloning of the beta subunit of the leukocyte adhesion proteins: homology to an extracellular matrix receptor defines a novel supergene family, *Cell*, 48, 681, 1987.
126. **Law, S. K., Gagnon, J., Hildreth, J. E. K., Wells, C. E., Willis, A. C., and Wong, A. J.,** The primary structure of the beta-subunit of the cell surface adhesion glycoproteins LFA-1, CR3 and p150,95 and its relationship to the fibronectin receptor, *EMBO J.*, 6, 915, 1987.
127. **Miller, L. J., Wiebe, M., and Springer, T. A.,** Purification and alpha subunit N-terminal sequences of human Mac-1 and p150,95 leukocyte adhesion proteins, *J. Immunol.*, 138, 2381, 1987.
128. **Poncz, M., Eisman, R., Heidenreich, R., Silver, S. M., Vilaire, G., Surrey, S., Schwartz, E., and Bennett, J. S.,** Structure of the platelet membrane glycoprotein IIb. Homology to the alpha subunits of the vitronectin and fibronectin membrane receptors, *J. Biol. Chem.*, 262, 8476, 1987.

129. **Suzuki, S., Argraves, W. S., Arai, H., Languino, L. R., Pierschbacher, M. D., and Ruoslahti, E.,** Amino acid sequence of the vitronectin receptor alpha subunit and comparative expression of adhesion receptor mRNAs, *J. Biol. Chem.,* 262, 14080, 1987.
130. **Takada, Y., Huang, C., and Hemler, M. E.,** Fibronectin receptor structures within the VLA family of heterodimers, *Nature (London),* 326, 607, 1987.
131. **Takada, Y., Strominger, J. L., and Hemler, M. E.,** The very late antigen family of heterodimers is part of a superfamily of molecules involved in adhesion and embryogenesis, *Proc. Natl. Acad. Sci. U.S.A.,* 84, 3239, 1987.
132. **Leptin, M., Aebersold, R., and Wilcox, M.,** Drosophila position-specific antigens resemble the vertebrate fibronectin-receptor family, *EMBO J.,* 6, 1037, 1987.
133. **Duband, J. L., Rocher, S., Chen, W. T., Yamada, K. M., and Thiery, J. P.,** Cell adhesion and migration in the early vertebrate embryo: location and possible role of the putative fibronectin receptor complex, *J. Cell Biol.,* 102, 160, 1986.
134. **Krotoski, D. M., Domingo, C., and Bronner-Fraser, M.,** Distribution of a putative cell surface receptor for fibronectin and laminin in the avian embryo, *J. Cell Biol.,* 103, 1061, 1986.
135. **Chen, W. T., Chen, J. M., and Mueller, S. C.,** Coupled expression and colocalization of 140k cell adhesion molecules, fibronectin, and laminin during morphogenesis and cytodifferentiation of chick lung cells, *J. Cell Biol.,* 103, 1073, 1986.
136. **Boucaut, J. C., Darribere, T., Poole, T. J., Aoyama, H., Yamada, K. M., and Thiery, J. P.,** Biologically active synthetic peptides as probes of embryonic development: a competitive peptide inhibitor of fibronectin function inhibits gastrulation in amphibian embryos and neural crest cell migration in avian embryos, *J. Cell Biol.,* 99, 1822, 1984.
137. **Naidet, C., Semeriva, M., Yamada, K. M., and Thiery, J. P.,** Peptides containing the cell-attachment recognition signal Arg-Gly-Asp prevent gastrulatin in *Drosophila* embryos, *Nature (London),* 325, 348, 1987.
138. **Darribere, T., Yamada, K. M., Johnson, K. E., and Boucaut, J. C.,** The 140-kDa fibronectin receptor complex is required for mesodermal cell adhesion during gastrulation in the amphibian *Pleurodeles waltlii, Dev. Biol.,* in press.
139. **Savagner, P., Imhof, B. A., Yamada, K. M., and Thiery, J. P.,** Homing of hemopoietic precursor cells to the embryonic thymus: characterization of an invasive mechanism induced by chemotactic peptides, *J. Cell Biol.,* 103, 2715, 1986.
140. **Bronner-Fraser, M.,** An antibody to a receptor for fibronectin and laminin perturbs cranial neural crest development in vivo, *Dev. Biol.,* 117, 528, 1986.
141. **Donaldson, D. J., Mahan, J. T., and Smith, G. N.,** Newt epidermal cell migration in vitro and in vivo appears to involve Arg-Gly-Asp-Ser receptors, *J. Cell Sci.,* 87, 525, 1987.
142. **Duband, J. L., Nuckolls, G. H., Ishihara, A., Hasegawa, T., Yamada, K. M., Thiery, J. P., and Jacobson, K.,** The fibronectin receptor exhibits high lateral mobility in embryonic locomoting cells but is immobile in focal contacts and fibrillar streaks in stationary cells, *J. Cell Biol.,* in press.
143. **Hynes, R. O.,** Alteration of cell-surface proteins by viral transformation and by proteolysis, *Proc. Natl. Acad. Sci. U.S.A.,* 70, 3170, 1973.
144. **Gahmberg, C. G. and Hakomori, S.,** Altered growth behavior of malignant cells associated with changes in externally labeled glycoprotein and glycolipid, *Proc. Natl. Acad. Sci. U.S.A.,* 70, 3329, 1973.
145. **Critchley, D. R.,** Cell surface proteins of NILI hamster fibroblasts labeled by a galactose oxidase, tritiated borohydride method, *Cell,* 3, 121, 1974.
146. **Gahmberg, C. G., Kiehn, D., and Hakomori, S.,** Changes in a surface-labelled galactoprotein and in glycolipid concentrations in cells transformed by a temperature-sensitive polyoma virus mutant, *Nature (London),* 248, 413, 1974.
147. **Hogg, N. M.,** A comparison of membrane proteins of normal and transformed cells by lactoperoxidase labeling, *Proc. Natl. Acad. Sci. U.S.A.,* 71, 489, 1974.
148. **Pearlstein, E. and Waterfield, M. D.,** Metabolic studies on ^{125}I-labeled baby hamster kidney cell plasma membranes, *Biochim, Biophys. Acta,* 362, 1, 1974.
149. **Robbins, P. W., Wickus, G. G., Branton, P. E., Gaffney, B. J., Hirschberg, C. B., Fuchs, P., and Blumberg, P. M.,** The chick fibroblast cell surface after transformation by Rous sarcoma virus, *Cold Spring Harbor Symp. Quant. Biol,* 39, 1173, 1974.
150. **Stone, K. R., Smith, R. E., and Joklik, W. K.,** Changes in membrane polypeptides that occur when chick embryo fibroblasts and NRK cells are transformed with avian sarcoma viruses, *Virology,* 58, 86, 1974.
151. **Vaheri, A. and Ruoslahti, E.,** Disappearance of a major cell-type specific surface glycoprotein antigen (SF) after transformation of fibroblasts by Rous sarcoma virus, *Int. J. Cancer,* 13, 579, 1974.
152. **Wartiovaara, J., Linder, E., Ruoslahti, E., and Vaheri, A.,** Distribution of fibroblast surface antigen. Association with fibrillar structures of normal cells and loss upon viral transformation. *J. Exp. Med.,* 140, 1522, 1974.

153. **Olden, K. and Yamada, K. M.,** Mechanism of the decrease in the major cell surface protein of chick embryo fibroblasts after transformation, *Cell,* 11, 957, 1977.
154. **Vaheri, A. and Ruoslahti, E.,** Fibroblast surface antigen produced but not retained by virus-transformed human cells, *J. Exp. Med.,* 142, 530, 1975.
155. **Vaheri, A., Ruoslahti, E., Westermark, B., and Ponten, J. A.** Common cell-type specific surface antigen in cultured human glial cells and fibroblasts: loss in malignant cells, *J. Exp. Med.,* 143, 64, 1976.
156. **Hynes, R. O. and Wyke, J. A.,** Alterations in surface proteins in chicken cells transformed by temperature-sensitive mutants for Rous sarcoma virus, *Virology,* 64, 492, 1975.
157. **Gallimore, P. H., McDougall, J. K., and Chen, L. B.,** Malignant behaviour of three adenovirus-2-transformed brain cell lines and their methylcellulose-selected sub-clones, *Int. J. Cancer,* 24, 477, 1979.
158. **Chen, L. B., Gallimore, P. H., and McDougall, J. K.,** Correlation between tumor induction and the large, external, transformation-sensitive protein on the cell surface, *Proc. Natl. Acad. Sci. U.S.A.,* 73, 3570, 1976.
159. **Hynes, R. O., Martin, G. S., Critchley, D. R., Shearer, M., and Epstein, C. J.,** Viral transformation of rat myoblasts: effects on fusion and surface properties, *Dev. Biol.,* 48, 35, 1976.
160. **Gallimore, P. H., McDougall, J. K., and Chen, L. B.,** In vitro traits of adenovirus-transformed cell lines and their relevance to tumorigenicity in nude mice, *Cell,* 10, 669, 1977.
161. **Rieber, M., Bacalao, J., and Alonso, G.,** Turnover of high-molecular-weight cell surface proteins during growth and expression of malignant transformation, *Cancer Res.,* 35, 2104, 1975.
162. **Palmieri, S., Kahn, P., and Graf, T.,** Quail embryo fibroblasts transformed by four v-myc-containing virus isolates show enhanced proliferation but are non tumorigenic, *EMBO J.,* 2, 2385, 1983.
163. **Jones, P. A., Laug, W. E., Gardner, A., Nye, C. A., Fink, L. M., and Benedict, W. F.,** In vitro correlates of transformation in C3H/10T1/2 clone 8 mouse cells, *Cancer Res.,* 36, 2863, 1976.
164. **Noonan, K. D., Wright, P. J., Bouck, N., and Di Mayorca, G.,** Leukemia virus infection of mammalian cells: effect on two "transformation-associated" surface properties, *J. Virol.,* 18, 1134, 1976.
165. **Clarke, S. M. and Fink, L. M.,** Studies on the iodinated surface membrane proteins and concanavalin A agglutination of transformed Syrian hamster cells, *Biochim. Biophys. Acta,* 464, 433, 1977.
166. **Clarke, S. M. and Fink, L. M.,** The effects of specific antibody-complement-mediated cytotoxicity on transformed and untransformed Syrian hamster cells, *Cancer Res.,* 37, 2985, 1977.
167. **Pearlstein, E., Hynes, R. O., Franks, L. M., and Hemmings, V. J.,** Surface proteins and fibrinolytic activity of cultured mammalian cells, *Cancer Res.,* 36, 1475, 1976.
168. **Klebe, R. J., Rosenberger, P. G., Naylor, S. L., Burns, R. L., Novak, R., and Kleinman, H.,** Cell attachment to collagen. Isolation of a cell attachment mutant, *Exp. Cell Res.,* 104, 119, 1977.
169. **Kahn, P. and Shin, S. I.,** Cellular tumorigenicity in nude mice: test of associations among loss of cell-surface fibronectin, anchorage independence and tumor-forming ability, *J. Cell Biol.,* 82, 1, 1979.
170. **Paraskeva, C. and Gallimore, P. H.,** Tumorigenicity and in vitro characteristics of rat liver epithelial cells and their adenovirus-transformed derivatives, *Int. J. Cancer,* 25, 631, 1980.
171. **Bober, F. J., Birk, D. E., and Raska, K.,** Expression of varying portions of the adenovirus 12 early region 1 in transformed cells affects tumorigenicity and interaction with extracellular matrix components, *Lab. Invest.,* 56, 37, 1987.
172. **Burridge, K.,** Changes in cellular glycoproteins after transformation: identification of specific glycoproteins and antigens in sodium dodecyl sulfate gels, *Proc. Natl. Acad. Sci. U.S.A.,* 73, 4457, 1976.
173. **Itaya, K. and Hakomori, S.,** Gangliosides and "galactoprotein A" ("LETS" - protein) of temperature-sensitive mutant of transformed 3T3 cells, *FEBS Lett.,* 66, 65, 1976.
174. **Keski-Oja, J., Gahmberg, C. G., and Alitalo, K.,** Pericellular matrix and cell surface glycoproteins of virus-transformed mouse epithelial cells, *Cancer Res.,* 42, 1147, 1982.
175. **Keski-Oja, J., Alitalo, K., Hautanen, A., and Rapp, U. R.,** Transformation of cultured epithelial cells by ethylnitrosourea. Altered expression of type I procollagen chains, *Biochim. Biophys. Acta,* 803, 153, 1984.
176. **Junker, J. L. and Wilson, M. J.,** Divergent expression of laminin and fibronectin in non-tumorigenic and transformed liver epithelial cells, *J. Cell Sci.,* 76, 213, 1985.
177. **Junker, J. L., Cottler-Fox, M., Wilson, M. J., Munoz, E. F., and Heine, U. I.,** Transformation-associated increase of adhesion, cellular fibronectin, and stress fiber development in a liver epithelial cell line, *J. Natl. Cancer Inst.,* 74, 173, 1985.
178. **Tsao, M. S., Grisham, J. W., and Nelson, K. G.,** Clonal analysis of tumorigenicity and paratumorigenic phenotypes in rat liver epithelial cells chemically transformed in vitro, *Cancer Res.,* 45, 5139, 1985.
179. **Wigley, C. G. and Summerhayes, I. C.,** Loss of LETS protein is not a marker for salivary gland or bladder epithelial cell transformation, *Exp. Cell Res.,* 118, 394, 1979.
180. **Blumberg, P. M., Driedger, P. E., and Rossow, P. W.,** Effect of a phorbol ester on a transformation-sensitive surface protein of chicken fibroblasts, *Nature (London),* 264, 446, 1976.
181. **Keski-Oja, J., Vaheri, A., and Ruoslahti, E.,** Fibroblast surface antigen (SF): the external glycoprotein lost in proteolytic stimulation and malignant transformation, *Int. J. Cancer,* 17, 261, 1976.

182. **Keski-Oja, J., Shoyab, M., Delarco, J. E., and Todaro, A. J.,** Rapid release of fibronectin from human lung fibroblasts by biologically active phorbol esters, *Int. J. Cancer,* 24, 218, 1979.
183. **Bolmer, S. D. and Wolf, G.,** Retinoids and phorbol esters alter release of fibronectin from enucleated cells, *Proc. Natl. Acad. Sci. U.S.A.,* 79, 6541, 1982.
184. **Zerlauth, G. and Wolf, G.,** Kinetics of fibronectin release from fibroblasts in response to 12-*O*-tetradecanoylphorbol-13-acetate and retinoic acid, *Carcinogenesis,* 5, 863, 1984.
185. **Zerlauth, G. and Wolf, G.,** Release of fibronectin is linked to tumor promotion: response of promotable and non-promotable clones of a mouse epidermal cell line, *Carcinogenesis,* 6, 73, 1985.
186. **Habliston, D. L., Birrell, G. B., Hedberg, K. K., and Griffith, O. H.,** Early phorbol ester induced release of cell surface fibronectin: direct observation by photoelectron microscopy, *Eur. J. Cell Biol.,* 41, 222, 1986.
187. **Zerlauth, G. and Wolf, G.,** Influence of tumor promoter on binding and release of fibronectin added to human lung fibroblasts, *Biochim. Biophys. Acta,* 927, 402, 1987.
188. **Hayashida, N., Kaneko, Y., Nishi, T., Miyazaki, J., Endo, Y., and Oda, T.,** Effects of the tumor promoter teleocidin on human hepatoma cells in tissue culture, *Gann,* 73, 534, 1982.
189. **Adams, S. L., Sobel, M. E., Howard, B. H., Olden, K., Yamada, K. M., de Crombrugghe, B., and Pastan, I.,** Levels of translatable mRNAs for cell surface protein, collagen precursors, and two membrane proteins are altered in Rous sarcoma virus-transformed chick embryo fibroblasts, *Proc. Natl. Acad. Sci. U.S.A.,* 74, 3399, 1977.
190. **Fagan, J. B., Sobel, M. E., Yamada, K. M., de Crombrugghe, B., and Pastan, I.,** Effects of transformation on fibronectin gene expression using cloned fibronectin cDNA, *J. Biol. Chem.,* 256, 520, 1981.
191. **Tyagi, J. S., Hirano, H., Merlino, G. T., and Pastan, I.,** Transcriptional control of the fibronectin gene in chick embryo fibroblasts transformed by Rous sarcoma virus, *J. Biol. Chem.,* 258, 5787, 1983.
192. **Tyagi, J. S., Hirano, H., and Pastan, I.,** Modulation of fibronectin gene activity in chick embryo fibroblasts transformed by a temperature-sensitive strain (ts68) of Rous sarcoma virus, *Nucleic Acids Res.,* 13, 8275, 1985.
193. **Leibovitch, S. A., Hillion, J., Leibovitch, M. P., Guillier, M., Schmitz, A., and Harel, J.,** Expression of extracellular matrix genes in relation to myogenesis and neoplastic transformation, *Exp. Cell Res.,* 166, 526, 1986.
194. **Hynes, R. O., Wyke, J. A., Bye, J. M., Humphreys, K. C., and Pearlstein, E. S.,** Are proteases involved in altering surface proteins during viral transformation?, in *Proteases and Biological Control,* Reich, E., Shaw, E., and Rifkin, D. B., Eds., Cold Spring Harbor Laboratory, Cold Spring Harbor, NY, 1975, 931.
195. **Blumberg, P. M. and Robbins, P. W.,** Effects of proteases on activation of resting chick embryo fibroblasts and on cell surface proteins, *Cell,* 6, 137, 1975.
196. **Zetter, B. R., Chen, L. B., and Buchanan, J. M.,** Effects of protease treatment on growth, morphology, adhesion, and cell surface proteins of secondary chick embryo fibroblasts, *Cell,* 7, 407, 1976.
197. **Hynes, R. O. and Pearlstein, E. S.,** Investigations of the possible role of proteases in altering surface proteins of virally transformed hamster fibroblasts, *J. Supramol. Struct.,* 4, 1, 1976.
198. **Mahdavi, V. and Hynes, R. O.,** Effects of cocultivation with transformed cells on surface proteins of normal cells, *Biochim. Biophys. Acta,* 542, 191, 1978.
199. **Hynes, R. O., Ali, I. U., Destree, A. T., Mautner, V., Perkins, M. E., Senger, D. R., Wagner, D. D., and Smith, K. K.,** A large glycoprotein lost from the surfaces of transformed cells, *Ann. N.Y. Acad. Sci.,* 312, 317, 1978.
200. **Hayman, E. G., Engvall, E., and Ruoslahti, E.,** Concomitant loss of cell surface fibronectin and laminin from transformed rat kidney cells, *J. Cell Biol.,* 88, 352, 1981.
201. **Wagner, D. D., Ivatt, R., Destree, A. T., and Hynes, R. O.,** Similarities and differences between fibronectins from normal and transformed hamster cells, *J. Biol. Chem.,* 256, 11708, 1981.
202. **Chen, W. T., Wang, J., Hasegawa, T., Yamada, S. S., and Yamada, K. M.,** Regulation of fibronectin receptor distribution by transformation, exogenous fibronectin, and synthetic peptides, *J. Cell Biol.,* 103, 1649, 1986.
203. **Virtanen, I., Lehto, V. P., and Vartio, T.,** Lack of fibronectin-binding plasma membrane proteins may explain defective pericellular matrix formation in transformed fibroblasts and fibrosarcoma cells, *Int. J. Cancer,* 39, 361, 1987.
204. **Ali, I. U.,** Structural analysis of fibronectin and its collagen-binding fragment from several cell lines, *Proc. Natl. Acad. Sci. U.S.A.,* 81, 28, 1984.
205. **Olden, K., Pratt, R. M., and Yamada, K. M.,** Role of carbohydrate in biological function of the adhesive glycoprotein fibronectin, *Proc. Natl. Acad. Sci. U.S.A.,* 76, 3343, 1979.
206. **Ali, I. U. and Hunter, T.,** Structural comparisons of fibronectins from normal and transformed cells, *J. Biol. Chem.,* 256, 7671, 1981.
207. **Teng, M. H. and Rifkin, D. B.,** Fibronectin from chicken embryo fibroblasts contains covalently bound phosphate, *J. Cell Biol.,* 80, 784, 1979.

208. **Yamada, K. M., Ohanian, S. H., and Pastan, I.,** Cell surface protein decreases microvilli and ruffles on transformed mouse and chick cells, *Cell,* 9, 241, 1976.
209. **Yamada, K. M., Yamada, S. S., and Pastan, I.,** Cell surface protein partially restores morphology, adhesiveness, and contact inhibition of movement to transformed fibroblasts, *Proc. Natl. Acad. Sci. U.S.A.,* 73, 1217, 1976.
210. **Willingham, M. C., Yamada, K. M., Yamada, S. S., Pouyssegur, J., and Pastan, I.,** Microfilament bundles and cell shape are related to adhesiveness to substratum and are dissociable from growth control in cultured fibroblasts, *Cell,* 10, 375, 1977.
211. **Ali, I. U., Mautner, V., Lanza, R., and Hynes, R. O.,** Restoration of normal morphology, adhesion and cytoskeleton in transformed cells by addition of a transformation-sensitive surface protein, *Cell,* 11, 115, 1977.
212. **Yamada, K. M. and Pastan, I.,** The relationship between cell surface protein and glucose and alpha-aminoisobutyric acid transport in transformed chick and mouse cells, *J. Cell. Physiol.,* 89, 827, 1977.
213. **Lage-Davila, A., Krust, B., Hofmann-Clerc, F., Torpier, G., and Montagnier, L.,** Bromodeoxyuridine-induced reversion of transformed characteristics in BHK21 cells: changes at the plasma membrane level, *J. Cell. Physiol.,* 100, 95, 1979.
214. **Pfeffer, L. M., Wang, E., and Tamm, I.,** Interferon effects on microfilament organization, cellular fibronectin distribution, and cell motility in human fibroblasts, *J. Cell Biol.,* 85, 9, 1980.
215. **Gerfaux, J., Rousset, S., Chany-Fournier, F., and Chany, C.,** Interferon effect on collagen and fibronectin distribution in the extracellular matrix of murine sarcoma virus-transformed cells, *Cancer Res.,* 41, 3629, 1981.
216. **Hayman, E. G., Engvall, E., and Ruoslahti, E.,** Butyrate restores fibronectin at cell surface of transformed cells, *Exp. Cell Res.,* 127, 478, 1980.
217. **Furcht, L. T., Mosher, D. F., Wendelschafer-Crabb, G., and Foidart, J. M.,** Reversal by glucocorticoid hormones of the loss of a fibronectin and procollagen matrix around transformed human cells, *Cancer Res.,* 39, 2077, 1979.
218. **Furcht, L. T., Mosher, D. F., Wendelschafer-Crabb, G., Woodbridge, P. A., and Foidart, J. M.,** Dexamethasone-induced accumulation of a fibronectin and collagen extracellular matrix in transformed human cells, *Nature (London),* 277, 393, 1979.
219. **Nielson, S. E. and Puck, T. T.,** Deposition of fibronectin in the course of reverse-transformation of Chinese hamster ovary cells by cyclic AMP, *Proc. Natl. Acad. Sci. U.S.A.,* 77, 985, 1980.
220. **Chen, L. B., Maitland, N., Gallimore, P. H., and McDougall, J. K.,** Detection of the large external transformation-sensitive protein on some epithelial cells, *Exp. Cell Res.,* 106, 39, 1977.
221. **Der, C. J. and Stanbridge, E. J.,** Lack of correlation between the decreased expression of cell surface LETS protein and tumorigenicity in human cell hybrids, *Cell,* 15, 1241, 1978.
222. **McCabe, R. P. and Evans, C. H.,** Plasminogen activator, fibronectin, lymphotoxin sensitivity, and natural skin reactivity relationships to guinea pig cell tumorigenicity, *J. Natl. Cancer Inst.,* 68, 329, 1982.
223. **Neri, A., Ruoslahti, E., and Nicolson, G. L.,** Distribution of fibronectin on clonal cell lines of a rat mammary adenocarcinoma growing in vitro and in vivo at primary and metastatic sites, *Cancer Res.,* 41, 5082, 1981.
224. **Marciani, D. J., Lyons, L. B., and Thompson, E. B.,** Characteristics of cell membranes from somatic cell hybrids between rat hepatoma and mouse L-cells, *Cancer Res.,* 36, 2937, 1976.
225. **Eun, C. K. and Klinger, H. P.,** Human chromosome 11 affects the expression of fibronectin (LETS protein) in human × mouse cell hybrids, *Cytogenet. Cell Genet.,* 25, 151, 1979.
226. **Klinger, H. P.,** Suppression of tumorigenicity in somatic cell hybrids. I. Suppression and reexpression of tumorigenicity in diploid human × D98AH2 hybrids and independent segregation of tumorigenicity from other cell phenotypes, *Cytogenet. Cell Genet.,* 27, 254, 1980.
227. **Der, C. J., Ash, J. F., and Stanbridge, E. J.,** Cytoskeletal and transmembrane interactions in the expression of tumorigenicity in human cell hybrids, *J. Cell Sci.,* 52, 151, 1981.
228. **Peehl, D. M. and Stanbridge, E. J.,** Characterization of human keratinocyte × Hela somatic cell hybrids, *Int. J. Cancer,* 27, 625, 1981.
229. **Larizza, L., Tenchini, M. L., Mottura, A., De Carli, L., Colombi, M., and Barlati, S.,** Expression of transformation markers and suppression of tumorigenicity in human cell hybrids, *Eur. J. Cancer Clin. Oncol.,* 18, 845, 1982.
230. **Lawrence, J. B. and Coleman, J. R.,** Analysis of myogenesis by somatic cell hybridization. II. Retention of myogenic competence and suppression of transformed properties in hybrids between differentiation competent and imcompetent rat L6 myoblasts, *J. Cell. Physiol.,* 114, 99, 1983.
231. **Tenchini, M. L., Larizza, L., Mottura, A., Colombi, M., Barlati, S., and De Carli, L.,** Studies on transformation markers and tumorigenicity in segregant clones from a human hybrid line, *Eur. J. Cancer Clin. Oncol.,* 19, 1143, 1983.
232. **Rampoldi, E., Larizza, L., Doneda, L., and Barlati, S.,** Fibronectin, laminin, in hybrids of Rous sarcoma virus transformed and normal mouse fibroblasts, *Tumori,* 71, 419, 1985.

233. **Der, C. J. and Stanbridge, E. J.,** Alterations in the extracellular matrix organization associated with reexpression of tumorigenicity in human cell hybrids, *Int. J. Cancer,* 26, 451, 1980.
234. **Stanbridge, E. J., Der, C. J., Doersen, C. J., Nishimi, R. Y., Peehl, D. M., Weissman, B. E., and Wilkinson, J. E.,** Human cell hybrids: analysis of transformation and tumorigenicity, *Science,* 215, 252, 1982.
235. **Labat-Robert, J., Birembaut, P., Robert, L., and Adnet, J. J.,** Modification of fibronectin distribution pattern in solid human tumours, *Diagn. Histopathol.,* 4, 299, 1981.
236. **Birembaut, P., Caron, Y., Adnet, J. J., and Foidart, J. M.,** Usefulness of basement membrane markers in tumoural pathology, *J. Pathol.,* 145, 283, 1985.
237. **Stenman, S. and Vaheri, A.,** Fibronectin in human solid tumors, *Int. J. Cancer,* 27, 427, 1981.
238. **Kochi, N., Tani, E., Morimura, T., and Itagaki, T.,** Immunohistochemical study of fibronectin in human glioma and meningioma, *Acta Neuropathol.,* 59, 119, 1983.
239. **Chronwall, B. M., McKeever, P. E., and Kornblith, P. L.,** Glial and nonglial neoplasms evaluated on frozen section by double immunofluorescence for fibronectin and glial fibrillary acidic protein, *Acta Neuropathol.,* 59, 283, 1983.
240. **Seitz, R. J. and Wechsler, W.,** Immunohistochemical demonstration of serum proteins in human cerebral gliomas, *Acta Neuropathol.,* 73, 145, 1987.
241. **Rutka, J. T., Myatt, C. A., Giblin, J. R., Davis, R. L., and Rosenblum, M. L.,** Distribution of extracellular matrix proteins in primary human brain tumours: an immunohistochemical analysis, *Can. J. Neurol. Sci.,* 14, 25, 1987.
242. **Linder, E., Stenman, S., Lehto, V. P., and Vaheri, A.,** Distribution of fibronectin in human tissues and relationship to other connective tissue components, *Ann. N.Y. Acad. Sci.,* 312, 151, 1978.
243. **Labat-Robert, J., Birembaut, P., Adnet, J. J.,** Mercantini, F., and Robert, L., Loss of fibronectin in human breast cancer, *Cell Biol. Int. Rep.,* 4, 609, 1980.
244. **Asch, B. B., Kamat, B. R., and Burstein, N. A.,** Interactions of normal, dysplastic, and malignant mammary epithelial cells with fibronectin in vivo and in vitro, *Cancer Res.,* 41, 2115, 1981.
245. **Burtin, P., Chavanel, G., Foidart, J. M., and Andre, J.,** Alterations of the basement membrane and connective tissue antigens in human metastatic lymph nodes, *Int. J. Cancer,* 31, 719, 1983.
246. **Lagace, R. Grimaud, J. A., Schurch, W., and Seemayer, T. A.,** Myofibroblastic stromal reaction in carcinoma of the breast: variations of collagenous matrix and structural glycoproteins, *Virchows Arch. A,* 408, 49, 1985.
247. **d'Ardenne, A. J., Kirkpatrick, P., Wells, C. A., and Davies, J. D.,** Laminin and fibronectin in adenoid cystic carcinoma, *J. Clin. Pathol.,* 39, 138, 1986.
248. **Gibert, M. A., Noel, P., Faucon, M., and Pavans de Caccatty, M.,** Comparative immunohistochemical localization of fibronectin and actin in human breast tumor cells in vivo and in vitro, *Virchows Arch. B,* 40, 99, 1982.
249. **Grigioni, W. F., Biagini, G., Errico, A. D., Milani, M., Villanacci, V., Garbisa, S., Mattioli, S., Gozzetti, G., and Mancini, A. M.,** Behaviour of basement membrane antigens in gastric and colorectal cancer. Immunohistochemical study, *Acta Pathol. Jpn.,* 36, 173, 1986.
250. **Burtin, P., Chavanel, G., Foidart, J. M., and Martin, E.,** Antigens of the basement membrane and the peritumoral stroma in human colonic adenocarcinomas: an immunofluorescence study, *Int. J. Cancer,* 30, 13, 1982.
251. **Niemczuk, P., Perkins, R. M., Talbot, I. C., and Critchley, D. R.,** Lack of correlation between metastasis of human rectal carcinoma and the absence of stromal fibronectin, *Br. J. Cancer,* 45, 500, 1982.
252. **Burtin, P., Chavanel, G., and Foidart, J. M.,** Immunofluorescence study of the antigens of the basement membrane and the peritumoral stroma in human colonic adenocarcinomas, *Ann. N.Y. Acad. Sci.,* 420, 229, 1983.
253. **d'Ardenne, A. J., Burns, J., Sykes, B. C., and Bennett, M. K.,** Fibronectin and type III collagen in epithelial neoplasms of gastrointestinal tract and salivary gland, *J. Clin. Pathol.,* 36, 756, 1983.
254. **Matsuda, T., Ogawa, M., Maruyama, J., Katsura, Y., and Yasutomi, M.,** A histochemical study of fibronectin (FN) of the large intestine by the immunoperoxidase technique, *Jpn. J. Clin. Oncol.,* 13, 327, 1983.
255. **Forster, S. J., Talbot, I. C., and Critchley, D. R.,** Laminin and fibronectin in rectal adenocarcinoma: relationship to tumour grade, stage and metastasis, *Br. J. Cancer,* 50, 51, 1984.
256. **Du Boulay, C. E.,** Demonstration of fibronectin in soft tissue tumours using the immunoperoxidase technique, *Diagn. Histopathol.,* 5, 283, 1982.
257. **d'Ardenne, A. J., Kirkpatrick, P., and Sykes, B. C.,** Distribution of laminin, fibronectin, and interstitial collagen type III in soft tissue tumours, *J. Clin. Pathol.,* 37, 895, 1984.
258. **Kumar, S., Marsden, H. B., and Calabuig, M. C.,** Childhood kidney tumours: in vitro studies and natural history, *Virchows Arch. A,* 405, 95, 1984.
259. **Kumar, S., Carr, T., Marsden, H. B., and Calabuig-Crespo, M. C.,** Study of childhood renal tumours using antisera to fibronectin, laminin, and epithelial membrane antigen, *J. Clin. Pathol.,* 39, 51, 1986.

260. **Sariola, H., Ekblom, P., Rapola, J., Vaheri, A., and Timpl, R.,** Extracellular matrix and epithelial differentiation of Wilm's tumor, *Am. J. Pathol.*, 118, 96, 1985.
261. **Szendroi, M. and Lapis, K.,** Fibronectin in differential diagnosis of primary hepatomas and carcinoma metastases in the liver, *Acta Morphol. Hung.*, 33, 101, 1985.
262. **Tabarin, A., Bioulac-Sage, P., Boussarie, L., Balabaud, C., de Mascarel, A., and Grimaud, J. A.,** Hepatocellular carcinoma developed on noncirrhotic livers. Sinusoids in hepatocellular carcinoma, *Arch. Pathol. Lab. Med.*, 111, 174, 1987.
263. **Tsumagari, J.,** Fibronectin in human hepatocellular carcinoma (HCC) and HCC cell lines, *Acta Pathol. Jpn.*, 37, 413, 1987.
264. **Nagai, H., Isemura, M., Arai, H., Abe, T., Shimoda, S., Motomiya, M., Sato, H., Hashimoto, K., Takusagawa, K., and Konno, K.,** Pattern of fibronectin distribution in human lung cancer, *J. Cancer Res. Clin. Oncol.*, 112, 1, 1986.
265. **Cam, Y., Caulet, T., Bellon, G., Poulin, G., Legros, M., and Pytlinska, M.,** Immunohistochemical localization of macromolecules of the basement membrane and the peritumoral stroma in human laryngeal carcinomas, *J. Pathol.*, 144, 35, 1984.
266. **Carbone, A., Manconi, R., Poletti, A., Colombatti, A., Tirelli, U., and Volpe, R.,** S-100 protein, fibronectin, and laminin immunostaining in lymphomas of follicular center cell origin, *Cancer*, 58, 2169, 1986.
267. **Kalimo, H., Lehto, M., Nanto-Salonen, K., Jalkanen, M., Risteli, L., Risteli, J., and Narva, E. V.,** Characterization of the perivascular reticulin network in a case of primary brain lymphoma. Immunohistochemical demonstration of collagen types I, III, IV, and V; laminin; and fibronectin, *Acta Neuropathol.*, 66, 299, 1986.
268. **Kochi, N., Budka, H., and Radaskiewicz, T.,** Development of stroma in malignant lymphomas of the brain compared with epidural lymphomas. An immunohistochemical study, *Acta Neuropathol.*, 71, 125, 1986.
269. **Truong, L. D., Harris, L., Mattioli, C., Hawkins, E., Lee, A., Wheeler, T., and Lane, M.,** Endodermal sinus tumor of the mediastinum. A report of seven cases and review of the literature, *Cancer*, 58, 730, 1986.
270. **Takagi, M.,** Ultrastructural and immunohistochemical characteristics of cardiac myxoma, *Acta Pathol. Jpn.*, 34, 1099, 1984.
271. **Ohmori, T., Arita, N., Uraga, N., Tabei, R., Tani, M., and Okamura, H.,** Malignant granular cell tumor of the esophagus. A case report with light and electron microscopic, histochemical, and immunohistochemical study, *Acta Pathol. Jpn.*, 37, 775, 1987.
272. **Meyer, J. R., Silverman, S., Daniels, T. E., Kramer, R. H., and Greenspan, J. S.,** Distribution of fibronectin and laminin in oral leukoplakia and carcinoma, *J. Oral Pathol.*, 14, 247, 1985.
273. **Mollenhauer, J., Roether, I., and Kern, H. F.,** Distribution of extracellular matrix proteins in pancreatic ductal adenocarcinoma and its influence on tumor cell proliferation in vitro, *Pancreas*, 2, 14, 1987.
274. **Peltonen, J., Aho, H., Halme, T., Nanto-Salonen, K., Lehto, M., Foidart, J. M., Duance, V., Vaheri, A., and Penttinen, R.,** Distribution of different collagen types and fibronectin in neurofibromatosis tumours, *Acta Path. Microbiol. Immunol. Scand. A*, 92, 345, 1984.
275. **Fleischmajer, R., Timpl, R., Dziadek, M., and Lebwohl, M.,** Basement membrane proteins, interstitial collagens, and fibronectin in neurofibroma, *J. Invest, Dermatol.*, 85, 54, 1985.
276. **Nelson, D. L., Little, C. D., and Balian, G.,** Distribution of fibronectin and laminin in basal cell epitheliomas, *J. Invest. Dermatol.*, 80, 446, 1983.
277. **Grimwood, R. E., Huff, J. C., Harbell, J. W., and Clark, R. A.,** Fibronectin in basal cell epithelioma: sources and significance, *J. Invest. Dermatol.*, 82, 145, 1984.
278. **Matalanis, G., Gardner, I. D., and Whitehead, R. H.,** Lectin binding patterns and monoclonal antibodies to epidermal antigens in tumours of the skin, *Pathology*, 18, 206, 1986.
279. **Inoue, A., Aozasa, K., Tsujimoto, M., Tamai, M., Chatani, F., and Ueno, H.,** Immunohistological study on malignant fibrous histiocytoma, *Acta Pathol. Jpn.*, 34, 759, 1984.
280. **Grimwood, R. E., Ferris, C. F., Nielsen, L. D., Huff, J. C., and Clark, R. A.,** Basal cell carcinomas grown in nude mice produce and deposit fibronectin in the extracellular matrix, *J. Invest. Dermatol.*, 87, 42, 1986.
281. **Kimata, K., Foidart, J. M., Pennypacker, J. P., Kleinman, H. K., Martin, G. R., and Hewitt, A. T.,** Immunofluorescence localization of fibronectin in chondrosarcoma cartilage matrix, *Cancer Res.*, 42, 2384, 1982.
282. **Brown, L. F., Chester, J. F., Malt, R. A., and Dvorak, H. F.,** Fibrin deposition in autochthonous Syrian hamster pancreatic adenocarcinomas induced by the chemical carcinogen *N*-nitroso-bis(2-oxopropyl)amine, *J. Natl. Cancer Inst.*, 78, 979, 1987.
283. **Kornblith, P. L.,** Role of tissue culture in prediction of malignancy, *Clin. Neurosurg.*, 25, 346, 1978.
284. **Paetau, A., Mellstrom, K., Westermark, B., Dahl, D., Haltia, M., and Vaheri, A.,** Mutually exclusive expression of fibronectin and glial fibrillary acidic protein in cultured brain cells, *Exp. Cell Res.*, 129, 337, 1980.

285. **Diserens, A. C., de Tribolet, N., Martin-Achard, A., Gaide, A. C., Schnegg, J. F., and Carrel, S.,** Characterization of an established human malignant glioma cell line: LN-18, *Acta Neuropathol.*, 53, 21, 1981.
286. **Koppel, H., Martin, J. M., Pilkington, G. J., and Lantos, P. L.,** Heterogeneity of a cultured neoplastic glial line. Establishment and characterization of six clones, *J. Neurol. Sci.*, 76, 295, 1986.
287. **McKeever, P. E., Hood, T. W., Varani, J., Taren, J. A., Beierwaltes, W. H., Wahl, R., Liebert, M., and Nguyen, P. K.,** Products of cells cultured from gliomas. V. Cytology and morphometry of two cell types cultured from glioma, *J. Natl. Cancer Inst.*, 78, 75, 1987.
288. **Tveit, K. M. and Pihl, A.,** Do cell lines in vitro reflect the properties of the tumours of origin? A study of lines derived from human melanoma xenografts, *Br. J. Cancer*, 44, 775, 1981.
289. **Persky, B., Thomson, S. P., Meyskens, F. L., and Hendrix, M. J.,** Methods for evaluating the morphological and immunohistochemical properties of human tumor colonies grown in soft agar, *In vitro*, 18, 929, 1982.
290. **Nakabayashi, H., Taketa, K., Miyano, K., Yamane, T., and Sato, J.,** Growth of human hepatoma cells lines with differentiated functions in chemically defined medium, *Cancer Res.*, 42, 3858, 1982.
291. **Glasgow, J. E. and Colman, R. W.,** Fibronectin synthesized by a human hepatoma cell line, *Cancer Res.*, 44, 3022, 1984.
292. **Matsuura, H. and Hakomori, S.,** The oncofetal domain of fibronectin defined by monoclonal antibody FDC-6: its presence in fibronectins from fetal and tumor tissues and its absence in those from normal adult tissues and plasma, *Proc. Natl. Acad. Sci. U.S.A.*, 82, 6517, 1985.
293. **Smith, H. S., Hackett, A. J., Riggs, J. L., Mosesson, M. W., Walton, J. R., and Stampfer, M. R.,** Properties of epithelial cells cultured from human carcinomas and nonmalignant tissues, *J. Supramol. Struct.*, 11, 147, 1979.
294. **Smith, H. S., Riggs, J. L., and Mosesson, M. W.,** Production of fibronectin by human epithelial cells in culture, *Cancer Res.*, 39, 4138, 1979.
295. **Neri, A. and Nicolson, G. L.,** Phenotypic drift of metastatic and cell-surface properties of mammary adenocarcinoma cell clones during growth in vitro, *Int. J. Cancer*, 28, 731, 1981.
296. **Stampfer, M. R., Vlodavsky, I., Smith, H. S., Ford, R., Becker, F. F., and Riggs, J.,** Fibronectin production by human mammary cells, *J. Natl. Cancer Inst.*, 67, 253, 1981.
297. **Taylor-Papadimitriou, J., Burchell, J., and Hurst, J.,** Production of fibronectin by normal and malignant human mammary epithelial cells, *Cancer Res.*, 41, 2491, 1981.
298. **Brown, K. W. and Parkinson, E. K.,** Alteration of the extracellular matrix of cultured human keratinocytes by transformation and during differentiation, *Int. J. Cancer*, 35, 799, 1985.
299. **Webb, K. S., Stone, K. R., Sharief, Y., and Paulson, D. F.,** Surface proteins of a transitional carcinoma cell line (KS-31E), *Urol. Res.*, 8, 77, 1980.
300. **Mosesson, M. W. and Umfleet, R. A.,** The cold insoluble globulin of human plasma. I. Purification, primary characterization, and relationship to fibrinogen and other cold insoluble fraction components, *J. Biol. Chem.*, 245, 5728, 1970.
301. **Mosher, D. F. and Williams, E. M.,** Fibronectin concentration is decreased in plasma of severely ill patients with disseminated intravascular coagulation, *J. Lab. Clin. Med.*, 91, 729, 1978.
302. **Plow, E. F., Birdwell, C., and Ginsberg, M. H.,** Identification and quantitation of platelet-associated fibronectin antigen, *J. Clin. Invest.*, 63, 540, 1979.
303. **Pearlstein, E. and Baez, L.,** A solid-phase radioimmunoassay for the determination of fibronectin levels in plasma, *Anal. Biochem.*, 116, 292, 1981.
304. **Saba, T. M., Albert, W. H., Blumenstock, F. A., Evanega, G., Staehler, F., and Cho. E.,** Evaluation of a rapid immunoturbidimetric assay for opsonic fibronectin in surgical and trauma patients, *J. Lab. Clin. Med.*, 98, 482, 1981.
305. **Stathakis, N. E., Fountas, A., and Tsianos, E.,** Plasma fibronectin in normal subjects and in various disease states, *J. Clin. Pathol.*, 34, 504, 1981.
306. **Todd-Kulikowski, H. D. and Parsons, R. G.,** A stable, sensitive assay for human fibronectin, *J. Immunol. Methods*, 44, 333, 1981.
307. **Boughton, B. J. and Simpson, A.,** Plasma fibronectin in acute leukaemia, *Br. J. Haematol.*, 51, 487, 1982.
308. **Eriksen, H. O., Clemmensen, I., Hansen, M. S., and Ibsen, K. K.,** Plasma fibronectin concentration in normal subjects, *Scand. J. Clin. Lab. Invest.*, 42, 291, 1982.
309. **Gonzalez-Calvin, J., Scully, M. F., Sanger, Y., Fok, J., Kakkar, V. V., Hughes, R. D., Gimson, A. E., and Williams, R.,** Fibronectin in fulminant hepatic failure, *Br. Med. J.*, 285, 1231, 1982.
310. **Matsuda, M., Yamanaka, T., and Matsuda, A.,** Distribution of fibronectin in plasma and liver in liver diseases, *Clin. Chim. Acta*, 118, 191, 1982.
311. **Scott, R. L., Sohmer, P. R., and MacDonald, M. B. G.,** The effect of starvation and repletion on plasma FN in man, *JAMA*, 248, 2025, 1982.
312. **Vuento, M., Korkolainen, M., and Stenman, U. H.,** Fibronectin binds to charge-modified proteins, *Adv. Exp. Med. Biol.*, 155, 623, 1982.

313. **Alexander, C. M., Lum, S. M., Rhodes, J., Boarman, C., Nicoloff, J. T., and Kumar, D.,** Rapid increase in both plasma fibronectin and serum triiodothyronine associated with treatment of diabetic ketoacidosis, *J. Clin. Endocrinol. Metab.*, 56, 279, 1983.
314. **Bowen, M. and Muller, T.,** Influence of sample preparation on estimates of blood fibronectin concentration, *J. Clin. Pathol.*, 36, 233, 1983.
315. **Choate, J. J. and Mosher, D. F.,** Fibronectin concentration in plasma of patients with breast cancer, colon cancer, and acute leukemia, *Cancer,* 51, 1142, 1983.
316. **Kawamura, K., Tanaka, M., Kamiyama, F., Higashino, K., and Kishimoto, S.,** Enzyme immunoassay of human fibronectin in malignant, collagen and liver diseases, *Clin. Chim. Acta,* 131, 101, 1983.
317. **Klingemann, H. G., Kosukavak, M., Hofeler, H., and Havemann, K.,** Fibronectin and factor VIII-related antigen in acute leukaemia, *Hoppe-Seyler's Z. Physiol. Chem.*, 364, 269, 1983.
318. **Norfolk, D. R., Bowen, M., Roberts, B. E., and Child, J. A.,** Plasma fibronectin in myeloproliferative disorders and chronic granulocytic leukaemia, *Br. J. Haematol.*, 55, 319, 1983.
319. **Siri, A., Carnemolla, B., Raffanti, S., Castellani, P., Balza, E., and Zardi, L.,** Fibronectin concentrations in pleural effusions of patients with malignant and non-malignant disease, *Cancer Lett.*, 22, 1, 1984.
320. **Zerlauth, G. and Wolf, G.,** Plasma fibronectin as a marker for cancer and other diseases, *Am. J. Med.*, 77, 685, 1984.
321. **de Russe, J., Colombat, P., Lavoix, X., and Bardos, P.,** Plasma fibronectin in various hemopathic diseases, *Clin. Chim. Acta,* 145, 49, 1985.
322. **Sawaya, R., Cummins, C. J., Smith, B. H., and Kornblith, P. L.,** Plasma fibronectin in patients with brain tumors, *Neurosurgery,* 16, 161, 1985.
323. **Boccardo, F., Guarneri, D., Zanardi, S., Castellani, P., Borsi, L., and Zardi, L.,** Fibronectin concentration in the plasma of patients with malignant and benign breast disease, *Cancer Lett.*, 33, 317, 1986.
324. **Eijan, A. M., Puricelli, L., Bal de Kier Joffe, E., Entin, D., Vuoto, D., Orlando, E., and de Lustig, E. S.,** Serial analysis of fibronectin concentration in plasma of patients with benign and malignant breast diseases, *Cancer,* 57, 1345, 1986.
325. **Parsons, R. G., Todd, H. D., and Kowal, R.,** Isolation and identification of a human serum fibronectin-like protein elevated during malignant disease, *Cancer Res.*, 39, 4341, 1979.
326. **Todd, H. D., Coffee, M. S., Waalkes, T. P., Abeloff, M. D., and Parsons, R. G.,** Serum levels of fibronectin and a fibronectin-like DNA-binding protein in patients with various diseases, *J. Natl. Cancer Inst.*, 65, 901, 1980.
327. **De Petro, G., Barlati, S., Vartio, T., and Vaheri, A.,** Transformation-enhancing activity of proteolytic fragments of fibronectin, *Proc. Natl. Acad. Sci. U.S.A.*, 78, 4965, 1981.
328. **Humphries, M. J. and Ayad, S. R.,** Stimulation of DNA synthesis by cathepsin D digests of fibronectin, *Nature (London),* 305, 811, 1983.
329. **Clemmensen, I. and Andersen, R.,** Different molecular forms of fibronectin in rheumatoid synovial fluid, *Arthritis Rheum.*, 25, 25, 1982.
330. **Bruhn, H. D. and Heimburger, N.,** Factor-VIII-related antigen and cold insoluble globulin in leukemias and carcinomas, *Haemostasis,* 5, 189, 1976.
331. **Blumenstock, F. A., Weber, P., and Saba, T. M.,** Purification of alpha-2-opsonic protein from human serum and its measurement by immunoassay, *J. Reticuloendothel. Soc.*, 23, 119, 1978.
332. **Ragni, M. V., Lewis, J. H., Spero, J. A., and Bontempo, F. A.,** Plasma fibronectin levels in clinical disease states and after cryoprecipitate infusion, *Thromb. Haemost.*, 52, 321, 1984.
333. **Saba, T. M. and Antikatzides, T. G.,** Humoral mediated macrophage response during tumour growth, *Br. J. Cancer,* 32, 471, 1975.
334. **Zardi, L., Cecconi, C., Barbieri, O., Carnemolla, B., Picca, M., and Santi, L.,** Concentration of fibronectin in plasma of tumor-bearing mice and synthesis by Ehrlich ascites tumor cells, *Cancer Res.*, 39, 3374, 1979.
335. **Saba, T. M., Gregory, T. J., and Blumenstock, F. A.,** Circulating immunoreactive and bioassayable opsonic plasma fibronectin during experimental tumor growth, *Br. J. Cancer,* 41, 956, 1980.
336. **Siri, A., Carnemolla, B., Castellani, P., Balza, E., Raffanti, S., and Zardi, L.,** Increased plasma fibronectin concentrations in tumor bearing mice, *Cancer Lett.*, 21, 117, 1983.
337. **Puricelli, L., Bal de Kier Joffe, E., Eijan, A. M., Entin, D., and de Lustig, E. S.,** Levels of plasmatic fibronectin in mice bearing adenocarcinomas of different metastasizing ability, *Cancer Lett.*, 29, 189, 1985.
338. **Humphries, M. J. and Yamada, K. M.,** Non-collagenous glycoproteins, in *Rheumatology. An Annual Review,* Vol. 10, Kuhn, K. and Krieg, T., Eds., Karger, Basel, 1986, 104.
339. **Fyrand, O. and Solum, N. O.,** Studies on cold insoluble globulin in dermatological patients. I. Immunochemical quantitation in citrated plasma from patients with increased amounts of heparin precipitable fraction (HPF), *Thromb. Res.*, 9, 447, 1976.
340. **Parenti, D., Carsons, S. E., Lavietes, B. B., Diamond, H. S., and Steinman, C. R.,** Fibronectin, a DNA binding protein, is elevated in systemic lupus erythematosus, *Arthritis Rheum.*, 25S, 41, 1982.

341. **Saba, T. M. and Jaffe, E.**, Plasma fibronectin (opsonic glycoprotein): its synthesis by vascular endothelial cells and role in cardiopulmonary integrity after trauma as related to reticuloendothelial function, *Am. J. Med.*, 68, 577, 1980.
342. **Blumenstock, F. A. and Saba, T. M.**, Fibronectin: a mediator of phagocytosis and its potential in treating septic shock, *Adv. Exp. Med. Biol.*, 155, 613, 1982.
343. **Goldman, A. S., Rudloff, H. B., McNamee, R., Loose, L. D., and Diluzio, N. R.**, Deficiency of plasma humoral recognition factor activity following burn injury, *J. Reticuloendothel. Soc.*, 15, 193, 1973.
344. **Saba, T. M., Kiener, J. L., and Holman, J. M.**, Fibronectin and the critically ill patient: current status, *Intensive Care Med.*, 12, 350, 1986.
345. **Saba, T. M.**, Fibronectin: role in phagocytic host defense and lung vascular integrity, in *Fibronectin in Health and Disease*, Carsons, S. E., Ed., CRC Press, Boca Raton, FL, 1989.
346. **Klockars, M., Pettersson, T., Vartio, T., Riska, H., and Vaheri, A.**, Fibronectin in exudative pleural effusions, *J. Clin. Pathol.*, 35, 723, 1982.
347. **Webb, K. S. and Lin, G. H.**, Urinary fibronectin: potential as a biomarker in prostatic cancer, *Invest. Urol.*, 17, 401, 1980.
348. **Kuusela, P., Vaheri, A., Palo, J., and Ruoslahti, E.**, Demonstration of fibronectin in human cerebrospinal fluid, *J. Lab. Clin. Med.*, 92, 595, 1978.
349. **Deverbizier, G., Beauchant, M., Chapron, A., Touchard, G., and Reiss, D.**, Fibronectin, a marker for malignant ascites, *Lancet*, 2, 1164, 1984.
350. **Hafter, R., Klaubert, W., Gollwitzer, R., von Hugo, R., and Graeff, H.**, Crosslinked fibrin derivatives and fibronectin in ascitic fluid from patients with ovarian cancer compared to ascitic fluid in liver cirrhosis, *Thromb. Res.*, 35, 53, 1984.
351. **Scholmerich, J., Volk, B. A., Kottgen, E., Ehlers, S., and Gerok, W.**, Fibronectin concentration in ascites differentiates between malignant and nonmalignant ascites, *Gastroenterology*, 87, 1160, 1984.
352. **Colli, A., Buccino, G., Cocciolo, M., Parravicini, R., Mariani, F., and Scaltrini, G.**, Diagnostic accuracy of fibronectin in the differential diagnosis of ascites, *Cancer*, 58, 2489, 1986.
353. **Runyon, B. A.**, Elevated ascitic fluid fibronectin concentration. A non-specific finding, *J. Hepatol.*, 3, 219, 1986.
354. **Fidler, I. J.**, General considerations for studies of experimental metastasis, *Methods Cancer Res.*, 15, 399, 1978.
355. **Fidler, I. J.**, Selection of successive tumour lines for metastasis, *Nature New Biol.*, 242, 148, 1973.
356. **Humphries, M. J., Olden, K., and Yamada, K. M.**, A synthetic peptide from fibronectin inhibits experimental metastasis of murine melanoma cells, *Science*, 233, 467, 1986.
357. **Humphries, M. J., Yamada, K. M., and Olden, K.**, Investigation of the biological effects of anti-cell adhesive synthetic peptides that inhibit experimental metastasis of B16-F10 murine melanoma cells, *J. Clin. Invest.*, in press.
358. **Ali, I. U. and Hynes, R. O.**, Effects of LETS glycoprotein on cell motility, *Cell*, 14, 439, 1978.
359. **Yamada, K. M., Olden, K., and Pastan, I.**, Transformation-sensitive cell surface protein: isolation, characterization, and role in cell morphology and adhesion, *Ann. N. Y. Acad. Sci.*, 312, 256, 1978.
360. **Couchman, J. R. and Rees, D. A.**, The behaviour of fibroblasts migrating from chick heart explants: changes in adhesion, locomotion and growth, and in the distribution of actomyosin and fibronectin, *J. Cell Sci.*, 39, 149, 1979.
361. **Couchman, J. R., Rees, D. A., Green, M. R., and Smith, C. G.**, Fibronectin has a dual role in locomotion and anchorage of primary chick fibroblasts and can promote entry into the division cycle, *J. Cell Biol.*, 93, 402, 1982.
362. **Nicolson, G. L., Irimura, T., Gonzalez, R., and Ruoslahti, E.**, The role of fibronectin in adhesion of metastatic melanoma cells to endothelial cells and their basal lamina, *Exp. Cell Res.*, 135, 461, 1981.
363. **Gehlsen, K. R., Argraves, W. S., Pierschbacher, M. D., and Ruoslahti, E.**, Inhibition of in vitro tumor cell invasion by Arg-Gly-Asp-containing synthetic peptides, *J. Cell Biol.*, 106, 925, 1987.
364. **Chen, W. T., Olden, K., Bernard, B. A., and Chu, F. F.**, Expression of transformation-associated protease(s) that degrade fibronectin at cell contact sites, *J. Cell Biol.*, 98, 1546, 1984.
365. **Fairbairn, S., Gilbert, R., Ojakian, G., Schwimmer, R., and Quigley, J. P.**, The extracellular matrix of normal chick embryo fibroblasts: its effect on transformed chick fibroblasts and its proteolytic degradation by the transformants, *J. Cell Biol.*, 101, 1790, 1985.
366. **Sas, D. F., McCarthy, J. B., and Furcht, L. T.**, Clearing and release of basement membrane proteins from substrates by metastatic tumor cell variants, *Cancer Res.*, 46, 3082, 1986.
367. **Chen, J. M. and Chen, W. T.**, Fibronectin-degrading proteases from the membranes of transformed cells, *Cell*, 48, 193, 1987.
368. **Pollanen, J., Hedman, K., Nielsen, L. S., Dano, K., and Vaheri, A.**, Ultrastructural localization of plasma membrane-associated urokinase-type plasminogen activator at focal contacts, *J. Cell Biol.*, 106, 87, 1988.

369. **Hirst, R., Horwitz, A., Buck, C., and Rohrschneider, L.,** Phosphorylation of the fibronectin receptor complex in cells transformed by oncogenes that encode tyrosine kinases, *Proc. Natl. Acad. Sci. U.S.A.,* 83, 6470, 1986.
370. **Duband, J. L., Dufour, S., Yamada, K. M., and Thiery, J. P.,** The migratory behavior of avian embryonic cells does not require phosphorylation of the fibronectin-receptor complex, *FEBS Lett.,* in press.
371. **Iwamoto, Y., Robey, F. A., Graf, J. Sasaki, M., Kleinman, H. K., Yamada, Y., and Martin, G. R.,** YIGSR, a synthetic laminin pentapeptide, inhibits experimental metastasis formation, *Science,* 238, 1132, 1987.
372. **Terranova, V. P., Liotta, L. A., Russo, R. G., and Martin, G. R.,** Role of laminin in the attachment and metastasis of murine tumor cells, *Cancer Res.,* 42, 2265, 1982.
373. **Barsky, S. H., Rao, C. N., Williams, J. E., and Liotta, L. A.,** Laminin molecular domains which alter metastasis in a murine model, *J. Clin. Invest.,* 74, 843, 1984.
374. **Malinoff, H. L., McCoy, J. P., Varani, J., and Wicha, M. S.,** Metastatic potential of murine fibrosarcoma cells is influenced by cell surface laminin, *Int. J. Cancer,* 33, 651, 1984.
375. **McCoy, J. P., Lloyd, R. V., Wicha, M. S., and Varani, J.,** Identification of a laminin-like substance on the surface of high-malignant murine fibrosarcoma cells, *J. Cell Sci.,* 65, 139, 1984.
376. **Haverstick, D. M., Cowan, J. F., Yamada, K. M., and Santoro, S. A.,** Inhibition of platelet adhesion to fibronectin, fibrinogen, and von Willebrand factor substrates by a synthetic tetrapeptide derived from the cell-binding domain of fibronectin, *Blood,* 66, 946, 1985.
377. **Humphries, M. J.,** unpublished.
378. **Thiery, J. P.,** personal communication.
379. **Chen, W. T.,** personal communication.
380. **Saga, S., Chen, W. T., and Yamada, K. M.,** submitted.
381. **Chen, W. T. and Yamada, K. M.,** unpublished data.
382. **Horwitz, A. F.,** personal communication.

Chapter 12

FIBRONECTIN IN THE SKIN

Richard A. F. Clark, M.D.

TABLE OF CONTENTS

I. INTRODUCTION

Our understanding of the role of fibronectin in the skin has steadily advanced since I last reviewed this topic in 1983.[1] For example, we now know that epidermal cells can deposit fibronectin in the extracellular matrix *in vivo*,[2,3] a function that was once relegated to mesenchymal cells. In addition, epidermal cells under certain circumstances can and do express fibronectin receptors *in vivo*[4] as well as *in vitro*.[5] Furthermore, the signals that might control *in vivo* epidermal cell expression of fibronectin are becoming elucidated.[6,7] Presumably the epidermal cells use fibronectin receptors to adhere to and migrate over fibronectin-coated surfaces[8,9] or to phagocytose fibronectin-coated particles,[10] Thus, recent data demonstrate that epidermal cells, like fibroblasts, can synthesize and deposit fibronectin in the extracellular matrix and use the endogenously produced fibronectin to facilitate reactive processes, such as migration and phagocytosis.

Besides advances in our understanding of fibronectin and keratinocyte pathobiology, over recent years the skin has continued to be used for wound repair investigations. Through observing the temporal and spatial relationships of fibronectin expression in a healing cutaneous wound with the expression of other molecules and with the influx and phenotypic modulation of a variety of inflammatory and tissue cells, many functions of and controls for fibronectin during tissue organization are being revealed.[11]

Having briefly highlighted some recent advances in our understanding of fibronectin in the skin, I will now attempt a more methodical review of the subject.

II. EMBRYOGENESIS AND MORPHOGENESIS

Much work demonstrating an association of fibronectin with migrating and proliferating cell populations during tissue organization has come from investigations in the field of developmental biology. A review of fibronectin in development is presented in Chapter 2. Here I will limit my review to those studies that pertain to the eventual development of the skin. Linder et al.[12] first noted that fibronectin occurred in various epithelial-mesenchymal boundaries during development. In many instances the expression of fibronectin seemed to be a property of primitive mesenchymal cells that was acquired early in morphogenesis and that often terminated during the differentiation of these cells. A parallel can be drawn between these observations and the appearance of fibronectin at the epidermal-stromal junction during wound repair reepithelialization. This is one of many examples where ontogeny appears to be recapitulated during wound repair.

Critchley et al.[13] studied the deposition of fibronectin in the ectoderm of gastrulating chick embryos. The authors proposed that fibronectin functioned to facilitate ectodermal translocation in early development. Similar observations have been noted in the mouse embryo[14] and the sea urchin embryo[15] and have been confirmed and extended in the chick embryo.[16] The concept that increased tissue fibronectin may have functional significance in these situations is substantiated by the fact that antibodies to fibronectin prevented the invagination of mesodermal cells during gastrulation in amphibian embryos.[17]

During development of the skin, fibronectin is observed at the epidermal-dermal junction[18] and around blood vessels[19] of human fetuses, in the feather tracts of chick embryos,[20,21] and around the developing hair shaft of mouse embryos.[22]

Fibronectin in the dorsal skin of the chick embryo is markedly concentrated in areas of the mesenchyme near the epidermal-dermal junction where prospective feather tracts are formed by the condensation of mesenchymal cells.[20,21] Interestingly, before the condensation, type I collagen is seen throughout this upper mesenchyme but disappears as condensation occurs. Thus, feather tract formation has three components: disappearance of type I collagen, appearance of fibronectin, and the localization and subsequent migration downward of mesenchymal cells. The invagination of epidermis to form feather rudiments follows with a continuing concentration of fibronectin around the follicular dermis and in the dermal core of the developing feather. [21] Although increased amounts of fibronectin occurs in the follicular dermis around the developing hair shaft in fetal rat skin,[22] no increase in fibronectin was noted under the hair germ, the area that probably corresponds to the feather tract in chick embryos.

Fine et al.[18] found that fibronectin occurred at the epidermal-dermal junction during the entire duration of fetal development examined (6 to 25 weeks.) Although they did not comment on this finding, the presence of fibronectin at the epidermal-dermal junction during skin development is consistent with the presence of fibronectin at many epithelial-stromal boundaries during development.[12] In contrast, fibronectin is not present in the basement membranes of normal adult skin.[23] The expression of fibronectin in the basement membrane zone of cutaneous blood vessels also varies dramatically with age.[19] During 14 to 18-weeks gestation, prominent amounts of fibronectin are observed in the microvasculature of human skin, while less fibronectin appears around newborn blood vessels, and little fibronectin is seen in normal adult dermal vessels. Thus, the presence of fibronectin at epithelial-stromal boundaries in fetal skin suggests that fibronectin may play a critical role in cutaneous development, perhaps through facilitating epithelial-mesenchymal interactions. This topic has been extensively reviewed by Sengel.[24]

III. WOUND REPAIR

Based on current information, fibronectin probably acts as a chemotactic factor for peripheral blood monocytes and as a nonspecific opsonin of debris during the inflammatory phase of wound repair. During granulation tissue formation, fibronectin provides a substrate for cell movement, a chemotactic or haptotic signal for such movement, and a nonspecific opsonin of debris that obstructs such movement. Epidermal cells and fibroblasts appear capable of phagocytizing fibronectin-coated material, and thus these tissue cells may be able to remove obstructive debris as they move. Fibronectin can act as a scaffold for new matrix deposition *in vitro* and likewise may act so for new matrix deposition *in vivo*. Finally, several authors have suggested that fibronectin may act as a growth factor during tissue repair. In the next five sections, I will briefly discuss the evidence supporting the contention that fibronectin has multiple roles during wound repair.

A. FIBRONECTIN AS A CHEMOTACTIC FACTOR

Monocytes are attracted to certain fragments of fibronectin containing the cell-binding domain. [25,26] The chemotactic activity of fibronectin fragments for monocytes is reminiscent of the chemotactic activity of collagen[27] and elastin[28] fragments for human monocytes. The fragments from all of these matrix proteins appear quite selective for peripheral blood monocytes over polymorphonuclear leukocytes or lymphocytes. Thus, connective tissue breakdown at a site of injury would call for the gradual accumulation of monocytes. The selective chemotactic activity of extracellular matrix fragments for peripheral blood monocytes might also explain why there is gradual accumulation of monocytes in areas of chronic inflammation.

Fibronectin and/or its fragments have also been found to stimulate fibroblast[29-31] and endothelial cell[32] movement by phagokinetic assays[29] and modified Boyden chamber assays.[30-32] Whether this stimulated movement is secondary to true chemotaxis (cells migrating in response to a chemical gradient in solution) or haptotaxis (cells migrating in response to a chemical gradient on a surface) has not been completely established. Nevertheless, it appears that fibronectin gradients in a wound might elicit the ingrowth of both fibroblasts and endothelial cells.

Fibronectin gradients would presumably derive from blood extravasation and coagulation[33] and from platelet release.[34] Fibrin probably acts *in vivo* as a lattice that incorporates fibronectin. Circulating plasma ^{125}I fibronectin concentrates in rat abdominal incisions within 15 to 20 min of wounding;[35] fibronectin appears by immunofluorescence in the extravascular clot of guinea pig skin wounds within 3 h after injury;[36] fibronectin drains from deep wounds in patients at levels exceeding that found in the plasma;[37] and plasma fibronectin concentrates in delayed hypersensitivity skin test sites with fibrin as determined by immunofluorescence and radioisotopic techniques.[38] In fact, the accumulation of fibrillar fibronectin in delayed hypersensitivity reactions is dependent of the deposition of fibrin as patients with afibrinogenemia fail to deposit either fibrin or fibrillar fibronectin in positive skin test sites.[39] Thus, gradients of fibronectin might form as blood is clotted and the coagulum nearest blood vessels lysed by proteases. Fibronectin fragments released by plasma- or cell-derived proteases such as plasmin,[40] cathepsin G,[41] or elastase,[42] respectively, might provide additional chemotactic signals for cell influx.

B. FIBRONECTIN AS AN OPSONIN

From the known *in vitro* binding properties of fibronectin, opsonization of collagen (especially denatured forms), fibrin, DNA, actin, complement coated particles (especially when coated with C1q which has a collagenous domain), and certain bacteria might be predicted. Fibronectin, in fact, opsonizes sheep erythrocytes, latex beads, or lipid emulsions for monocyte/macrophage phagocytoses, if the test particles are first coated with gelatin (denatured collagen).[43-46] Gelatin-coated latex beads can also be opsonized with fibronectin for phagocytoses by human fibroblasts[47] and human epidermal cells.[4,5] In addition, intact human plasma fibronectin binds directly to a variety of live microorganisms without cofactors such as gelatin. Microorganisms to which fibronectin binds includes: protein-A containing strains of *Staphylococcus aureus*;[48] clinical strains of *S. epidermidis* and protein A-deficient *S. aureus*,[49] *Streptococcus pneumoniae*,[50] some strains of *Pseudomonas aeruginosa*,[51] *Treponema pallidum*,[52] and influenza, parainfluenza, and mumps viruses.[53] However, since subsequent incubation of some fibronectin-coated microbes with various human phagocytes promotes relatively little phagocytosis,[49,54] the opsonic value of fibronectin-microorganism interactions is questionable.

Although fibronectin appears to have a limited capacity to act directly as an opsonin, fibronectin has been demonstrated to promote opsonic activity of other systems. For instance, Czop[55] has shown that a 180-kDa fragment of fibronectin greatly enhances human peripheral blood monocyte phagocytoses of particles which bind to β-glucan receptors of monocytes. Yeasts exemplify particles with surface β-glucan.[55] In 1980 Bevilacqua et al.[56] showed that fibronectin-adherent monocytes phagocytosed IgG-opsonized erythrocytes much more avidly than did glass-adherent monocytes. In addition, Pommier et al.[57] demonstrated that fibronectin activates monocytes to phagocytose complement-coated (C3b-coated) particles as well as IgG-coated particles. Monocytes either binding to a fibronectin-coated surface or in a fibronectin-containing medium developed the capacity to phagocytose complement-coated particles. Thus, incubation of fibronectin with monocytes has a direct effect on the cells which leads to increased phagocytosis.[58] The mechanisms(s) by which fibronectin enhances monocyte phagocytoses of particles having β-glucan, IgG, or complement on their surface is unknown. Nevertheless, monocytes infiltrating tissue sites of inflammation, such as delayed-type hypersensitivity reaction, are coated with fibronectin[39] and thereby might more readily phagocytose β-glucan containing particles and IgG and/or complement-coated particles within these inflammatory sites. Thus, fibronectin may facilitate debridement of a wound by its ability to promote monocyte/macrophage, fibroblast, and epidermal cell phagocytoses.

C. FIBRONECTIN AS A SUBSTRATUM FOR CELL MOVEMENT

Fibronectin has the capacity to bind, simultaneously, cells through the RGDS binding domain and extracellular matrix through heparin-, fibrin-, and collagen-binding domains. Many *in vitro* studies have shown that fibronectin can mediate the binding of human fibroblasts,[59] epidermal cells,[5,8,9] endothelial cells,[60-62] and monocyte/macrophages[56,63] to various extracellular matrix substrata. Fibronectin is a component of the initial plasma clot in wounds, and is cross-linked both to itself and to fibrin by plasma transglutaminase (coagulation factor XIIIa).[64,65] Monocyte/macrophage penetration of a fibrin gel *in vitro* seems to be regulated by transglutaminase cross-linking and the presence or absence of fibronectin.[66,67] In addition, fibronectin is required for fibroblast attachment to fibrin, and cross-linking of fibronectin to fibrin by factor XIIIa augments fibroblast attachment.[68]

Cutaneous wound repair models have provided much information regarding associations between fibronectin and migrating cell populations.[4,11,19,36,69-79] Whether full-thickness wounds are made in the skin of guinea pigs or domestic pigs (References 36 and 79, respectively) or sponges are implanted subcutaneously in mice,[69] fibronectin appears early in granulation tissue formation together with the invading fibroblasts. Types III and I collagen appear 2 to 7 d later.[69] Fibronectin matrix is assembled first on the surface of randomly oriented fibroblasts in 5-d wounds and then over the next 2 d the fibronectin fibrils become linked.[79] During the same period of time the fibroblasts become aligned in a parallel radial array across the wound and assume the actin bundle-rich myofibroblast phenotype. In this model, wound contraction commences between days 7 and 10 after injury. After wound contraction is accomplished and as collagen is organized into larger bundles in wounds older than 2 weeks, both the fibroblasts and fibronectin disappear.[69]

In ultrastructural studies we have also noted a striking coordinance in 7- and 9-day-old guinea pig wounds between fibronectin fibrils and myofibroblasts.[75] The temporal and spatial coincidence of fibronectin and myofibroblasts in wound healing has led us to speculate that a functional relationship exists between these components of granulation tissue resulting in wound contraction. Since well-spread fibroblasts in tissue culture exhibit a colinear distribution of extracellular fibronectin fibrils and intracellular actin microfilaments,[80-82] both of which insert at the plasma membrane into a vinculin-containing plaque,[83] this same colinear distribution was sought and found in 7- and 9-d granulation tissue of healing excisional wounds in guinea pig skin.[75,76] Thus, the fibronectin attachment to the plasma membranes of fibroblasts, their colinear distribution with intracellular actin bundles, their linkages to adjacent fibroblasts and collagen bundles, and the synchrony of fibronexus assembly with the initiation of wound contraction suggest that the fibronexus complex may serve as a granulation tissue microtendon that transmits the collective forces generated by myofibroblast throughout the granulation tissue to the wound margins and thereby affect wound contraction.

Additional investigations have demonstrated that fibronectin is markedly accentuated in blood vessels around the perimeter of excisional wounds that contain proliferating endothelial cells and in the neovasculature of the granulation tissue.[71] Furthermore, using reciprocal species-specific antibodies, we demonstrated that the increased vessel wall fibronectin observed in proliferating blood vessels around wounded cutaneous rat grafts implanted on the flanks of athymic mice was rat, not mouse, fibronectin.[73] In other words, the increased microvascular fibronectin was produced *in situ* presumably by proliferating endothelial cells, and not passively absorbed plasma fibronectin. Since fibronectin has been shown to mediate the adherence of cultured microvascular cells,[62] perhaps blood vessel wall fibronectin, which appears subjacent to the proliferating endothelium of the granulation tissue neovasculature, facilitates the proliferation and movement of the endothelial cells.[84]

Fibronectin and fibrin also appear to provide a provisional matrix for epidermal cell migration during guinea pig,[70] rat,[73] and domestic pig[136] cutaneous wound repair. Once wounds are reepithelialized in any of the animal models studied, the subepithelial fibronectin and fibrin progressively disappear. Using the same xenograft system described above and reciprocal species-specific antibodies,[73] the fibronectin under the migrating epithelium was shown to be plasma-derived early in reepithelialization (1 to 4 d) and both plasma- and *in situ*-derived later in reepithelialization (4 to 10 d). More recent studies have shown not only that cultured epidermal cells deposit fibronectin between themselves and the culture dish,[85,86] but also that human keratinocytes are capable of depositing fibronectin in the extracellular matrix *in vivo*.[3] In the latter experiments, cultured human keratinocytes were injected subcutaneously into athymic mice. Using anti-human fibronectin antibodies that did not recognize mouse fibronectin, we found that human fibronectin was deposited immediately adjacent to organizing aggregates of human keratinocytes. Since keratinocytes were the only human cells present in this system as judged by Hoechst dye nuclear staining patterns and keratin immunofluorescence, the human fibronectin must have been deposited by the epidermal cells. After the epidermal cell aggregates organized into a epidermal inclusion cyst (2 to 5 weeks after injection), the fibronectin was replaced with the normal basement membrane proteins, laminin and type IV collagen. We believe that the organization of epidermal inclusion cysts in this model parallels the reepithelialization of a cutaneous wound. Thus, the *in situ*-derived fibronectin beneath the migrating epithelial tongue probably derived from the overlying epidermal cells.

Although several laboratories have demonstrated that cultured keratinocytes can adhere to and translocate on fibronectin-coated surfaces[8,9] and phagocytoses fibronectin-coated particles,[10] Grinnell and his co-workers have shown that freshly isolated keratinocytes from normal skin do not in fact adhere to fibronectin-coated surfaces until held in culture for 2 to 5 d.[5] In contrast, keratinocytes freshly isolated from the migrating epidermal tongue of a healing wound do adhere to fibronectin coated dishes.[4] More recently, Toda et al.[87] have shown formally that keratinocytes freshly isolated from normal skin do not express fibronectin receptors on their surface but do develop plasma membrane fibronectin receptors after 2 d in culture. Thus, resident epidermal cells do not express fibronectin receptors. Perhaps some signal arises after cutaneous injury which induces keratinocytes to express fibronectin receptors as part of transforming into a migrating cell phenotype. Once epidermal cells express fibronectin receptors, acquire a f-actin motor apparatus, and become detached from neighboring cells, they would be capable of moving over fibronectin-coated surfaces and phagocytosing fibronectin-coated debris in their path as they repave a wound surface.

These data taken together suggest that fibronectin may provide a provisional matrix for cell translocation. A more direct test of this hypothesis was provided by Nishida et al.[88] and by Donaldson et al.[77] These authors have shown that fibronectin augments epithelial cell movement across a denuded rabbit cornea or across a glass coverslip juxtaposed to a reepithelializing newt limb wound, respectively, and that either antibodies to the fibronectin cell-binding site or RGDS peptides interfere with such movement. In fact, Nishida et al.[89] have demonstrated that autologous plasma fibronectin applied to chronic corneal ulcers potentiates corneal reepithelialization. Thus, fibronectin may play a critical role in wound repair by providing a matrix over which cells can translocate.

D. FIBRONECTIN AS A MATRIX SCAFFOLD

In addition to attracting cells to the wound site and providing migratory cells with a provisional matrix, fibronectin may also play a direct role in initial matrix deposition and assembly. This speculation follows from both immunohistochemical studies on healing wounds and *in vitro* fibroblast culture experiments. Light microscopic studies demonstrate that fibroblasts in dermal wound healing are surrounded by a fibronectin-rich matrix during granulation tissue formation.[36,74] Electron microscopic studies of healing experimental dermal wounds demonstrated myofibroblasts with abundant surface fibronectin aligned with intracellular stress fiber.[75,76] Although double-label immunoelectron microscopy to localize collagen types I or III and fibronectin has not been performed in healing wounds, the normal dermis contains fibers staining with antibodies to both fibronectin and the aminoterminal propeptide of collagen type I.[90]

The matrix deposited by cultured fibroblasts contains composite fibrils containing fibronectin, collagen types I, III, VI, sulfated proteoglycans, including heparan and chondroitin sulfate, and other components.[91,94] Preventing formation of a fibronectin matrix by human fibroblasts *in vitro* inhibits deposition of collagen types I and III.[95] Of course, the fibroblast requirement for a fibronectin matrix to deposit types I and III collagen could be an artifact of tissue culture. However, the unfilled space of a tissue culture dish, and the serum supplement typically added to fibroblast culture medium, which contains platelet-derived polypeptide growth factors,[96] mimic rather well an open wound bed suggesting that the fibroblast tissue culture model is probably relevant to wound repair.

Fibronectin deposition in fibroblast culture appears to occur in at least two stages. Initially, fibronectin is deposited between the freshly plated fibroblasts and the underlying tissue culture dish where it appears to promote cell attachment and spreading.[59] Fibronectin is subsequently organized beneath the spreading cells in the form of "footprints" and linear strands closely associated with the fibronectin receptor.[97-101] Later, following cell proliferation, an extensive three-dimensional pericellular matrix surround the fibroblasts.[91,102,103] Because the pericellular matrix is between and above the cells and not simply interposed between the cells and a planar surface, the pericellular matrix, topologically if not biochemically, closely resembles the matrix deposited during early wound repair.

The fibronectin matrix of cultured fibroblasts appears to serve as a scaffold for the deposition of types I and III collagen[95] and type VI collagen.[94] The evidence that fibronectin is pivotal in early extracellular matrix deposition has prompted studies on the molecular features of fibronectin matrix assembly. Large fibronectin fragments lacking the aminoterminus and added exogenously to an animal or to fibroblast cultures are not incorporated into extracellular matrix.[104,105] In addition, endogenous fibroblast-fibronectin matrix assembly is effectively inhibited by both antibodies to the aminoterminal 25-kDa domain and fragments containing this sequence.[106] The RGDS fibroblastic cell-binding domain may be critical for initiating matrix assembly,[106] but once matrix assembly has been initiated, the cell-binding domain appears unnecessary for further matrix development.[107] Other investigators, however, have presented evidence that suggests that initial fibronectin fibril formation is dependent on cell surface gangliosides rather than the RGDS-containing cell-binding domain.[108,109]

E. FIBRONECTIN AS A GROWTH PROMOTER

Plasma fibronectin has been noted to facilitate fibroblast growth *in vitro* and in this capacity has appeared to act as a progression factor.[110] Recent data, however, have demonstrated that most preparations of plasma fibronectin are contaminated with TGF-beta.[111] Since TGF-beta augments fibroblast growth in the presence of PDGF,[112] the findings of Bitterman et al.[110] may be an artifact of TGF-beta contaminating their plasma fibronectin preparations. In addition, since numerous studies [113-115] have shown that TGF-beta promotes wound repair, studies showing that plasma fibronectin promotes wound repair[116] must be interpreted with caution unless the authors demonstrate that their fibronectin preparations contain no growth factors.

In conclusion, fibronectin may have several important roles in wound repair as listed above. Whereas, adding fibronectin to the cornea may promote wound repair through a direct fibronectin effect since the tears and cornea lack appreciable quantities of fibronectin; the addition of fibronectin to wounds that have access to ample blood supply seems like taking coal to Newcastle. Not only does the blood contain 300 μg /ml of fibronectin but, in addition, connective tissue cells and epidermal cells appear to synthetize and deposit great quantities of fibronectin at sites of wound repair.[36,72,73] Nevertheless, in certain disease states a systemic or local depletion of fibronectin may occur and the addition of fibronectin may be beneficial.[117,118]

IV. INFLAMMATION

At sites of psoriatic skin lesions, fibronectin occured in circumscribed areas of the epidermis.[119] The epidermal areas which contained fibronectin overlaid enlarged dermal papillae that actually projected up into the stratum corneum. In contrast, normal epidermis contained no fibronectin.[120,121] The deposition pattern of fibronectin in the stratum corneum suggested that fibronectin occurred both intra- and extracellularly.[119] Fibronectin was not prominent along the epidermal-dermal junction except in areas of elongated papillae. Here a marked increase in junctional fibronectin was noted and was associated with intensification of fibronectin around papillary capillaries. In lichen planus[122] and discoid lupus[123] lesions, a marked increase in fibronectin was noted at the epidermal-dermal junction where basal cell degeneration had occurred. Once again, fibronectin intensified in the microvasculature near lesional skin. Fibronectin also increased markedly in the involved papillae of dermatitis herpetiformis skin lesions.[124] Since fibrin deposition occurred in conjunction with the increased fibronectin, the latter was probably plasma derived. The same might be true for fibronectin in the dermal papillae of psoriatic lesions and along the epidermal-dermal junction in lichen planus and lupus lesions although concomitant staining was not done for fibrin in these studies.

In our own studies on fibronectin deposition at sites of delayed-type hypersensitivity skin reactions,[38] we observed that during the first 2 d after intradermal antigen challenge in sensitized guinea pigs, a large quantity of fibronectin appeared as fibrillar strands and minute nodular foci in association with fibrin in the extravascular space. By radioisotopic technique this interstitial accumulation of fibronectin was plasma derived. In contrast, fibronectin accumulation within blood vessels began 3 d after challenge when interstitial deposition of plasma fibronectin and fibrin had ceased and endothelial cells within the blood vessels at these sites had begun to proliferate. From these data, we postulated and later confirmed[39,72] that blood-vessel-associated fibronectin was produced *in situ* by endothelial cells.

In the first study as described above we made an excisional wound in a healed rat skin graft implanted on the back of an athymic mouse. Then using reciprocal species-specific antibodies, we demonstrated that an increased vessel wall fibronectin observed in proliferating blood vessels around the wound was rat, not mouse, fibronectin. In other words the increased microvascular fibronectin was produced *in situ*. The second study investigated the fibronectin distribution in delayed-type hypersensitivity (DTH) reactions of normals and patients with afibrinogenemia. At 48 h after challenge, large quantities of fibronectin and fibrin were deposited throughout the interstitium in DTH reaction sites of normals. However, little fibronectin and no fibrin were observed in the interstitium of afibrinogenemia DTH reaction sites. The little fibronectin present was associated with the surfaces of monocyte / macrophages. In contrast, increased blood vessel wall fibronectin was observed in sites of both normals and patients. Thus plasma-derived, interstitial fibronectin seems dependent on fibrin for its deposition, while increases in vessel wall fibronectin are independent of fibrin deposition. These data are consistent with the vessel wall fibronectin being derived from *in situ*, possible endothelial cell, production.

Extrapolating these data to the skin disorders listed above, it is likely that the increased fibronectin noted within the microvasculature near sites of inflammation is deposited by activated endothelial cells. Likewise, the marked increase of fibronectin within the thickened walls of blood vessels in skin of patients with erythropoietic protoporphyria[125] are probably secondary to multiple episodes of endothelial cell injury, subsequent activation / proliferation of these cells, and a concomitant increased synthesis and deposition of fibronectin by the rejuvenating endothelium. Even the usual amounts of fibronectin found around normal [19] and ectactic[126] blood vessels may be derived from the endothelial cells lining the vessel lumen rather than the plasma.

V. TUMORS

Basal cell epithelioma[127,128] and squamous cell carcinoma[129] have increased quantities of fibronectin. In fact, much of the fibronectin associated with basal cell epithelioma is deposited by the cancer cells themselves.[2] This information stemmed from studies in which human basal carcinoma cells were implanted into athymic mice and several days thereafter the tissue was stained for human fibronectin using an antibody that did not recognize mouse fibronectin. Large amounts of fibronectin were seen both within and surrounding the tumor aggregates. At first pass this increase seems in opposition to the extensive studies showing that loss of cell surface fibronectin is a common, though not universal, feature of transformed cells.[130,131] However, additional studies suggest that a better correlation may exist between loss of fibronectin and acquisition of metastatic potential.[132,133] Since basal cell epithelioma and cutaneous squamous cell carcinoma arising in sun-exposed areas have a low metastatic potential, the increase in fibronectin in really not discordant with previous work.

In basal cell epithelioma, fibronectin not only is present within the tumor lobule, unlike the normal epidermis where it is absent,[120,121] but also is markedly increased at the dermal-epidermal junction and in the surrounding stroma.[127,128] This finding is actually reminiscent of the increase in fibronectin that occurs along the ectodermal basement membrane just preceding lens invagination[134] and neural plate invagination.[135]

By electron microscopy technique, epithelial cell-substratum adhesion complexes have been demonstrated between basal epithelioma cells and the underlying basement membrane, as well as between human squamous carcinoma cells and their underlying basement membrane.[129] These adhesion complexes (epinexus), which are similar to fibronexus,[82] seem to replace half-desmosomes which occur along the normal epidermal-dermal junction. The epinexus is easy to distinguish from a hemidesmosome since actin filaments run into the former while tonofilments slip into the latter.[129] The epinexus, therefore, might provide a tether along which these epidermal tumors could invaginate during dermal invasion.

VI. CONCLUSION

Fibronectin in the skin appears most prominent when adjacent cells are migrating or proliferating. This is true whether embryogenesis, morphogenesis, wound repair, inflammatory conditions, or tumors are being investigated. These observed associations between fibronectin and migrating and proliferating cells are consistent with the *in vitro* data that fibronectin contains distinct and specific binding sites for cells and for extracellular matrix materials such as collagen, fibrin, glycosaminoglycans, and fibronectin itself. Since fibronectin can be rapidly deposited and degraded around the perimeter of cells, fibronectin might act as a transient, cell-regulated, tether for cell-substratum binding during processes that require cell motility such as migration, proliferation, and tissue contraction. Evidence has accumulated in skin studies that, during the tissue organization processes listed above, the cell-matrix tether function of fibronectin might apply to monocyte / macrophages, epidermal cells, endothelial cells, and fibroblasts.

ACKNOWLEDGMENTS

We thank Marialyce Austin for her expert typing assistance and are deeply grateful to the American Medical Association for allowing us to reprint parts of the article entitled

REFERENCES

1. **Clark, R. A. F.,** Fibronectin in the skin, *J. Invest. Dermatol.,* 81, 475, 1983.
2. **Grimwood, R. E., Ferris, C. F., Nielsen, L. D., Huff, J. C. and Clark, R. A. F.,** Basal cell carcinomas grown in nude mice produce and deposit fibronectin in the extracellular matrix, *J. Invest. Dermatol.* 87, 42, 1986.
3. **Grimwood, R. E., Baskin, J. B., Nielsen, L. D., Ferris, C. F., and Clark, R. A. F.,** Fibronectin extracellular matrix assembly by human epidermal cells implanted into athymic mice, *J. Invest. Dermatol.,* 90, 434, 1988.
4. **Takashima, A., Billingham, R. E., and Grinnell, F.,** Activation of rabbit keratinocyte fibronectin receptor function *in vivo* during wound healing, *J. Invest. Dermatol.,* 86, 585, 1986.
5. **Takashima, A. and Grinnell, F.,** Fibronectin-mediated keratinocyte migration and initiation of fibronectin receptor function *in vitro, J. Invest. Dermatol.,* 85, 304, 1985.
6. **Nickoloff, B. J., Mitra, R. S., Riser, B. L., Dixit, V. M., and Varani, J.,** Modulation of keratinocyte motility, *Am. J. Path.,* 132, 543, 1988.
7. **Wikner, N. E., Baskin, J. B., Nielsen, L. D., and Clark, R. A. F.,** Transforming growth factor-beta stimulates the expression of fibronectin by human keratinocytes, *J. Invest. Dermatol.,* 91, 207, 1988.
8. **Clark, R. A. F., Folkvord, J. M., and Wertz, R. L.,** Fibronectin as well as other extracellular matrix proteins mediate human keratinocyte adherence, *J. Invest. Dermatol.,* 84, 378, 1985.
9. **O'Keefe, E. J., Payne, R. E., Jr., Russell, N., and Woodley, D. T.,** Spreading and enhanced motility of human keratinocytes on fibronectin, *J. Invest. Dermatol.,* 85, 125, 1985.
10. **Takashima, A. and Grinnell, F.,** Human keratinocyte adhesion and phagocytosis promoted by fibronectin, *J. Invest. Dermatol.,* 83, 352, 1984.
11. **Clark, R. A. F.** Potential roles of fibronectin in cutaneous wound repair, *Arch. Dermatol.,* 124, 201, 1988.
12. **Linder, E., Vaheri, A., Ruoslahti, E., and Wartiovaara, J.,** Distribution of fibroblast surface antigen in the developing chick embryo, *J. Exp. Med.,* 142, 41, 1975.
13. **Critchley, D. R., England, M. A., Wakely, J., and Hynes, R. O.,** Distribution of fibronectin in the ectoderma of gastrulation chick embryos, *Nature (London),* 280, 498, 1979.
14. **Wartiovaara, F., Leivo, I., and Vaheri, I.,** Expression of cell surface associated glycoprotein fibronectin in the early mouse embryo, *Dev. Biol.,* 69, 247, 1979.
15. **Spiegel, E., Burger, M., and Spiegel, M.,** Fibronectin in the developing sea urchin embryo, *J. Cell. Biol.,* 8787, 309, 1980.
16. **Duband, J. L. and Thiery, J. P.,** Appearance and distribution of fibronectin during chick embryo gastrulation and neuralation, *Dev. Biol.,* 94, 337, 1982.
17. **Boucaut, J. C., Darribere, T., Boulekbache, H., and Thiery, J. P.,** Prevention of gastrulation but not neurulation by antibodies to fibronectin in amphibian embryos, *Nature (London),* 307, 364, 1984.
18. **Fine, J-D., Smith, L. T., Holbrook, K. A., and Katz, S. I.,** The appearance of four basement membrane zone antigens in developing human fetal skin, *J. Invest. Dermatol.,* 83, 66, 1984.
19. **Tonnesen, M. G., Jenkins, D., Jr., Siegal, S. L., Lee, L. A., Huff, J. C., and Clark, R. A. F.,** Expression of fibronectin, laminin, and factor VIII-related antigen during development of the human cutaneous microvasculature, *J. Invest. Dermatol.,* 85, 564, 1985.
20. **Kitamura, K.,** Distribution of endogenous galactoside-specific lectin, fibronectin and type I and III collagens during dermal condensation in chick embryos, *J. Embryol. Exp. Morphol.,* 65, 41, 1981.
21. **Manger, A., Demarchex, M., Herbage, D., Grimaud, J. A., Drugnet, M., Hartmann, D., and Sengel, P.,** Immunofluorescent localization of collagen types I and III, and of fibronectin during feather morphogenesis in the chick embryo, *Dev. Biol.,* 94, 93, 1982.
22. **Gibson, W. T. and Couchman, J. R.,** Fibronectin distribution during the development of fetal rat skin, *J. Invest. Dermatol.,* 81, 480, 1983.
23. **Woodley, D. T. and Briggaman, R. A.,** Reformation of the dermal-epidermal junction during wound healing, in *Molecular and Cellular Biology of Wound Repair,* Clark, R. A. F. and Henson, P. M., Eds., Plenum Press, New York, 1988, 559.
24. **Sengel, P.,** Role of extracellular matrix in the development of skin and cutaneous appendages, *Prog. Clin. Biol. Res.,* 171, 123, 1985.

25. **Norris, D. A., Clark, R. A. F., Swigart, L. M. et al.**, Fibronectin fragment(s) are chemotactic for human peripheral blood monocytes, *J. Immunol.*, 129, 1612, 1982.
26. **Clark, R. A. F., Wikner, N. E., Doherty, D. E., and Norris, D. A.**, Cryptic chemotactic activity of fibronectin for human monocytes resides in the 120 kDa fibroblastic cell-binding fragment, *J. Biol. Chem.*, 263, 12115, 1988.
27. **Postlewaite, A. E. and Kang, A. H.**, Collagen and collagen peptide-induced chemotaxis of human blood monocytes, *J. Exp. Med.*, 143, 1299, 1976.
28. **Senior, R. M., Griffin, G. L., and Mecham, R. P.**, Chemotactic activity of elastin-derived peptides, *J. Clin. Invest.*, 66, 859, 1980.
29. **Ali, U. and Hynes, R. O.**, Effect of LETS glycoprotein on cell motility, *Cell,* 14, 439, 1978.
30. **Postlethwaite, A. E., Keski-Oja, J., Balian, G., and Kang, A. H.**, Induction of fibroblast chemotaxis by fibronectin. Localization of the chemotactic region to a 140,000 molecular weight non-gelatin-binding fragment, *J. Exp. Med.*, 15, 494, 1981.
31. **Seppa, H. E. J., Yamada, K. M., Seppa, S. T., et al.**, The cell binding fragment of fibronectin is chemotactic for fibroblasts, *Cell. Biol. Int. Rep.*, 5, 813, 1981.
32. **Bowersox, J. C. and Sorgente, N.**, Chemotaxis of aortic endothelial cells in response to fibronectin, *Cancer Res.*, 42, 2547, 1982.
33. **Mosesson, M. W., and Umfleet, R. A.**, The cold insoluble globulin of human plasma. I. Purification, primary characterization, and relationship to fibrinogen and other cold insoluble fraction components, *J. Biol. Chem.*, 245, 5728, 1970.
34. **Ginsberg, M. H., Painter, R. G., Birdwell, C., and Plow, E. F.**, The detection, immunofluorescent localization, and thrombin induced release of human platelet-associated fibronectin antigen, *J. Supramol. Struct.*, 11, 167, 1979.
35. **Kaplan, J. E., Molnar, J., Saba, T. M., and Allen, C.**, Comparative disappearance and localization of isotopically labelled opsonic protein and soluble albumin following surgical trauma, *J. Reticuloendothel. Soc.*, 20, 375, 1976.
36. **Grinnell, F., Billingham, R. E., and Burgess, L.**, Distribution of fibronectin during wound healing *in vivo, J. Invest. Dermatol.*, 76, 181, 1981.
37. **Robbins, A. B., Doran, J. E., Reese, A. C., and Mansburger, A. R., Jr.**, Cold insoluble globulin levels in operative trauma: serum depletion, wound sequestration, and biological activity: an experimental and clinical study, *Ann. Surg.*, 46, 663, 1980.
38. **Clark, R. A. F., Dvorak, H. F., and Colvin, R. B.**, Fibronectin in delayed-type hypersensitivity skin reactions: associations with vessel permeability and endothelial cell activation, *J. Immunol.*, 126, 787, 1981.
39. **Clark, R. A. F., Horsburgh, C. R., Hoffman, A. A., Dvorak, H. F., Mosesson, M. W., and Colvin, R. B.**, Fibronectin deposition in delayed-type hypersensitivity reactions of normals and a patient with afibrinogenemia, *J. Clin. Invest.*, 74, 1011, 1984.
40. **Petersen, T. E., Thogersen, H. C., Skorsjengaard, K., et al.**, Partial primary structure of bovine plasma fibronectin: three types of internal homology, *Proc. Natl. Acad. Sci. U.S.A.* 80, 137, 1983.
41. **DePetro, G., Barlati, S., Vartio, T., and Vaheri, A.**, Transformation-enhancing activity of gelatin-binding fragments of fibronectin, *Proc. Natl. Acad. Sci. U.S.A.*, 78, 4965, 1981.
42. **McDonald, J. A. and Kelley, D. G.**, Degradation of fibronectin by human leukocyte elastase. Release of biologically active fragments, *J. Biol. Chem.*, 255, 8848, 1980.
43. **Rourke, F. J., Blumenstock, F. A., and Kaplan, J. E.**, Effect of fibronectin fragments on macrophage phagocytosis of gelatinized particles, *J. Immunol.*, 132, 1931, 1984.
44. **Gudewicz, P. W., Molnar, J., Lai, M. Z., et al.**, Fibronectin-mediated uptake of gelatin-coated latex particles by peritoneal macrophages, *J. Cell. Biol.*, 87, 427, 1980.
45. **van de Water, L., Schoeder, S., Crenshaw, E. B., et al.**, Phagocytosis of gelatin-latex particles by a murine macrophage line is dependent on fibronectin and heparin, *J. Cell. Biol.* 90, 32, 1981.
46. **Doran, J. E., Mansberger, A. R., Edmondson, H. T., and Reese, A. C.**, Cold insoluble globin and heparin interactions in phagocytosis by macrophage monolayers: lack of heparin requirement, *J. Reticuloendothel. Soc.*, 29, 275, 1981.
47. **Grinnell, F., and Bennett, M. H.**, Fibroblast adhesion on collagen substrata in the presence and absence of plasma fibronectin, *J. Cell. Sci.*, 48, 19, 1981.
48. **Doran, J. E. and Raynor, R. H.**, Fibronectin binding to protein A-containing staphylococci, *Infect. Immunol.*, 33, 683, 1981.
49. **Verbrugh, H. A., Peterson, P. K., Smith, D. F., et al.**, Human fibronectin binding to Staphylococcal surface protein and its relative inefficiency in promoting phagocytosis by human polymorphonuclear leukocytes, monocytes and alveolar macrophages, *Infect. Immunol.*, 33, 811, 1981.
50. **Hof, D. G., Repine, J. E., Peterson, P. K., and Hoidal, J. R.**, Phagocytosis by human alveolar macrophages and neutrophils: qualitative differences in the opsonic requirements for uptake of *Staphylococcus aureus* and *Streptococcus pneumoniae in vitro, Am. Rev. Respir. Dis.*, 121, 65, 1980.

51. **Niehaus, G. D., Schumacker, P. R., and Saba, T. M.,** Reticuloendothelial clearance of blood-borne particulates, *Ann. Surg.*, 191, 479, 1980.
52. **Peterson, K. M., Baseman, J. B., and Alderete, J. F.,** *Treponema pallidum* receptor binding protein interact with fibronectin, *J. Exp. Med.*, 157, 1958, 1983.
53. **Julkunen, I., Hautanen, A., and Keski-Oja, J.,** Interaction of viral envelope glycoproteins with fibronectin, *Infect. Immunol.*, 40, 876, 1983.
54. **van de Water, L., Destree, A. T., and Hynes, R. O.,** Fibronectin binds to some bacteria but does not promote their uptake by phagocytic cells, *Science*, 220, 201, 1983.
55. **Czop, J. K.,** Phagocytosis of particulate activators of the alternative complement pathway: effects of fibronectin, *Adv. Immunol.*, 38, 361, 1986.
56. **Bevilacqua, M. P., Amrani, D., Mosesson, M. W., and Bianco, C.,** Receptors for cold-insoluble globulin (plasma fibronectin) on human monocytes, *J. Exp. Med.*, 153, 42, 1981.
57. **Pommier, C. G., Inada, S., Fried, L. F. et al.,** Plasma fibronectin enhances phagocytosis of opsonized particles by human peripheral blood monocytes, *J. Exp. Med.*, 157, 1844, 1983.
58. **Brown, E. J.,** The role of extracellular matrix proteins in the control of phagocytosis, *J. Leuk. Biol.*, 39, 579, 1986.
59. **Grinnell, F. and Feld, M. K.,** Initial adhesion of human fibroblasts in serum-free medium: possible role of secreted fibronectin, *Cell*, 17, 117, 1979.
60. **Macarak, E. J. and Howard, P. S.,** Adhesion of endothelial cells to extracellular matrix proteins, *J. Cell. Physiol.*, 116, 76, 1983.
61. **Palotie, A., Tryggvason, K., Peltonen, L., and Seppa, H.,** Components of subendothelial aorta basement membrane. Immunohistochemical localization and role in cell attachment, *Lab. Invest.*, 49, 362, 1983.
62. **Clark, R. A. F., Folkvord, J. M., and Nielsen, L. D.,** Either exogenous or endogenous fibronectin can promote adherence of human endothelial cells, *J. Cell. Sci.*, 82, 263, 1986.
63. **Horsburgh, C. R., Clark, R. A. F., and Kirpatrick, C. H.,** Lymphokines and platelets promote human monocyte adherence to fibrinogen and fibronectin *in vitro*, *J. Leuk. Biol.*, 41, 14, 1987.
64. **Mosher, D. F.,** Cross-linking of cold-insoluble globulin by fibrin-stabilizing factor, *J. Biol. Chem.*, 250, 6614, 1975.
65. **Mosher, D. F. and Johnson, R. B.,** Specificity of fibronectin-fibrin cross-linking, *Ann. N.Y. Acad. Sci.*, 408, 583, 1983.
66. **Ciano, P. S., Colvin, R. B., Dvorak, A. M., et al.,** Macrophage migration in fibrin gel matrices, *Lab. Invest.*, 54, 62, 1986.
67. **Lanir, N., Ciano, P. S., van de Water, L., McDonagh, J., Dvorak, A. M., and Dvorak, H. F.,** Macrophage migration in fibrin gel matrices. II. Effects of clotting factor XIII, fibronectin and glycosaminoglycan content on cell migration, *J. Immunol.*, 140, 2340, 1988.
68. **Grinnell, F., Feld, M., and Minter, D.,** Fibroblast adhesion to fibrinogen and fibrin substrata: requirement
69. **Kurkinen, M., Vaheri, A., Roberts, P. R., and Steinman, S.,** Sequential appearance of fibronectin and collagen in experimental granulation tissue, *Lab. Invest.*, 43, 47, 1980.
70. **Clark, R. A. F., Lanigan, J. M., DellaPelle, P., Manseau, E., Dvorak, F. G., and Colvin, R. B.,** Fibronectin and fibrin provide a provisional matrix for epidermal cell migration during wound healing, *J. Invest. Dermatol.*, 70, 264, 1982.
71. **Clark, R. A. F., DellaPelle, P., Manseau, E., Lanigan, J. M., Dvorak, H. F., and Colvin, R. B.,** Blood vessel fibronectin increases in conjunction endothelial cel proliferation and capillary ingrowth during wound healing, *J. Invest. Dermatol.*, 79, 269, 1982.
72. **Clark, R. A. F., Quinn, J. H., Winn, H. J., Lanigan, J. M., DellaPelle, P., and Colvin, R. B.,** Fibronectin is produced by blood vessels in response to injury, *J. Exp. Med.*, 156, 646, 1982.
73. **Clark, R. A. F., Winn, H. J., Dvorak, H. F., and Colvin, R. B.,** Fibronectin beneath reepithializing epidermis *in vivo:* sources and significance, *J. Invest. Dermatol.*, 80 (Suppl), 26, 1983.
74. **Repesh, L. A., Fitzgerald, T. J., and Furcht, L. T.,** Fibronectin involvement in granulation tissue and wound healing in rabbits, *J. Histochem. Cytochem.*, 30, 351, 1982.
75. **Singer, I. I., Kawka, D. W., Kazazis, D. M., and Clark, R. A. F.,** *In vivo* codistribution of fibronectin and actin fibers in granulation tissue: immunofluorescence and electron microscope studies of the fibronexus at the myofibroblast surface, *J. Cell. Biol.*, 98, 2091, 1984.
76. **Singer, I. I., Kawka, D. W., and Kazazis, D. M.,** Localization of the fibronexus at the surface of granulation tissue myofibroblasts using double-label immunogold electron microscopy on ultrathin frozen sections, *Eur. J. Cell. Biol.*, 38, 94, 1985.
77. **Donaldson, D. J., Mahan, J. T., Hasty, D. L, and Furcht, L. T.,** Location of a fibronectin domain involved in newt epidermal cell migration, *J. Cell. Biol.*, 101, 73, 1985.
78. **Martin, D. E., Reese, M. C., Maher, J. E., and Reese, A. C.,** Tissue debris at the injury site is coated by plasma fibronectin and subsequently removed by tissue macrophages, *Arch. Dermatol.*, 124, 226, 1988.
79. **Welch, M. P., Folkvord, J. M., and Clark, R. A. F.,** The relationship of fibroblast phenotype and matrix assembly to wound contraction, *Clin. Res.*, 36, 254A, 1988.

80. **Hynes, R. O. and Destree, A. T.,** Relationships between fibronectin (LETS protein) and actin, *Cell,* 15, 875, 1978.
81. **Hynes, R. O., Destree, A. T., and Wagner, D. D.,** Relationships between microfilaments, cell-substratum, adhesion and fibronectin, *Cold Spring Harbor Symp. Quant. Biol.,* 46, 659, 1981.
82. **Singer, I. I.,** The fibronexus: a transmembrane association of fibronectin-containing fibers and bundles of 5 nm microfilaments in hamster and human fibroblasts, *Cell,* 16, 675, 1979.
83. **Singer, I. I. and Paradiso, P. R.,** A transmembrane relationship between fibronectin and vinculin (130 kd): serum modulation in normal and transformed hamster fibroblasts, *Cell,* 24, 481, 1981.
84. **Ausprunk, D. H. and Folkman, J.,** Migration and proliferation of endothelial cells in preformed and newly formed blood vessels during tumor angiogenesis, *Microvasc. Res.,* 14, 53, 1977.
85. **O'Keefe, E. J., Woodley, D. T., Castillo, G., Russell, N., and Payne, R. E., Jr.,** Production of soluble and cell-associated fibronectin by cultured keratinocytes, *J. Invest. Dermatol.* 82, 150, 1984.
86. **Kubo, M., Norris, D. A., Howell, S. E., Ryan, S. R., and Clark, R. A. F.,** Human keratinocytes synthesize, secrete, and deposit fibronectin in the pericellular matrix, *J. Invest. Dermatol.,* 82, 580, 1984.
87. **Toda, K-I., Tuan, T-L., Brown, P. J., and Grinnell, F.,** Fibronectin receptors of human keratinocytes and their expression during cell culture, *J. Cell. Biol.,* 105, 3097, 1987.
88. **Nishida, T., Nakagawa, S., Awate, T., et al.,** Fibronectin promotes epithelial migration of cultured rabbit cornea *in situ, J. Cell. Biol.,* 97, 1653, 1983.
89. **Nishida, T., Ohashi, Y., Awata, T., et al.,** Fibronectin. A new therapy for corneal trophic ulcer, *Arch. Ophthalmol.,* 101, 1046, 1983.
90. **Fleischmajer, R., Timpl, R., Tuderman, L., et al.,** Ultrastructural identification of extension amino-peptides of type I and III collagens in human skin, *Proc. Natl. Acad. Sci. U.S.A.,* 78, 7360, 1981.
91. **Hedman, K., Kurkinen, M., Alitalo, K., et al.,** Isolation of the pericellular matrix of human fibroblast cultures, *J. Cell. Biol.,* 81, 83, 1979.
92. **Hedman, K., Johansson, S., Vartio, T., et al.,** Structure of the pericellular matrix: association of heparan and chondroitin sulfate with fibronectin-procollagen fibers, *Cell,* 28, 663, 1982.
93. **Hedman, K., Vartio, T., Johansson, S., et al.,** Integrity of the pericellular fibronectin matrix of fibroblasts is independent of sulfated glycosaminoglycans, *EMBO J.,* 3, 581, 1984.
94. **Carter, W. G.,** The role of intermolecular disulfide bonding in deposition of GP 140 in the extracellular matrix, *J. Cell. Biol.,* 99, 105, 1984.
95. **McDonald, J. A., Kelley, D. G., and Broekelmann, T. J.,** Role of fibronectin in collagen deposition: Fab[1] to the gelatin-binding domain of fibronectin inhibits both fibronectin and collagen organization in fibroblast extracellular matrix, *J. Cell. Biol.,* 92, 485, 1982.
96. **Ross, R., Raines, E. W., and Bowen-Pope, D. F.,** The biology of platelet-derived growth factor, *Cell,* 46, 155, 1986.
97. **Avnur, Z. and Geiger, B.,** The removal of extracellular fibronectin from areas of cell-substrate contact, *Cell,* 25, 121, 1981.
98. **Chen, W. T., Hasegawa, E., Hasegawa, T., et al.,** Development of cell surface linkage complexes in cultured fibroblasts, *J. Cell. Biol.,* 100, 1103, 1985.
99. **Damskey, C. H., Knudsen, K. A., Bradley, D., et al.,** Distribution of the cell substratum attachment (CSAT) antigen on myogenic and fibroblastic cells in culture, *J. Cell. Biol.,* 100, 1528, 1985.
100. **Chen, W-T., Wang, J., Hasegawa, T., Yamada, S. S., and Yamada, K. M.,** Regulation of fibronectin receptor distribution by transformation, exogenous fibronectin and synthetic peptides, *J. Cell. Biol.,* 103, 1649, 1986.
101. **Singer, I. I., Scott, S., Kawka, D. W., Kazazis, D. M., Gailit, J., and Ruoslahti, E.,** Cell surface distribution of fibronectin and vitronectin receptors depends on substrate composition and extracellular matrix accumulation, *J. Cell. Biol.,* 106, 2171, 1988.
102. **Furcht, L. T., Wendelschafer, C., Mosher, D. F., and Foidart, J. M.,** An axial periodic fibrillar arrangement of antigenic determinants for fibronectin and procollagen on ascorbate treated human fibroblasts, *J. Supramol. Struct.,* 13, 15, 1980.
103. **Irish, P. S. and Hasty, D. L.,** Immunocytochemical localization of fibronectin in human fibroblast cultures using a cell surface replica technique, *J. Histochem. Cytochem.,* 31, 69, 1983.
104. **Oh, E., Pierschbacher, M., and Rouslahti, E.,** Deposition of plasma fibronectin in tissues, *Proc. Natl. Acad. Sci. U.S.A.,* 78, 3218, 1981.
105. **Millis, A. J., Hoyle, M., Mann, D. M., and Brennan, M. J.,** Incorporation of cellular and plasma (fibronectin) into smooth muscle cell extracellular matrix *in vitro, Proc. Natl. Acad. Sci. U.S.A.* 82, 2746, 1985.
106. **McDonald, J. A., Quade, B. J., Broekelmann, T. J., et al.,** Both the cell adhesive domain and an aminoterminal matrix assembly domain participate in fibronectin assembly into fibroblast pericellular matrix, *J. Biol. Chem.,* 262, 2957, 1987.
107. **McKeown-Longo, P. J. and Mosher, D. F.,** Interaction of the 70,000-mol-wt aminoterminal fragment of fibronectin with the matrix-assembly receptor of fibroblasts, *J. Cell. Biol.,* 100, 364, 1985.

108. **Spiegel, S., Yamada, K. M., Hom, B. E., Moss, J., and Fishman, P. H.,** Fluorescent gangliosides as probes for the retention and organization of fibronectin by ganglioside-deficient mouse cells, *J. Cell. Biol.,* 100, 721, 1985.
109. **Spiegel, S., Yamada, K. M., Hom, B. E., Moss, J., and Fishman, P. H.,** Fibrillar organization of fibronectin is expressed coordinately with cell surface gangliosides in a variant murine fibroblast, *J. Cell. Biol.,* 102, 1896, 1986.
110. **Bitterman, P. B., Rennard, S. I., Aderbert, S., and Crystal, R. G.,** Role of fibronectin as a growth factor for fibroblasts, *J. Cell Biol.,* 97, 1925, 1983.
111. **Fava, R. A. and McClure, D. B.,** Fibronectin-associated transforming growth factor, *J. Cell. Physiol.,* 131, 184, 1987.
112. **Roberts, A. B., Anzano, M. A., and Wakefield, L. M.,** Type β-transforming growth factor: a bifunctional regulator of cellular growth, *Proc. Natl. Acad. Sci. U.S.A.,* 82, 119, 1985.
113. **Sporn, M. B., Roberts, A. B., Shull, J. H., et al.,** Polypeptide transforming growth factors isolated from bovine sources and used for wound healing *in vivo, Science,* 219, 1329, 1983.
114. **Roberts, A. B., Sporn, M. B., Assoian, R. K., et al.,** Transforming growth factor type β: rapid induction of fibrosis and angiogenesis *in vivo* and stimulation of collagen formation *in vitro, Proc. Natl. Acad. Sci. U.S.A.,* 83, 4167, 1986.
115. **Lawrence, W. T., Norton, J. A., Sporn, M. B., et al.,** The reversal of an Adriamycin induced healing impairment with chemoattractants and growth factors, *Ann. Surg.,* 203, 142, 1986.
116. **Cheng, C. Y., Martin, D. E., Leggett, C. G., et al.,** Fibronectin enhances healing of excised wounds in rats, *Arch. Dermatol.,* 124, 221, 1988.
117. **Saba, T. M., Blumenstock, F. A., and Bernard, H.,** Cryoprecipitate reversal of opsonic alpha-2-surface binding glycoprotein deficiencies in septic surgical trauma patients, *Science,* 201, 622, 1978.
118. **Lundsgaard-Hansen, P., Doran, J. E., Ruli, E., et al.,** Purified fibronectin administration to patients with severe abdominal infections: a controlled study, *Ann. Surg.,* 202, 745, 1985.
119. **Fyrand, O.,** Studies on fibronectin in the skin. II. Indirect immunofluorescence studies in psoriasis vulgaris, *Arch. Dermatol. Res.,* 226, 33, 1979.
120. **Fyrand, O.,** Studies of fibronectin in the skin. I. Indirect immunofluorescence studies in normal human skin, *Br. J. Dermatol.,* 101, 263, 1979.
121. **Couchman, J. R., Gibson, W. T., Thom, D., Weaver, A. C., Rees, D. A., and Parrish, W. E.,** Fibronectin distribution in epithelial and associated tissues of the rat, *Arch. Dermatol. Res.,* 266, 295, 1979.
122. **Fyrand, O.,** Studies on fibronectin in the skin. III. Indirect immunofluorescence studies in lichen planus, *Acta Dermatol. Venereol.,* 59, 487, 1979.
123. **Fyrand, O.,** Studies on fibronectin in the skin. IV. Indirect immunofluorescence studies in lupus erythematosus, *Br. J. Dermatol.,* 102, 167, 1980.
124. **Reitamo, S., Reunala, R., Konttinen, Y. T., Saksela, O., and Salo, O. P.,** Inflammatory cells, IgA C3, fibrin, and fibronectin in skin lesions in dermatitis herpetiformis, *Br. J. Dermatol.,* 105, 167, 1981.
125. **Breathnach, S. M., Bhogal, B., de Beer, F. C., Melrose, S. M., Black, M. M., and Pepys, M. B.,** Immunohistochemical studies of amyloid P component and fibronectin in erthyropoietic protoporphyria, *Br. J. Dermatol.,* 108, 267, 1983.
126. **Finley, J., Clark, R. A. F., Colvin, R. B., Blackman, R., Noe, J., and Rosen, S.,** Patterns of immunofluorescent staining with antibodies to factor VIII related antigen, fibronectin and collagenous basement membrane protein in normal human skin and port wine stains, *Arch. Dermatol.,* 118, 971, 1982.
127. **Nelson, D. L., Cottle, C. D., and Balian, G.,** Distribution of fibronectin and laminin in basal cell epitheliomas, *J. Invest. Dermatol.,* 80, 446, 1983.
128. **Grimwood, R. E., Huff, J. C., Harbell, J. W., and Clark, R. A. F.,** Fibronectin in basal cell epithelioma: sources and significance, *J. Invest. Dermatol.,* 82, 145, 1984.
129. **Singer, I. I.,** Fibronexus-like adhesion sites are present at the invasive zone of human epidermoid carcinomas, in *40th Annu. Proc. Electron Microscopy Soc. Am.,* Bartory, G. W., Ed., Electron Microscopy Society of America, Washington, D.C., 106.
130. **Yamada, K. and Olden, K.,** Fibronectins — adhesive glycoproteins of cell surface and blood, *Nature*
131. **Hynes, R. O. and Yamada, K. M.,** Fibronectins: multi-functional modular glycoproteins, *J. Cell. Biol.,* 95, 369, 1982.
132. **Smith, H. S., Riggs, J. L., and Mosesson, M. W.,** Production of fibronectin by human epithelial cells in culture, *Cancer Res.,* 39, 4138, 1979.
133. **Labat-Robert, J., Birembaut, P., Adnet, J. J., Mercantin, F., and Robert, L.,** Loss of fibronectin in human breast cancer, *Cell. Biol. Int. Rep.,* 4, 608, 1988.
134. **Kurkinen, M., Alitalo, K., Vaheri, A., Stenman, S., and Saxen, L.,** Fibronectin in the development of embryonic chick eye, *Dev. Biol.,* 69, 589, 1979.
135. **Newgreen, D. and Thiery, J. P.,** Fibronectin in early avian embryos. Synthesis and distribution along the migration pathways of neutral crest cells, *Cell. Tissue Res.,* 211, 269, 1980.
136. **Welch, and Clark, R. A. F.,** personal observations.

Chapter 13

FIBRONECTIN IN THE LUNG

Jiro Fujita and Stephen I. Rennard

TABLE OF CONTENTS

I. INTRODUCTION

Reviews of the structure and function of the fibronectins are given elsewhere in this volume. This chapter will provide an overview of fibronectin in the lung. Fibronectin functions in the lung in three separate, interacting compartments: the pulmonary parenchymal interstitium, the intravascular space of the pulmonary vascular bed, and the epithelial surface lining the airspaces of the lung. By virtue of its multiple binding sites, fibronectin can mediate a variety of interactions between cells and components of the extracellular milieu. As in other organs, a variety of roles have been suggested for fibronectin in the pulmonary parenchyma including the organization of extracellular matrix and the regulation of cellular recruitment, proliferation and differentiation both during normal development and during repair processes following injury. Fibronectin present in the vascular lumen, by virtue of its activity as an opsonin, protects the lung from microembolic injury. In addition, plasma fibronectin may be an important determinant of lung vascular membrane integrity, particularly at times of stress. Fibronectin present on the epithelial surface of the lung appears to aid in host defense against bacterial infection. Fibronectin, therefore, is likely involved in a number of diverse pathologic conditions affecting the lung.

Since the lung is an organ that can be relatively easily sampled, a number of studies have been possible investigating the functional role of fibronectin in the lung. For example, the flexible fiberoptic bronchoscope makes it readily simple to sample the epithelial surface of the lung.[1] As a result, evidence from both man and animals has been obtained suggesting a role for fibronectin in the fibrotic lung diseases. In addition, the lung has its own circulation, and chornically cannulated animal models are available.[2,3] As a result, it is possible to suggest a role for fibronectin in maintenance of lung vascular function. This growing body of evidence is of interest in its own right. Moreover, since the lung is easier to sample than many other organs, the lung can serve as a useful model for the role of fibronectin in disease processes in other organs.

II. THE PULMONARY INTERSTITIUM

A. LOCATION AND SOURCES

The interstitial space contains both soluble fibronectin, a component of interstitial fluid, and tissue fibronectin. While the latter is defined by its relative insolubility, a number of techniques including denaturation and partial proteolytic degradation are able to extract insoluble fibronectin from the lung tissue.[4,5] The presence of both soluble and tissue fibronectin in the lung interstitium complicates immunohistochemical localization of fibronectin in the lung.[6] Methods of preparation which allow soluble fibronectin to be fixed into the tissue specimen yield slightly different immunohistochemical staining patterns from techniques in which soluble fibronectin is extensively washed away prior to fixation.[7] Nevertheless, most studies of adult mammalian lung reveal fibronectin along the alveolar and capillary basal laminae[7-9] as well as co-distribution with reticulin fibers in larger vessels[10] and its presence in the basement membranes of airways.[7] Ultrastructural studies[7] indicated uniform immunohistochemical staining of basal laminae of airway epithelia and capillaries. Staining was also noted between epithelial cell junctions and in pinocytotic vesicles suggesting fironectin "traffic" in the lung. The pattern was generally similar, although much less extensive in the adult chicken lung.[11]

It is likely that there are multiple sources of interstitial fibronectin. Many of the more than 40 specialized cell types found in the lung can produce fibronectin *in vitro*. The *in vitro* culture conditions used to grow cells, however, likely have profound effects on the synthesis of fibronectin. Thus, while *in vitro* data provide evidence that pulmonary cells may produce fibronectin, it is not conclusive evidence that fibronectin is produced *in vivo*.

Cells for which *in vitro* evidence of fibronectin production has been obtained include: fibroblasts,[12-14] endothelial cells,[15,16] smooth muscle cells,[17] mesothelial cells,[18] type II alveolar epithelial cells,[19,20] bronchial epithelial cells,[21,22] as well as macrophages[23-25] and, in small amounts, neutrophils.[26] For fibroblasts, both fetal[12,13] and adult,[13,14] fibronectin represents a major *in vitro* secreted product. It can represent as much as 5% of total protein synthesis for these cells.[12,13] Fibronectin synthesis by fibroblasts can be increased by factors such as TGF-β.[27,28] The presence of such factors in the media used for *in vitro* cultures may be responsible for the large *in vitro* production rates consistently observed. Endothelial cells, including these isolated from large[29] and small pulmonary vessels[16] also can synthesize large amounts of fibronectin. Also in these cells, culture conditions are important determinants of fibronectin production.[30,31] For both of these cell types, culture conditions in which cells are rapidly growing in response to serum or growth factors are very unlike the normal adult lung where cell replication is a relatively rare event. In this respect, the culture conditions in which marked fibronectin production is observed resembles a healing wound or (more vaguely) a developing fetal tissue. The immunohistologic observations using a monoclonal antibody specific for cellular fibronectin demonstrate little staining for cellular fibronectin in normal adult tissue, but notable staining in fetal and in "injured" adult lung.[6,32] This supports a role for fibronectin produced locally in the lung during normal development and in repair of the lung following injury (see below).

Alveolar type II cells[19,20] and bronchial epithelial cells also produce fibronectin in culture.[21,22] Both of these cell types can produce an extracellular matrix *in vitro*.[21,33,34] Fibronectin appears able to mediate the attachment of type II cells to an extracellular matrix substrate.[35] Moreover, the presence of the rich extracellular matrix of fibronectin appears to augment type II cell proliferation and to alter the differentiated state of these cells.[33] While studies of the several different types of epithelial cells present in the airway are unavailable, studies with mixed cells indicate both fibronectin production[21,22] and production of an extracellular matrix to which these cells attach preferentially.[34] Although fibronectin does not appear to be the preferred atttachment molecule, these cells can respond to fibronectin as a chemoattractant,[36] and fibronectin included as a culture matrix may augment the growth of these cells as well.[37,38] Thus, while less is known about fibronectin production by epithelial cells in the lung than by mesenchymal cells, the observations are similar: fibronectin is produced by proliferating cells, and it appears to augment the attachment, proliferation, and differentiation of these cells. Thus it is reasonable to suggest a similar role for locally produced epithelial derived fibronectin; i.e., that it functions during division and orientation, processes which are required during development and during repair.

At least 20 different species of fibronectin may be produced due to differential splicing of the gene into mRNA.[39-42] Which of these species are produced in the lung and under what conditions is not currently known. Moreover, several pieces of evidence suggest that much of the insoluble tissue fibronectin present in the adult lung may be derived from soluble plasma fibronectin and thus orginate in the liver. First, infused heterologous fibronectin can be incorporated into basement membranes.[43,44] Second, *in vitro* studies of cultured fibroblasts demonstrate that these cells can both produce fibronectin[12-14] and incorporate the fibronectin present in the culture medium into an insoluble pericellular matrix.[45] Third, extraction of adult human lung with cathepsin D released fibronectin fragments characteristic of the plasma-type fibronectin.[46] There may be important differences in the source of fibronectin during different stages of lung development. Thus, the fibronectin extracted from fetal tissues possessed the extra domain (ED) characteristic of cellular-type fibronectin while this was lacking in fibronectin extracted from adult tissues.[46] Moreover, staining of adult human lung with a monoclonal antibody specific for the EIIIA region of cellular fibronectin reveals little staining in contrast to fetal tissues which stain well.[32] Together, these data suggest that fibronectin derived from different sources may have varying functional roles at various times in the lung in health and disease.

III. FUNCTIONAL SIGNIFICANCE OF FIBRONECTIN IN THE PULMONARY INTERSTITIUM

A. DEVELOPMENT

Fibronectin and its cellular receptor are thought to play major roles during fetal development both for migrating and nonmigrating cells undergoing differentiation[47] (see Chapter 2). While studies are incomplete, immunohistochemical and biochemical studies in the rat,[48,49] mouse,[50] and rabbit[51] demonstrate increased fibronectin, and studies in chicken[11] demonstrate increased fibronectin and its receptor at sites associated with active morphogenesis. Fibronectin is generally associated with basal laminae, although pericellular staining of mesenchymal cells and interstitial fibers has also been observed. The three stages of lung development[52] generally show analagous patterns of fibronectin deposition.

The first stage, the glandular stage, during which the lungs bud from the embryonic gut and form the larger airways by branching morphogenesis shows fibronectin associated with basement membranes. The second stage, the canalicular stage, during which terminal airways develop, demonstrates fibronectin most prominently in the basement membranes of the distal developing sites. During the final alveolar stage, fibronectin immunochemical staining is most intense at sites of alveolar septal formation. All these observations are consistent with the concept that fibronectin provides a ''preliminary matrix'' over which cells can migrate during development.[47] In addition, deposition of fibronectin in neonatal pig lung is more extensive than that of interstitial collagens (types I, III, V).[10,11] This supports the concept that fibronectin may serve as a template for collagen fiber formation.[53]

The source of fibronectin deposited at sites of active morphogenesis is unclear. Biochemical extraction has revealed significantly more ''cellular'' fibronectin containing the ED domain in fetal tissues than in adult tissues.[46] Moreover, quantitative estimates suggest an increased content of fibronectin in lung during development compared to the adult.[11,51] Both these observations suggest local production during development, but whether the source is epithelial cells or mesenchymal cells, and the extent to which plasma fibronectin plays a role is unknown. Fibronectin is not the only matrix membrane protein deposited in a developmentally regulated pattern. Laminin and type IV collagen also appear to be deposited early in the development of the lung.[10,11] Collagenase, presumably by disrupting extracellular collagen, blocks salivary gland branching morphogenesis,[54] and inhibitors of collagen synthesis[55,56] can block early lung morphogenesis. Finally, it is likely that cell-cell contacts[50,57] also play a role in the mesenchymal-epithelial interactions that characterize lung development.

Fibronectin, nevertheless, particularly locally produced fibronectin, may have several functions in developing lung. While the suggested roles for fibronectin during development are speculative, they are consistent with known *in vitro* functions of fibronectin. Moreover, these putative roles for fibronectin are analogous to the suggested role of fibronectin in the fibrotic process (see below). Specifically, (1) fibronectin may help regulate cellular migration by providing a framework over which cells may migrate; (2) fibronectin may help orient cells in the newly developing structure; (3) fibronectin may regulate cell proliferation; and, (4) fibronectin may participate in the organization of the newly deposited extracellular matrix.

B. MAINTENANCE AND REPAIR — CONNECTIVE TISSUE DEPOSITION

It has been suggested that fibronectin plays a role in normal lung maintenance and repair. Three lines of evidence, all indirect, support this suggestion: (1) *in vitro* studies indicate a role for fibronectin in organizing newly deposited connective tissue collagen; (2) the observation that depletion of plasma fibronectin in several pathological states is associated with evidence of lung dysfunction presumably due to altered lung vascular membrane function (see next section below); and, (3) observations that fibronectin appears to play a role in fibrosis which can be considered an ''extreme'' form of tissue repair.

The mechanisms by which connective tissue fiber formation is regulated are only partially understood, but involve interactions among various connective tissue components. Fibronectin may play a key role in fiber function. Not only is fibronectin a major secreted product of fibroblasts,[12-14] it can polymerize to form fibers.[58] Immunohistochemical localization of type I, III, and VI collagen demonstrates colocalization of the collagens with the fibronectin fibers.[59,60] Fiber formation is also regulated by conversion of procollagen to collagen by procollagen peptidase and, perhaps, by the ratio of collagen types present.[62] Thus regulation of collagen fiber formation is likely to be a complex process in which fibronectin will probably play a role. Inasmuch as some collagen turnover takes place in normal lung tissue, this may be a ''normal'' function of fibronectin.

C. PULMONARY FIBROSIS

Fibrosis is a process in which fibroblasts and the dense connective tissue produced by fibroblasts replace the normal architectural elements of a tissue.[63] Because the normal tissue structures are disrupted, organ function is usually disturbed. Fibrosis can develop in the lung as a consequence of a number of disease processes. For example, the interstitial lung diseases are characterized by an inflammatory process of the alveolar structures which can lead to the formation of dense fibrosis.[64,65] As a result of the destruction of the pulmonary alveoli, respiratory failure can ensue. Obliterative fibrosis of the terminal airways can be a consequence of bronchiolitis.[66,67] The resulting bronchiolitis obliterans can also have major clinical impact. Fibrosis is also a feature associated with the larger airways in bronchitis,[68] and condensations of collagen which may represent areas of localized fibrosis have been reported in pulmonary emphysema.[69] In these latter two conditions, the physiologic consequences of the fibrotic changes are thought to be minimal. Thus, the physiologic impact of fibrosis is thought to depend on the location and extent of the fibrotic changes.

The various pathologic processes which lead to fibrosis are generally characterized by inflammation.[64,65,70] It has been suggested that fibrosis develops as a consequence of the repair process following injury and inflammation.[71] This process is thought to involve a number of distinct stages including: (1) injury of the normal lung structures; (2) recruitment of mesenchymal fibroblasts; (3) orientation of mesenchymal fibroblasts at sites of injury; (4) proliferation of fibroblasts; (5) production of new fibrous connective tissue macromolecules. Considerable circumstantial evidence suggests that fibronectin can play a role at all stages of this process. As such, fibronectin may function as a ''wound-healing hormone'' in the lung. Fibrosis may, in this context, represent relatively ineffective healing.

The evidence which supports a role for fibronectin in the fibrotic process includes: (1) laboratory evidence demonstrates that fibronectin is capable of participating in all five stages of the fibrotic process; (2) quantitative measures indicate increased fibronectin in the lung in fibrotic processes in human disease and in animal models; (3) quantitative measures indicate increased production by cells in the lung during the fibrotic process; (4) histologic studies indicate increased deposition of fibronectin in tissues during fibrotic processes; (5) increases in fibronectin appear to precede the development of fibrosis in both human and animal studies; (6) patients with fibrotic diseases who have increased levels of fibronectin production by alveolar macrophages appear to be at risk for the development of progressive fibrosis. While this mass of data shows an association of fibronectin with the fibrotic process, it does not show a ''causative'' role. Nevertheless, the weight of evidence suggests that fibronectin plays a crucial role during this form of ''wound healing'' which frequently follows inflammation in the lung.

Laboratory evidence supports the potential role of fibronectin at all stages of the fibrotic process. Fibronectin is sensitive to proteolytic degradation.[72,73] As such, the inflammatory processes which frequently lead to fibrosis can degrade local fibronectin. In support of this, release of tissue fibronectin from basement membrane has been reported in an *in vitro* model system[74] and directly following lung injury.[75,76] Because the interactions between the parenchymal cells of the lung and the connective tissue framework which determine tissue architecture depend, in part, on interactions mediated by fibronectin, degradation of fibronectin by inflammatory proteases may lead to tissue disruption. In addition, fragments of fibronectin retain certain biological activities. Thus, it is possible that fibronectin fragments released early in the inflammatory process could mediate subsequent stages of the development of fibrosis.[77]

Fibroblast recruitment is a crucial feature of fibrotic processes.[78,79] Fibroblasts which accumulate in the lung in fibrosis are thought to differ from the fibroblasts present in the normal pulmonary parenchyma. For example, in idiopathic pulmonary fibrosis, the fibroblasts which accumulate have features which resemble smooth muscle cells. As a result, these fibrotic fibroblasts have been termed "myofibroblasts".[80] Fibronectin is a potent chemoattractant capable of mediating fibroblast recruitment.[77,81,82] In addition, fibronectin is chemokinetic for fibroblasts.[77] As a result, it may augment the chemotactic recruitment of fibroblasts by other chemoattractants.

Once recruited to sites of inflammation, fibroblasts must attach to the extracellular matrix prior to proliferation. While fibroblasts are capable of attaching to the extracellular matrix through other factors such as vitronectin or laminin,[83] attachment through fibronectin is thought to be a major means of interaction between fibroblasts and the extracellular matrix. The ability of fibronectin to interact with a variety of extracellular matrix components raises the possibility that fibroblasts can attach in "abnormal" orientations. For example, fibronectin can bind to fibrin and can be covalently cross-linked to it by factor XIII.[84] In an inflammatory process in the lung, which may be accompanied by leakage of fibrinogen from the plasma and deposition of fibrin in the tissue, this could result in deposition of fibronectin within the tissue.[85] Fibronectin attached to polymerized fibrin in a tissue could then serve as a nidus for fibroblast attachment.[86] In support of this, activation of coagulation pathways have been observed in several pulmonary fibrotic processes.[87,88]

Fibronectin can also mediate the attachment of fibroblasts to immune complexes. Immune complexes can bind the first component of complement, C1qrs.[89] After activation of this component, the C1q remains bound to immune complexes while the C1rs component is separated. Fibronectin can then bind to the C1q which in turn is bound to the immune complex.[90-92] This bound fibronectin can serve as a nidus for fibroblast attachment.[93] In support of this, diseases characterized by the deposition of immune complexes within the lung such as hypersensitivity pneumonitis,[94] idiopathic pulmonary fibrosis,[95] and systemic vasculitis[96] are frequently associated with the development of fibrosis of the lung.[64]

Fibronectin can also bind to denatured collagens.[97] Thus, in an inflammatory milieu where tissue destruction is taking place and collagen fragments may be generated, fibronectin may bind at unusual sites within tissues. Fibronectin bound in such locations could then serve as a nidus for fibroblast attachment. The ability of fibronectin to bind to unusual sites within tissues is thought to be crucial to the development of the fibrotic process. A key aspect of this process is the formation of dense fibrous connective tissue at abnormal locations within the lung. The ability of fibronectin to serve as a nidus for deposition of fibrous tissue may play a crucial role in the ultimate morphologic outcome.

Once fibroblasts have been recruited to a site of inflammation and have attached to components of the extracellular matrix, their subsequent proliferation is regulated by the presence of growth factors in the local milieu.[98-100] Regulation of fibroblast growth is a complex process subject to both stimulatory factors and to inhibitors. Optimal fibroblast growth requires the pressence of at least two classes of growth stimulators termed competence factors and progression factors. Competence factors, the paradigm of which is platelet-derived growth factors (PDGF), act early during the G1 phase of the growth cycle and allow the cells to respond to subsequent exposure to progression factors.[98] Progression factors, the paradigm of which is insulin-like growth factor, act later in the G1 phase of the cell cycle and allow competence-factor-primed cells to proceed with cell division.[99] Fibronectin has, in this regard, been shown to function as a competence factor and can replace PDGF in stimulating fibroblast growth.[100] Interestingly, fibroblast growth can be inhibited by the presence of the prostaglandin PGE_2.[101] This may play an important role in regulating fibroblast proliferation *in vivo* situations since PGE_2 has been estimated to be present in lung epithelial lining fluid at concentrations at which it is an effective inhibitor of fibronectin stimulated fibroblast growth.[102,103] Thus, fibronectin may interact with other growth stimulators and inhibitors in determining the extent of fibroblast proliferation within the lung.

Following fibroblast proliferation, new connective tissue must be synthesized. Fibronectin represents a major secreted product of lung fibroblasts.[12-14] As discussed above, the production of fibronectin by fibroblasts and its subsequent polymerization in the extracellular matrix may serve as a template for the formation of collagen fibers. In this regard, it is of interest that both fibronectin and collagen production by fibroblasts can be strongly stimulated by TGF-β, a mediator potentially important in the development of fibrosis. [27,28,104] This may account for the altered phenotype which appears to characterize fibroblasts from patients with interstitial lung disease.[105]

Thus, fibronectin is capable of participating in all stages of the fibrotic process. Importantly, each stage of the fibrotic process can also be mediated by other factors. Not only is fibronectin disrupted during the early phase of an inflammatory process, but collagen, elastin, and other tissue glycoproteins and proteoglycans can also be disrupted by inflammatory processes.[95,106-108] In this context, it has been suggested that destruction of basement membranes may be the crucial feature which determines whether or not effective healing can follow an inflammatory process.[109] Whether disruption of fibronectin is crucial in this regard remains to be evaluated. Numerous other chemotactic factors for fibroblasts have been described besides fibronectin. These include PDGF,[110,111] LTB4,[112] TGF-β,[113] fragments of the connective tissue components collagen and elastin,[114-116] fibrin fragments,[117] and activated complement components.[118] In addition, a fibroblast chemoattractant released by activated lymphocytes has been described.[119] Thus, fibroblast recruitment to sites of inflammation could result from mediators other than fibronectin. Finally, fibroblast recruitment may also be mediated by inhibitors of chemotaxis which can include interferon[120] and a neutrophil-derived factor.[120a] Similarly, fibroblast attachment can be mediated by fibronectin, but fibroblasts also possess binding sites for other attachment factors including laminin and vitronectin.[83] Thus, fibroblasts could attach and become oriented in tissues through fibronectin-independent mechanisms. In addition, while fibronectin has activity as a fibroblast growth factor,[100] other factors present in the inflammatory milieu such as the platelet-derived growth factor may serve as competence factors. In this regard, increased PDGF production has been reported by macrophages associated with the fibrotic pulmonary disease, idiopathic pulmonary fibrosis.[110] Thus, fibroblast proliferation can also occur independently of a role for fibronectin. Finally, while fibronectin is a major secreted product of fibroblasts and while it may serve as a template for collagen fiber formation, collagen fibers may be able to form independent of fibronectin.[61] Thus, the specific role of fibronectin in the inflammatory process remains problematic. It appears likely that fibronectin will play a role in the development of fibrosis in many conditions. It is likely that the various fibrotic disorders which affect the lung, however, will be heterogeneous in their pathophysiology. That is, while fibronectin may be important in some disorders at certain stages, other factors may substitute for fibronectin in other disorders.

Processes leading to fibrosis in the lung are associated with increased fibronectin content in both lung tissue and in fluid recovered from the epithelial surface of the lung. These findings have been consistent features in a number of animal models of fibrosis. Increased extractable fibronectin has been observed following bleomycin-induced fibrosis in dog,[121] in paraquat-induced fibrosis in the rat,[122] and in radiation-induced fibrosis in the mouse.[123] Bronchoalveolar lavage has revealed increased fibronectin content in bronchoalveolar lavage fluid in paraquat-induced fibrosis in rats[122] and monkeys,[124] and in asbestos-exposed sheep.[125] The fibronectin content was not increased in a model of hypersensitivity pneumonitis in sheep in which fibrosis did not develop.[126] Thus, in a number of animal systems, increaseed fibronectin content both in lung tissue and bronchoalveolar lavage fluid is associated with the fibrotic process. Similar observations have been made in human studies. Patients with a variety of interstitial lung disorders have increased content of fibronectin in their bronchoalveolar lavage fluid.[127] In addition, in a rabbit model of acute lung injury resembling the adult respiratory distress syndrome (ARDS), there is also an increased content of fibronectin in bronchoalveolar lavage fluid.[75,76] In this regard, ARDS frequently leads to the development of extensive pulmonary fibrosis. Finally, in newborns with respiratory distress syndrome, low fibronectin levels were thought to contribute to the capillary leak of this condition (see bellow), while elevated levels seemed to predict the development of bronchopulmonary dysplasia, a fibrotic process of the terminal airways.[128] Taken together, these observations suggest that fibronectin may play a role early in the fibrotic process.

Much of the fibronectin contained in bronchoalveolar lavage fluid represents plasma-type fibronectin.[76,129] Nevertheless, increased local fibronectin production appears to play an important role in the fibrotic process. Alveolar macrophages normally produce low levels of fibronectin.[23-25] Fibronectin production by these cells increases dramatically in fibrotic lung disease.[25] Increased production of fibronectin by alveolar macrophages has also been noted in several animal models of pulmonary fibrosis including paraquat-induced fibrosis in monkeys,[124] asbestos-exposed sheep[125] and guinea pigs,[130] and increased content of fibronectin messenger RNA was noted in bleomycin-exposed hamsters.[131] In this latter model, it is likely that fibronectin was produced by a variety of cells in the lung including macrophages, fibroblasts, and perhaps other cells.

Increased production of fibronectin by alveolar macrophages has been noted in normal human volunteers undergoing exposure to 100 % oxygen.[132] Such a brief exposure results in a reversible increase in fibronectin production but does not result in fibrosis. Prolonged exposure to oxygen can result in the development of pulmonary fibrosis. This suggests that the increased production by alveolar macrophages may be an early event in the development of pulmonary fibrosis. In an animal model of hyperoxia in rats, increased cell surface fibronectin on alveolar macrophages was noted,[133] but no increase in bronchoalveolar lavage fibronectin was observed. It is noteworthy that in this study, also, pulmonary fibrosis was not observed. Thus, increased production of fibronectin by cells in the lung appears to be an early feature of pulmonary fibrosis. The alveolar macrophage may play a central role in this regard. The alveolar macrophage is the major mononuclear phagocyte present in the lower respiratory tract under normal circumstances. While the majority of fibronectin present in the lung is likely derived from other sources, specifically from the plasma for example,[43,44,75,76,129] macrophage-derived fibronectin may still have an important role. It has been suggested that macrophage fibronectin is considerably more active as a fibroblast chemoattractant than is plasma fibronectin, for example.[134] As such, activation of alveolar macrophages to produce increased amounts of fibronectin may be a crucial feature in the early phases of fibroblast recruitment.

In support of a role for fibronectin in human disease are observations that production of both fibronectin and the alveolar-macrophage-derived growth factor (AMDGF or IGF-1a) precede the clinical progression of fibrosis in human fibrotic diseases.[135] Patients whose macrophages were not producing fibronectin did not demonstrate progressive disease. Importantly, patients who deteriorated had macrophages which were producing both AMDGF and fibronectin suggesting that fibronectin production alone is not a sufficient stimulus to lead to progressive fibrosis. Additionally, unaffected individuals with a form of familial pulmonary fibrosis have been noted to have alveolar macrophages spontaneously releasing increased levels of fibronectin.[136] While it is unknown if these individuals will develop fibrosis, the production of fibronectin has clearly preceded the clinical syndrome.

Histologic studies of pulmonary fibrosis have consistently demonstrated increased fibronectin deposition within tissues. Fibronectin deposition is a prominent feature of hypoxia-induced fibrosis in the rat[137] where fibronectin deposition was prominent in the alveolar interstitium. In guinea pigs,[133] fibronectin deposition within the alveolar interstitium was noted as early as 6 h following oxygen exposure and was thought to be related to the formation of protein-rich plasma exudates. Analogous findings were observed in paraquat-induced fibrosis in monkeys where fibronectin was deposited not only within the pulmonary interstitium but was also deposited in the intra-alveolar spaces during the formation of hyaline membranes.[138] The deposition of fibronectin within the alveolar space was thought to be a potent stimulus for the development of intra-alveolar fibrosis. Similar results have been observed in human disease. Fibronectin deposition is a prominent feature in the alveolar interstitium and basement membranes of fibrotic human lungs.[6,7] In addition, prominent immunohistochemical staining for fibronectin is observed in the fibrotic lesions of pulmonary pneumoconiosis.[139] Finally, fibronectin is present in the exudative lesions characterizing early ARDS in humans prior to the development of fibrosis.[138] Taken together, these observations are consistent with a role for fibronectin in directing the morphologic process of fibrosis. Specifically, fibronectin appears to attach to components of the extracellular matrix either within the interstitial space of within the alveolar lumen prior to the development of fibrosis. This is consistent with a role for fibronectin in orienting newly recruited fibroblasts in the lung as part of the fibrotic process.

Taken together, these observations suggest that fibronectin plays a role in the fibrotic process analogous to a ''wound-healing hormone''. Specifically, following injury, fibronectin can be deposited in tissues as a result of exudation of plasma proteins together with binding of fibronectin to components of the extracellular milieu, e.g., polymerized fibrin, immune complexes, or denatured connective tissue macromolecules. Increased fibronectin production by local alveolar macrophages can then result in fibroblast recruitment to sites of inflammation. These fibroblasts can then bind to fibronectin attached at tissue sites, proliferate in response to fibronectin and other growth factors, and subsequently synthesize new connective tissue matrix. Such a scheme can allow for repair of tissues following injury. In situations where the initial damage is not extensive, this process may lead to tissue restoration. Consistent with this are the observations that extensive intra-alveolar fibrosis can develop following the adult respiratory distress syndrome[138] and that this process might be reversible.[140,141] It has been suggested that the reversibility depends on the preservation of the original basal laminae.[109,138,142] Should the recruitment of fibroblasts and the deposition of new connective tissue take place in a situation where tissue destruction has occurred, however, the morphologic disruption may be irreversible.

IV. THE PULMONARY VASCULATURE

Fibronectin is a major protein component of plasma, and is, therefore, present within the intravascular space of the lung. Because the lung capillary membranes are relatively ''leaky'', pulmonary lymphatics are relatively rich in plasma proteins including fibronectin.[143,144] Thus, it is likely that a considerable amount of plasma-derived fibronectin passes from the intravascular space of the lung capillaries, into the pulmonary interstitium, and into pulmonary lymphatics. Evidence for a major role for this plasma fibronectin in the maintenance of lung function stems, in part, from the extensive work of Saba and colleagues[145] (see Chapter 4).

Plasma fibronectin is a major circulating opsonin.[146-148] It mediates the phagocytic clearance of particulate debris such as collagen fragments or circulating fibrin. Plasma fibronectin is consumed during this opsonic process.[147,149] Thus, intravascular generation of fibrin as occurs in disseminated intravascular coagulation[150] or experimental infusion of gelatin (denatured collagen)-coated particles[148] results in both clearance of the circulating debris and in depletion of plasma fibronectin. The depletion of plasma fibronectin is associated with a loss of opsonin function.[147,148,151] Subsequent challenges with additional particulate results in microembolization of the lung with secondary physiological impairment.[151a,152] A similar depletion of plasma fibronectin occurs following burns, sepsis, major trauma, and major surgery and is also thought to be due to fibronectin consumption due to the generation of intravascular debris.[147,152,153-156]

Fibronectin depletion induced by the intravascular infusion of trypsin results in a marked increase in lung lymph flow.[157] Fibronectin depletion caused by injection of gelatin-coated particles alone, however, causes no change in lung lymph flow. A second challenge with live pseudomonas, however, results in a marked increase in lung lymph flow in the fibronectin-depleted animals.[158] Repletion of fibronectin with cryoprecipitate prior to bacterial challenge can block the increase in lung lymph flow.[158,159] This fibronectin-depletion-associated increase in lung lymph flow has been suggested to play a role in the noncardiogenic pulmonary edema of ARDS that frequently accompanies severe illness.[147,152]

The mechanicisms by which plasma fibronectin depletion may lead to increased lung lymph flow may invlove cell matrix interactions in the lung. Plasma fibronectin exchanges with lung basement membrane fibronectin.[44,45] Depletion of plasma fibronectin in a rat model system caused a rapid decrease in extracellular lung fibronectin.[160] Aggregation of activated neutrophils in the small vessels of the lung, a frequent finding in ARDS, is associated with detachment of endothelial cells.[138] *In vitro* studies support a role for neutrophil degradation of basement membrane fibronectin with loss of endothelial cells.[74,160a] In an acute lung injury model in rabbits, tissue fibronectin was released into plasma and bronchoalveolar lavage fluid.[75] In this setting an influx of plasma fibronectin which may increase with injury[153] may help maintain lung structure by preserving a matrix for endothelial cell attachment. Plasma fibronectin depletion could lead to depletion of fibronectin in the alveolar basement membranes and makes the endothelial cell-matrix interaction more susceptible to disruption.

The possibility that plasma fibronectin contributes to the pathogenesis of ARDS has created considerable interest in the use of fibronectin both prognostically and therapeutically.[161-163] Fibronectin levels did not appear to identify those patients who would develop ARDS among a population at risk.[164] In another study of 98 critically ill surgical patients, fibronectin levels were lower in patients with sepsis but did not correlate with outcome.[165] Thus, while fibronectin appears to correlate generally with severity of illness,[166-168] it is not a clear predictor of outcome. As a result, it is difficult to assess therapeutic interventions using fibronectin. This is particularly true as fibronectin is usually given as cryoprecipitates which contains fibrinogen and other bioactive substances.[169]

Fibronectin administration as cryoprecipitate, on the other hand, can raise plasma fibronectin levels.[158,161,162,169] The increment in fibronectin level, moreover, is due to the fibronectin component of the cryoprecipitate,[158] but an effect of other components on the response of critically ill patients is certainly not excluded.[162] Physiologic benefits following fibronectin administration have been reported including improved lung function, improved chest X-ray, and improved gas exchange.[169-173] Improvement in peripheral circulation and in renal function have also been reported.[170,171] This could be due to an effect of fibronectin on vascular beds other than the lung, or could be due to other supportive measures. Overall, in studies done to date, survival in critically ill patients does not appear to be improved by cryoprecipitate infusion.[161,162,162a] Such negative results, however, must be interpreted with caution. Survival of patients with respiratory failure secondary to severe underlying illness depends on control of the underlying cause, management of nonpulmonary secondary problems, and management of respiratory failure. The ability of fibronectin to effect physiologic improvement, even in selected patients, offers promise as a potential addition to the supportive care of such patients. As the case of these complex patients continues to change, the importance of a specific intervention such as fibronectin replacement will have to be reassessed.

V. THE PULMONARY EPITHELIAL SURFACE

Fibronectin is also present on the epithelial surface of the lung. It is present in alveolar lining fluid at concentrations in excess of those expected by leakage from plasma alone.[127,174] Moreover, it is present in mucus secretions at very high relative concentrations.[175] Interestingly, the majority of this epithelial surface fibronectin appears to be of the plasma-type.[75,76,129] Several mechanisms could explain the high local concentration of plasma-type fibronectin: (1) secretion of plasma fibronectin onto the epithelial surface; (2) "trapping" of fibronectin on the epithelial surface during fluid secretion and reabsorption; (3) synthesis of "plasma-type" fibronectin by lung cells *in situ*.

The recent observation that airway serosal epithelial cells can produce an albumin-like species[176] supports the concept that species thought to originate exculsively in the liver and to circulate in the blood can also be produced in the lung. Perhaps this locally produced fibronectin functions as an opsonin and aid in host defenses (see below). Alveolar macrophages, the normal mononuclear phagocyte found on the epithelial surface of the lower respiratory tract, can produce fibronectin.[23-25] Moreover, production of fibronectin by alveolar macrophages is increased in a variety of lung disorders, and a role for macrophage production of fibronectin has been suggested in the response of the lung to injury (see above). It may be that the epithelial surface of the lung will contain several forms of fibronectin derived from several sources with several differing functional roles.

VI. HOST DEFENSE

The epithelial fluid lining the lung contains a number of macromolecules thought to protect against infection including immunoglobulins, lactoferin, lysozyme, complement, and fibronectin. [177] While the precise roles of these macromolecules in host defense molecules are speculative, fibronectin may function in three ways: (1) to reduce bacterial adherence to epithelial surfaces; (2) to aid in phagocytic clearance of bacteria, and (3) in recruitment of neutrophils.

Adherence of bacteria to the airway surface is thought to be an initial stage in the infection process leading to Gram-negative pneumonias.[178,179] Some bacteria such as staphylococci possess binding sites for fibronectin, and fibronectin may be able to bind these bacteria and prevent their subsequent adhesion to epithelial surfaces.[180] Fibronectin bound to buccal epithelial surface, moreover, appears to inhibit binding of certain Gram-negative bacteria.[181,182] The fibronectin covering the buccal surface can be lost due to proteolytic activity resulting in better bacterial binding and, presumably, more efficient colonization. Such a process may play a role in the chronic bacterial colonization of the airway in cystic fibrosis or in chronic bronchitis. The recent finding that the gel phase of mucus contains high livels of fibronectin raises the possibility that fibronectin may help entrap bacteria for subsequent mechanical removal by ciliary clearance and cough.[175]

Fibronectin, moreover, is an opsonin.[148,183] Both fibronectin[24] and certain of its fragments[184,185] can mediate opsonization of particulate and bacteria by alveolar macrophages. Thus, fibronectin directly bound to bacteria could aid in the clearance of bacteria from the lung.[186] Fibronectins may also indirectly mediate bacterial opsonization through its ability to bind Clq[90-92] and, perhaps, to the collagenous sequences present in surfactant apoproteins.[187] Clq is present in the lung[177] and will become bound to immune complexes.[89] Subsequent binding of fibronectin may mediate opsonization. It has also been suggested that surfactant apoprotein may help in clearance of particles from the lower respiratory tract.[188,189] These proteins are thought to coat hydrophillic particulates and, by virtue of their collagenous sequence, could, theoretically, allow fibronectin-mediated opsonization.

Fibronectin can also alter neutrophil adhesion [190] and monocyte chemotaxis.[191] As such, fibronectin may play a role in leukocyte recruitment into the lung in host defenses.

VII. SUMMARY AND FUTURE DIRECTIONS

Fibronectin appears to play three interconnected roles in the lung. Plasma fibronectin, by its ability to mediate opsonization of intravascular particulates and by its ability to exchange with fibronectin deposited in pulmonary basement membranes, appears to be important in the maintenance of lung vascular and endothelial function. Loss of this function of fibronectin may play an important role in the development of pulmonary capillary leak syndromes. Replacement of fibronectin may prove to be a therapeutic modality for patients with a variety of critical illnesses. Second, tissue fibronectin appears to play an important role in tissue recruitment orientation and differentiated function within the lung. This appears to be true both during development and in repair following lung injury. In particular, fibronectin appears to play a role in the development of fibrosis which often represents an ineffective form of repair and can be associated with loss of pulmonary function. Modulation of the role of fibronectin in the fibrotic process may be a possible therapeutic strategy in fibrotic lung disorders. Finally, fibronectin is present on the epithelial surface of the lung where it may play a role in protection of the host from bacterial infection.

Recent studies of the structure and genetics of fibronectin indicate that there are a multiplicity of forms of fibronectin. These forms of fibronectin will differ in their various functional domains and, thus, will differ in the roles they may play in normal and disease processes. Clearly, a major important goal for the near future will be to characterize the forms of fibronectin associated with various disease states. It seems reasonable to speculate that different forms of fibronectin will be involved in the different pathogenetic processes noted above. As the role of specific forms of fibronectin in disease processes becomes clearer, therapeutic strategies can be evaluated. Specifically, in situations where fibronectin depletion or degradation plays a role, as appears to be the case in acute lung injury, restoration of fibronectin can be evaluated. In conditions where fibronectin plays a pathogenetic role by mediating the recruitment of cells and the subsequent disruption of normal tissue architecture, as in the development of fibrosis, inhibition of fibronectin release can be evaluated as a therapeutic strategy. Thus, it is likely that better classification of the specific function of fibronectin in disease processes will allow for better testing of fibronectin in clinical situations.

At a more fundamental level, an understanding of the role of fibronectin and its specific interaction with individual types of lung cells and lung matrix components promises important understanding of normal lung function. Thus, it is likely that advances will continue in understanding the role of fibronectin in protecting the lung from bacterial infection including both the forms and the sources of the fibronectin involved. It is likely that the various functions of fibronectin in the repair and fibrotic processes also involve heterogeneous forms and sources of fibronectin. These need to be defined. Finally, the role of fibronectin "traffic" needs to be elucidated. It seems reasonable to speculate that by virtue of its ability to mediate a host of interactions between cells and the extracellular environment and by virtue of the fact that many cells within the lung can both produce and interact with fibronectin that fibronectin will turn out to be one of the major forms by which cells communicate. Additional understanding of the forms of fibronectin and its receptors on specific lung cells will undoubtedly greatly improve the understanding of the cell biology of the lung.

ACKNOWLEDGMENT

I would like to acknowledge the assistance of Dr. John A. McDonald for his assistance regarding the content of this review, Dr. John Peters for the use of his as yet unpublished observations, and Ms. Lillian Richards for her assistance in the preparation of the manuscript.

REFERENCES

1. **Reynolds, H. Y.,** Bronchoalveolar lavage, *Am. Rev. Resp. Dis.*, 135, 250, 1987.
2. **Staub, N. C., Bland, R. D., Brigham, K. L., Demling, R., Erdmann, A. J., and Woolverton, W. C.,** Preparation of chronic lung lymph fistulas in sheep, *J. Surg. Res.*, 19, 135, 1975.
3. **Brigham, K. L., Woolverton, W. C., Blake, L. A., and Staub, N. C.,** Increased sheep lung vascular permeability caused by pseudomonas bacteremia, *J. Clin. Invest.*, 54, 792, 1974.
4. **Bray, B. A.,** Cold-insoluble globulin (fibronectin) in connective tissues of adult human lung and in trophoblast basement membrane, *J. Clin. Invest.*, 62, 745, 1978.
5. **Bray, B. A., Mandl, I., and Turino, G. M.,** Heparin facilitates the extraction of tissue fibronectin, *Science*, 214, 793, 1981.
6. **McDonald, J. A.,** Fibronectin in the lung, *Biology of Extracellular Matrix: Fibronectin*, Mosher, D. F., Ed., Academic Press, San Francisco, 1989, 363.
7. **Torikata, C., Villiger, B., Kuhn, C., III., and McDonald, J. A.,** Ultrastructural distribution of fibronectin in normal and fibrotic human lung, *Lab. Invest.*, 52, 399, 1985.
8. **Stenman, S. and Vaheri, A.,** Distribution of a major connective tissue protein, fibronectin, in normal human tissues, *J. Exp. Med.*, 147, 1054, 1978.
9. **Gil, J. and Martinez-Hernandez, A.,** The connective tissue of the rat lung, *J. Histochem. Cytochem.*, 32, 230, 1984.
10. **Mills, A. N. and Haworth, S. G.,** Pattern of connective tissue development in swine pulmonary vasculature by immunolocalization, *J. Pathol.*, 153, 171, 1987.
11. **Chen, W. T., Chen, J. M., and Mueller, S. C.,** Coupled expression and colocalization of 140K cell adhesion molecules, fibronectin, and laminin during morphogenesis and cytodifferentiation of chick lung cells, *J. Cell Biol.*, 103, 1073, 1986.
12. **Baum, B. J., McDonald, J. A., and Crystal, R. G.,** Metabolic fate of the major cell surface protein of normal human fibroblasts, *Biochem. Biophys. Res. Commun.*, 79, 8, 1977.
13. **McDonald, J. A., Baum, B. J., Rosenberg, S. M., Kelman, J. A., Brin, S. C., and Crystal, R. G.,** Destruction of a major extravascular adhesive glycoprotein (fibronectin) of human fibroblasts by neutral proteases from polymorphonuclear leukocyte granules, *Lab. Invest.*, 40, 350, 1979.
14. **Rennard, S. I., Berg, R., Martin, G. R., Foidart, J. M., and Gehron-Robey, P.,** Enzyme linked immunoassay (ELISA) for connective tissue components, *Anal. Biochem.*, 104, 205, 1980.
15. **Jaffe, E. A. and Mosher, D. F.,** Synthesis of fibronectin by cultured human endothelial cells, *J. Exp. Med.*, 147, 1779, 1978.
16. **Habliston, D. L., Whitaker, C., Hart, M. A., Ryan, U. S., and Ryan, J. W.,** Isolation and culture of endothelial cells from the lungs of small animals, *Am. Rev. Respir. Dis.*, 119, 853, 1979.
17. **Bornstein, P. and Ash, J. F.,** Cell surface-associated structural proteins in connective tissue cells, *Proc. Natl. Acad. Sci. U.S.A.*, 74, 2480, 1977.
18. **Rennard, S. I., Jaurand, M. C., Bignon, J., Kawanami, O., Ferrans, V. J., Davidson, J., and Crystal, R. G.,** Role of pleural mesothelial cells in the production of the submesothelial connective tissue matrix of lung, *Am. Rev. Respir. Dis.* 130, 267, 1984.
19. **Sage, H., Farin, F. M., Striker, G. E., and Fisher, A. B.,** Granular pneumocytes in primary culture secrete several major components of the extracellular matrix, *Biochemistry*, 22, 2148, 1983.
20. **Crouch, E. and Longmore, W.,** Collagen binding proteins secreted by tye II pneumocytes in culture, *Biochim. Biophys. Acta*, 924, 81, 1987.
21. **Stoner, G. D., Katoh, Y., Foidart, J. M., Trump, B. F., Steinert, P. M., and Harris, C. C.,** Cultured human bronchial epithelial cells: blood group antigens, keratin, collagens and fibronectin, *In Vitro*, 17, 577, 1981.
22. **Shoji, S., Ertl, R. F., Rickard, K. A., and Rennard, S. I.,** Bronchial epithelial cells produce chemotactic activity for lung fibroblasts, *Am. Rev. Resp. Dis.*, 134, 14, 1988.

23. **Alitalo, K., Hovi, T., and Vaheri, A.,** Fibronectin is produced by human macrophages, *J. Exp. Med.*, 151, 602, 1980.
24. **Villiger, B., Kelley, D. G., Engleman, W., Kuhn, C., III., and McDonald, J. A.,** Human alveolar macrophage fibronectin: synthesis, secretion, and ultrastructural localization during gelatin-coated latex particle binding, *J. Cell Biol.*, 90, 711, 1981.
25. **Rennard, S. I., Hunninghake, G. W., Bitterman, P. B., and Crystal, R. G.,** Production of fibronectin by the human alveolar macrophage: mechanism for the recruitment of fibroblasts to sites of tissue injury in interstitial lung diseases, *Proc. Natl. Acad. Sci. U.S.A.* 78, 7147, 1981.
26. **Beaulieu, A. D., Menard, C., and Parent, C.,** Fibronectin synthesis by polymorphonuclear cells in synovial fluids, *Clin. Res.*, 31, 338A, 1983.
27. **Varga, J., Rosenbloom, J., and Jimenez, S. A.,** Transforming growth factor (TGF) causes a persistent increase in steady-state amounts of type I and type III collagen and fibronectin mRNAs in normal human dermal fibroblasts, *Biochem. J.*, 247, 597, 1987.
28. **Ignotz, R. A. and Massague, J.,** Transforming growth factor-beta stimulates the expression of fibronectin and collagen and their incorporation into the extracellular matrix, *J. Biol. Chem.*, 261, 4337, 1986.
29. **Mecham, R. P., Madaras, J., McDonald, J. A., and Ryan U.,** Elastin production by cultured calf pulmonary artery endothelial cells, *J. Cell Physiol.* 116, 282, 1983.
30. **Vlodavsky, I., Johnson, I. K., Greenburg, G., and Gospodarowicz, D.,** Vascular endothelial cells maintained in the absence of fibroblast growth factor undergo structural and functional alterations that are incompatible with their in vivo differentiated properties, *J. Cell Biol.*, 83, 468, 1979.
31. **Clark, R. A., DellaPelle, P., Manseau, E., Lanigan, J. M., Dvorak, H. F., and Colvin, R. B.,** Blood vessel fibronectin increases in conjunction with endothelial cell proliferation and capillary ingrowth during wound healing, *J. Invest. Derm.*, 9, 269, 1982.
32. **McDonald, J. A.,** personal communication, 1988.
33. **Rannels, S. R., Stinson-Fisher, C., Heuser, L. J., and Rannels, D. E.,** Culture of type II pneumocytes on a type II cell-derived fibronectin-rich matrix, *Physiol. Soc.*, C759, 1987.
34. **Rickard, K. A., Shoji, S., and Rennard, S. I.,** Bronchial epithelial cells exhibit greater attachment to biosynthesized extracellular matrix than to purified components of connective tissue, *Clin. Res.*, 36, 510A, 1988.
35. **Clark, R. A. F., Mason, R. J., Folkvord, J. M., and McDonald, J. A.,** Fibronectin mediates adherence of rat alveolar type II epithelial cells via the fibroblastic cell-attachment domain, *J. Clin. Invest.*, 77, 1831, 1986.
36. **Shoji, S., Rickard, K. A., Ertl, R. F., and Rennard, S. I.,** Lung fibroblasts produce chemotactic activity which stimulates bronchial epithelial cells, *Clin. Res.*, 35, 868A, 1987.
37. **Lechner, J. F., Haugen, A., Autrup, H., McClendon, I. A., Trump, B. F., and Harris, C. C.,** Clonal growth of epithelial cells from normal adult human bronchus, *Cancer Res*, 41, 2294, 1981.
38. **Jetten, A. M., Shirley, J. E., and Stoner, G.,** Regulation of proliferation and differentiation of respiratory tract epithelial cells by TGF β, *Exp. Cell Res.*, 167, 539, 1986.
39. **Paul, J. I., Schwarzbaner, J. E., Tamkun, J. W., and Hynes, R. O.,** Cell-type specific fibronectin subunits generated by alternative splicing, *J. Biol. Chem.*, 261, 12258, 1986.
40. **Zardi, L., Carnemolla, B., Siri, A., Petersen, T. E., Paoletta, G., Sebastio, G., and Baralle, F. E.,** Transformed human cells produce a new fibronectin isoform by preferential alternative splicing of a previously unobserved exon, *EMBO J.*, 6, 2337, 1987.
41. **Kornblihtt, A. R., Umezawa, K., Vibe-Pedersen, K., and Baralle, F. E.,** Primary structure of human fibronectin: differential splicing may generate at least 10 polypeptides from a single gene, *EMBO J.*, 4, 1755, 1985.
42. **Schwarzbauer, J. E., Tamkun, J. W., Lemischka, I. R., and Hynes, R. O.,** Three different fibronectin mRNAs arise by alternative splicing within the coding region, *Cell,* 35, 421, 1983.
43. **Oh, E., Pierschbacher, M., and Ruoslahti, E.,** Deposition of plasma fibronectin in tissues, *Proc. Natl. Acad. Sci. U.S.A.*, 78, 3218, 1981.
44. **Deno, D. C., Saba, T. M., and Lewis, E.,** Kinetics of endogenously labelled plasma fibronectin: incorporation into tissues, *Am. J. Physiol.*, 245, R564, 1983.
45. **Hayman, E. G. and Ruoslahti, E.,** Distribution of fetal bovine serum fibronectin and endogenous rat cell fibronectin in extracellular matrix, *J. Cell Biol.*, 83, 255, 1979.
46. **Sekiguchi, K., Klos, A., Hirohashi, S., and Hakomori, S.,** Human tissue fibronectin: expression of different isotypes in the adult and fetal tissues. *Biochem. Biophys. Res. Commun.*, 141, 1012, 1986.
47. **Duband, J. L., Rocher, S., Chen, W. T., Yamada, K. M., and Thiery, J. P.,** Cell adhesion and migration in the early vertebrate embryo: location and possible role of the putative fibronectin receptor complex, *J. Cell Biol.*, 102, 160, 1986.
48. **Rosenkrans, W. A., Jr., Albright, J. T., Hausman, R. E., and Penney, D. P.,** Light-microscopic immunocytochemical localization of fibronectin in the developing rat lung, *Cell Tissue Res.*, 233, 113, 1983.

49. **Rosenkrans, W. A., Jr., Albright, J. T., Hausman, R. E., and Penney, D. P.,** Ultrastructural immunocytochemical localization of fibronectin in the developing rat lung, *Cell Tissue Res.*, 234, 165, 1983.
50. **Jaskoll, T. F. and Slavkin, H. C.,** Ultrastructural and immunofluorescence studies of basal-lamina alterations during mouse-lung morphogenesis, *Differentiation*, 28, 36, 1984.
51. **Snyder, J. M., O'Brien, J. A., and Rodgers, H. F.,** Localization and accumulation of fibronectin in rabbit fetal lung tissue, *Differentiation*, 34, 32, 1987.
52. **Arey, L. B.,** *Developmental Anatomy*, W. B. Saunders, Philadelphia 1966, 263.
53. **McDonald, J. A., Broekelmann, T. J., Kelley, D. G., and Villiger, B.,** Gelatin-binding domain-specific anti-human plasma fibronectin Fab ′ inhibits fibronectin-mediated gelatin binding but not cell spreading, *J. Biol. Chem.* 256, 5583, 1981.
54. **Grobstein, C. and Cohen, J.,** Collagenase: effect on the morphogenesis of embryonic salivary epithelium in vitro, *Science*, 150, 626, 1965.
55. **Alescio, T.,** Effect of proline analogue, azetidine-2-carboxylic acid, on the morphogenesis in vitro of mouse embryonic lung, *J. Embryol. Exp. Morphol.*, 29, 439, 1973.
56. **Spooner, B. S. and Faubion, J. M.,** Collagen involvement in branching morphogenesis of embryonic lung and salivary gland, *Dev. Biol.*, 77, 84, 1980.
57. **Reddi, A. H.,** *Extracellular Matrix and Development in Extracellular Matrix Biochemistry*, Piez, K. and Reddi, A. H., Eds., Elsevier, 1984, 375.
58. **Chen, L. B., Murray, A., Segal, R. A., Bushnell, A., and Walsh, M. L.,** Studies on intercellular LETS glycoprotein matrices, *Cell*, 14, 377, 1978.
59. **McDonald, J. A., Kelley, D. G., and Broekelmann, T. J.,** Role of fibronectin in collagen deposition: Fab′ to the gelatin-binding domain of fibronectin inhibits both fibronectin and collagen organization in fibroblast extracellular matrix, *J. Cell Biol.*, 92, 485, 1982.
60. **Carter, W. G.,** The role of intermolecular disulfide bonding in deposition of GP140 in the extracellular matrix, *J. Cell Biol.*, 99, 105, 1984.
61. **Miyahara, M., Hayashi, K., Berger, J., Tanzawa, K., Njieha, E. K., Tvelstad, R. L., and Prockop, D. J.,** Formation of collagen fibrils by enzymic cleavage of precursors of type I collagen in vitro, *J. Biol. Chem.*, 259, 9891, 1984.
62. **Lapiere, C. M., Nusgens, B., and Pierard, G. E.,** Interaction between collagen type I and type III in conditioning bundles organization, *Connect. Tissue Res.*, 5, 21, 1977.
63. **Agelli, M. and Wahl, S. M.,** Cytokines and fibrosis, *Clin. Exp. Rheumatol.* 4, 379, 1986.
64. **Davis, W. B. and Crystal, R. G.,** Chronic interstitial lung disease, *Curr. Pulmonol.*, 5, 347, 1984.
65. **Crystal, R. G., Bitterman, P. B., Rennard, S. I., Hanoe, A., and Keogh, B. A.,** Interstitial lung disease of unknown etiology: disorders characterized by chronic inflammation of the lower respiratory tract, *N. Engl. J. Med.*, 310, 154, 1984.
66. **Katzenstein, A. A., Myers, J. L., Prophet, W. D., Corley, L. S., and Shin, M. S.,** Bronchiolitis obliterans and usual interstitial pneumonia, *Am. J. Surg. Pathol.*, 10, 373, 1986.
67. **McLoud, T. C., Epler, G. R., Colby, T. V., Gaensler, E. A., and Carrington, C. B.,** Bronchiolitis obliterans, *Radiology*, 159, 1, 1986.
68. **Reid, L.,** The pathology of chronic bronchitis, in *Recent Trends in Chronic Bronchitis*, Oswald, N. C., Ed., Lloyd-Luke, London, 1958, 26.
69. **Pierce, J. A. and Ebert, R. V.,** Fibrous network of the lung and its change with age, *Thorax*, 20, 469, 1965.
70. **Keogh, B. A. and Crystal, R. G.,** Alveolitis: the key to the interstitial lung disorders, *Thorax*, 37, 1, 1982.
71. **Rennard, S. I., Bitterman, P. B., and Crystal, R. G.,** Current concepts of the pathogenesis of fibrosis: lessons from pulmonary fibrosis, in *Myelofibrosis and the Biology of Connective Tissue*, Berk, P., Ed., Alan Liss, New York, 1984, 359.
72. **McDonald, J. A. and Kelley, D. G.,** Degradation of fibronectin by human leukocyte elastase, *J. Biol. Chem.*, 255, 8848, 1980.
73. **Vartio, T.,** Characterization of the binding domains in the fragments cleaved by cathepsin G from human plasma fibronectin, *Eur. J. Biochem.*, 123, 223, 1982.
74. **Sibille, Y., Lwebuga-Mukasa, J. S., Polomski, L., Merrill, W. W., Ingbar, D. H., and Gee, J. B. L.,** An in vitro model for polymorphonuclear-leukocyte-induced injury to an extracellular matrix. *Am. Rev. Respir. Dis.*, 134, 134, 1986.
75. **Peters, J. H., Ginsberg, M. H., Case, C. M., and Cochrane, C. G.,** Release of soluble fibronectin containing an extra type III domain (EDI) during acute pulmonary injury mediated by oxidants or leukocytes in vivo, *Am. Rev. Resp. Dis.*, 138, 167, 1988.
76. **Peters, J. H., Ginsberg, M. H., Bohl, B. P., Sklar, I. A., and Cochrane, C. G.,** Intravascular release of intact cellular fibronectin during oxidant-induced injury of the in vitro perfused rabbit lung, *J. Clin. Invest.*, 78, 1596, 1986.

77. **Postlethwaite, A. E., Keski-Oja, J., Balian, G., and Kang, A. H.,** Induction of fibroblast chemotaxis by fibronectin: localization of the chemotactic region to an 140,000 molecular weight non-gelatin binding fragment, *J. Exp. Med.*, 153, 494, 1980.
78. **Crapo, J. D.,** Morphologic changes in pulmonary oxygen toxicity, *Annu. Rev. Physiol.*, 48, 721, 1986.
79. **Crystal, R. G., Bradley, K. H., Baum, B. J., et al.,** Cells, collagen and idiopathic pulmonary fibrosis, *Lung*, 155, 199, 1978.
80. **Woodcock-Mitchell, J., Adler, K. B., and Low, R. B.,** Immunohistochemical identification of cell types in normal and in bleomycin-induced fibrotic rat lung. Cellular origins of interstitial cells, *Am. Rev. Resp. Dis.*, 130, 910, 1984.
81. **Tsukamoto, Y., Helsel, W. E., and Wahl, S. M.,** Macrophage production of fibronectin, a chemoattractant for fibroblasts, *J. Immunol.*, 127, 673, 1981.
82. **Rennard, S. I., Hunniunghake, G. W., Bitterman, P. B., and Crystal, R. G.,** Production of fibronectin by the human alveolar macrophage: a mechanism for the recruitment of fibroblasts to sites of tissue injury in interstitial lung disease, *Proc. Natl. Acad. Sci.*, 78, 7147, 1981.
83. **Rouslahti, E., Hayman, E. G., and Pierschbacher, M. D.,** Extracellular matrices and cell adhesion, *Arteriosclerosis, 1985*, 5, 581, 1985.
84. **Mosher, D. F.,** Crosslinking of cold-insoluble globulin by fibrin stabilizing factor, *J. Biol, Chem.*, 250, 6614, 1975.
85. **Fukuda, Y., Ferrans, V. J., Schoenberger, C. I., Rennard, S. I., and Crystal, R. G.,** Patterns of pulmonary structural remodeling after experimental paraquat toxicity, *Am. J. Pathol.*, 118, 452, 1985.
86. **Grinnell, F., Feld, M., and Minter, D.,** Fibroblast adhesion to fibrinogen and fibrin substrata: requirement for cold insoluble globulin, *Cell*, 19, 517, 1980.
87. **Chapman, H. A., Jr., Allen, C. L., Stone, O. L., and Fair, D. S.,** Human alveolar macrophages synthesize factor VII in vitro. Possible role in interstitial lung disease, *J. Clin. Invest.*, 75, 2030, 1985.
88. **Chapman, H. A., Jr., Bertozzi, P., and Reilly, J. J., Jr.,** Role of enzymes mediating thrombosis and thrombolysis in lung disease, *Chest*, 93, 1256, 1988.
89. **Atkinson, J. P. and Frank, M. M.,** *Complement in Clinical Immunology*, Parker, C. W., Ed., Saunders, Philadelphia, 1980, 219.
90. **Bing, D. H., Almeda, S., Lahav, J., Isliker, H., and Hynes, R. O.,** Fibronectin binds to the Clq component of complement, *Proc. Natl. Acad. Sci. U.S.A.* 79, 4198, 1982.
91. **Pearlstein, F., Sorvillo, E., and Gigli, I.,** Interaction of human plasma fibronectin with a subunit of the first component of complement Clq, *J. Immunol.*, 128, 2036, 1982.
92. **Menzel, E. J., Smolen, J. S., Liotta, L., and Reid, K. B.,** Interaction of fibronectin with Clq and its collagen-like fragment (CLF), *FEBS Lett.*, 129, 188, 1981.
93. **Rennard, S. I., Chen, Y. F., Robbins, R. A., Gadek, J. E., and Crystal, R. G.,** Fibronectin mediates cell attachment to Clq: a mechanism for the localization of fibrosis in inflammatory disease, *Clin. Exp. Immunol.*, 97, 1925, 1983.
94. **Kawanami, O., Basset, F., Barrios, R., Lacronique, J. G., Ferrans, V. J., and Crystal, R. G.,** Hypersensitivity pneumonitis in man, *AJP*, 110, 275, 1983.
95. **Crystal, R. G., Fulmer, J. D., Roberts, W. C., Moss, M. L., Line, B. R., and Reynolds, H. Y.,** Idiopathic pulmonary fibrosis: clinical, histologic, radiographic, physiologic, scintigraphic, cytologic and biochemical aspects, *Ann. Int. Med.*, 85, 769, 1976.
96. **Leavitt, R. Y. and Fauci,, A. S.,** Pulmonary vasculitis, *Am. Rev. Respir. Dis.*, 134, 149, 1986.
97. **Engvall, E. and Ruoslahti, E.,** Binding of soluble form of fibroblast surface protein, fibronectin, to collagen, *Int. J. Cancer.* 20, 1, 1977.
98. **Ross, R.,** The biology of platelet derived growth factor, *Cell*, 46, 155, 1986.
99. **Harrington, M. A. and Pledger, W. J.,** Characterization of growth factor modulated events regulating cellular proliferation, *Methods Enzymol.*, 147, 400, 1987.
100. **Bitterman, P. B., Rennard, S. I., Adelberg, S., and Crystal, R. G.,** Role of fibronectin as a growth factor for fibroblasts, *J. Cell Biol.*, 97, 1925, 1983.
101. **Bitterman, P. B., Wewers, M. D., Rennard, S. I., Adelberg, S., and Crystal, R. G.,** Modulation of alveolar macrophage-driven fibroblast proliferation by alternative macrophage mediators, *J. Clin. Invest.*, 77, 700, 1986.
102. **Ozaki, T., Rennard, S. I., and Crystal, R. G.,** Cyclooxygenase metabolites are compartmentalized in the human lower respiratory tract, *J. Appl. Physiol.*, 62, 219, 1987.
103. **Ozaki, T., Rennard, S. I., and Crystal, R. G.,** Arachidonic acid cyclooxygenase metabolites in lung epithelial lining fluid, *Clin. Res.*, 31, 165A, 1983.
104. **Fine, A. and Goldstein, R. H.,** The effect of transforming growth factor-beta on cell proliferation and collagen formation by lung fibroblasts, *J. Biol. Chem.* 262, 3897, 1987.
105. **McDonald, J. A., Broekelmann, T. J., Matheke, M. L., Crouch, E., Koo, M., and Kuhn, C.,** A monoclonal antibody to the carboxyterminal domain of procollagen type I visualizes collagen-synthesizing fibroblasts. Detection of an altered fibroblast phenotype in lungs of patients with pulmonary fibrosis, *J. Clin. Invest*, 78, 1237, 1986.

106. **Hance, A. J. and Crystal, R. G.,** The connective tissue of lung, *Am. Rev. Respir. Dis,* 112, 657, 1975.
107. **Senior, R. M. and Campbell, E. J.,** Neutral proteinases from human inflammatory cells. A critical review of their role in extracellular matrix degradation, *Clin. Lab. Med.,* 3, 345, 1983.
108. **Rennard, S. I., Ferrans, V. J., Bradley, K. H., and Crystal, R. G.,** Lung connective tissue, in: *Mechanisms in Respiratory Toxicology,* Witschi, H. P., Ed., Vol. 2, CRC Press, Boca Raton, FL, 1982, 115,
109. **Vrako, R.,** Significance of basal lamina for regeneration of injured lung, *Virchows Arch. Pathol. Anat.,* 355, 264, 1972.
110. **Martinet, Y., Rom, W. N., Grotendorst, G. R., Martin, G. R., and Crystal, R. G.,** Exaggerated spontaneous release of platelet-derived growth factor by alveolar macrophages from patients with idiopathic pulmonary fibrosis, *N. Eng. J. Med.,* 317, 202, 1987.
111. **Seppa, H., Grotendorst, G., Seppa, S., Schiffman, E., and Martin, G. R.,** The platelet-derived growth factor is chemotactic for fibroblasts, *J. Cell. Biol.,* 92, 584, 1982.
112. **Mensing, H. and Czarnetzki, B. M.,** Leukotriene B_4 induces in vitro fibroblast chemotaxis, *J. Invest. Dermatol.,* 82, 9, 1984.
113. **Postlethwaite, A. E., Keski-Oja, J., Moses, H. L., and Kang, A. H.,** Stimulation of the chemotactic migration of human fibroblasts by transforming growth factor, *J. Exp. Med.,* 165, 251, 1987.
114. **Postlethwaite, A. E., Seyer, J. M., and Kang, A. H.,** Chemotactic attraction of human fibroblasts to type I, II and III collagen and collagen-derived peptides, *Proc. Natl. Acad. Sci. U.S.A.* 75, 871, 1978.
115. **Albini, A. and Adelmann-Grill, B. C.,** Collagenolytic cleavage products of collagen type I as chemattractants for human dermal fibroblasts, *Eur. J. Cell. Biol.,* 36, 104, 1985.
116. **Senior, R. M., Griffin, G. L., and Mechan, R. P.,** Chemotactic response of fibroblasts to tropoelastin and elastin-derived peptides, *J. Clin. Invest.,* 70, 614, 1982.
117. **Senior, R. M., Skogen, W. F., Griffin, G. L., and Wilner, G. D.,** Effects of fibrinogen derivatives upon the inflammatory response, *J. Clin. Invest.,* 77, 1014, 1986.
118. **Postlethwaite, A. E., Snyderman, R., and Kang, A. H.,** Generation of a fibroblast chemotactic factor in serum by activation of complement, *J. Clin. Invest., 64,* 1379, 1979.
119. **Postlethwaite, A. E., Snyderman, R., and Kang, A. H.,** The chemotactic migration of human fibroblasts to a lymphocyte-derived factor, *J. Exp. Med.,* 144, 1188, 1976.
120. **Martinet, Y., Rom, W. N., Grotendorst, G. R., Martin, G. R., and Crystal, R. G.,** Exaggerated spontaneous release of platelet-derived growth factor by alveolar macrophages from patients with idiopathic pulmonary fibrosis, *N. Engl. J. Med.,* 317, 202, 1987.
120a. **Adelmann-Grill, B. C., Hein, R., Wach, F., and Kreis, T.,** Inhibition of fibroblast chemotaxis by recombinant human interferon gamma and interferon alpha, *J. Cell. Physiol.,* 130, 270, 1987.
121. **Bray, B. A., Osman, M., Ashtyani, H., Mandl, I., and Turino, G. M.,** The fibronectin content of canine lungs is increased in bleomycin-induced fibrosis, *Exp. Mol. Pathol.,* 44, 353, 1986.
122. **Dubaybo, B. A. and Thet, L. A.,** Changes in lung tissue and lavage fibronectin after paraquat injury in rats, *Res. Commun. Chem. Pathol. Pharmacol.,* 51, 211, 1986.
123. **Rosenkrans, W. A., Jr. and Penney, D. P.,** Cell-cell matrix interactions in induced lung injury, *Radiat. Res.,* 109, 127, 1987.
124. **Schoenberger, C. I., Rennard, S. I., Bitterman, P. B., Fukuda, Y. F., Ferrans, V. J., and Crystal, R. G.,** Paraquat induced pulmonary fibrosis: role of the alveolitis in modulating the development of fibrosis, *Am. Rev. Respir. Dis.,* 1129, 168, 1984.
125. **Begin, R., Martel, M., Desmarais, Y., Drapeau, G., Boileau, R., Rola-Pleszczynski, M., and Masse, S.,** Fibronectin and procollagen 3 levels in bronchoalveolar lavage of asbestos-exposed human subjects and sheep, *Chest,* 89, 237, 1986.
126. **Bosse, J., Boileau, R., and Begin, R.,** Chronic allergic airway disease in the sheep model: functional and lung-lavage features, *J. Allergy Clin. Immunol.,* 79, 339, 1987.
127. **Rennard, S. I. and Crystal, R. G.,** Fibronectin in human bronchopulmonary lavage fluid, *J. Clin. Invest.,* 69, 113, 1981.
128. **Gerdes, J. S., Yoder, M. C., Douglas, S. D., Paul, M., Harris, M. C., and Polin, R. A.,** Tracheal lavage and plasma fibronectin: relationship to respiratory distress syndrome and development of bronchopulmonary dysplasia, *J. Pediatr.,* 108, 601, 1986.
129. **Peters, J. H.,** personal communication, 1988.
130. **Schoenberger, C. I., Bitterman, P. B., Rennard, S. I., Fukuda, Y., Ferrans, V. J., and Crystal, R. G.,** Role of fibronectin and alveolar macrophage derived fibroblast growth factor in experimental pulmonary fibrosis, *Am. Rev. Resp. Dis.,* 125 (2), 217, 1982.
131. **Raghow, R., Lurie, S., Seyer, J. M., and Kang, A. H.,** Profiles of steady state levels of messenger RNA's coding for type 1 procollagen, elastin, and fibronectin in hamster lungs undergoing bleomycin-induced interstitial pulmonary fibrosis, *J. Clin. Invest.,* 76, 1733, 1985.

132. **Davis, W. B., Rennard, S. I., Bitterman, P. B., and Crystal, R. G.,** Pulmonary oxygen toxicity: early reversible biologic changes in human alveolar structures induced by hyperoxia, *N. Eng. J. Med.,* 309, 879, 1983.
133. **Kradin, R. L., Zhu, Y., Hales, C. A., Bianco, C., and Colvin, R. B.,** Response of pulmonary macrophages to hyperoxic pulmonary injury, *AJP,* 125, 349, 1986.
134. **Rennard, S. I., Hunninghake, G., Davis, W., Moritz, E., and Crystal, R. G.,** Macrophage fibronectin is 1000-fold more potent as a fibroblast chemoattractant than plasma fibronectin, *Clin. Res.,* 30, 356A, 1982.
135. **Bitterman, P. B., Rennard, S. I., Keogh, B., Adelberg, S., and Crystal, R. G.,** Chronic alveolar macrophage release of fibronectin and alveolar macrophage derived growth factor correlates with functional deterioration in fibrotic lung diseases, *Clin. Res.,* 31, 414A, 1983.
136. **Bitterman, P. B., Rennard, S. I., Keogh, B. A., Hunninghake, G. W., Wewers, M. D., Adelberg, S., and Crystal, R. G.,** Familial pulmonary fibrosis: evidence of lung inflammation in unaffected family members, *N. Engl. J. Med.,* 314, 1343, 1986.
137. **Durr, R. A., Dubaybo, B. A., and Thet, L. A.,** Repair of chronic hyperoxic lung injury: changes in lung ultrastructure and matrix, *Exp. Mol. Pathol.,* 47, 219, 1987.
138. **Fukuda, Y., Ishizaki, M., Masuda, Y., Kimura, G., Kawanami, O., and Masugi, Y.,** The role of intraalveolar fibrosis in the process of pulmonary structural remodeling in patients with diffuse alveolar damage, *AJP,* 126, 171, 1987.
139. **Wagner, J. C., Burns, J., Munday, D. E., and McGee, JO'D.,** Presence of fibronectin in pneumoconiotic lesions, *Thorax,* 37, 54, 1982.
140. **Alberts, W. M., Priest, G. R., and Moser, K. M.,** The outlook for survivors of ARDS, *Chest,* 84, 272, 1983.
141. **Rinaldo, J. E. and Rogers, R. M.,** Adult respiratory distress syndrome. Changing concepts of lung injury and repair, *N. Engl. J. Med.,* 306, 900, 1982.
142. **Fukuda, Y., Ferrans, V. J., Schoenberger, C. I., Rennard, S. I., and Crystal, R. G.,** Patterns of pulmonary structural remodeling after experimental paraquat toxicity, *AJP,* 118, 452, 1985.
143. **Cohler, L. F., Saba, T. M., Lewis, E., Vincent, P. A., and Charash, W. E.,** Plasma fibronectin therapy and lung protein clearance with bacteremia after surgery, *J. Appl. Physiol,* 63, 623, 1987.
144. **Saba, T. M., Niehaus, G. D., Scovill, W. A., Blumenstock, F. A., Newell, J. C., Holman, J., Jr., and Powers, S. R., Jr.,** Lung vascular permeability after reversal of fibronectin deficiency in septic sheep, *Ann. Surg.,* 198, 654, 1983.
145. **Saba, T., Kiener, J. L., and Holman, J. M., Jr.,** Fibronectin and the critically ill patient: current status, *Int. Care Med.,* 12, 350, 1986.
146. **Saba, T. M. and DiLuzio, N. R.,** Reticuloendothelial blockade and recovery as a function of opsonic activity, *Am. J. Physiol.,* 216, 197, 1969.
147. **Saba, T. M. and Jaffe, E.,** Plasma fibronectin (opsonic glycoprotein): its synthesis by vascular endothelial cells and role in cardiopulmonary integrity after trauma as related to reticuloendothelial function, *Am. J. Med.,* 68, 577, 1980.
148. **Blumenstock, E. A., Saba, T. M., Weber, P., and Laffin, R.,** Biochemical and immunological characterization of human opsonic a_2-SB glycoprotein: its identity with cold insoluble globulin, *J. Biol. Chem.,* 253, 4287, 1978.
149. **Kiener, J. L., Cho, E., Saba, T. M.,** Reticuloendothelial function following infusion of bacterial and non-bacterial particulates: relation to plasma fibronectin, *Curr. Surg.,* 43, 37, 1986.
150. **Mosher, D. F. and Williams, E. M.,** Fibronectin concentration is decreased in plasma of severely ill patients with disseminated intravascular coagulation, *J. Lab. Clin. Med.* 91, 729, 1978.
151. **Blumenstock, F. A., and Saba, T. M., Roccario, E., Cho., E., and Kaplan, J. E.,** Opsonic fibronectin after trauma and particle injection as determined by a peritoneal macrophage monolayer assay, *J. Reticuloendothel. Soc.,* 30, 61, 1981.
151a. **Kiener, J. L., Cho, E., and Saba, T. M.,** Comparative effect of circulating bacterial or nonbacterial particulates on plasma fibronectin: relationship to lung deposition of blood-borne foreign particles, *Cir. Shock,* 19, 357, 1986.
152. **Saba, T. M.,** Plasma fibronectin and hepatic Kupffer cell function, *Progr. Liver Dis.,* 7, 109, 1982.
153. **Deno, D. C., McCafferty, M. H., Saba, T. M., and Blumenstock, F. A.,** Mechanism of acute depletion of plasma fibronectin following thermal injury in rats, *J. Clin. Invest.,* 73, 20, 1984.
154. **Ekindjian, O. G., Marien, M., Wassermann, D., Bruxelle, J., Cazalet, C., Konter, E., and Yonger, J.,** Plasma fibronectin time course in burned patients: influence of sepsis, *J. Trauma.,* 24, 214, 1984.
155. **Scovill, W. A., Saba, T. M., Blumenstock, F. A., Bernard,, H., and Powers, S. R., Jr.,** Opsonic a_2 surface binding glycoprotein therapy during sepsis, *Ann. Surg.,* 188, 521, 1978.
156. **Stevens, L. E., Clemmer, T. P., Laub, R. M., Miya, F., and Robbins, L.,** Fibronectin in severe sepsis, *Surg. Gynecol. Obstetr.,* 162, 222, 1986.

157. **Cohler, L. F., Saba, T. M., and Lewis, E. P.**, Lung vascular injury with protease infusion, *Ann. Surg.*, 202, 240, 1985.
158. **Saba, T. M., Niehaus, G. D., Scovill, W. A., Blumenstock, F. A., Newell, J. C., Holman, J., Jr., and Powers, S. R., Jr.**, Lung vascular permeability after reversal of fibronectin deficiency in septic sheep, *Ann. Surg.*, 198, 654, 1983.
159. **Holman, J. M., Jr., Saba, T. M., Niehaus, G. D., and Lewis, E.**, Lung fluid and protein flux during postoperative sepsis, *Cir. Shock.*, 17, 121, 1985.
160. **Saba, T. M., Cho, E., and Blumenstock, F. A.**, Effect of acute plasma fibronectin depletion on tissue fibronectin levels: analysis by a new fluorescent immunoassay, *Exp. Mol. Pathol.*, 41, 81, 1984.
160a. **Richards, P. S., Saba., T. M., Del Vecchio, P. J., Vincent, P. A., and Gray, V. C.**, Matrix fibronectin disruption in association with altered endothelial cell adhesion induced by activated polymorphonuclear leukocytes, *Exp. Mol. Pathol.*, 45, 1, 1986.
161. **Snyder, E. L. and Luban, N. L. C.**, Fibronectin: applications to clinical medicine, *CRC Crit. Rev. Clin. Lab. Sci.*, 23, 15, 1986.
162. **Doran, J. E., Lundsgaard-Hansen, P., and Rubli, E.**, Plasma fibronectin: relevance for anesthesiology and intensive care, *Int. Care Med.*, 12, 340, 1986.
162a. **Hesselvik, F., Brodin, B., Carlsson, C., Cedergren, B., Jorfeldt, L., and Lieden, G.**, Cryoprecipitate infusion fails to improve organ function in septic shock, *Crit. Care Med.*, 15, 475, 1987.
163. **Czop, J. K.**, Plasma fibronectin and the critically ill, *Int. Care Med.*, 12, 337, 1986.
164. **Maunder, R. J., Harlan, J. M., Pepe, P. E., Paskell, S., Carrico, C. J., and Hudson, L. D.**, Measurement of plasma fibronectin in patients who develop the adult respiratory distress syndrome, *J. Lab. Clin. Med.*, 104, 583, 1984.
165. **Rublik, E., Bussard, S., Frei, E., Lundsgaard-Hansen, P., and Pappova, E.**, Plasma fibronectin and associated variables in surgical intensive care patients, *Ann. Surg.*, 197, 310, 1983.
166. **Richards, W. O., Scovill, W. A., and Baekhyo, S.**, Opsonic fibronectin deficiency in patients with intra-abdominal infection, *Surgery,* 94, 210, 1983.
167. **Dietch, F. A., Gelder, F., and McDonald, J. C.**, The relationship between CIg depletion and peripheral neutrophil function in rabbits and man, *J. Trauma.*, 22, 469, 1982.
168. **Dietch, E. A., Gelder, F., and McDonald, J. C.**, Sequential prospective analysis of the nonspecific host defense system after thermal injury, *Arch. Surg.*, 119, 83, 1984.
169. **Saba, T. M., Blumenstock, F. A., Scovill, W. A., and Bernard, H.**, Cryoprecipitate reversal of opsonic α_2 SB glycoprotein deficiency in septic surgical and trauma patients, *Science,* 201, 622, 1978.
170. **Scovill, W. A., Saba, T. M., Blumenstock, F. A., et al.**, Opsonic a_2 surface binding glycoprotein therapy during sepsis, *Ann. Surg.*, 188, 521, 1978.
171. **Brodin, B., Berghem, L., Frigerg-Nielson, S., Nordstrom, H., and Schildt, B.**, Fibronectin the treatment of septicemia — a preliminary report, in *7th World Congr. Anesthesiol.*, Rugheimer, E., Wawersik, J., Zindler, M., Eds., Exerpta Medica, Hamburg, 1980, 504.
172. **Robbins, A. B., Doran, J. E., Reese, A. C., and Mansberger, A. R.**, Cold-insoluble globulin levels in operative trauma: serum depletion, wound sequestration and biological activity, *Am. Surg.*, 46, 663, 1980.
173. **Lanser, M. E. and Saba, T. M.**, Decreased resistance to Staphylococcus aureus peritonitis by opsonic fibronectin depletion, *Surg. Forum,* 32, 54, 1981.
174. **Villiger, B., Broekelmann, T., Kelley, D., Heymach, G. J., and McDonald, J. A.**, Bronchoalveolar fibronectin in smokers and nonsmokers, *Am. Rev. Respir. Dis.*, 124, 652, 1981.
175. **Haag, M., Morin, R., Ertl, R. F., Von Essen, S., and Rennard, S. I.**, Fibronectin is concentrated in the gel sub-phase of human sputum, *Am. Rev. Respir. Dis.*, 133, A256, 1986.
176. **Jacquot, J., Benali, R., Sommerhoff, C. P.,, Finkbeiner, W. E., Goldstein, G., Puchelle, E., and Basbaum, C. B.**, Identification of albumin-like protein released by cultured bovine tracheal serous cells, *Am. Rev. Resp. Dis.*, 137, 11A, 1988.
177. **Reynolds, H. Y. and Chretien, J.**, Respiratory tract fluids: analysis of content and contemporary use of understanding lung diseases, *DM,* 30, 1, 1984.
178. **Johanson, W. G., Jr., Pierce, A. K., Sanford, J. P., and Thomas, G. D.**, Nosocomial respiratory infections with gram-negative bacilli: the significance of colonization of the respiratory tract, *Ann. Int. Med.*, 77, 701, 1972.
179. **Johanson, W. G., Jr., Woods, D. E., and Chaudhuri, T.**, Association of respiratory tract colonization with adherence of gram-negative bacilli to epithelial cells, *J. Infect. Dis.*, 139, 667, 1979.
180. **Murphy, S. and Florman, A. L.**, Lung defenses against infection: a clinical correlation, *Pediatrics,* 72, 1, 1983.
181. **Woods, D. E., Bass, J. A., Johanson, W. G., Jr., Straus, D. C.**, Role of adherence in the pathogenesis of Pseudomonas aeruginosa lung infection in cystic fibrosis patients, *Infect. Immun.*, 30, 694, 1980.

182. **Woods, D. E., Straus, D. C., Johanson, W. G., Jr., Bass, J. A.,** Role of salivary protease activity in adherence of gram-negative bacilli to mammalian buccal epithelial cells in vivo, *J. Clin. Invest.*, 68, 1435, 1981.
183. **Hormann, H.,** Fibronectin and phagocytosis, *Blut,* 51, 307, 1985.
184. **Czop, J. K., Kadish, J. L., and Austen, K. F.,** Augmentation of human monocyte opsonin-independent phagocytosis by fragments of human plasma fibronectin, *Proc. Natl. Acad. Sci. U.S.A.*, 78, 3649, 1981.
185. **Czop, J. K., McGowan, S. E., and Center, D. M.,** Opsonin-independent phagocytosis by human alveolar macrophages: augmentation by human plasma fibronectin, *Am. Rev. Respir. Dis.*, 125, 607, 1982.
186. **Deitch, E. A., Gelder, F., and McDonald, J. C.,** The role of plasma fibronectin as a nonantibody, noncomplement opsonin for *Staphylococccus aureus, J. Trauma,* 24, 208, 1984.
187. **Benson, B., Hawgood, S., Schilling, J., Clements, J., et al.,** Structure of canine pulmonary surfactant apoprotein: cDNA and complete amino acid sequence, *Proc. Natl. Acad. Sci. U.S.A.*, 82, 6379, 1985.
188. **Green, G. M.,** Alveolo-bronchiolar clearance mechanisms, in *Pulmonary Surfactant System,* Cosmi, E. V. and Scarpelli, E. M., Eds., Elsevier, Amsterdam, 1983, 313.
189. **Reifenrath, R.,** Surfactant action in bronchial mucus transport, in *Pulmonary Surfactant System,* Cosmi, E. V. and Scarpelli, E. M., Eds., Elsevier, Amsterdam, 1983, 313.
190. **Vercellotti, G. M., McCarthy, J., Furcht, L. T., Jacob, H. S., and Moldow, C. F.,** Inflamed fibronectin: an altered fibronectin enhances neutrophil adhesion, *Blood,* 62, 1063, 1983.
191. **Norris, D. A., Clark, R. A. F., Swigart, L. F., Huff, C., Weston, W. L., and Howell, S. E.,** Fibronectin fragments are chemotactic for human peripheral blood monocytes, *J. Immunol.*, 129, 1612, 1982.

Chapter 14

FIBRONECTIN IN LIVER DISEASE

Marcos Rojkind and Patricia Greenwel

TABLE OF CONTENTS

I. INTRODUCTION

Fibronectin is an ubiquitous protein. In many tissues it is associated with basement membranes as well as with interstitial collagens. However, it is not an integral component of basement membranes.[1] In the liver, fibronectin is part of the loose connective tissue, and it has a similar distribution to that of type I and type III collagens.[2] It is also present in the space of Disse where it forms clusters of amorphous material.[1] Fibronectin is in close contact with hepatocytes and endothelial cells. With the former, it establishes direct interaction with the plasmalemma, and with the latter, the contact is established with the nonvascular surface of the sinusoidal lining cells.[3]

Fibronectin synthesis and distribution varies in physiological as well as pathological states. It has been shown that fibronectin biosynthesis is increased during regeneration and in liver cirrhosis. However, while in the former, fibronectin distribution tends to resemble that present in the normal tissue, in cirrhosis, excess deposition of fibronectin occurs in abnormal locations. In general it is mixed with other connective tissue components in fibrous septa. It is also increased in the space of Disse during its capillarization.[2] These modifications in matrix distribution and organization may bear a direct relationship to alterations in liver cell function. It has been shown that hepatocytes attach to different substrata, and that attachment to a given matrix will vary under different physiological states.[4] Normal adult hepatocytes attach with high efficiency to fibronectin-coated dishes. However, hepatocytes obtained from partially hepatectomized animals attach preferentially to laminin.[5] It has been shown also that cell phenotype may vary according to the substrata used for attachment. While fibronectin stimulates the production of fibronectin and collagens by cultured hepatocytes,[6] laminin favors the survival of the cultured cells[7] and glycosaminoglycans stimulate the transcription of liver specific genes.[8]

Many cells in culture are capable of producing fibronectin.[9-18] However, the liver in general, and the hepatocytes in particular, are the main contributors of plasma fibronectin.[19] Therefore, alterations in liver function, such as those observed in acute and chronic liver diseases, may be accompanied by alterations in plasma fibronectin levels. Fibronectin is also present in several body fluids[20-24] and may appear in variable concentration in ascites.[25-27]

In this chapter, we shall analyze two specific topics: (1) fibronectin as an acute phase protein, and (2) fibronectin alterations in acute or chronic liver diseases.

II. FIBRONECTIN AS AN ACUTE PHASE REACTANT

Although it was suggested several years ago that fibronectin could be an acute phase protein,[19,28] this suggestion was confirmed only recently.[29,30] In chicken and rats treated with turpentine, fibronectin plasma levels increase. In the rat a maximal increase is obtained after 24 h (412 ± 59 μg/ml vs. 150 ± 50 μg/ml), and values decline to normal by 72 h.[30] Rat hepatocytes obtained 24 h after turpentine administration synthesize three times more protein than control hepatocytes. Fibronectin production follows the same pattern.[30] However, the fractional synthesis of fibronectin is the same in controls and turpentine-treated animals, suggesting that in the latter, increased fibronectin production is associated with the induction of several liver-produced proteins generically named acute phase reactants.

The actual mechanism of stimulation in fibronectin production is not known. However, it has been suggested that glucocorticoids (dexamethasone) and a cytokine produced by circulating monocytes (HSF, hepatocyte stimulating-factor) may play an important role in the activation of fibronectin synthesis by the liver.[29,30]

Serum corticosterone and HSF were measured in the turpentine-treated chicken.[29] While HSF values peaked at 8 to 12 h, corticosteroids reached their peak value by 24 h. The combined administration of HSF and corticosterone to hepatocyte cultures produce an additive effect. It was suggested from these data that both factors are important, although they may act by different but related mechanisms, i.e., increasing the cytosolic levels of fibronectin mRNA.[31,32] HSF may act at the transcriptional level,[33] while glucocorticoids could have an effect on the stability and half-life of mRNAs.[34]

From the kinetic data, mainly the serum increases in HSF and corticosteroids, it has been suggested that HSF could stimulate the production of corticosteroids. It has been shown already that HSF stimulates cultured pituitary cells to release ACTH.[35] Recent studies have shown that HSF is identical with IFN-β2[36] and has been renamed as IL-6.[37]

Studies from several laboratories have suggested that IL-1 and tumor necrosis factor (TNF) may play an important role in inducing the acute phase response.[38-42] Accordingly, the actual mechanisms of activation of fibronectin production during the acute phase could be more complex. It is possible that several cytokines are required for the production of all the acute phase proteins, including fibronectin. However, it is also possible that some of the original preparations of IL-1 or TNF used for stimulation of an acute phase response were contaminated with HSF, and that the latter stimulated the production of acute phase proteins. Recombinant TNF and IL-1 have a limited activity in stimulating the acute phase reaction in cultured hepatocytes or hepatoma cells.[38,43-44]

It has been reported that plasma fibronectin levels after trauma and severe burns vary with the animal species investigated. While in rats fibronectin values increase, in patients it decreases. It is thus possible that fibronectin is not an acute phase protein in man, while it is in rat,[30] rabbit,[28] and chicken.[29]

III. PLASMA LEVELS

Fibronectin plasma levels have been measured in controls as well as in patients with a large variety of liver diseases. In normal controls, values may range from 300 to 400 μg/ml of plasma.[45] However, different groups, perhaps depending on the methods used for quantification of fibronectin, have reported other values.[46]

Fibronectin plasma levels vary with age and sex. They are lower in newborns[47-48] and increase with age. Values are particularly elevated in females over 50 years of age.[49] Accordingly, changes that occur in disease states must be compared with properly age- and sex-matched controls. A proper matching will facilitate the analysis and interpretation of the data, since wide variation in values have been reported among homogeneous populations analyzed by a single group. Variation could be higher than 20 to 30%.[49-53]

Plasma fibronectin values are modified in the presence of renal disease. Although fibronectin plasma levels do not differ from normal values in patients with chronic renal disease, patients with proteinuria contain plasma fibronectin levels which are 2SDs above controls.[51] There is an inverse correlation between fibronectin plasma levels and total plasma protein concentration.[51] Thus, to properly evaluate alterations in plasma fibronectin that occur in patients with liver disease, one must rule out possible abnormalities in renal function.

An additional factor to consider is the nutritional status of the subject. It has been reported that plasma fibronectin is elevated in rats with vitamin A deficiency.[54] The increase in fibronectin is associated with an increased production (synthesis and secretion) of the protein. Vitamin A appears to regulate fibronectin gene expression at the transcriptional level. Both transcription rates and steady-state levels of cytoplasmic mRNA are increased.[54] The increase in plasma fibronectin values observed in vitamin-A deficient animals is fully reversible upon administration of vitamin A.

Progesterone inhibits the stimulation of fibronectin production induced by glucocorticoids during the acute phase reaction in rats.[55] Therefore, progesterone could also play a role in regulating fibronectin plasma levels in normal subjects. This could explain the variations observed in females: the decrease in plasma fibronectin after puberty and the increase observed after 50 years of age.[49]

Fibronectin is associated with cell surfaces and is also deposited into the extracellular matrix. It has been shown that factor XIII, a transglutaminase involved in fibrin cross-linking, may also participate in establishing cross-links between fibronectin and other components of the connective tissue[56] as well as those formed between fibronectin and fibrin.[57] Therefore, it is implied that factor XIII could play an important role in fibronectin clearance from the circulation.[58] In experimental animals, clearance of exogenous administered fibronectin after bacteremia is delayed, and this delay is associated with a decrease in circulating levels of factor XIII.[59] Accordingly, modifications in factor XIII could also influence plasma fibronectin levels.

IV. FIBRONECTIN AND THERAPY

Fibronectin has been shown to play an important role in phagocytosis. It is considered to be the main plasma modulator of phagocytosis.[60] Depletion of fibronectin can result in greater incidence of infection[61] and in increased susceptibility to the vascular manifestation of shock.[62] Therefore, fibronectin administration to patients with low plasma fibronectin values could result in clinical improvement. There is experimental evidence to suggest that fibronectin administration to rats ameliorates liver damage induced by galactosamine.[63] In patients with sepsis, in which plasma fibronectin is decreased, the administration of cryoprecipitate can increase plasma fibronectin by 52% ± 18%.[64] In other patients with low fibronectin values such as those with surgical or traumatic shock[65,66] or severely burned patients,[67] the administration of fibronectin in a cryoprecipitate form improves opsonic function as well as the clinical status of the patients.[65,68] Similar results have been obtained using purified fibronectin.[69] These results are of interest because it has been suggested that prolongation of fibronectin depression correlates with increased mortality and morbidity.[70]

The above results should be taken with caution. Under some experimental conditions such as the induction of a generalized Schwartzman reaction in rabbits or the production of intraabdominal abscesses in rats, plasma fibronectin levels may be higher than controls.[28] This could be due in part to the induction of an acute phase reaction in these animal species. However, as suggested above, fibronectin may not be an acute phase protein in man, and therefore, fibronectin plasma levels may not increase during the production of an acute phase situation induced by traumatic shock or severe burns. The administration of fibronectin may not be free of side reactions. It could produce particulate aggregation and potential microembolization of debris to capillary beds.[71] In addition, great care should be taken in ruling out possible contamination of the cryoprecipitates with hepatitis and HIV viruses.

V. PLASMA FIBRONECTIN MEASUREMENTS AS A DIAGNOSTIC TOOL

A. OBSTRUCTIVE JAUNDICE

Plasma fibronectin levels were measured in 16 patients that underwent surgery for obstructive jaundice. It was observed that even in patients that developed fatal septicemia and general system failure, there was no significant decrease in circulating fibronectin.[72] These results indicate that the decrease in fibronectin reported by others in posttraumatic septicemic patients may be associated with trauma and not with septicemia.

B. ACUTE AND CHRONIC LIVER DISEASE

Measurements of plasma fibronectin have been performed in patients with various liver diseases. Although it is generally agreed that fibronectin levels are decreased in patients with acute fulminant hepatitis,[58,73] results in other acute and chronic liver diseases are conflicting. In fulminant hepatic failure, a mean of 68 μg/ml of fibronectin has been reported. However, values range from 0 to 158 μg/ml of plasma. These values are significantly lower than controls which show a mean of 268 μg/ml (range 178 to 380 μg/ml).[58] No correlation was found between fibronectin plasma levels and the degree of encephalopathy. The authors suggested that the drop in fibronectin could be explained by consumption during the active phagocytic process. However, the presence of aggregates, containing fibrin cross-linked to fibronectin, similar to those found in patients with leukemia,[74] suggest a possible activation of factor XIII with formation of nonfunctional complexes. In this regard, it has been already shown that activation of the coagulation system occurs in patients with fulminant hepatic failure.[75]

Fibronectin values are decreased in decompensated cirrhotic patients (208.8 μg/ml ± 104.3 μg/ml; controls 401.3 ± 85.8 μg/ml).[50,53] In compensated cirrhotics, fibronectin values can be normal,[76] higher,[50] or lower than normal.[52-53] In all the studies, there is a common feature: they show a wide spread of plasma fibronectin values. Thus, although the means may be different, it is difficult, on the basis of a plasma fibronectin value alone, to establish a diagnosis or suggest a prognosis. However, others have claimed that fibronectin values correlate with severity of fibrotic, inflammatory, and necrotic changes of the liver, and thus reflect the severity of tissue injury.[52]

C. ASCITES

It has been suggested that fibronectin measurements in ascites fluid could help in distinguishing between ascites associated with malignancy and that resulting from complication of liver damage. Fibronectin values are higher in the former than in the latter.[25-26] However, this is still a controversial topic. Different groups of investigators have reported contradictory results.[27,77]

VI. FIBRONECTIN IN PRIMARY AND METASTATIC NEOPLASIAS

Immunocytochemical studies have suggested that qualitative estimations of liver fibronectin could be of help in establishing differential diagnoses among hepatic malignancies. Fibronectin is a common feature of hepatocellular carcinomas, as determined by immunocytochemical methods. However, fibronectin is missing in cholangiocarcinomas. Similarly, metastatic carcinomas do not contain fibronectin in their parenchyma, it is only present in the reactive stroma. However, metastases with strong fibroplastic character and with pericellular fibrosis could give a strong positive reaction with antifibronectin antibodies.[78]

VII. CONCLUSIONS

Our interpretation of the current literature differs from that presented by others. We believe that measurements of plasma fibronectin are of interest, although greater care in evaluation and interpretation of the data is necessary. These measurements, with a few exceptions (i.e., acute liver failure) should not be applied in the general practice. In many instances, this test will increase the cost of patient care, without providing more information than that obtained with a good clinical history or adequate evaluation of routine laboratory data.

In this computer era, some people have tried very hard to find possible correlations between fibronectin plasma levels and clinical, biochemical, and histological parameters. Depending on the quality and sophistication of the programs used, it is possible to find the desired correlation. Although the application of statistical methods could be a good intellectual exercise, the biological and clinical relevance of these data remains to be determined.

REFERENCES

1. **Martínez-Hernández, A.,** The hepatic extracellular matrix. I. Electron immunohistochemical studies in normal rat liver, *Lab. Invest.*, 51, 57, 1984.
2. **Hahn, E., Wick, G., Pencev, D., and Timpl, R.,** Distribution of basement membrane proteins in normal and fibrotic human liver: Collagen type IV, laminin and fibronectin *Gut,* 21, 63, 1980.
3. **Clement, B., Grimaud, J. A., Campion, J. P., Deugnier, Y, and Guillouzo, A.,** Cell types involved in collagen and fibronectin production in normal and fibrotic human liver, *Hepatology,* 6, 225, 1986.
4. **Mourelle, M., Cordero-Hernández, J., Ponce-Noyola, P., and Rojkind, M.,** Abnormal matrix recognition by Morris hepatomas correlates with low glucagon binding capacity, *Hepatology,* 3, 303, 1983.
5 **Carlsson, R., Engvall, E., Freeman, A., and Rouslahti, E.,** Laminin and fibronectin in cell adhesion: enhanced adhesion of cells from regenerating liver to laminin, *Proc. Natl. Acad. Sci. U.S.A.*, 78, 2403, 1981.
6. **Foidart, J. M., Berman, J. J., Paglia, L., Rennard, S., Abe, S., Perantoni, A., and Martin, G. R.,** Synthesis of fibronectin, laminin and several collagens by a liver-derived epithelial cell line, *Lab. Invest.*, 42, 525, 1980.
7. **Bissell, D. M., Arenson, D. M., Maher, J. J., and Roll, F. J.,** Support of cultured hepatocytes by a laminin-rich gel. Evidence for a functionally significant subendothelial matrix in normal rat liver, *J. Clin. Invest.*, 79, 801, 1987.
8. **Fujita, M., Spray, D. C., Choi, H., Saez, J. C., Watanabe, T., Rosenberg, L., Hertzberg, E. L., and Reid, L. M.,** Glucosaminoglycans and proteoglycans induce gap junction expression and restore transcription of tissue-specific mRNAs in primary liver cultures, *Hepatology,* 7, 1S, 1987.
9. **Zetter, B. R., Martin, G. R., Birdwell, C. R., and Gospodarowicz, D.,** Role of high molecular weight glycoprotein in cellular morphology, adhesion and differentiation, *Ann. N.Y. Acad. Sci.*, 312, 299, 1978.
10. **Chen, L. B., Maitland, N., Gallimore, P. H., and McDougall, J. K.,** Detection of the large external transformation-sensitive protein on some epithelial cells, *Exp. Cell. Res.*, 106, 39, 1977.
11. **Yamada, K. M. and Olden, K.,** Fibronectins: adhesive glycoproteins of cell surface and blood, *Nature (London),* 275, 179, 1978.
12. **Zucker, M. B., Mosseson, M. W., Broekman, M. J., and Kaplan, K. L.,** Release of platelet fibronectin (cold insoluble globulin) from alpha granules induced by thrombin or collagen; lack of requirement for plasma fibronectin in ADP-induced platelet aggregation, *Blood,* 54, 8, 1979.
13. **Hoffstein, S. T., Weissmann, G., and Pearlstein, E.,** Fibronectin is a component of the surface coat of human neutrophils, *J. Cell. Sci.*, 50, 315, 1981.
14. **Alitalo, K. I., Hovi, T., and Vaheri, A.,** Fibronectin is produced by human macrophages, *J. Exp. Med.*, 151, 602, 1980.
15. **Vaheri, A., Ruoslahti, E., Westermark, B., and Ponten, J. A.,** A common cell-type surface antigen in cultured human glial cells and fibroblasts lost in malignant cells, *J. Exp. Med.*, 143, 64, 1976.
16. **Burke, J. M., Balian, G., Ross, R., and Bornstein, P.,** Synthesis of types I and III procollagen and collagen by monkey aortic smooth muscle cells in vitro, *Biochemistry,* 16, 3243, 1977.
17. **Hassell, J. R., Pennypacker, J. P., Yamada, K., and Pratt, R. M.,** Changes in cell surface proteins during normal Vitamin A-inhibited chondrogenesis in vitro, *Ann. N.Y. Acad. Sci.*, 312, 406, 1978.
18. **Voss, B., Allam, S., Rauterberg, J., Ullrich, K., Gieselmann, K., and Von Figura, K.,** Primary cultures of rat hepatocytes synthesize fibronectin, *Biochem. Biophys. Res. Commun.*, 90, 1348, 1979.
19. **Owens, M. R. and Cimino, C. D.,** Synthesis of fibronectin by the isolated perfused rat liver, *Blood,* 59, 1305, 1982.
20. **Crouch, E., Balian, G., Holbrook, K., Duskin, D., and Bornstein, P.,** Amniotic fluid fibronectin, *J. Cell Biol.*, 78, 701, 1978.
21. **Vuento, M., Salonen, E., Koskimeis, A., and Stenman, U. H.,** High concentrations of fibronectin-like antigens in human seminal plasma, *Hoppe Seyler's Z. Physiol. Chem.*, 361, 1453, 1980.
22. **Scott, D. L., Wainwright, A. C., Walton, K. W., and Williamson, N.,** Significance of fibronectin in rheumatoid arthritis and osteoarthrosis, *Ann. Rheum. Dis.*, 40, 142, 1981.
23. **Carsons, S., Lavietes, B. B., and Diamond, H. S.,** Factors influencing the incorporation of fibronectin into synovial fluid cryoprotein, *J. Lab. Clin. Med.*, 102, 722, 1983.
24. **Kuusela, P., Vaheri, A., Palo, J., and Ruoslahti, E.,** Demonstration of fibronectin in human cerebrospinal fluid, *J. Lab. Clin. Med.*, 92, 595, 1978.
25. **Deverbizier, G., Beauchant, M., Chapron, A., Touchard, G., and Reiss, D.,** Fibronectin, a marker for malignant ascites, *Lancet,* p. 1104, November 1984.
26. **Colli, A., Buccino, G., Cocciolo, M., Parravicini, R., Mariani, F., and Scaltrini, G.,** Diagnostic accuracy of fibronectin in the differential diagnosis of ascites, *Cancer,* 58, 2489, 1986.
27. **Runyon, B. A.,** Elevated ascitic fluid fibronectin concentration, a non specific finding, *J. Hepatol.*, 3, 219, 1986.

28. **Grossman, J., Pohlman, T., Koerner, F., and Mosher, M. D.,** Plasma fibronectin concentration in animal models of sepsis and endotoxemia, *J. Surg. Res.*, 34, 145, 1983.
29. **Amrani, D. L., Mauzy-Melizz, D., and Mosesson, M. W.,** Effect of hepatocyte-stimulating factor and glucocorticoids on plasma fibronectin levels, *Biochem. J.*, 238, 365, 1986.
30. **Pick-Kober, K. H., Munker, D., and Gressner, A. M.,** Fibronectin is synthesized as an acute phase reactant in rat hepatocytes, *J. Clin. Chem. Clin. Biochem.*, 24, 521, 1986.
31. **Baumann, H. and Eldredge, D.,** Dexamethasone increases the synthesis and secretion of partially active fibronectin in rat hepatoma cells, *J. Cell Biol.*, 95, 29, 1982.
32. **Oliver, N., Newby, R. F., Furcht, L. T., and Bourgeois, S.,** Regulation of fibronectin biosynthesis by glucocorticoids in human fibrosarcoma cells and normal fibroblasts, *Cell*, 33, 287, 1983.
33. **Baumann, H., Hill, R. E., Sauder, D. N., and Jahreis, G. P.,** Regulation of major acute-phase plasma proteins by hepatocyte-stimulating-factors of human squamous carcinoma cells, *J. Cell Biol.*, 102, 370, 1986.
34. **Jefferson, D. M., Reid, L. M., Giambrone, M. A., Shafritz, D. A., and Zern, M. A.,** Effects of dexamethasone on albumin and collagen gene expression in primary cultures of adult rat hepatocytes, *Hepatology*, 5, 14, 1985.
35. **Woloski, B. M. R. N. J., Smith, E. M., Meyer, W. J., Fuller, G. M., and Blalock, J. E.,** Corticotropin-releasing activity of monokines, *Science*, 230, 1035, 1985.
36. **Gauldie, J., Richards, C., Harnish, D., Lansdorp, P., and Baumann, H.,** Interferonβ_2/B-cell stimulatory factor type 2 shares identity with monocyte-derived hepatocyte-stimulating factor and regulates the major acute phase protein response in liver cells, *Proc. Natl. Acad. Sci. U.S.A.*, 84, 7251, 1987.
37. **Poupart, P., Vandenabeele, P., Cayphas, S., Van Snick, J., Haegeman, G., Kruys, V., Fiers, W., and Content, J.,** B cell growth modulating and differentiating activity of recombinant human 26-kDa protein (BSF-2, HuIFN β_2, HPGF), *EMBO J.*, 6, 1219, 1987.
38. **Gauldie, J., Sauder, D. N., McAdam, K. P. W. J., and Dinarello, C. A.,** Purified interleukin-1 (IL-1) from human monocytes stimulates acute-phase protein synthesis by rodent hepatocytes in vitro, *Immunology*, 60, 203, 1987.
39. **Perlmutter, D. H., Goldberger, G., Dinarello, C. A., Mizel, S. B., and Colten, H. R.,** Regulation of class III major histocompatibility complex gene products by interleukin-1, *Science*, 232, 850, 1986.
40. **Ramadori, G., Sipe, J. D., Dinarello, C. A., Mizel, S. B., and Colten, H. R.,** Pretranslational modulation of acute phase hepatic protein synthesis by murine recombinant interleukin-1 (IL-1) and purified IL-1, *J. Exp. Med.*, 162, 930, 1985.
41. **Darlington, G. J., Wilson, D. R., and Lachman, L. B.,** Monocyte-conditioned medium, interleukin-1 and tumor necrosis factor stimulate the acute phase response in human hepatoma cells in vitro, *J. Cell Biol.*, 103, 787, 1986.
42. **Perlmutter, D. H., Dinarello, C. A., and Colten, H. R.,** Cachectin/tumor necrosis factor regulates hepatic acute-phase gene expression, *J. Clin. Invest.*, 78, 1349, 1986.
43. **Koj, A., Kurdowska, A., Magielska-Zero, D., Rokita, H., Sipe, J. D., Dayer, J. M., Demczuk, S., and Gauldie, J.,** Limited effects of recombinant human and murine interleukin-1 and tumor necrosis factor on production of acute phase proteins by cultured rat hepatocytes, *Biochem. Int.*, 14, 553, 1987.
44. **Baumann, H., Onorato, V., Gauldie, J., and Jahreis, G. P.,** Distinct sets of acute phase plasma proteins are stimulated by separate human hepatocyte-stimulating factors and monokines in rat hepatoma cells, *J. Biol. Chem.*, 262, 9756, 1987.
45. **Mosesson, M. W. and Umfleet, R. A.,** The cold-insoluble globulin of human plasma, *J. Biol. Chem.*, 245, 5328, 1970.
46. **Zerlauth, G. and Wolf, G.,** Plasma fibronectin as a marker for cancer and other diseases, *Am. J. Med.*, 77, 685, 1984.
47. **McCafferty, M. H., Lepow, M., Saba, T. M., Cho, E., Meuwissen, H., White, J., and Zuckerbrod, S. F.,** Normal fibronectin levels as a function of age in the pediatric population, *Pediatr. Res.*, 17, 482, 1983.
48. **Barnard, D. R. and Arthur, M. M.,** Fibronectin (cold insoluble globulin) in the neonate, *J. Pediatr.*, 102, 453, 1983.
49. **De Russe, J., Colombat, P., Lavoix, X., and Bardos, P.,** Plasma fibronectin in various hemopathic diseases, *Clin. Chim. Acta*, 145, 49, 1985.
50. **Matsuda, M., Yamanaka, T., and Matsuda, A.,** Distribution of fibronectin in plasma and liver diseases, *Clin. Chim. Acta*, 118, 191, 1982.
51. **Cosio, F. G. and Bakaletz, A. P.,** Abnormal plasma fibronectin levels in patients with proteinuria, *J. Lab. Clin. Med.*, 104, 867, 1984.
52. **Jitoku, M., Koide, N., and Nagashima, H.,** Decreased plasma fibronectin in liver diseases correlated to the severity of fibrotic, inflammatory and necrotic changes of liver tissue, *Acta Med. Okayama*, 40, 189, 1986.

53. **Gabrielli, G. B., Casaril, M., Bonazzi, L., Baracchino, F., Bellisola, G., and Corrocher, R.,** Plasma fibronectin in liver cirrhosis and its diagnostic value, *Clin. Chim. Acta,* 160, 289, 1986.
54. **Kim, H. Y. and Wolf, G.,** Vitamin A deficiency alters genomic expression for fibronectin in liver and hepatocytes, *J. Biol. Chem.,* 262, 365, 1987.
55. **Nimmer, D., Bergtrom, G., Hirano, H., and Amrani, D. L.,** Regulation of plasma fibronectin biosynthesis by glucocorticoids in chick hepatocyte cultures, *J. Biol. Chem.,* 262, 10369, 1987.
56. **Upchurch, H. F., Conway, E., Patterson, M. K., Birckbichler, P. J., and Maxwell, M. D.,** Cellular transglutaminase has affinity for extracellular matrix, In Vitro, *Cell. Dev. Biol.,* 23, 795, 1987.
57. **Mosher, D. F.,** Cross-linking of cold insoluble globulin by fibrin stabilizing factor, *J. Biol. Chem.,* 250, 6614, 1975.
58. **Almasio, P. L., Hughes, R. D., and Williams, R.,** Characterization of the molecular forms of fibronectin in fulminant hepatic failure, *Hepatology,* 6, 1340, 1986.
59. **Kiener, J. L., Cho, E., and Saba, T. M.,** Factor XIII as a modulator of plasma fibronectin alterations during experimental bacteremia, *J. Trauma,* 26, 1013, 1986.
60. **Snyder, E. L. and Luban, N. L. C.,** Fibronectin: applications to clinical medicine, *CRC Crit. Rev. Clin. Lab. Sci.,* 23, 15, 1986.
61. **Pardy, B. J., Spencer, R. C., and Dudley, H. A. F.,** Hepatic reticuloendothelial protection against bacteremia in experimental hemorrhagic shock, *Surgery,* 8, 193, 1977.
62. **Zweifach, B. W., Benacerraf, B., and Thomas, L.,** Relationship between the vascular manifestations of shock produced by endotoxin, trauma and hemorrhage. II. The possible role of the RES in resistance to each type of shock, *J. Exp. Med.,* 106, 403, 1957.
63. **Moriyama, T., Aoyama, H., Ohnishi, S., and Imawari, M.,** Protective effects of fibronectin in galactosamine-induced liver failure in rats, *Hepatology,* 6, 1334, 1986.
64. **Hesselvick, J. F.,** Plasma fibronectin levels in sepsis: influencing factors, *Crit. Care Med.,* 15, 1092, 1987.
65. **Scovill, W. A., Saba, T. M., Blumenstock, F. A., Bernard, H., and Powers, S. R.,** Opsonic alpha-2-surface binding glycoprotein therapy during sepsis, *Ann. Surg.,* 188, 521, 1978.
66. **Saba, T. M., Blumenstock, F. A., Scovill, W. A., and Bernard, H.,** Cryoprecipitate reversal of opsonic alpha-2-A B Glycoprotein deficiency in septic surgical and trauma patients, *Science,* 201, 622, 1978.
67. **Lanser, M. E., Saba, T. M., and Scovill, W. A.,** Opsonic glycoprotein (plasma fibronectin) levels after burn injury. Relationship to extent of burn and development of sepsis, *Ann. Surg.,* 192, 776, 1980.
68. **Ninnemann, J. L.,** Immunologic defense against infection. Alterations following thermal injuries. *J. Burn. Care Rehabil.,* 3, 355, 1982.
69. **Saba, T. M., Blumenstock, F. A., Shah, D. M., Landaburu, R. H., Hrinda, M. E., Deno, D. C., Holman, J. M., Jr., Cho, E., Dayton, C., and Cardarelli, P. M.,** Reversal of opsonic deficiency in surgical, trauma and burn patients by infusion of purified human plasma fibronectin. Correlation with experimental observations, *Am. J. Med.,* 80, 229, 1986.
70. **Scovill, W. A., Saba, T. M., Kaplan, J. E., Bernard, H., and Powers, S., Jr.,** Deficits in reticuloendothelial humoral control mechanisms in patients after trauma, *J. Trauma,* 16, 898, 1976.
71. **Saba, T. M.,** Reversal of plasma fibronectin deficiency in septic-injured patients by cryoprecipitate infusion, in, *Massive Transfusion in Surgery and Trauma,* Collins, J. A., Murawski, K., and Shafer, A. W., Eds., Alan R. Liss, New York, 1982, 129.
72. **Hadjis, N. S., Griffin, S., Blumgart, L. H., and Knox, P.,** Outcome of surgery for obstructive jaundice is not associated with reduced levels of plasma fibronectin, *Clin. Sci.,* 70, 73, 1986.
73. **González Calvin, J., Scully, M. F., Sanger, Y., Fok, J., Kakkar, V. V., Hughes, R. D., Gimson, A. E. S., and Williams, R.,** Fibronectin in fulminant hepatic failure, *Br. Med. J.,* 285, 1231, 1982.
74. **Reilly, J. T., Galloway, M. J., Mackie, M. J., et al.,** Fibronectin C in acute leukemia, *Br. J. Haematol.,* 58, 83, 1984.
75. **Hughes, R. D., Lane, D. A., Ireland, H., et al.,** Fibrinogen derivatives and platelet activation products in acute and chronic liver disease, *Clin. Sci.,* 68, 701, 1985.
76. **Gluud, C., Dejgaard, A., and Clemmensen, I.,** Plasma fibronectin concentrations in patients with liver diseases, *Scand. J. Clin. Lab. Invest.,* 43, 533, 1983.
77. **Villar, M., García-Bragado, F., Vilardell, M., Brosca, M., and Rodrigo, M. J.,** Fibronectin concentration in ascites does not differentiate between malignant and non malignant ascites, *Gastroenterology,* 94, 556, 1988.
78. **Szendroi, M. and Lapis, K.,** Fibronectin in differential diagnosis of primary hepatomas and carcinoma matastases in the liver, *Acta Morphol. Hung.,* 33, 101, 1985.

Chapter 15

FIBRONECTIN IN CARTILAGE

Nancy Burton-Wurster and George Lust

TABLE OF CONTENTS

I. INTRODUCTION

There is both immunohistochemical and biochemical evidence for the presence of fibronectin in articular cartilage. Its importance to arthritic disease is suggested by reports that the fibronectin content of degenerated cartilage is markedly increased in human osteoarthritis and in animal models of the disease.

II. FIBRONECTIN IN ARTICULAR CARTILAGE — IMMUNOHISTOCHEMICAL EVIDENCE

The presence of fibronectin in articular cartilage has been documented now in several laboratories by immunohistochemical techniques although findings of the precise distribution of fibronectin within the cartilage have not been consistent. Weiss and Reddi[1] reported that fibronectin in rat embryonic cartilage was masked by proteoglycans but could be detected after the cartilage had been treated with hyaluronidase. Clemmensen et al.[2] demonstrated fibronectin in human articular cartilage, and this detection also required prior treatment with hyaluronidase. They identified fibronectin both within the matrix and pericellularly, but the highest intensity of staining was around the chondrocytes. Evans et al.[3] found fibronectin in the interterritorial matrix of adult bovine nasal cartilage without the need to use hyaluronidase treatment, but observed no staining around the chondrocytes. In contrast, Glant et al.[4] reported that fibronectin in normal adult human cartilage was limited to the pericellular area. They stated that fibronectin can also appear in the territorial and interterritorial areas of the surface layer of human articular cartilage obtained from arthritic joints. Their tissues were pretreated with chondroitinase ABC and collagenase prior to immunohistochemical staining. Burton-Wurster et al.[5] detected fibronectin throughout the matrix in both canine and lapine articular cartilage whether or not the cartilage was disease free or osteoarthritic. Subsequently, they demonstrated that the duration and intensity of hyaluronidase pretreatment of cartilage sections was critical in revealing the location of fibronectin within the articular cartilage. Milder treatment revealed fibronectin pericellularly while masking detection in the matrix.[6] Shiozawa et al.,[7] possibly because they used tissue which was fixed in paraformaldehyde rather than frozen sections, did not detect fibronectin in normal or osteoarthritic tissue, but did find fibronectin on the surface of articular cartilage in rheumatoid arthritis. Nevertheless, in those areas where fibronectin was detected, they extended their observation to the ultrastructural level and observed fibronectin in association with collagen fibrils as well as with an unidentified amorphous substance in the matrix. Recently, Jones et al.[50] detected the presence of fibronectin within the matrix of human osteoarthritic cartilage, but not in disease-free human cartilage. The fibronectin was mainly in the surface zones but also in the mid and deep zones. It was found pericellularly as well as within the matrix. Rees et al.[51] also made observations at the ultrastructural level and found fibronectin within an amorphous substance, and most importantly, in the endoplasmic reticulum and Golgi of some chondrocytes. This is evidence that fibronectin is synthesized by chondrocytes *in vivo*. Silbermann et al.[8] reported an enhanced appearance of fibronectin in cartilage from the mandibular joint of newborn mice after explants were treated with dexamethasone.

The consensus to be drawn from these studies is that fibronectin is present in articular cartilage; however, inferences from immunohistochemical studies as to specific localization within the matrix or to quantitative differences in cartilage fibronectins must be made carefully in light of the strong dependence on the methods used to prepare and probe the tissue. Biochemical studies on cartilage explants confirm the presence of fibronectin in cartilage matrix and, in conjunction with cell culture experiments, provide additional evidence that the chondrocyte itself is capable of fibronectin biosynthesis. Unfortunately, the role of fibronectin in cartilage matrix has not been elucidated. A possible role, that of cell attachment, has not been established; rather it was reported that chondronectin, not fibronectin, is the cell attachment factor for chondrocytes.[9] But the importance of fibronectin as a marker for, or modulator of, chondrocyte differentiation/dedifferentiation or phenotype modulation is consistent with much of the information obtained from chondrocytes in cultures. In addition, there is evidence that fibronectin content of cartilage is strikingly increased in osteoarthritic lesions. These topics are discussed now.

III. FIBRONECTIN IN ARTICULAR CARTILAGE — BIOCHEMICAL EVIDENCE

Methods of Extraction and Quantitation

It was reported that urea, in conjunction with heparin, solubilized fibronectin from tissues.[10] This has worked well for cartilage fibronectin, yielding a product which released subunits of approximately 220,000 Da upon reduction. This material could be quantitated in an ELISA assay[11,12] using available polyclonal and monoclonal antibodies to fibronectin with little or no interference from coextracted matrix molecules, or could be purified easily on a gelatin affinity column.[11,13] Since extraction with urea did not extract all of the cartilage fibronectin, it was important to determine that the quantity of extracted fibronectin was representative of the total fibronectin. Extraction with guanidinium chloride released additional fibronectin from the tissue, but proteoglycans had to be removed before detection with antibody was possible.[14,15] Treatment with testicular hyaluronidase also released additional fibronectin from cartilage.[5] Nevertheless, the treatment of cartilage with guanidinium chloride or hyaluronidase extracted more fibronectin from osteoarthritic cartilage than from disease-free cartilage, substantiating the finding that the increased fibronectin in osteoarthritic joints represented a real increase in content, rather than increased extractability of fibronectin from arthritic cartilage.[14,5] Fibronectin can also be extracted from cartilage by heating at 80 to 100°C with 0.2% $NaDodSO_4$ and 2% mercaptoethanol. This material has been suitable for gel electrophoresis but not for ELISA assay. A report by Bowness[16] suggested the presence of insoluble structural glycoproteins in cartilage matrix which cross reacted with antibodies to fibronectin but were of lower molecular weight. These have not been further identified. In our laboratory, cartilage has been extracted with 4 *M* urea in phosphate buffer (pH 7.2; 0.05 *M*) to obtain fibronectin for biochemical studies on the structure of cartilage fibronectin and on the increases in fibronectin in cartilage in several animal models of arthritis.

IV. THE STRUCTURE OF CARTILAGE FIBRONECTIN

Recent work suggested that there is a single gene for fibronectin, but alternate splicing of the primary transcript from this gene results in multiple mRNAs encoding for different fibronectin subunits. Posttranslational modifications contribute further to the heterogeneity of this molecule.[17] For human fibronectin, alternate splicing has now been documented in three regions of the pre-mRNA. These are ED-A and ED-B, both type III homology repeats, and IIICS, a 120-amino acid segment which may be omitted or partially or completely inserted between the last two type III homology repeats (Figure 1.)[17-19] The presence or absence of IIICS accounts for the difference in molecular weight of the two subunits of plasma fibronectin. Of the Human plasma fibronectin subunits, 40% are of the beta type, containing no IIICS. In contrast, 75% of the subunits of fibronectin derived from cultured fibroblasts are IIICS containing α subunits and this figure increases to 90% for fibronectin from tumor-derived or transformed fibroblasts.[20] ED-A and ED-B are not detected on plasma fibronectin. ED-A is expressed in the "cellular" fibronectin of cultured fibroblasts and expression increases tenfold in tumor-derived or transformed human fibroblasts.[21] ED-B is barely detectible in fibronectin from normal cells but was first identified in fibronectin

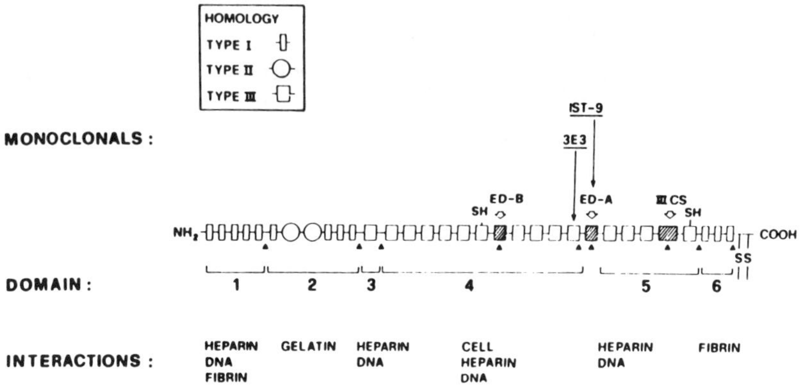

FIGURE 1. Model of the domain structure of a subunit of human FN. White arrows indicate the regions of variability due to alternative splicing of the FN mRNA precursors. Arrowheads indicate the thermolysin cleavage sites. Arrows indicate the sites where the two Mabs 3E3 and IST-9, react. The figure also indicated the internal homologies. (From Zardi, L. et al., *EMBO J.*, 6, 2337, 1987. With permission of IRL Press Limited, Oxford, England.)

produced by tumor cells.[18] Although ED-A was identified in platelet-derived fibronectin,[22] in general, little is known about the distribution of these extra domains in tissues *in vivo*. Indeed, the work of Sekiguchi et al.[23] suggests that the fibronectins derived from cultured fibroblasts are similar to fibronectins derived from fetal tissues while fibronectins from adult tissue are similar to plasma fibronectin but not necessarily derived from plasma. The functional significance of the various fibronectin isoforms is at present unknown but remains an important question.

The fibronectin extracted from canine articular cartilage could be distinguished from canine plasma fibronectin by several criteria including molecular weight, glycosylation, and reactivity with a monoclonal antibody raised in our laboratory and designated 13G3B7. However, the cartilage fibronectin also differed from the "cellular" fibronectin produced by fibroblasts in culture in that very little of the fibronectin found in cartilage (< 2%) was of the isotype which contains the extra domain A (ED-A). Fibronectin which was extracted from cartilage and reduced with mercaptoethanol migrated on sodium dodecyl sulfate polyacrylamide gels predominantly as a doublet, as did canine plasma fibronectin. The larger subunit of cartilage fibronectin was 3- to 5-kDa larger than the upper subunit of plasma fibronectin; the lower subunit of cartilage fibronectin was about 10-kDa smaller than the lower molecular weight subunit of plasma fibronectin. Monoclonal antibody 13G3B7 reacted with both subunits of plasma fibronectin, but failed to recognize the small subunit of cartilage fibronectin (Figure 2), and had reduced overall affinity for cartilage fibronectin in an enzyme-linked immunosorbent assay. The epitope recognized by 13G3B7 was located near the C-terminal end of the molecule in the last type III homology repeat. A subunit of intermediate molecular weight was also observed in cartilage fibronectin but was present at markedly reduced levels. Cartilage fibronectin also differed from plasma fibronectin with respect to glycosylation. As was reported for human synovial fluid fibronectin,[24,25] canine cartilage and synovial fluid fibronectins contained less sialic acids (but did not completely lack sialic acids) than plasma fibronectin and reacted more strongly with wheat germ agglutinin.[26] Preliminary results suggested that fibronectin from equine cartilage is similar to fibronectin from canine cartilage. Additional work needs to be done to characterize fibronectin from the cartilage of humans and other species.

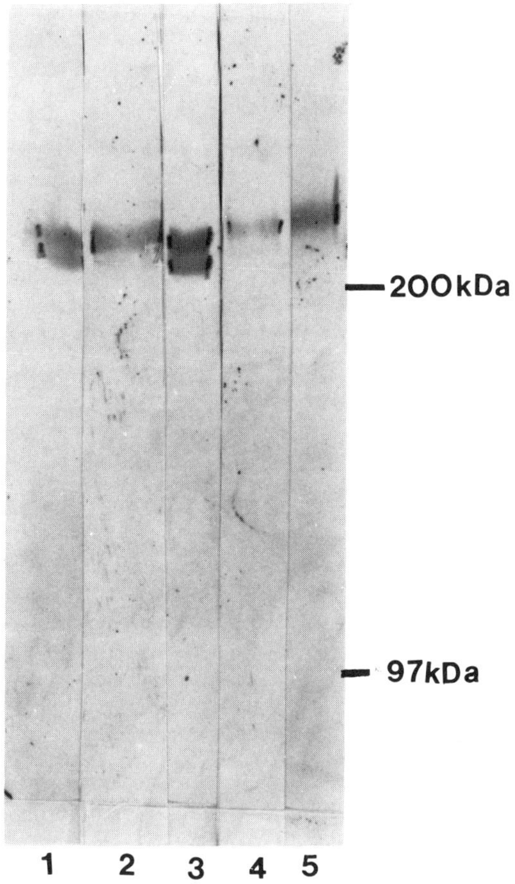

FIGURE 2. Western blot of canine plasma with extracts of canine cartilage separated on 4% $NaDodSO_4$-PAGE and probed with monoclonal antibody 13G3B7. Lanes 1,3 — Canine plasma. Lane 2 — Osteoarthritic cartilage from a hip joint (urea extract). Lane 4 — Disease-free cartilage from normal shoulder joints (urea extract). Lane 5 — Cartilage without visible abnormality from the surrounding osteoarthritic cartilage in a hip joint (direct extraction with $NaDodSO_4$ and mercaptoethanol). (From Burton-Wurster, N. and Lust, G., *Trans. 34th Annu. Meet. Orthop. Res. Soc.*, 13, 492, 1988. With permission.)

V. FIBRONECTIN PRODUCTION BY CHONDROCYTES IN CELL CULTURE

Chondrocytes in cell culture do synthesize fibronectin, albeit the total amount of fibronectin produced can vary with the particular culture.[27,28] This has been the experience of investigators working with chondrocytes isolated from embryonic chick sterna, and our recent experience with chondrocytes isolated from canine articular cartilage obtained from dogs between 6 to 18 months old has been similar.

Fibronectin production by cultured chondrocytes has been reported to parallel other changes that are considered to reflect chondrocyte dedifferentiation. For example, Gionti et al.[28] reported that Rous sarcoma virus (RSV) transformation of chick embryo chondrocytes resulted in a switch to the spindle shaped morphology characteristic of fibroblasts, a decrease in type II collagen and chondronectin synthesis, but an increase in fibronectin synthesis. Allebach et al[29] showed that these changes occurred at the level of increased production of mRNA for fibronectin and decreased production of mRNA for type II collagen. Dessau et al.[27] reported that as chondrocytes round up, total fibronectin associated with them is reduced. Gerstenfeld et al.[30] treated chondrocytes from chick sternum with phorbol-12-myristate-13-acetate (PMA) and found that round, floating cells will attach, spread, and accumulate increased levels of B-actin mRNA and fibronectin mRNA but reduced levels of type II collagen mRNA.

Hiraki et al.[31] demonstrated that rabbit costal chondrocytes treated with dibutyryl cyclic AMP will synthesize greater amounts of cartilage-specific proteoglycans and that those synthesized are of higher molecular weight than untreated. We have looked for a possible relationship between cyclic AMP, synthesis of cartilage-specific macromolecules, fibronectin synthesis, and cell shape in canine articular chondrocytes in monolayer cultures. We have observed that addition of dibutyryl cyclic AMP to the cultures favored retention of a rounded morphology and decreased fibronectin synthesis.[32] Experiments suggested that the isotype of fibronectin synthesized in culture may also be affected by the addition of dibutyryl cyclic AMP to the culture medium. The untreated canine chondrocytes in culture assumed a polygonal morphology and expressed, at a relatively high level (10 to 25%), an isotype of fibronectin in which the type III homology termed ED-A is present. This isotype was detected in cartilage fibronectin only at very low levels (1 to 2%).[33] The proportion of fibronectin containing the ED-A sequence was reduced when chondrocytes remained round in the presence of dibutyryl cAMP. At the same time, the cultures treated with dibutyryl cAMP increased production of keratin sulfate, probably reflecting increased synthesis of the cartilage-specific proteoglycan.

Hassell et al.[34] and Bernard et al.[35] reported that retinoic acid, which caused cultured chick chondrocytes to assume a fibroblastic, spindle-shaped morphology and inhibited proteoglycan synthesis, also mildly increased total fibronectin production, increased fibronectin accumulation in the cell layer, and shifted glycosylation patterns from high mannose oligosaccharides to the complex type oligosaccharides.

Finally, adding fibronectin itself to chick chondrocytes in culture induced the spindle-shaped morphology characteristic of fibroblasts, inhibited proteoglycan synthesis, and increased fibronectin synthesis.[36,37] More recently, West et al.[38] reported that, in their hands, only cellular fibronectin was able to effect these changes. Gibson et al.[39] reported that fibronectin in the culture medium of chick embryo sternal cartilage chondrocytes promoted the synthesis of two low molecular weight cartilage collagens probably related to what is now called type IX collagen.

The phenotypic modulation of chondrocytes in culture and the possible relationship of this to arthritic diseases is an interesting and important topic but further discussion will not be attempted here. The reader is referred to articles by Wurster and Lust[40] and Benya.[41]

VI. FIBRONECTIN IN OSTEOARTHRITIS

A. THE SPONTANEOUS OSTEOARTHRITIS WHICH ACCOMPANIES CANINE HIP DYSPLASIA

Two lines of Labrador retrievers are maintained at the James A. Baker Institute at the College of Veterinary Medicine at Cornell University. They differ in susceptibility to canine hip dysplasia and the osteoarthritis which accompanies it. Dogs from the line which is prone to develop hip dysplasia usually develop the disease from 4 to 9 months of age and at necropsy will have characteristic degenerative focal lesions on the cartilage of one or both hip joints, and occasionally also on the shoulders and knees. The fibronectin content is markedly increased in the deteriorating cartilage, as much as 40-fold when compared to

TABLE 1
Fibronectin Content of Canine Articular Cartilages

Source of cartilage	Fibronectin (ng/mg wet cartilage): Surrounding area		Site of lesion predilection	
Hip joint disease free[a]	17 ± 32	(7)	30 ± 41	(9)
Shoulder joint, disease free	35 ± 45	(16)	42 ± 60	(17)
	Surrounding area		**Site of degeneration**	
Hip joint, osteoarthritic	274 ± 348	(9)	3192 ± 3781	(8)
Shoulder joint, osteoarthritic	121	(1)	1931	(1)

Note: The 11 dogs used in this study were between the ages of 5 and 11 months. Cartilage was removed from the appropriate region of the joint surfaces and extracted with a buffer containing heparin/urea and analyzed for fibronectin by an enzyme-linked immunosorbent assay. Data are expressed as means ± S.D. Numbers in parentheses are the number of cartilage samples analyzed.

[a] Some samples of disease-free cartilage were below the limit of detection for fibronectin; these were assigned a value of zero.

Reproduced in part from Wurster, N. and Lust, G., *Biochem. Biophys. Acta*, 800, 52, 1984, with permission of Elsevier Science Publishers B.V.

disease-free cartilage. Cartilage from the area surrounding a focal lesion will often also exhibit elevated fibronectin content. It is not unusual to be able to extract as much as 2 to 4 μg of fibronectin from 1 mg of lesion cartilage. Corresponding values for normal cartilage are of the order of 100 ng/mg of wet cartilage, and values for surrounding area cartilage could vary from 300 to 1000 ng/mg cartilage (Table 1).[12,42]

It is interesting to consider the macroscopically "normal" cartilage from osteoarthritic joints. As the disease progresses, these areas will eventually degenerate, and so may be thought to be in an early stage of the disease. Our data suggested that fibronectin was increased in cartilage in these early osteoarthritic stages while the glycosaminoglycan content was still in the normal (disease free) range (Table 2). For the data in Table 2, 56% of samples of normal cartilage with a fibronectin content greater than 300 ng/mg wet weight were from a joint with an osteoarthritic lesion. In contrast, only 11% of samples with a fibronectin content less than 100 ng/mg wet weight were from joints with osteoarthritis. This trend was found to be significant by a test for linear trends in proportions ($p < 0.017$).[43]

The source of the increased fibronectin, especially in the early stages, is of interest. The differences in structure between plasma and cartilage fibronectin makes it extremely unlikely that the plasma is the source of the accumulated cartilage fibronectin. At this point, evidence that the chondrocyte is the source of the increased fibronectin in osteoarthritic cartilage includes the following observations: cartilage explants were capable of fibronectin biosynthesis; explants of lesion cartilage synthesized more fibronectin than normal and then retained more of this newly synthesized fibronectin within the matrix;[42] fibronectin synthesized by cartilage explants was of similar molecular weight to the bulk of the fibronectin which accumulated in degenerated cartilage.[26] On the other hand, synovial fluid, which may well contain a mixture of fibronectins from the synovium and the cartilage, cannot be ruled out as a source of cartilage fibronectin, especially in the lesion when fibrillation is present. This was made clear by studies which looked at the uptake by cartilage explants of purified

TABLE 2
Comparison of Glycosaminoglycan Levels in Cartilage with Low or Moderately Elevated Fibronectin Content

	Fibronectin content		Fibronectin content	
Experiment	low level (μg/mg)	GAG (μg/mg)	Elevated (μg/mg)	GAG (μg/mg)
1	0.05 ± 0.05 (12)	68.0 ± 20.0 (12)	0.67 ± 0.28 (7)	75.8 ± 10.3 (7)
2	0.08 ± 0.04 (26)	52.5 ± 9.3 (26)	0.25 ± 0.05 (19)	51.3 ± 10.0 (19)

Note: Experiment 1: cartilage was harvested and stored frozen until analyzed for fibronectin in an ELISA, or for glycosaminoglycan content, by reactivity with the dye dimethymethylene blue. Experiment 2: cartilage was held at 4°C for various times (days), then incubated at 37°C for 24 h. Fibronectin and GAG contents were determined on same sample. The percentage of water in these samples was 71%. Cartilage obtained from femoral and humeral heads of 20 dogs. GAG, glycosaminoglycan.

From Burton-Wurster, N. and Lust, G., Incorporation of purified plasma fibronectin into explants of articular cartilage from disease-free and osteoarthritic canine joints, *J. Orthop. Res.*, 4, 1437, 1986. Copyright Orthopaedic Research Society. With permission of Raven Press.

plasma fibronectin labeled with biotin or ^{125}I-iodine. Degenerated cartilage accumulated as much as tenfold more fibronectin than disease-free cartilage although penetration occurred only from the articular surface.[44] Disease-free cartilage maintained a barrier to fibronectin penetration from the articular surface which was sustained even after the proteoglycan content was markedly reduced by incubation of the cartilage with catabolin or lipopolysaccharide.[44,6] The caveats relevant to these experiments include the possible modifications of fibronectin by iodination and biotinylation procedures and the fact that plasma fibronectin is not identical to cartilage or synovial fluid fibronectin and therefore would not normally be the fibronectin seen by the cartilage.

B. SURGICALLY INDUCED OSTEOARTHRITIS IN THE RABBIT

Osteoarthritis can be induced surgically in the rabbit as described by Colombo et al.[45] In this procedure, the collateral and sesamoid ligaments of the right knee were sectioned and 405 mm of the anterior lateral meniscus was removed from New Zealand white male rabbits. This resulted in the appearance of osteoarthritic lesions predominantly on the right lateral femoral condyle. The unoperated left knee, or a sham operation on the right knee of control animals, served as controls. At 6 weeks post surgery, moderate to severe lesions on the right lateral femoral condyle exhibited loss of Safranin O stain, flaking, and loss of the superficial layer, fibrillation, loss of chondrocytes, and multicellular clusters, as observed by histochemical techniques. Such lesions contained up to 20-fold more fibronectin than normal cartilage from the contralateral or sham-operated knees as determined in an ELISA assay on 4-*M* urea extracts of the cartilage (Table 3).[5]

C. INDUCED MODELS OF CARTILAGE DEGENERATION IN CANINE JOINTS

In a preliminary study,[46] the fibronectin content of cartilage in two different models of joint degeneration was determined. In the first model, osteoarthritis was induced in adult mongrel dogs by anterior cruciate ligament transections as described by Palmoski et al.[47] After 8 to 24 weeks, animals were sacrificed and articular cartilage was obtained from the habitually loaded areas of the femoral condyles. The disease was present at an early stage when macroscopic lesions were not observed, reduction in Safranin O staining was mild, and chondroitin sulfate levels were unchanged or increased, but some fissuring and loosening

TABLE 3
Fibronectin and GAG Content in Articular Cartilage Extracts from Rabbit Knee Joints

Source of cartilage Femoral condyles	Fibronectin (μg/mg wet cartilage)		GAG (μg/mg wet cartilage)	
	Medial	Lateral	Medial	Lateral
Right knee — operated	0.15 ± 0.08 (12)	1.18 ± 0.71 (12)	32.0 ± 10.5 (8)	24.5 ± 7.7 (8)
Left knee — unoperated	0.09 ± 0.06 (8)	0.04 ± 0.04 (8)		
Right knee — sham operated	0.15 ± 0.08 (4)	0.08 ± 0.08 (4)		

Note: Rabbits were sacrificed at 6 weeks after surgery and joints frozen. After thawing, cartilage was removed from the articular surface, extracted and analyzed for fibronectin content or for GAG content. Data reported as means and standard deviations. The number in parentheses represents the number of individual joints that were analyzed.

From Burton-Wurster, N. et al., *J. Rheumatol.*, 13, 175, 1986. With permission.

of the fibrous subsurface network as well as chondrocyte cloning could be seen. Mild elevations of fibronectin, up to 593-ng/mg wet weight) were observed in the cartilage taken from the operated joint. The second model was a model of cartilage atrophy produced by immobilization of one hind limb for 6 weeks. As described by Palmoski et al.,[48] this model does not go on to develop osteoarthritis. Articular cartilage from the habitually loaded areas of the femoral condyles of the immobilized limb showed reduction of Safrinin O staining but no fissuring, fibrillation, or cloning. Fibronectin levels in this atrophic cartilage were normal.

The data discussed so far are consistent with the notion that increased fibronectin content in cartilage is associated with osteoarthritis, but that proteoglycan loss, which is a prominent feature of osteoarthritic cartilage, especially in the latter stages, is not necessary or sufficient to trigger fibronectin accumulation.

D. FIBRONECTIN IN HUMAN OSTEOARTHRITIS AND RHEUMATOID ARTHRITIS

Miller et al.[15] examined specimens of osteoarthritic cartilage from 17 femoral heads immediately after surgical removal. They were able to identify fibronectin in guanidine hydrochloride extracts of osteoarthritic cartilage, and occasionally, in some preparations of rheumatoid cartilage specimens. In addition, they demonstrated that explants of human osteoarthritic cartilage were able to synthesize fibronectin. The authors estimated the fibronectin content of osteoarthritic samples to be 0.5 to 1% of the total dry weight, or approximately 2.5 μg/mg wet weight of cartilage, comparable to values obtained for osteoarthritic dog cartilage.

Fibronectin levels in normal human cartilage were below the levels of detectability by their assay, a double immunodiffusion assay. Using the more sensitive ELISA assay, we were able to detect low levels (80 ng/mg wet) of fibronectin in normal human cartilage supplied to us by Dr. Miller.[52]

E. FIBRONECTIN IN AN ANTIGEN-INDUCED MODEL OF RHEUMATOID ARTHRITIS IN THE RABBIT

Dr. Arnold Rubin of the Ciba-Geigy Corporation in Summit, NJ and ourselves have begun to look at the fibronectin content in articular cartilage in an antigen-induced model of rheumatoid arthritis.

In this model of rheumatoid arthritis described by Rubin and Roberts,[49] New Zealand white rabbits were sensitized intradermally with 1 ml of complete Freund's adjuvant. After 4 weeks, they were challenged with 25 μg of mycobacterium butyricum intraarticularly in one knee. This challenge was repeated at 6 weeks. Joints were harvested at 7, 9, and 12 weeks after the initial sensitization. The contralateral knee, as well as untreated animals, served as a control. Cartilage was sampled from both the tibial and femoral condyles. This model produced mild to moderate villous hyperplasia of the synovial membrane with a thickening of the synovial lining cell layer, lymphocytic nodule formation, and a generalized infiltration of plasma cells reminiscent of rheumatoid arthritis. In the cartilage, a loss of Safranin O staining and decreased cellularity was observed. In a preliminary experiment, fibronectin content increased twofold after 7 d and continued to increase to as much as eightfold after 42 d. At the same time, loss of glycosaminoglycan was about 32% by day 7 with a maximum loss of 35% by 42 d.

VII. SUMMARY

In conclusion, the evidence is strong that adult articular cartilage contains fibronectin. The content of fibronectin is strikingly increased in osteoarthritic lesions in both spontaneous and surgically induced animal models of osteoarthritis, and also in a rabbit model of rheumatoid arthritis. The same observations have been made in human tissue, confirming the relevance for the human disease. Recent studies describing structural differences in cartilage, and also in synovial fluid fibronectins, when compared to plasma fibronectin, suggest that the accumulated fibronectin is produced locally within the joint. The existence of a unique subunit of canine cartilage fibronectin which differs from plasma fibronectin at an epitope in the last type III homology raises an important question. What is the relationship between the structure of fibronectin and its function in cartilage? Also, what are the effects of the increased quantities of fibronectin on the biology of diseased tissue? On the basis of histochemical evidence and the known biochemical interactions of fibronectin with collagen, proteoglycans, and hyaluronic acid, we expect fibronectin to be able to interact with components in the matrix and in the pericellular lacuna, and perhaps also with the chondrocytes. But what really are the important interactions of cartilage fibronectin *in vivo*? Information obtained in cell culture suggests it may be possible to relate increased fibronectin to cell dedifferentiation. But to what extent does phenotypic modulation of the chondrocyte influence or reflect cartilage degeneration *in vivo*? The focus of further research will be to answer some of these questions.

REFERENCES

1. **Weiss, R. E. and Reddi, A. H.** Appearance of fibronectin during the differentiation of cartilage, bone, and bone marrow, *J. Cell Biol.*, 88, 630, 1981.
2. **Clemmensen, I., Hølund, B., Johansen, N., and Andersen, R. B.**, Demonstration of fibronectin in human articular cartilage by an indirect immunoperoxidase technique, *Histochemistry*, 76, 51, 1982.
3. **Evans, H. B., Ayad, S., Abden, M. Z., Hopkins, S., Morgan, K., Walton, K. W., Weiss, J. B., and Hold, P. J. L.**, Localization of collagen types and fibronectin in cartilage by immunofluorescence, *Ann. Rheum. Dis.*, 42, 575, 1983.
4. **Glant, T. T., Hadházy, Cs., Mikecz, K., and Sipos, A.**, Appearance and persistence of fibronectin in cartilage specific interaction of fibronectin with collagen type II, *Histochemistry*, 82, 149, 1985.
5. **Burton-Wurster, N., Butler, M., Harter, S., Colombo, C., Quintavalla, J., Swartzendurber, D., Arsenis, C., and Lust, G.**, Presence of fibronectin in articular cartilage in two animal models of osteoarthritis, *J. Rheumatol.*, 13, 175, 1986.
6. **Burton-Wurster, N., Horn, V. J., and Lust, G.**, Immunohistochemical localization of fibronectin and chondronectin in canine articular cartilage, *J. Histochem. Cytochem.* 36, 581, 1988.

7. **Shiozawa, S. and Ziff, M.**, Immunoelectron microscopic demonstration of fibronectin in rheumatoid pannus and at the cartilage-pannus junction, *Ann. Rheum. Dis.*, 42, 254, 1983.
8. **Silbermann, M., von der Mark, K., Maor, G., and van Menxel, M.**, Dexamethasone impairs growth and collagen synthesis in condylar cartilage in vitro, *Bone Miner.* 2, 87, 1987.
9. **Hewitt, A. T., Kleinman, H. K., Pennypacker, J. P., and Martin, G. R.**, Identification of an adhesion factor for chondrocytes, *Proc. Natl. Acad. Sci. U.S.A.*, 77, 385, 1980.
10. **Bray, B. A., Mandl, I., and Turino, G. M.**, Heparin facilitates the extraction of tissue fibronectin, *Science,* 214, 793, 1981.
11. **Engvall, E. and Ruoslahti, E.**, Binding of soluble form of fibroblast surface protein, fibronectin, to collagen, *Int. J. Cancer,* 20, 1, 1977.
12. **Wurster, N. B. and Lust, G.**, Fibronectin in osteoarthritic canine articular cartilage, *Biochem. Biophys. Res. Commun.*, 109, 1094, 1982.
13. **Pena, S. D. J., Mills, G., Hughes, R. C., and Aplin, J. C.**, Polypeptide heterogeneity of hamster and calf fibronectins, *Biochem. J.*, 189, 337, 1980.
14. **Burton-Wurster, N. and Lust, G.**, Deposition of fibronectin in articular cartilage of canine osteoarthritic joints, *Am. J. Vet. Res.*, 46, 2542, 1985.
15. **Miller, D. R., Mankin, H. J., Shoji, H., and D'Ambrosia, R. D.**, Identification of fibronectin in preparations of osteoarthritic human cartilage, *Connect. Tissue Res.*, 12, 267, 1984.
16. **Bowness, J. M.**, Comparison of cartilage structural glycoproteins with matrix proteins and fibronectin, *Can. J. Biochem.*, 59, 181, 1981.
17. **Hynes, R. O.**, Molecular biology of fibronectin., *Annu. Rev. Cell. Biol.*, 1, 67, 1985.
18. **Zardi, L., Carnemolla, B., Siri, A., Petersen, T. E., Paolella, G., Sebastio, G., and Baralle, E.**, Transformed human cells produce a new fibronectin isoform by preferential alternative splicing of a previously unobserved exon, *EMBO J.*, 6, 2337, 1987.
19. **Carnemolla, B., Borsi, L., Zardi, L., Owens, R. J., and Baralle, F. E.**, Localization of the cellular-fibronectin-specific epitope recognized by the monoclonal antibody IST-9 using fusion proteins expressed in E. coli, *FEBS,* 215, 269, 1987.
20. **Castellani, P., Siri, A., Rosellini, C., Infusini, E., Borsi, L., and Zardi, L.**, Transformed human cells release different fibronectin variants than do normal cells, *J. Cell Biol.*, 103, 1671, 1986.
21. **Borsi, L., Carnemolla, B., Castellani, P., Rosellini, C., Vecchio, D., Allemanni, G., Chang, S. E., Taylor-Papdimitriou, J., Pande, H., and Zardi, L.**, Monoclonal antibodies in the analysis of fibronectin isoforms generated by alternative splicing of mRNA precursors in normal and transformed human cells, *J. Cell Biol.*, 104, 595, 1987.
22. **Paul, J. I., Schwarzbauer, J. E., Tamkun, J. W., and Hynes, R. O.**, Cell-type-specific fibronectin subunits generated by alternative splicing, *J. Biol. Chem.*, 261, 12258, 1986.
23. **Sekiguchi, K., Klos, A. M., Hirohashi, S., and Hakomori, S.**, Human tissue fibronectin: expression of different isotypes in the adult and fetal tissues, *Biochem. Biophys. Res. Commun.*, 141, 1012, 1986.
24. **Carsons, S. E., Lavietes, B. B., Diamond, H. S., and Berkowitz, E.**, Carbohydrate heterogeneity of fibronectins: identification of plasma and synovial forms by lectin binding, *Clin. Res.*, 33, 504A, 1985.
25. **Carnemolla, B., Castellani, P., Cutolo, M., Borsi, L., and Zardi, L.**, Lack of sialic acid in synovial fluid fibronectin, *FEBS,* 171, 285, 1984.
26. **Burton-Wurster, N. and Lust, G.**, Fibronectins from articular cartilage and plasma differ in molecular weight, glycosylation, and in reactivity with a monoclonal antibody, *Trans. 34th Annu. Meet. Orthopaed. Res. Soc.*, 13, 492, 1988.
27. **Dessau, W., Sasse, J., Timpl, R., Jilek, F., and von der Mark, K.**, Synthesis and extracellular deposition of fibronectin in chondrocyte cultures, *J. Cell. Biol.*, 79, 342, 1978.
28. **Gionti, E., Capasso, O., and Cancedda, R.**, The culture of chick embryo chondrocytes and the control of their differentiated functions in vitro, *J. Biol. Chem.*, 258, 7190, 1983.
29. **Allebach, E. S., Boettiger, D., Pacifici, M., and Adams, S. L.**, Control of types I and II collagen and fibronectin gene expression in chondrocytes delineated by viral transformation, *Mol. Cell. Biol.*, 5, 1002, 1985.
30. **Gerstenfeld, L. C., Finer, M. H., and Boedtker, H.**, Altered β-actin gene expression in phorbol myristate acetate-treated chondrocytes and fibroblasts, *Mol. Cell. Biol.*, 5, 1425, 1985.
31. **Hiraki, Y., Yutani, Y., Takigawa, M., Kato, Y., and Suzuki, F.**, Differential effects of parathyroid hormone and somatomedin-like growth factors on the sizes of proteoglycan monomers and their synthesis in rabbit costal chondrocytes in culture, *Biochim. Biophys. Acta,* 845, 445, 1985.
32. **Leipold, H. R., Burton-Wurster, N., and Lust, G.**, Dibutyryl cyclic AMP reduces fibronectin production in cultured chondrocytes, *Trans. 33rd Annu. Meet. Orthop. Res. Soc.*, 12, 156, 1987.
33. **Burton-Wurster, N., Leipold, H. R., and Lust, G.**, Dibutryryl cyclic AMP decreases expression of ED-A fibronectin by canine chondrocytes, *Biochem. Biophys. Res. Comm.*, 154, 1088, 1988.
34. **Hassell, J. R., Pennypacker, J. P., Kleinman, H. K., Pratt, R. M., and Yamada, K. M.**, Enhanced cellular fibronectin accumulation in chondrocytes treated with Vitamin A, *Cell,* 17, 821, 1979.

35. **Bernard, B. A., De Luca, L. M., Hassell, J. R., Yamada, K. M., and Olden, K.,** Retinoic acid alters the proportion of high mannose to complex type oligosaccharides on fibronectin secreted by cultured chondrocytes, *J. Biol. Chem.,* 259, 5310, 1984.
36. **West, C. M., Lanza, R., Rosenbloom, J., Lowe, M., Holtzer, H., and Avdalovic, N.,** Fibronectin alters the phenotypic properties of cultured chick embryo chondroblasts, *Cell,* 17, 491, 1979.
37. **Pennypacker, J. P., Hassell, J. R., Yamada, K. M., and Pratt, R. M.,** The influence of an adhesive cell surface protein on chondrogenic expression in vitro, *Exp. Cell Res.,* 121, 411, 1979.
38. **West, C. M., de Weerd, H., Dowdy, K., and de la Paz, A.,** A specificity for cellular fibronectin in its effect on cultured chondroblasts, *Differentiation,* 27, 67, 1984.
39. **Gibson, G. J., Kielty, C. M., Garner, C., Schor, S. L., and Grant, M. E.,** Identification and partial characterization of three low-molecular-weight collagenous polypeptides synthesized by chondrocytes cultured within collagen gels in the absence and in the presence of fibronectin, *Biochem. J.,* 211, 417, 1983.
40. **Wurster, N. B. and Lust, G.,** Fibronectin in osteoarthritic cartilage — a possible indication of phenotype modulation of the chondrocyte?, in *Degenerative Joints,* Vol. 2, Verbruggen, G. and Veys, E. M., Eds., Excerpta Medica, Amsterdam, 1985, 141.
41. **Benya, P. D. and Brown, P. D.,** Modulation of the chondrocyte phenotype in vitro, in *Articular Cartilage Biochemistry Workshop Conference Hoechst-Werk Albert,* Kuettner, K. E., Schleyerbach, R., and Hascall, V. C., Eds., Raven Press, New York, 1986, 219.
42. **Wurster, N. B. and Lust, G.,** Synthesis of fibronectin in normal and osteoarthritic articular cartilage, *Biochim. Biophys. Acta,* 800, 52, 1984.
43. **Burton-Wurster, N. and Lust, G.,** Fibronectin and water content of articular cartilage explants after partial depletion of proteoglycans, *J. Orthopaed. Res.,* 4, 437, 1986.
44. **Burton-Wurster, N. and Lust, G.,** Incorporation of purified plasma fibronectin into explants of articular cartilage from disease-free and osteoarthritic canine joints, *J. Orthopaed. Res.,* 4, 409, 1986.
45. **Colombo, C., Butler, M., O'Byrne, E., et al.,** A new model of osteoarthritis in rabbits. I. Development of knee joint pathology following lateral meniscectomy and section of the fibular collateral and sesamoid ligaments, *Arthritis Rheum.,* 26, 875, 1983.
46. **Brandt, K. D., Wurster, N. B., and Lust, G.,** Fibronectin in degenerating cartilage: relationship to chondrocyte cloning, *Trans. 32nd Annu. Meet. Orth. Res. Soc.,* 11, 255, 1986.
47. **Palmoski, M. J., Colyer, R. A., and Brandt, K. D.,** Marked suppression by salicylate of the augmented proteoglycan synthesis in osteoarthritic cartilage, *Arthritis Rheum.,* 23, 83, 1980.
48. **Palmoski, M., Perricone, E., and Brandt, K. D.,** Development and reversal of a proteoglycan aggregation defect in normal canine knee cartilage after immobilization, *Arthritis Rheum.,* 22, 508, 1979.
49. **Rubin, A. S. and Roberts, E. D.,** Morphometric quantitation of histopathologic changes in articular cartilage in an immunologically-induced rabbit model of rheumatoid arthritis, *Lab Invest.,* 57, 1342, 1987.
50. **Jones, K. L., Brown, M., Ali, S. Y., and Brown, R. A.,** An immunohistochemical study of fibronectin in human osteoarthritic and disease free articular cartilage, *Ann. Rheum. Dis.,* 46, 809, 1987.
51. **Rees, J. A., Ali, S. Y., and Brown, R. A.,** Ultrastructural localisation of fibronectin in human osteoarthritic articular cartilage, *Ann. Rheum. Dis.,* 46, 816, 1987.
52. **Burton-Wurster, N. and Lust, G.,** unpublished observations, 1984.

Chapter 16

FIBRONECTIN AND RHEUMATIC DISEASE

David L. Scott, Paul Mapp, and K. E. Herbert

TABLE OF CONTENTS

I. INTRODUCTION

Rheumatic diseases cover a wide range of different pathological entities. They include inflammatory arthritis, degenerative arthritis, muscle disorders, systemic connective tissue diseases, spinal diseases, and nonarticular rheumatic problems. Some of the principal inflammatory arthropathies, such as rheumatoid arthritis, have significant extra-articular features. There are diverse pathological mechanisms involved. In these circumstances it is difficult to make a brief and definitive assessment of the overall role of fibronectin in arthritis. However, there are many similarities between the major inflammatory arthropathies, and by limiting ourselves to this group of disorders we can draw general conclusions on the role of fibronectin in joint disease.

Rheumatoid arthritis is the central joint disease on which to focus any review of the pathology of synovial inflammation. It is a chronic progressive inflammatory arthritis which involves about 1 to 2% of the adult population. The way in which diagnostic criteria are applied influences the apparent incidence of rheumatoid arthritis, principally because of the recognition of varying numbers of mild cases. The joints in rheumatoid arthritis demonstrate inflammatory symptoms of pain, swelling, and stiffness. Histopathologically there is synovial hyperplasia, marginal pannus formation, and inflammatory cell infiltration of the synovium with follicular aggregation of lymphocytes. In some cases, fibrin deposition, and a marked increase in vascularity is seen.[1] Many joints have large synovial effusions, with depolymerization of hyaluronic acid and a mixed variable polymorphonuclear and lymphocytic cell content. Rheumatoid nodules with their characteristic central area of fibrinoid necrosis are found at many sites, including, on occasion, the synovium. The presence of nodules, serositis, and vasculitis represent extra-articular disease.

There are several reasons why a connective tissue protein such as fibronectin may have a pathogenic role in the rheumatic disease. Connective tissue proliferation characterizes inflammatory arthritis, and fibronectin is involved both in cell-cell interactions in the hyperplastic synovium and as part of the immature connective tissue matrix. It is also a component of synovial fluid, and its opsonic functions may be important in the removal of debris from the joint cavity. Finally, fibronectin may interact with immune complexes. These functional roles imply that fibronectin is more important in the response of the joints and their related structures to inflammation than as a primary component of synovial inflammation. This is also typical of the involvement of fibronectin in wound healing and tissue repair in other pathological situations.

A number of different pathogenic mechanisms have been proposed as the basis of rheumatoid arthritis. These include autoimmunity, mycoplasma,[2] and viral infection.[3] They are not mutually exclusive; for example, a viral infection could trigger an autoimmune disease in a genetically susceptible subject. The actual cause of rheumatoid arthritis, as with other connective tissue diseases, remains unknown. There is little to suggest that a connective tissue protein such as fibronectin is involved in the initiation of rheumatic diseases. On the other hand, there is considerable evidence that fibronectin may influence the course of the disease, and at present, this may be of prognostic significance; if we can affect the course of rheumatic diseases we may be able to favorably affect their outcome.

II. PLASMA AND SYNOVIAL FLUID FIBRONECTIN

Fibronectin is not an acute phase protein in humans. Plasma levels do not rise in most inflammatory diseases, and studies in rheumatoid arthritis have invariably shown concentrations to be normal.[4-6] By contrast, synovial fluid fibronectin levels are significantly elevated in comparison to plasma. A summary of these results from comparable studies is shown in Table 1. Others have reported similar results.[10-12] The finding of elevated synovial fluid

TABLE 1
Fibronectin Concentrations in Synovial Fluid and Plasma

Sample	Study	Mean fibronectin (g/l)	Ref.
RA synovial fluid	Scott et al.	0.70	4
	Carsons et al.	0.70	5
	Clemmenson and Anderson	0.81	6
	Vartio et al.	0.44	7
RA plasma	Scott et al.	0.32	4
	Clemmenson and Anderson	0.33	6
Normal plasma	Matsuda et al.	0.28	8
	Mosher and Williams	0.33	9
	Scott et al.	0.32	

levels is quite distinct from that of other proteins which show lower concentrations in synovial fluid than plasma. Synovial fluid protein levels are partially related to protein size with smaller proteins having in many instances, higher relative levels than larger proteins. This is in keeping with the concept that they are present in the synovial fluid by "filtration" from the plasma. The molecular weight of fibronectin (450,000 Da) suggests that high levels found in synovial fluid cannot arise by such a mechanism; local production is the most likely explanation. This is suggested by studies of paired plasma and synovial fluid samples from rheumatoid patients.[4-6] At the same time, there is a wider scatter of synovial fluid fibronectin levels in rheumatoid arthritis and not all patients have high levels; in some cases it is below the plasma concentration.

In contrast, there is evidence that plasma fibronectin levels are altered in systemic lupus erythematosus (SLE). Carsons et al.[13] showed elevated mean plasma fibronectin levels in 22 patients with systemic lupus attending a rheumatology clinic. Similarly elevated levels have been described in two other studies of lupus patients.[12,14] Subgroups of patients with SLE have shown a relationship between the elevated plasma fibronectin levels and clinical disease activity. The reasons for this acute phase reaction of plasma fibronectin in SLE is uncertain, though it could be related to an involvement in the pathophysiology of immune-complex mediated vasculitis.

A. BIOCHEMICAL HETEROGENEITY OF SYNOVIAL FLUID FIBRONECTIN

Synovial fluid fibronectin is a soluble form of the molecule. Immunochemically it has many similarities to the plasma form, but it is quite distincet from the insoluble cellular fibronectin seen in fibroblast cultures. There is, however, some evidence to suggest that synovial fluid fibronectin is different from the plasma molecule. There are oligosaccharide component variations in its gelatin-binding regions and it is often partially degraded (see Chapter 1). Fibronectin from different sources is known to have biochemical differences due to posttranslational modifications.[15] Should synovial fluid fibronectin differ substantially from the plasma form, this could affect its role in immune complex disease. Clemmensen and Andersen,[6] Carsons et al.,[15] and ourselves[18] have suggested that fibronectin is biochemically different in synovial fluid. Of synovial fluids obtained from patients with active arthritis, 86% contained fragments of the native molecule and 39% of the fluids no longer showed the dimeric, native form. Compared with native fibronectin, the synovial fluid peptides were as active in promoting synoviocyte chemotaxis and in glycosaminoglycan binding, but displayed lower affinity for fibrin and gelatin. Additional results which support the concept of biochemical differences in synovial fluid fibronectin, especially in relation to protein complexes, are the binding of synovial fluid fibronectin to immunoglobulin G

(16); evidence of fibronectin complexes in synovial fluid by two-dimensional immunoelectrophoresis;[16] and evidence of greater microheterogeneity of synovial fluid fibronectin as shown by immunoblotting.[17,18]

B. EXPERIMENTAL MODELS OF FIBRONECTIN IN INFLAMMATION

How can synovial fluid fibronectin be examined experimentally? The induction of a subcutaneous air pouch in rats gives rise to a mesenchymal cavity with similarities to the synovial membrane.[19] This can be used to model the pathophysiology of fibronectin in chronic inflammation in the synovial space. For these studies, animals are sensitized with bovine serum albumin; after 13 d they are challenged with antigen injected directly into the pouch. This is repeated every 7 d to maintain chronicity.[20] There is one species difference which complicates studies of fibronectin in rats; recent results have shown that fibronectin exhibits an acute phase response in arthritic rats.[21] The air-pouch model has shown that there is a rapid increase in rat plasma fibronectin levels following antigenic challenge into the pouch. This is similar to other results in animal systems showing fibronectin to be an acute phase protein; the level gradually falls with time. By contrast, pouch fluid fibronectin gradually rose and was above control plasma levels after 15 d. Similar to what has been demonstrated in human RA, there was no relationship between pouch fluid fibronectin levels and those of plasma fibronectin. Nor was there any correlation between pouch fluid volume or cell content and fibronectin levels. This supports the concept that the fibronectin level is not a measure of inflammation at a localized site such as the air pouch, but most likely represents the proliferative response of mesenchymal tissue.

There has been one reported clinical study showing a correlation between clinical improvement and synovial fluid fibronectin levels in arthritic patients treated with orgotein.[22] This led Stecher and her colleagues to investigate in detail whether fibronectin is a disease marker in animal models of arthritis. In a series of studies[21,23-26] (and see Chapter 17 in this volume) they have shown that quantification of fibronectin in plasma and tissue fluid differentiates between the effect of slow-acting antirheumatic drugs and nonsteroidal anti-inflammatory drugs. The differential effects have been shown in carrageenan-induced pleurisy, autoimmune disease in MRL/1 mice, and adjuvant arthritis in rats. In these situations slow-acting remittive drugs such as gold, penicillamine, and hydroxychloroquine reduce the elevated fibronectin levels; nonsteroidal agents like indomethacin and naproxen have no effect. It is of interest that clinical use of these slow-acting drugs may result in actual remission of RA accomplished by reduction in the proliferation of synovial tissue (pannus).

C. RICE BODIES

Rice bodies are particulate material of variable size found in the synovial fluid of most patients with rheumatoid effusions. Their removal may be of therapeutic value. Most rice bodies contain large amounts of immunoreactive fibronectin. Fibrous rice bodies and those showing early fibrotic change contain more fibronectin. Areas of rice bodies which show "mature" collagenous tissue react negatively for fibronectin. The origin of rice bodies is controversial; they may be entirely derived from degenerate synovial membrane or show significant progressive development in the synovial fluid itself. In either case, fibronectin is implicated in their formation and their natural history.[27]

III. SYNOVIAL MEMBRANE FIBRONECTIN

The distribution of fibronectin in the synovial membrane and pannus has been extensively studied using immunohistology with both indirect immunofluorescence and immunoperoxidase methods and also immunoelectron-microscopy.[7,28-33] The main feature of its distribution in rheumatoid arthritis can be summarized as follows.

A. THE SYNOVIAL LINING CELL LAYER

Fibronectin is present intracellularly within the hyperplastic synovial lining cells; it is thought to be predominantly within the fibroblastic B-type cell.[30] It is also seen closely surrounding the lining cells in their immediate extracellular matrix. Examples of the distribution of fibronectin are shown in Figures 1 and 2.

B. SUBINTIMAL RETICULAR CONNECTIVE TISSUE

Fibronectin is present in a predominantly extracellular distribution beneath the synovial lining cell layer. It forms small reticular fibrils with a complex matrix pattern between cells; it is present in the basement membranes and perivascular connective tissue of the small blood vessels; and it is seen in small micronodules at some sites. Fibronectin shows no relationship to the inflammatory cell infiltrate in this area.

C. SUPERFICIAL FIBRIN DEPOSITS

Fibrin deposits on the surface of the synovium invariably contain immunoreactive fibronectin. The same is true of smaller amounts of fibrin seen at slightly deeper sites of the synovium. However, the distribution of fibrin-related antigen is extremely limited when compared to the overall distribution of immunoreactive fibronectin.

D. VASCULAR BASEMENT MEMBRANE

In addition to its localization in the perivascular connective tissue, immunoreactive fibronectin is also found in the basement membranes of capillaries and small blood vessels. These vessels are mainly distributed in the subintimal connective tissues but are also seen in slightly deeper sites. Examples are shown in Figures 3 and 4.

E. THE RHEUMATOID PANNUS

The marginal pannus is a site of proliferating connective tissue with marked vascularity. It contains large amounts of extracellular immunoreactive fibronectin forming a dense fibrillar network surrounding the mesenchymal cells. Fibronectin is also closely related to the blood vessels which are present in the pannus, in a distribution similar to that at other sites of the synovium. The pannus immediately adjacent to cartilage itself mainly reacts negatively for fibronectin, though this negative reaction is the subject of some debate (see Chapter 15). Our most recent studies on a large number of samples show that in some areas underlying cellular pannus, the cartilage matrix does contain significant quantities of immunoreactive fibronectin, as does the cytoplasm of more deeply lying chondrocytes (Figure 5 and 6). The question of the distribution of fibronectin in normal cartilage is controversial. Some authors report large amounts,[34] while others consider very little fibronectin is present.[35]

The underlying fibrotic areas of the synovium react negatively for fibronectin. Similarly there is no relationship between lymphocytic infiltration or lymphocyte aggregate formation and fibronectin; areas with many lymphocytes react negatively for fibronectin.

There is no evidence that changes in the distribution of fibronectin are specific for rheumatoid arthritis.[28] Its distribution in seronegative inflammatory arthritis, such as in the peripheral synovitis of anklylosing spondylitis and psoriasis, is virtually indistinguishable from that seen in rheumatoid disease. However, no histopathological features of synovitis have been found to show diagnostic specificity for rheumatoid arthritis.[36]

In osteoarthritis there is little immunoreactive synovial fibronectin, though in those osteoarthritic biopsies with marked synovial proliferation it is present in a similar distribution to that seen in rheumatoid biopsies. In control synovial biopsies from noninflammatory conditions, such as torn menisci, there are relatively small amounts of immunoreactive fibronectin in the lining-cell layer, around the sparse small blood vessels, and in the thin layer of subintimal connective tissue.

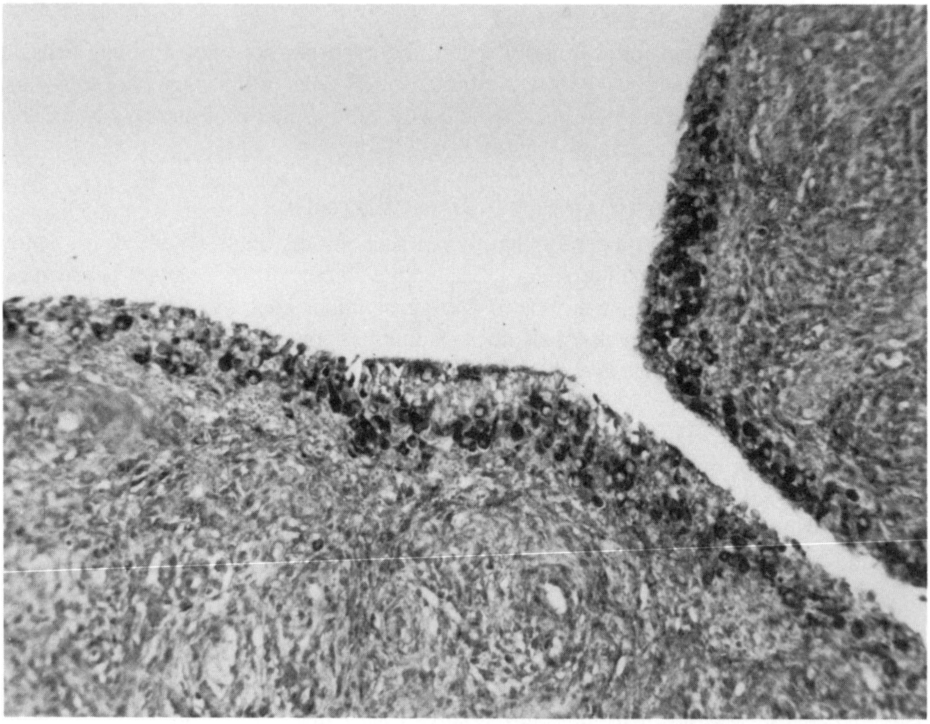

FIGURE 1. Fibronectin in rheumatoid synovium. The fibronectin immunoreactivity is predominantly in the proliferating synovial lining cell layer. PAP method; magnification × 20.

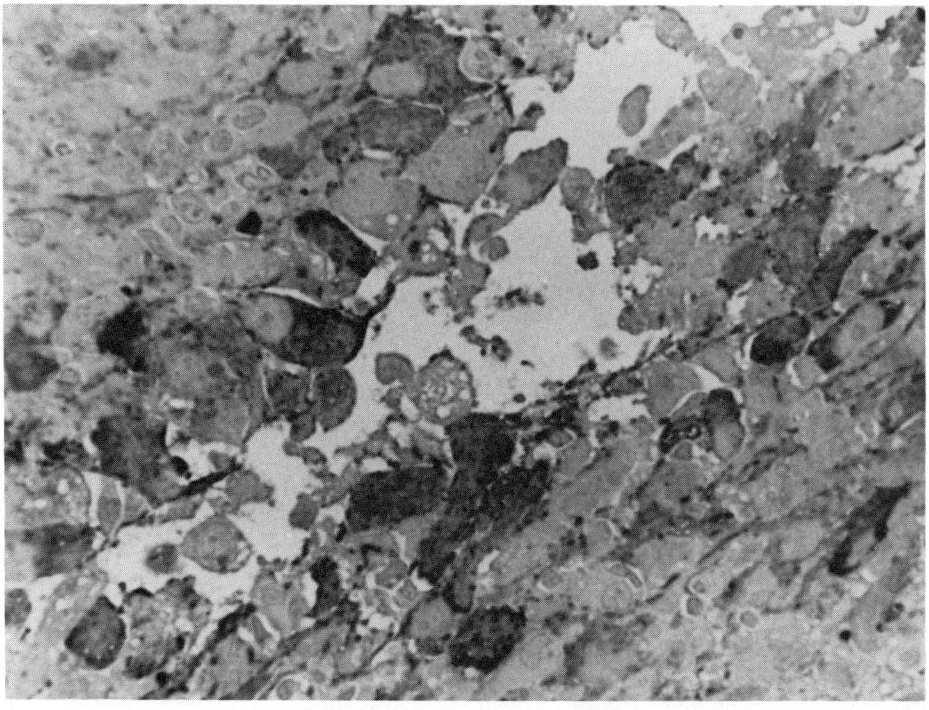

FIGURE 2. High power view of intracellular fibronectin in synovial lining cell layer from rheumatoid patient. PAP method on etched resin section; magnification × 800.

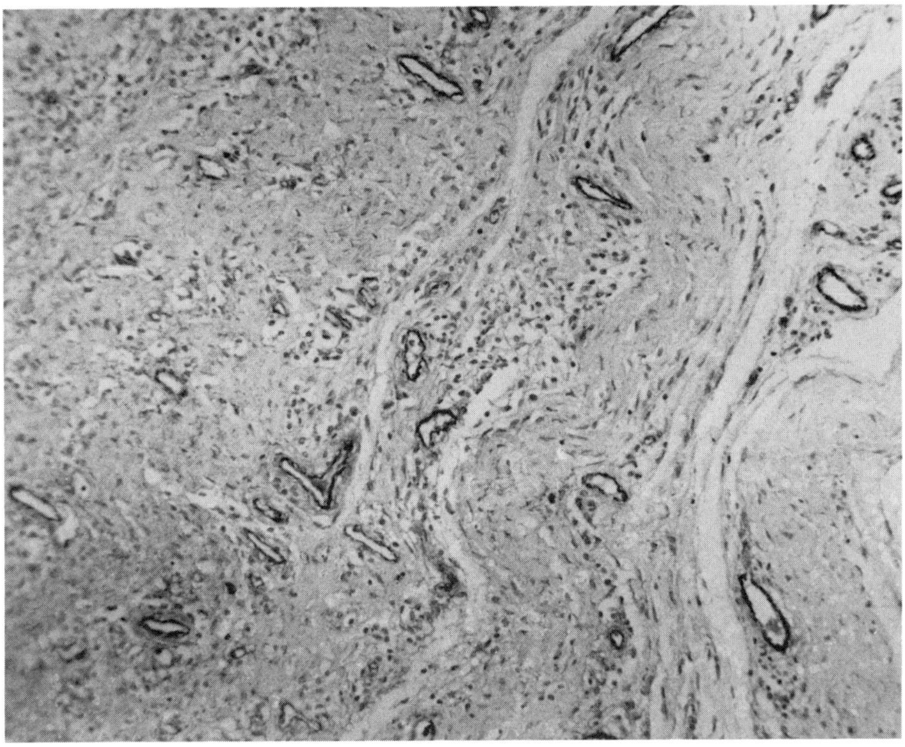

FIGURE 3. Fibronectin in vascular basement membranes of rheumatoid synovium. PAP method with trypsin digestion to unmask staining; magnification × 120.

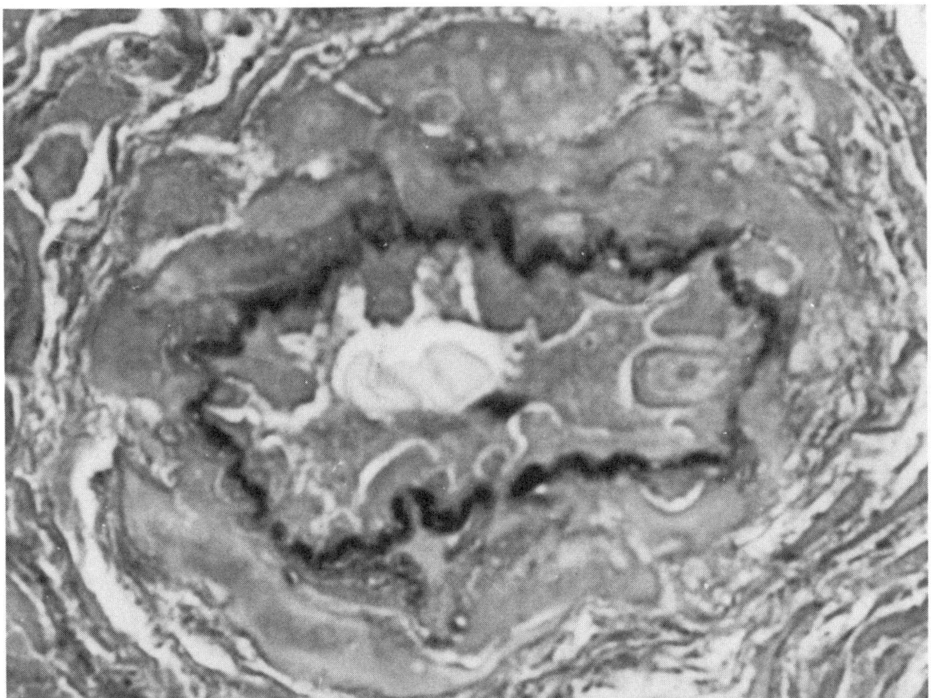

FIGURE 4. High power view of fibronectin in vascular basement membranes shown by PAP method with trypsin digestion to unmask staining; magnification × 800.

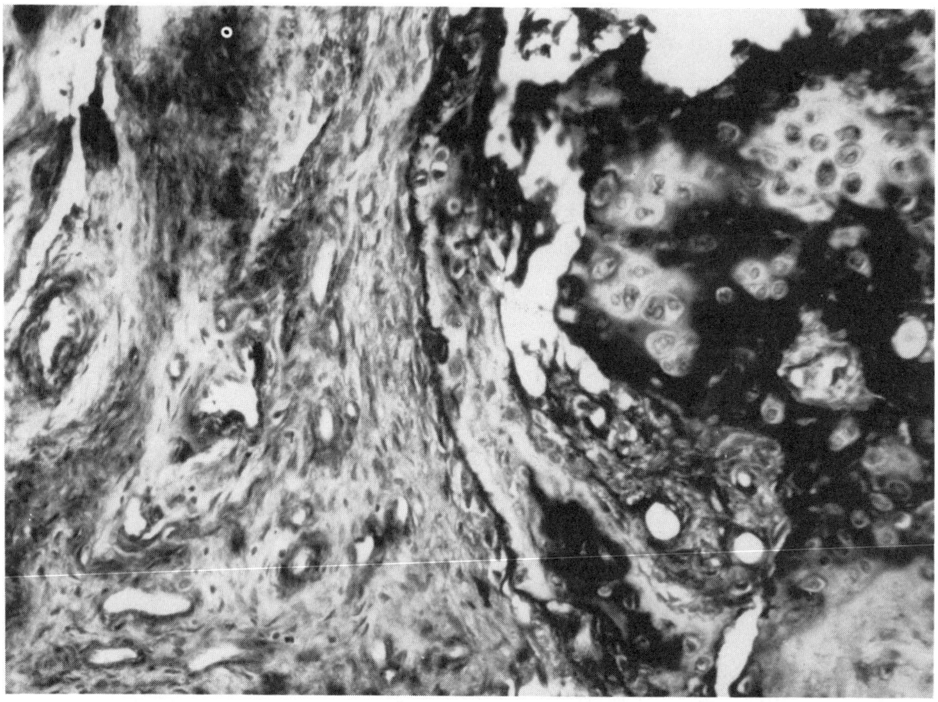

FIGURE 5. Cartilage-pannus junction showing distribution of fibronectin in specimen from a rheumatoid patient. Note the extracellular reactivity for fibronectin in the proliferating pannus tissue and the variable reactivity in the cartilage. Some areas of cartilage show dense fibronectin immunoreactivity. PAP method; magnification × 100.

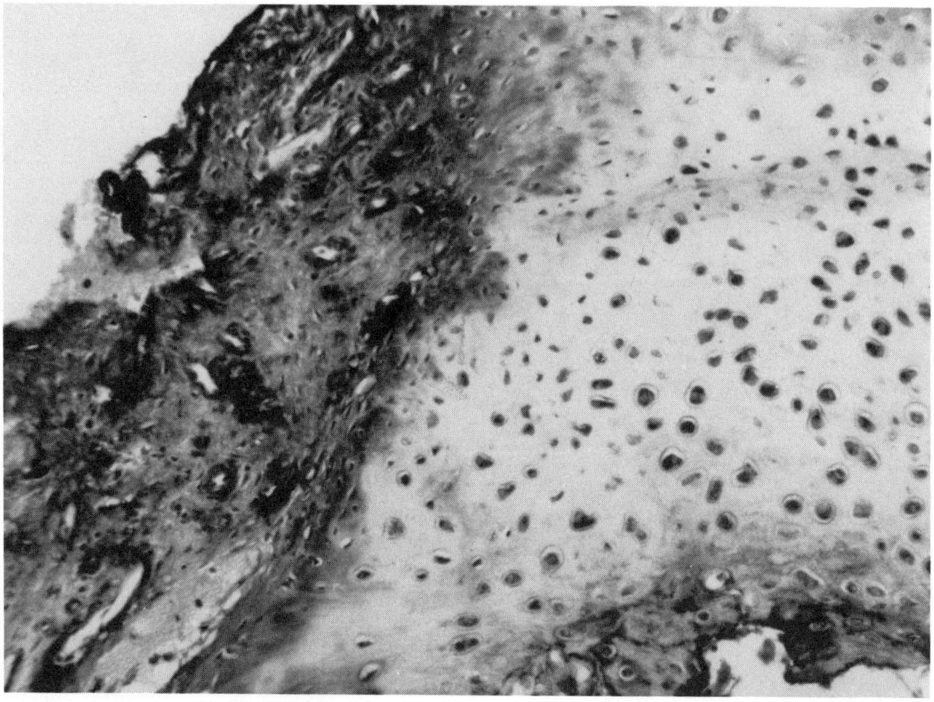

FIGURE 6. Another area of cartilage-pannus junction showing fibronectin in the pannus tissue with weak staining of the cartilage. Most areas of cartilage matrix are negative, but the chondrocytes show significant reactivity; magnification × 100.

The clinical studies have been supplemented with evaluations of experimental arthritis. Examination of the distribution of synovial fibronectin in rabbit antigen-induced chronic synovitis showed an identical pattern of deposition to that of human rheumatoid disease.[37,38] Studies on the distribution of fibronectin during the initial acute phase of the disease (2 to 3 d after intra-articular challenge), showed that there is widespread deposition of fibronectin, probably reflecting the presence of fibronectin from plasma exudates which characterizes this acute phase. In established chronic disease, fibronectin codistributes with fibrin and is present in the synovial lining cell layer. In synovia undergoing organization, it is present on immature collagen fibrils and is observed intracellularly in several types of mesenchymal cells, suggesting enhanced local synthesis.

IV. RELATIONSHIP OF FIBRONECTIN TO OTHER CONNECTIVE TISSUE COMPONENTS OF THE SYNOVIUM

Reticulin is not a single discrete entity. Argyrophilic reticulin fibers result from the interaction of at least three proteins — fibronectin, a noncollagenous reticulin component,[39] and collagen type III.[31] Reticulin fibers are a significant component of the rheumatoid synovium.[40] In acutely inflamed synovia they are predominantly found beneath the hyperplastic lining cell layer. The reticulin fibers of the rheumatoid synovium contain fibronectin codistributed with the noncollagenous reticulin component and collagen type III. With developing chronicity, ''mature'' collagen is more prominent, which does not contain immunoreactive fibronectin. The ''mature'' thick collagen fibers are composed of collagen type I and react specifically with antisera against this instead of collagen type III. This *in vivo* specificity in the relationship of fibronectin to collagen type III is not reflected by it *in vitro*, for it can bind to both collagen types.

Laminin, a high molecular weight basement membrane structural glycoprotein, is a marker for vascular proliferation. It codistributes with collagen type IV.[41-43] There is a close relationship between the distribution of laminin, noncollagenous reticulin component, and fibronectin in blood vessel walls in the inflamed synovium.[44] However, compared to the other proteins, laminin has a relatively restricted distribution.

V. FIBRONECTIN AND IMMUNE COMPLEXES

Does fibronectin interact with the immune system in rheumatic disease? One way in which this could occur is through an interaction with immune complexes. Immune complexes have been implicated in the pathogenesis of many arthropathies. Both circulating and localized (synovial fluid) complexes may play a role in the pathogenesis of RA.

The concept that fibronectin may be involved in immune complexes stems from its known opsonic role. The binding of fibronectin to gelatin is a well-recognized characteristic.[45] Fibronectin is responsible for the phagocytosis and removal of gelatin-coated particles by the reticuloendothelial system. Its mechanism of action lies within the properties of its domain structure and, in particular, the domains binding gelatin (denatured collagen) and cells. The binding of plasma fibronectin to a gelatin-coated substrate may cause conformational changes in the tertiary structure of fibronectin. This enables recognition by phagocytic cells and is followed by internalization of the complex. Binding to collagen results in a significant change in the antigenic structure of the fibronectin molecule.[46] Specific fibronectin receptors are present on monocytes, and fibronectin mediates binding of gelatin to macrophages.[47] The presence of fibronectin alone is not sufficient for removal of tissue debris because true phagocytosis which requires internalization of the substrate as well as its binding was not evident in studies of fibronectin using liver slices, though in peritoneal macrophages phagocytosis it has been demonstrated.[47] Phagocytosis of gelatin-coated particles by neutrophils has also been demonstrated but interestingly did not induce microbial killing activity as indicated by the lack of respiratory burst.

The question whether fibronectin affects immune complex formation has been looked at in an experimental *in vivo* model of its opsonic role.[48] Complexes of a charged colloid (dextran sulfate) and plasma fibronectin were used to simulate immune complexes. High molecular weight dextran sulfate (500,000 Da) forms fibronectin-containing complexes in rat plasma *in vitro* and lowers plasma fibronectin acutely in a dose-related fashion when given parenterally to rats. The plasma changes are accomplished by deposition of dextran sulfate (shown histochemically as metachromatic material) and of fibronectin (shown by specific immunofluorescence) in an identical distribution within reticuloendothelial cells of rat liver and spleen. The protein and polysaccharide components of the complex are disposed of by the reticuloendothelial system at markedly different rates. If the deposition of charged colloid can be extrapolated to immune complexes, it becomes clear that fibronectin has a major potential to modify the clearance of immune complexes and thus control, to a greater or lesser extent, many of the features of immunologically mediated diseases.

The first evidence suggesting interaction of fibronectin with mechanisms of specific immune opsonization arose from work on C3b receptors on red blood cells. Although not necessary for the initial binding of C3b-coated red cells to monocytes, fibronectin was compulsory for their internalization.[49,50] Similarly, substrates capable of activating the alternative pathway of complement were recognized but not internalized by monocytes unless in the presence of certain fibronectin fragments (180 kDa). Direct evidence of the association of fibronectin with components of immune complexes, such as complement components and immunoglobulins, was first demonstrated for C1q[51] (see Chapter 5). The complexing of fibronectin with immunoglobulins and complement could aid opsonization of the immune complex by the concerted actions of Fc, complement, and fibronectin receptors on phagocytic cells. *In vitro*, fibronectin did not bind to solid-phase bovine serum albumin — anti-bovine serum albumin complexes, but a great potentiation in binding was observed for complexes also containing C1q. In the disease situation, the interaction between fibronectin and C1q is suggested to facilitate attachment of fibroblast to immune complexes and to provide a nidus for fibroblast attachment and activation in tissues where immune complexes are deposited. C1q is not the only complement protein to interact directly with fibronectin; C3b is also recognized by a site on the fibronectin molecule. Indeed, fibronectin has been shown to enhance the phagocytosis of C3b/C3bi-coated immune complexes by human monocytes.[49,50]

Complement-fixing immune complexes are often thought to be of particular pathogenic significance. The ability of fibronectin to participate in such complexes was evaluated by the use of conglutinin binding.[52] Radiolabeled antifibronectin bound to complement-fixing complexes in a majority of rheumatoid arthritis serum and synovial fluid samples; there were significant differences between the findings in rheumatoid sera and controls. These interactions suggested for the first time that fibronectin is an integral part of circulating immune complexes with important consequences for immune-complex deposition in tissues and removal by phagocytic cells. Immunoglobulin-fibronectin interactions have received less attention than complement-fibronectin binding. Ferraccioli et al.,[53] showed a correlation between fibronectin and the presence of IgM rheumatoid factor. *In vitro* they showed fibronectin interacted directly with IgM rheumatoid factor and heat-aggregated IgG at the Fc part of the immunoglobulin molecule. This suggests that fibronectin forms part of rheumatoid factor-containing immune complexes in rheumatic diseases.

VI. FIBRONECTIN IN CRYOPRECIPITATES AND POLYETHYLENE GLYCOL PRECIPITATES

Historically, fibronectin was isolated as cold insoluble globulin.[54] This led to its initial characterization as a coprecipitant of plasma together with fibrinogen.[55] Since these studies, a number of investigations have shown that fibronectin is a cryoprecipitable plasma protein. The mechanism of its insolubilization is complex because a number of different experimental approaches lead to its cryoprecipitation. These include the production of cryofibrinogen,[56] the heparin precipitable fraction of plasma,[57] and serum, plasma, and synovial fluid cryoprecipitates.[58-61] Cryoglobulins are the major constituents of most cryoprecipitates. They can be divided into monoclonal, mixed monoclonal-polyclonal, and mixed polyclonal types. Most studies have examined fibronectin in mixed polyclonal cryoglobulins from patients with inflammatory connective tissue diseases. However, fibronectin also coprecipitates with monoclonal immunoglobulins. Hautenen and Keski-Oja[62] have shown there is an affinity of fibronectin for myeloma IgG. The implication is that fibronectin will bind specifically to IgG in cryoprecipitates. The simplest explanation for the finding of fibronectin in cryoprecipitates is that it binds to immune complexes and is coprecipitated with them in the cold. However, there are multiple potential interactions leading to precipitation, and fibronectin is found in high concentrations in the factor VIII-rich cryoprecipitate fractions produced as a product of normal plasma.

The formation of cryoprecipitates is too variable to allow a reasonable appreciation of the role of fibronectin in the formation of immune complexes. Using polyethylene glycol (PEG), mol. wt. 6000, to precipitate protein complexes is a more suitable approach. Native fibronectin is not insolubilized by PEG to any significant extent; radiolabeling studies show less than 1% of purified fibronectin is precipitated by 2 to 4% PEG.[52] However, when IgM-anti-IgM immune complexes are formed *in vitro*, fibronectin is coprecipitated with them in the presence of PEG.[17] This occurs whether or not serum is present, and therefore may be independent of the presence of complement. Other studies suggest, however, that the amount of fibronectin present is directly related to the C1q content.

Studies of PEG precipitates formed at room temperature (20°C) and 4°C from the serum of patients with systemic lupus erythematosus and rheumatoid arthritis showed that more fibronectin was present in precipitates formed in the cold. This suggests that factors involved in PEG precipitation of fibronectin may be different from those involved in cold precipitation.

Very small amounts of PEG-precipitable fibronectin are present in normal serum, but some patients with rheumatoid arthritis and systemic lupus erythematosus have large amounts of PEG-precipitable fibronectin. The amount of PEG-precipitable fibronectin is not related to the serum fibronectin levels. However, there are significant correlations between fibronectin and immunoglobulins in PEG precipitates. Similar results are seen in rheumatoid synovial fluid. In summary, the data suggest that fibronectin is a component of immune complexes precipitated by PEG.

VII. CONCLUSIONS

The pathological changes of chronic arthritis are complex, and fibronectin is involved in several of them. We have shown that it is implicated in the most important features of rheumatoid arthritis and related diseases. It is present in synovial fluid in high concentrations, and is a component of both immune complexes and particulate rice bodies. Its presence in immune complexes may be important for their opsonization and removal. Large amounts of fibronectin are found in the synovial membrane, especially when it is undergoing rapid proliferative change. Fibronectin also appears to play a role in the destruction of cartilage and bone by proliferating pannus. These changes are not specific for any one joint disease, but they are related to the nature of the pathological process in the joint and chronic joint injury.

How may fibronectin influence the treatment of joint diseases? Fibronectin levels in synovial fluid may indicate whether or not synovial proliferation is likely to continue and may have prognostic value. There is some evidence that fibronectin levels are influenced by treatment with antirheumatic drugs. Genetically engineered fibronectin and its fragments will soon become available, and these could be injected into the joints as a novel local treatment. Analogues of the binding site tetrapeptide (Arg-Gly-Asp-Ser)[63] may be suitable for administration and could influence the proliferative aspects of synovitis.

The most likely impact of investigations on fibronectin in rheumatic disease is an indirect one; by understanding the principal pathological changes of synovial proliferation, we may be able to fundamentally influence this process by therapeutic intervention. This, together with alteration of the inflammatory component of the disease, may allow us to alter the clinical course of synovitis, and reduce the amount of joint damage which is usually the end result of arthritic disease.[64]

REFERENCES

1. **Gardner, D. L.,** General pathology of the peripheral joints, in *The Joints and Synovial Fluid,* Vol 2, Sokoloff, L., Ed., Academic Press, New York, 1980, 316.
2. **Bartholomew, L. E.,** Isolation and characterization of mycoplasmas (PPLO) from patients with rheumatoid arthritis, systemic lupus erythematosus and Reiters syndrome, *Arthritis Rheum.,* 8, 377, 1965.
3. **Smith, R. W., McGinty, L., Smith, C. A., Godzoski, C. W., and Boyd, R. J.,** Association of parvoviruses with rheumatoid arthritis in humans, *Science,* 223, 1425, 1984.
4. **Scott, D. L., Farr, M., Crockson, A. P., and Walton, K. W.,** Synovial fluid and plasma fibronectin levels in rheumatoid arthritis, *Clin. Sci.,* 62, 71, 1982.
5. **Carsons, S., Mosesson, M. W., and Diamond, H. S.,** Detection and quantitation of fibronectin in synovial fluid of patients with rheumatoid disease, *Arthritis Rheum.,* 24, 1261, 1981.
6. **Clemmensen, I. and Andersen, R. B.,** Different molecular forms of fibronectin in rheumatoid synovial fluid, *Arthritis Rheum.,* 25, 25, 1982.
7. **Vartio, T., Vaheri, A., Von Essen, R., Isomaki, H., and Stenman, S.,** Fibronectin in synovial fluid and tissue in rheumatoid arthritis, *Eur. J. Clin. Invest.,* 11, 207, 1981.
8. **Matsuda, B., Yoshida, N., Aoki, N., and Wakabayashi, K.,** Distribution of cold-insoluble globulin in plasma tissues, *Ann. N.Y. Acad. Sci.,* 312, 56, 1978.
9. **Mosher, D. F. and Williams, E. M.,** Fibronectin concentration is decreased in plasma of severely ill patients with disseminated intravascular coagulation, *J. Lab. Clin. Med.,* 91, 729, 1978.
10. **Lu-Steffes, M., Iammartino, A. J., Schmid, C. R., Castor, C. W., Davis, L., Entwistle, R., and Anderson, B.,** Fibronectin in rheumatoid and non-rheumatoid arthritis synovial fluids and in synovial fluid cryoproteins, *Ann. Clin. Lab. Sci.,* 72, 178, 1982.
11. **Gressner, X. A. M. and Wallraff, P.,** Der Einsatz der Lasernephelometrie zur Bestimmung und rechnerunterstutzten Auswertung der Fibronectinkonzentration in verschiedenen korper Fussigkeiten, *J. Clin. Chem. Biochem.,* 18, 797, 1980.
12. **Fyrand, O., Munthe, E., and Solum, N.O.,** Studies on cold insoluble globulin. I. Concentrations in citrated plasma in rheumatic disorders, *Ann. Rheum. Dis.,* 37, 347, 1978.
13. **Carsons, S., Parenti, D., and Lavietes, B. B.,** Plasma fibronectin in systemic lupus erythematosus: relationship to clinical activity, DNA binding and acute phase proteins, *J. Rheum.,* 12, 1088, 1985.
14. **Kawarmura, K., Tanaka, M., Kamiyama, F., Higashino, K., and Kishimoto, S.,** Enzyme immunoassay of plasma fibronectin in malignant collagen, and liver diseases, *Clin. Chem. Acta,* 131, 101, 1983.
15. **Carsons, S., Lavietes, B. B., and Diamond, H. S.,** The immunoreactivity, ligand, and cell binding characteristics of rheumatoid synovial fluid fibronectin, *Arthritis Rheum.,* 28, 601, 1985.
16. **Scott, D. L., Carter, S. D., Coppock, J. S., and Robinson, M. W.,** Difference between plasma and synovial fluid fibronectin, *Rheumatol. Int.,* 5, 49, 1985.
17. **Herbert, K. E., Coppock, J. S., Griffiths, A. M., and Walton, K. W.,** Fibronectin and immune complexes in rheumatic diseases, *Ann. Rheum. Dis.,* 46, 734, 1987.
18. **Griffiths, A. M., Herbert, K. E., Perrett, D., and Scott, D. L.,** Fibronectin fragments in synovial fluid, *Clin. Chem. Acta,* in press.
19. **Yoshino, S., Bacon, P. A., Blake, D. R., Scott, D. L., Wainwright, A., and Walton, K. W.,** A model of persistent antigen-induced chronic inflammation in the rat air pouch, *Br. J. Exp. Path.,* 65, 191, 1984.
20. **Scott, D. L., Robinson, M. W., and Yoshino, S.,** Fibronectin in chronic inflammation: studies using the rat air pouch model of chronic allergic inflammation, *Br. J. Exp. Path.,* 66, 519, 1985.

21. **Stecher, V. J., Kaplan, J. E., Connolly, K., Meilens, Z., and Saelens, J. K.**, Fibronectin in acute and chronic inflammation, *Arthritis Rheum.*, 29, 394, 1986.
22. **Goebel, K. M. and Storch, U.**, Effect of intra-articular orgotein versus a corticosteroid on rheumatoid arthritis of the knees, *Am. J. Med.*, 74, 124, 1983.
23. **Mielens, Z. E., Connolly, K., and Stecher, V. J.**, Effect of disease modifying anti-rheumatic drugs and non-steroidal anti-inflammatory drugs upon cellular and fibronectin responses in a pleurisy model, *J. Rheumatol.*, 12, 1083, 1985.
24. **Connolly, K., Stecher, V. J., Saelens, J. K., and Kaplan, J. E.**, The relationship between plasma fibronectin levels and auto-immune disease activity in MRL/1 mice, *Proc. Soc. Exp. Biol. Med.*, 180, 149, 1985.
25. **Connolly, K., Stecher, V. J., Kaplan, J. E., Mielens, Z., Rostami, H. J., and Saelens, J. K.**, The effect of anti-inflammatory drugs on plasma fibronectin, *J. Rheum.*, 12, 758, 1985.
26. **Stecher, V. J., Connolly, K. M., and Speight, P. T.**, Fibronectin and macrophages as parameters of disease modifying anti-rheumatic activity, *Br. J. Clin. Pract.*, in press.
27. **Popert, A. J., Scott, D. L., Wainwright, A. C., Walton, K. W., Williamson, N., and Chapman, A. J.**, The frequency of occurrence, mode of development and significance of rice-bodies in rheumatoid joints, *Ann. Rheum. Dis.*, 41, 109, 1982.
28. **Scott, D. L., Wainwright, A., Walton, K. W., and Williamson, N.**, Significance of fibronectin in rheumatoid arthritis and osteoarthritis, *Ann. Rheum. Dis.*, 40, 142, 1981.
29. **Mayston, V., Mapp, P. I., Davies, P. G., and Revell, P. A.**, Fibronectin in the synovium in chronic inflamed joint disease, *Rheumatol. Int.*, 4, 129, 1984.
30. **Mapp, P. I., and Revell, P. A.**, Fibronectin production by synovial intimal cells, *Rheumatol. Int.*, 5, 229, 1985.
31. **Matsubara, T., Spycher, M. A., Ruttner, A. T., and Fehr, K.**, Ultrastructural localization of fibronectin in the lining layer of the rheumatoid synovium, *Rheumatol. Int.*, 3, 75, 1983.
32. **Scott, D. L., Delamere, J. P., and Walton, K. W.**, The distribution of fibronectin in the rheumatoid pannus, *Br. J. Exp. Pathol.*, 62, 362, 1981.
33. **Shiozawa, S. and Ziff, M.**, Immunoelectron microscopic demonstration of fibronectin in rheumatoid pannus and at the cartilage-pannus junction, *Ann. Rheum. Dis.*, 42, 51, 1982.
34. **Clemmensen, I., Holund, B., Johansen, N. and Anderson, R. B.**, Demonstration of fibronectin human articular cartilage by an indirect immunoperoxidase technique, *Histochemistry*, 76, 51, 1982.
35. **Evans, H. B., Ayad, S., Abedin, M. Z., Hopkins, S., Morgan, K., Walton, K. W., Weiss, J. B., and Holt, P. J.**, Localization of collagen types and fibronectin in cartilage by immunofluorescence, *Ann. Rheum. Dis.*, 42, 575, 1983.
36. **Goldenberg, D. L., Egan, M. S., and Cohen, A. S.**, Inflammatory synovitis in degenerative joint disease, *J. Rheumatol.*, 9, 204, 1982.
37. **Scott, D. L., Almond, T. J., Walton, K. W., and Hunneyball, I. M.**, Fibronectin in antigen-induced arthritis in the rabbit, *J. Pathol.*, 141, 143, 1983.
38. **Holund, B., Clemmenson, I., and Wanning, M.**, Sequential appearance of fibronectin and collagen fibres in experimental arthritis in rabbits, *Histochemistry*, 80, 39, 1984.
39. **Unsworth, D. J., Scott, D. L., Almond, T. J., Beard, H. K., Holborow, E. J., and Walton, K. W.**, Studies on reticulin I: serological and immunological investigation of the occurrence of collagen type III, fibronectin and non-collagenous glycoprotein of Pras and Glynn in reticulin, *Br. J. Exp. Pathol.*, 63, 154, 1982.
40. **Scott, D. L., Salmon, M., and Walton, K. W.**, Reticulin and its related structural connective tissue proteins in the rheumatoid synovium, *Histopathology*, 8, 469, 1984.
41. **Rohde, H., Wick, G., and Timple, R.**, Immunochemical characterization of the basement membrane glycoprotein laminin, *Eur. J. Biochem.*, 102, 195, 1979.
42. **Mieltinen, M., Foidart, J. M., and Ekblom, P.**, Immunohistochemical demonstration of laminin, the major glycoprotein of basement membranes as an aid to the diagnosis of soft tissue tumors, *Am. J. Clin. Pathol.*, 79, 306, 1983.
43. **Laurie, G. E., Lebland, C. P., and Martin, G. R.**, Light microscope immunolocalization of type IV collagen, laminin, heparin sulphate glycoprotein, and fibronectin in the basement membranes of a variety of rat organs, *Am. J. Anat.*, 167, 71, 1983.
44. **Scott, D. L., Salmon, M., Morris, C. J., Wainwright, A., and Walton, K. W.**, Laminin and vascular proliferation in rheumatoid arthritis, *Ann. Rheum. Dis.*, 43, 551, 1984.
45. **Mosesson, M. W. and Amrani, D. L.**, The structure and biologic activities of plasma fibronectin, *Blood*, 56, 145, 1980.
46. **Cierniewski, C. S., Babinska, A., Niewiarowska, J., and Augustyniak, W.**, Alteration of the antigenic structure of human fibronectin caused by complexing with collagen, *Hoppe-Seyler's Z. Physiol. Chem.*, 364, 515, 1983.

47. **Gudewicz, P. W., Molnar, J., Lai, M. Z., Beezhold, D. W., Siegrieg, G. E., Credo, R. B., and Lorand, L.,** Fibronectin-mediated uptake of gelatin-coated latex particles by peritoneal macrophages, *J. Cell Biol.*, 87, 427, 1980.
48. **Walton, K. W., Almond, T. J., Robinson, M. W., and Scott, D. L.,** An experimental model for the study of opsonic activity of fibronectin in the clearance of intravascular complexes, *Br. J. Exp. Path.*, 65, 191, 1984.
49. **Pommier, C. G., Inada, S., Fries, L. F., Takahashi, T., Frank, M. M., and Brown, E. J.,** Plasma fibronectin enhances phagocytosis of opsonized particles by human peripheral blood monocytes, *J. Exp. Med.*, 157, 1844, 1983.
50. **Wright, S. D., Craigmyle, L. S., and Silverstein, S. C.,** Fibronectin and serum amyloid P component stimulate C3b-mediated phagocytosis in cultured human monocytes, *J. Exp. Med.*, 158, 1338, 1983.
51. **Menzel, E. J., Smolen, J. S., Liotta, I., and Reid, K. B.,** Interaction of fibronectin with C12 and its collagen like fragment (CLF), *FEBS Lett.*, 129, 188, 1981.
52. **Carter, S. D., Scott, D. L., and Elson, C. J.,** Fibronectin associated with immune complexes from sera and synovial fluids in rheumatoid arthritis, *Br. J. Rheumatol.*, 25, 353, 1986.
53. **Ferraccioli, G., Karsh, J., and Osterland, C. K.,** Interaction between fibronectin, rheumatoid factor and aggregated gammaglobulins, *J. Rheumatol.*, 12, 680, 1985.
54. **Edsall, J. T., Gilbert, G. A., and Scheraga, H. A.,** The non-clotting component of the human plasma fraction I-I (cold insoluble globulin), *J. Am. Chem. Soc.*, 77, 157, 1955.
55. **Edsall, J. T.,** Some early history of cold insoluble globulin, *Ann. N.Y. Acad. Sci.*, 312, 1, 1978.
56. **Stathakis, N. E., Mosesson, M. W., Chen, A. B., and Galanakis, D. K.,** Cryoprecipitation of fibrin-fibrinogen complexes induced by cold insoluble globulin of plasma, *Blood*, 51, 1211, 1978.
57. **Stathakis, N. E. and Mosesson, M. W.,** Interactions among heparin, cold insoluble globulin and fibrinogen in formation of the heparin-precipitable fraction, *J. Clin. Invest.*, 60, 855, 1977.
58. **Wood, G., Rucker, M., Davies, J. W., Entwistle, R., and Anderson, B.,** Interaction of plasma fibronectin with selected cryoglobulins, *Clin. Exp. Immunol.*, 40, 358, 1980.
59. **Anderson, B., Rucker, M., Entwistle, R., Schmid, F. R., and Wood, G. W.,** Plasma fibronectin is a component of cryoglobulins from patients with connective tissue and other diseases, *Ann. Rheum. Dis.*, 40, 50, 1981.
60. **Scott, D. L., Almond, T. J., Naqvi, S. N. H., Lea, D. J., Stone, R., and Walton, K. W.,** The significance of fibronectin in cryoprecipitation in rheumatoid arthritis and other disorders, *J. Rheum.*, 9, 514, 1982.
61. **Strevey, K., Beaulieu, A. D., Menard, C., Valet, J. P., Latulippe, L., and Herbert, J.,** The role of fibronectin in the cryoprecipitation of monoclonal cryoglobulins, *Clin. Exp. Immunol.*, 55, 340, 1984.
62. **Hautenen, A. and Keski-Oja, J.,** Affinity of myeloma IgG proteins for fibronectin, *Clin. Exp. Immunol.*, 53, 233, 1983.
63. **Pierschbacher, M., Hayman, E. G., and Ruoslahti, E.,** Synthetic peptide with cell attachment activity of fibronectin, *Proc. Natl. Acad. Sci. U.S.A.*, 80, 1224, 1983.
64. **Scott, D. L., Coulton, B. L., Symmons, D. P. M., and Popert, A. J.,** Longterm outcome of treating rheumatoid arthritis: results after 20 years, *Lancet*, i, 1108, 1987.
65. **Scott, D. L., Mapp, P. A., and Herbert, K. E.,** unpublished observations.

Chapter 17

FIBRONECTIN IN ANIMAL MODELS OF INFLAMMATION AND AUTOIMMUNE DISEASES

Vera J. Stecher and Kevin M. Connolly

TABLE OF CONTENTS

I. INTRODUCTION

Fibronectin (Fn) has several biologic activities which may be of importance in the pathogenesis of the chronic inflammatory response. Fn provides a link between inflammatory cells and connective tissue[1,2] and enhances cell adhesion,[3] chemotaxis,[4] and phagocytosis.[5] Since these biologic processes are key features of chronic inflammation, Fn may provide a biochemical marker of disease activity. A comparison of the effect on Fn of disease-modifying antirheumatic drugs (DMARDs) as opposed to nonsteroidal anti-inflammatory drugs (NSAIDs) and steroids, supports this hypothesis.[6]

II. CHRONIC INFLAMMATION RESULTING FROM AUTOIMMUNE DISEASE

Adjuvant-induced arthritis in rats has many of the characteristics of rheumatoid arthritis.[7] It is a chronic, progressive, deforming arthritis of the peripheral joints, with a primary mononuclear cell response consisting of bone and joint space invasion by pannus.[8] In the rat adjuvant arthritis model, chronic, systemic inflammation is preceded by an increase in the level of plasma fibronectin (Fn).[9] Within 24 h of adjuvant injection, rat plasma Fn levels rise from normal levels of approximately 350 μg/ml to almost 700 μg/ml. The twofold increase in plasma Fn in adjuvant-induced arthritic rats is sustained throughout the course of the disease, indicating that plasma Fn levels have an association with disease activity. Fn levels remained high for more than 6 months. Clinically, Scott et al.[10] and other investigators[11] have postulated that the high Fn concentration in rheumatoid synovium is a result of *in situ* production, and that increased levels of Fn may be evident in the plasma only in cases of patients with severe, systemic disease. Indeed, Carsons et al.[12] found that patients with systemic lupus erythematosus (SLE) had increased plasma Fn levels and that the highest levels of Fn occurred in patients with severe disease activity. Furthermore, Fn levels in certain patients were found to parallel disease activity longitudinally. Administration of Plaquenil (400 mg/d) and prednisone (20 mg/d) to a SLE patient who had flared resulted in decreases in plasma Fn levels and serum DNA binding which coincided with abatement of all clinical activity. In another report, a correlation between clinical improvement and synovial fluid Fn levels was found in arthritic patients treated with orgotein.[13] Fn levels in joint fluid in the case of arthritics and plasma Fn levels in the case of SLE patients may be a relevant indicator of disease activity.

In the rat, however, a relatively small blood volume, combined with severe, systemic disease, may result in the liberation of a significant amount of Fn into the plasma. Liver synthesis accounts for most of the circulating plasma Fn in normal animals, but currently it has not been possible to determine what part liver synthesis, as opposed to joint and connective tissue production, plays in elevated plasma Fn levels of arthritic rats.

A. EFFECT OF NSAIDs AND DMARDs ON PLASMA FIBRONECTIN LEVELS AND PAW SWELLING IN ARTHRITIC RATS

During the development of adjuvant-induced arthritis in the rat, the disease becomes systemic and the noninjected or contralateral paw becomes inflamed and swollen by day 17 after induction of disease. The systemic nature of the disease was assessed by measuring noninjected paw swelling. Rocket immunoelectrophoresis, using purified goat anti-rat Fn, provided a specific and sensitive means of measuring plasma Fn in rats.[14] Two days after the induction of disease by injection of adjuvant, NSAIDs or DMARDs were administered daily on a mg/kg regimen for 15 d at their anti-inflammatory doses. Table 1 illustrates the efficacy in this assay of the NSAIDs: aspirin (55, 100, 200 mg/kg), naproxen (1, 3, 10 mg/kg), ibuprofen (30, 55, 100 mg/kg), and phenylbutazone (10, 30, 55 mg/kg). Each drug

TABLE 1
Activity of NSAIDs in the Developing Arthritis and Fibronectin Assays

Group	Dose mg base/kg	Noninjected paw % inhibition	Injected paw % inhibition	Fibronectin % reduction
Aspirin	55	37[a]	21[a]	31
	100	59[a]	32[a]	32
	200	86[a]	68[a]	32
Naproxen	1	35	27[a]	0
	3	61[a]	45[a]	19
	10	86[a]	69[a]	20
Ibuprofen	30	50[a]	42[a]	31
	55	65[a]	51[a]	32
	100	69[a]	57[a]	25
Phenylbutazone	10	55[a]	35[a]	0
	30	74[a]	55[a]	30
	55	83[a]	64[a]	22

[a] $p \leq 0.01$ compared to arthritic controls, n = 10 animals per group.

caused a significant inhibition of swelling in both the injected and noninjected paws. However, the NSAIDs had no significant effect on plasma Fn levels. Dose-response studies using normal animals treated with NSAIDs established that normal Fn levels were not changed by these drugs.[14]

Since NSAIDs used clinically control swelling and symptomatic relief but not the underlying progression of the disease,[15] local production of Fn by the invasive pannus tissue[16] would be expected to continue, thereby keeping Fn levels high. In addition, since Fn is a membrane-bound protein of both leukocytes and synovial cells,[17] Fn may be released into the synovial fluid and blood plasma as these cells are destroyed. The relative importance of new synthesis as opposed to cell breakdown as contributors to plasma Fn is difficult to quantify due to the systemic nature of arthritis.

The mechanisms by which DMARDs exert their antirheumatic activity and slow joint deterioration are not well understood. Some of the proposed modes of action involve a reduction in macrophage function, including modulation of lymphocytes via macrophage production of interleukin-1.[18,19] In addition, DMARDs are known to reduce the phagocytic activity of synovial macrophages[20] and stabilize their lysosomes which contain enzymes capable of degrading cartilage.[21,22] Furthermore, the macrophage is a site of synthesis of fibronectin,[23] and DMARDs may also modulate its production.[24]

Table 2 illustrates the effect of DMARDs on plasma Fn levels and inhibition of swelling of both the injected and noninjected paws of adjuvant-induced arthritic rats. Three classes of drugs which are generally accepted as DMARDs were included in this study: antimalarials, typified by Plaquenil (hydroxychloroquine sulfate); gold, both oral (auranofin) and injectable (gold sodium thiomalate), and D-penicillamine. All four drugs were able to reduce systemic inflammation but unlike the pattern seen for the NSAIDs (Table 1), treatment with DMARDs resulted in significant reductions in the elevated plasma Fn levels characteristic of untreated arthritic control animals. DMARDs administered to normal animals had no effect on normal plasma Fn levels.

Since there is clinical[10-11,13,25-26] as well as preclinical[9,27] evidence indicating an association between high Fn levels and arthritic disease, the *in vivo* differential effects of DMARDs as opposed to NSAIDs on plasma Fn levels in arthritic animals may be relevant to the mechanism of action of these drugs in affecting the pathogenesis of rheumatic disease.

TABLE 2
Activity of DMARDs in the Developing Arthritis and Fibronectin Assays

Group	Dose mg base/kg	Noninjected paw % inhibition	Injected paw % inhibition	Fibronectin % reduction
Plaquenil	55	10	16	21
	75	25	26[a]	33[b]
	100	60[a]	30[a]	96[a]
Oral gold	3	69[a]	54[a]	29
	10	82[a]	67[a]	77[a]
	30	107[a]	75[a]	86[a]
IM gold	3	44[a]	30[a]	41[a]
	10	46[a]	36[a]	75[a]
	30	38[a]	41[a]	65[a]
D-penicillamine	30	24	27[a]	6
	55	24	27[a]	63[a]
	100	37[a]	28[a]	38[a]

[a] $p \leq 0.01$ vs. arthritic controls, n = 10 animals per group.
[b] $p \leq 0.05$ vs. arthritic controls, n = 10 animals per group.

B. EFFECT OF GLUCOCORTICOIDS ON PLASMA FIBRONECTIN LEVELS IN NORMAL AND ARTHRITIC RATS

The glucocorticoids are another class of drugs used as anti-inflammatory agents in the treatment of arthritis.[28] *In vitro* studies have demonstrated that dexamethasone enhances Fn production in cultures of chick hepatocytes,[29] rat hepatocytes,[30] rat hepatomas,[31] and SV40 transformed human fibroblasts.[32] In an *in vivo* study, the normal and adjuvant-induced arthritic rats were given daily doses of dexamethasone, methylprednisolone, or corticosterone in order to determine how paw inflammation and Fn levels are affected by these drugs.[33] In all normal groups, glucocorticoids significantly enhanced plasma Fn levels, paralleling the manner in which dexamethasone increased Fn production *in vitro*.[29-32] In all arthritic groups, glucorticoids significantly decreased paw inflammation in a dose-response manner, similar to the anti-inflammatory effect of NSAIDs.[14] Of the three glucocorticoids, only corticosterone did not significantly enhance Fn levels in arthritic rats, possibly due to its comparative lack of potency and narrow therapeutic window.

Endogenous levels of glucocorticoids appear to increase plasma Fn levels, since adrenalectomy of normal rats resulted in a significant decrease in the concentration of plasma Fn.[33] Elevated levels of Fn in adjuvant-induced arthritic animals may, therefore, be related to increased levels of endogenous glucocorticoids.

The significant increase in rat plasma Fn levels following steroid treatment may be due to an effect upon liver metabolism since normal plasma Fn levels are maintained by liver synthesis.[34] In view of the possible involvement of Fn in the actual progression of disease, the induction by steroids of high concentrations of Fn may exacerbate the arthritic process. It has been hypothesized that the Fn-mediated bond between leucocyte and cartilage allows formation of a sheltered microenvironment conducive to tissue degradation by proteolytic enzymes which normally would be rendered inactive by endogenous inhibitors.[2] A certain basal level of Fn may be desirable since Fn may be involved in matrix repair or modulation of cells active in the repair process. Although it is intriguing to speculate upon the involvement of Fn in the pathophysiology of rheumatoid arthritis, at this time we can only state that Fn appears to be a reliable marker associated with some autoimmune diseases.

III. ACUTE INFLAMMATION

If sustained high plasma Fn levels are partially the result of synthesis in inflamed joints, then models of inflammation which are not chronic and do not exhibit systemic joint disease would not be expected to be accompanied by sustained high levels of plasma Fn. This was indeed the case in two animal models.

A. CARRAGEENAN-INDUCED PAW EDEMA

A 1% solution of carrageenan[9] was injected into the right hind paws of rats and the difference in volume between injected and uninjected paws was recorded at various time intervals. Six hours following carrageenan injection, the right hind paws were measurably swollen, but no significant rise in plasma Fn levels was measured. After 24 h, plasma Fn levels were significantly ($p \leq 0.001$) higher than normal and remained high on days 2 and 4 as did paw swelling. By day 7, Fn levels returned to normal and paw swelling decreased to less than values recorded during the first 6 h following carrageenan injection.

B. CARRAGEENAN-INDUCED PLEURISY

A 0.25% solution of carrageenan was injected intrapleurally into rats, and plasma samples were taken sequentially thereafter for Fn analysis. Carrageenan in this model induces leukocyte accumulation in the pleural cavity but does not induce chronic or systemic joint disease.[35] There was a sevenfold increase in the number of leukocytes in the pleural cavity 72 h following injection. Differential cell counts revealed a predominantly mononuclear cell infiltrate. There was neither an increase in plasma Fn levels nor evidence of disseminated inflammation.[9]

The interchange of proteins between pleural fluid and blood is complex and influenced by factors such as intrapleural pressure.[36] Movement of macromolecules from the pleural cavity is restricted[37] and proteins appear able to leave the pleural space only via the lymphatics.[38] In contrast, there is extensive movement of molecules between the peritoneal cavity and the vascular circulation. This may explain why intraperitoneal inflammation has been reported to result in a temporary increase in serum Fn levels.[39]

1. Fibronectin Levels within the Pleural Cavity

Although plasma levels of Fn were not elevated during the course of carrageenan-induced pleurisy in the rat, intrapleural levels of Fn increased significantly and paralleled the rise in intrapleural mononuclear cell counts.[24] The data indicated that 3 d after injection of carrageenan was the optimal time interval for the study of mononuclear cells and Fn. Mononuclear cells constituted 90% of total cells.

An association between high Fn levels and arthritic disease[17] may relate to the increased numbers of macrophages at sites of chronic inflammation such as the rheumatoid synovium.[40] The macrophage is known to produce Fn[23] which is present in high concentrations in the synovial fluid[10,25,26] and pannus tissue of arthritics.[11] Furthermore, when activated, macrophages produce numerous secretory products, including neutral proteases and collagenase which play a destructive role in eroding cartilage during the course of arthritis.[41,42]

2. Activity of DMARDs and NSAIDs in the Pleurisy Assay

Evaluation of the accumulation of macrophages in the pleural cavity of rats 72 h after the introduction of the inflammatory irritant, carrageenan, provides an assay which distinguishes the known NSAIDs from DMARDs.[35,24] Nonsteroidal anti-inflammatory drugs are able to affect neutrophil accumulation (acute inflammation) but do not inhibit macrophage infiltration, whereas DMARDs reduce marcophage but not neutrophil accumulation. The DMARDs, Plaquenil (10 mg/kg), D-penicillamine (10 mg/kg), oral gold (10 mg/kg), and IM gold (10 mg/kg), all cause a significant inhibition in the accumulation of macrophages: 61, 49, 35 and 83%, respectively.[6] NSAIDs, including ibuprofen (100 mg/kg), aspirin (100 mg/kg), naproxen (55 mg/kg), phenylbutazone (100 mg/kg), and indomethacin (1 mg/kg) have no macrophage inhibitory activity.

The pleural washes of normal, untreated rats contain 10 to 20 μg Fn. Unmedicated control animals with carrageenan-induced pleurisy develop an exudate which contains a total average of 123 μg of Fn. The DMARDs such as Plaquenil cause a highly significant reduction in the concentration of Fn per milliliter of exudate as well as a decrease in the total volume of exudate.[6] All the DMARDs inhibited Fn production in a dose-related manner.[24] These results suggest that the mechanisms of action of DMARDs which relate to their antirheumatic profile includes inhibition of Fn.

IV. FIBRONECTIN IN ANIMAL MODELS OF OSTEOARTHRITIS (OA)

Adjuvant arthritis in the rat is not the only disease model in which Fn has been studied. Fn has been used as a disease marker in animal models of OA, antigen-induced arthritis, and lupus.

In models of OA, Fn measurements were made at the site of joint inflammation, in contrast to quantitation of plasma Fn in AA rats. Burton-Wurster and Lust have done extensive studies on spontaneously occurring OA in dogs by culturing cartilage from normal and OA animals[43,44] (and Chapter 15). Incubating these tissue explants with ^{3}H-phenylalanine, they were able to quantitate Fn synthesis using a gelatin-Sepharose column to isolate labeled-Fn from tissue supernatants. Tissue from OA dogs synthesized about four times more Fn than normal cartilage.[45] Measurement of Fn extracted from articular tissue using a urea/heparin/phosphate buffer[46] indicated that cartilage from OA dogs contained 10 to 40 times more Fn than cartilage from normal dogs.[47] Even the synovial fluid from OA animals had a twofold increase in Fn compared to normal controls.[6] The Fn in the OA joints appears to be a result of chondrocyte synthesis since histochemical and electron microscopic studies failed to reveal the presence of fibroblasts or macrophages in the tissue explants.[45]

Spontaneous canine OA is not the only example of a model in which high levels of Fn could be isolated from arthritic joints. Cartilage tissue was removed from the joints of rabbits with surgically induced OA.[49] The tissue from the OA rabbits contained an increased quantity of Fn compared to normal joint tissue.[47] Rabbits have also been used in an antigen-induced arthritis assay.[50,51] This model also resulted in an increased depostion of Fn in synovial tissue.[52,53] Since the degree of inflammation in antigen-induced arthritis is greater than in OA, the source of the Fn in the antigen-induced model may be the macrophages, fibroblasts, and endothelial cells involved in the inflammatory process,[52] in addition to Fn synthesis by chondrocytes in OA.[45] As a structural protein, Fn presence can be viewed positively, assisting in tissue repair by cross-linking and orienting collagen molecules. However, as an immunologically active molecule, it can exacerbate inflammation via its opsonic, adhesive, and chemotactic properties.

Although deposition of Fn in arthritic joints has been most extensively examined in the dog and the rabbit, Livne et al. have described a model of spontaneously occurring osteoarthritis in the temporomandibular joint of aged (22 months) ICR mice. Histological and electron microscopic studies revealed a greater abundance of Fn on the articular surface of specimens from aged ICR mice compared to young controls.[54]

V. FIBRONECTIN IN ANIMAL MODELS OF SYSTEMIC LUPUS ERYTHEMATOSUS (SLE)

Mice have been used in the characterization of autoimmune disorders such as SLE. The MRL/lpr mouse spontaneously develops a lupus-like syndrome which is fatal within 5 months of birth.[55] Although we have not examined Fn deposition in these mice, we have demonstrated that the increase in plasma Fn levels is associated with a rise in mortality and proteinuria in male and female MRL/lpr mice.[27] The high plasma Fn concentration is probably due to both elevated hepatocyte synthesis and synthesis from activated macrophages, fibroblasts, and endothelial cells participating in the systemic inflammatory process. As in other models of autoimmune disease, it is difficult to determine if Fn in MRL/lpr mice is just a marker of disease, or if it is involved in the pathophysiology of the disease due to enhancement of phagocytosis, chemotaxis, and cell adhesion.

VI. FIBRONECTIN IN ANIMAL MODELS OF CHRONIC INFLAMMATION

The assay systems in this section are models of chronic inflammation which lack an autoimmune component; in other words, the inflammatory process does not induce immunological breakdown of self-recognition. Inflammation in these models was defined in the most general of terms. Tissue damage due to contact with activated inflammatory cells (i.e., macrophages and neutrophils) or their soluble factors, was considered "inflammation". Chronic refers to the time of measurement. Inflammation was considered chronic if the endpoint was measured in terms of days not hours after stimulation. If measurements were made hours as well as days after initial stimulus, then the model was viewed as having an acute (short-term) and chronic (long-term) phase.

In the rat air pouch model of chronic inflammation, air was injected under the skin of a rat, forming a subcutaneous bubble into which an antigenic agent could be injected.[56,57] Within 24 h of antigen injection, plasma Fn levels were twofold higher than in uninjected controls, and were maintained at that high level over the 19 d duration of the experiment. In contrast, Fn levels in the exudate from the pouch increased gradually, but were significantly above normal plasma levels by day 13.[58] Histological studies of tissue from the inflamed pouch indicated a progressive rise in tissue-bound Fn over the time course of the assay.[59]

Buildup of tissue-bound Fn in the air pouch model of chronic inflammation is similar to the accumulation of Fn in the granulomata of animals experiencing chronic inflammation due to *Schistosoma mansoni*.[60] Fn has been shown to deposit around the parasite eggs forming a granulomatous extracellular matrix. Fn contributes to the pathophysiology of schistosomiasis by binding macrophages to the Fn-coated schistosoma eggs, forming hepatic granulomas.[61] There is also evidence that activated macrophages in the granuloma release Fn which acts as a chemoattractant for fibroblasts.[62] The recruitment of fibroblasts by Fn is another way in which Fn advances the pathogenesis of the schistosomiasis lesions.

Fibronectin also accumulates in experimental granulation tissue formed by subcutaneous implantation of cellulose sponges into rodents.[63,64] In this model of chronic inflammation, fibroblasts are not only attracted by Fn, but also synthesize it.[63] As in the schistosomiasis model of granuloma formation, Fn in this sponge model accumulates and forms an extracellular matrix, around which is constructed an orderly array of collagen fibers.[64] Thus, Fn can also be an important factor in tissue repair. In the inflammatory process, Fn has a dual function; on the one hand, it attracts and binds inflammatory cells (i.e., macrophages and neutrophils), and on the other hand, it is a structural protein which binds collagen and participates in the repair process.

The involvement of Fn in the repair process can easily be demonstrated by examining various models of wound healing. Histological examination of sections from full thickness skin wounds in guinea pigs indicated that Fn deposited around the fibrin clot and was bound to collagen fibers in the granulation tissue. In this model, as in the experimental granulation model,[64] Fn forms a matrix upon which collagen can be built.[65] It also preferentially binds to denatured collagen and as an opsonin, aids the wound-healing process by facilitating the removal of tissue debris by macrophages.[65] Studies involving mechanical trauma of striated muscle in rats also showed that Fn accumulated at the site of the wound.[66] In this study,[125]

I-labeled Fn was injected i.v. into rats suffering muscle trauma. The labeled Fn went preferentially to the location of the wound, probably due to its affinity for denatured collagen, fibrin, and fibrinogen found at the site of inflammation.[66]

Fn accumulation at the site of inflammation has also been measured in a delayed-type hypersensitivity skin reaction. Using radioisotopic and immunofluorescence techniques, Clark et al.[67] quantitated the deposition of Fn in the dermal extravascular space and the dermal vessel wall of guinea pigs. Initial accumulation was presumed to be from the pool of Fn in the plasma; later deposits were the result of Fn synthesis by endothelial cells in the vessel walls.

Paraquat-induced lung injury in monkeys is another model of chronic inflammation.[68] When cynomologous monkeys were administered paraquat for 2 weeks, they developed a pulmonary fibrosis with elevated levels of Fn in lung lavage fluid.[69] Since the activated alveolar macrophages rapidly synthesized and released Fn, the Fn which accumulated in the inflamed lung was presumed to be ''cellular'' as opposed to plasma Fn. Release of cellular Fn contributed to the pathology of the injury by attracting fibroblasts, which in turn spurred the development of pulmonary fibrosis.

VII. FIBRONECTIN IN ANIMAL MODELS OF INFLAMMATION WITH AN ACUTE COMPONENT

In other experimental models of lung injury, Fn accumulates at the site of injury within hours of the insult.[70] Lung tissue from rabbits, when exposed to superoxide anions, released Fn into the circulation within 6 h. The source of Fn was traced to fibroblasts (as opposed to hepatocytes), based on their characteristic electrophoretic mobility[71] and reaction with antibody specific for the type III Fn domain associated with cellular fibroblast Fn but not plasma hepatocyte Fn.[72]

Heretofore, every animal model of inflammation discussed has been accompanied by an elevation in the level of Fn. It would, however, be misleading to imply that inflammation invariably is followed by an abnormal increase in Fn. The Fn level will vary depending on the site and time of sampling. Fn is a plasma protein with unique properties. It is an opsonic molecule with affinity for denatured collagen and fibrin. With such an affinity, it accumulates at the site of inflammation, where macrophages, denatured collagen, and fibrin-fibrinogen complexes are found in abundance. Thus, in the very early (acute) states of inflammation, plasma Fn levels are actually reduced, since the plasma Fn has localized at the site of inflammation. However, like other more typical acute phase proteins, Fn is synthesized by hepatocytes triggered to produce Fn (and other acute phase proteins) as a response to an inflammatory stimulus. Thus, inflammation causes a transient drop in plasma Fn levels which is subsequently restored and often surpassed by hepatocyte synthesis. Another factor which separates Fn from most other acute phase proteins is that its synthesis is not solely dependent on the activity of hepatocytes. Macrophages, fibroblasts, and endothelial cells are all capable of synthesizing Fn, and this capability contributes to the elevated levels of Fn observed locally and systemically in all but the earliest stages of the inflammatory process.

An example of plasma Fn deficiency in the early stages of inflammation is a model of acute lung vascular injury in sheep.[73] When protease was infused into the lungs of sheep, depression in the plasma Fn level was significant within 2 h. The authors speculated that the plasma Fn was sequestered in the lungs or in the reticuloendothelial system (RES).

In several models of peritoneal inflammation, there is an early, transient fall in plasma Fn concentration followed by Fn rebound and overshoot in the later stages of the inflammation.[74-76] In a model of peritoneal sepsis, cecal ligation induced a decrease in plasma Fn levels at the 2-h timepoint due to sequestration of Fn at the focus of intraperitoneal infection.[77] Within 24 h, levels of plasma Fn were significantly higher than normal controls due to a

50% rise in the synthesis of Fn following inflammation. In animals with sterile peritonitis resulting from i.p. injection of casein, plasma Fn levels were elevated 24 h after initiation of the inflammation.[39] Peritoneal Fn was also elevated, and appeared to coincide with the increase in the number of peritoneal macrophages drawn to the site of the inflammation. Because of its opsonic activity, elevation of Fn enhanced clearance of colloidal particles, e.g., gelatin-coated beads injected into rats 72 h after induction of inflammation. However, in the initial stages of peritoneal inflammation, the ability of the reticuloendothelial system (RES) to remove blood-borne particulates was depressed, presumably due to sequestration of opsonic Fn in the peritoneal cavity.[78] Although it appears that plasma Fn enhances the ability of the RES to remove sterile particulate matter from the circulation, the evidence that plasma Fn aids in the removal of bacteria from the circulation is ambiguous.[79,80]

In a different model of peritoneal inflammation, Grossman et al. induced peritoneal inflammation in rats by either injecting endotoxin or surgically inducing an abdominal abscess.[81] Depression of plasma Fn occurred in both inflammatory models between 1 and 4 h after induction of peritoneal inflammation. However, the level of plasma Fn rebounded rapidly, so that 24 h after endotoxin injection, plasma Fn levels were significantly above normal. In rats subjected to surgically induced inflammation, elevated plasma Fn levels occurred 72 h after trauma. Clearance of particulate matter was not measured, so no conclusions could be made regarding the opsonic activity of the plasma during the various stages of the inflammation.

Although Loegering and Schneidkraut[82] did not directly measure the absolute concentration of plasma Fn in their rat model of endotoxin-induced inflammation, they did determine that the opsonic activity of the plasma was unchanged over time as measured in the rat liver slice bioassay.[83] However, in another model of acute endotoxemia, a reduction in the plasma Fn concentration was noted. In this model, Modig and Borg administered an i.v. infusion of endotoxin into pigs over a 6-h period.[84] In that interval, they saw a significant decrease in plasma Fn level although they never measured the opsonic activity of the plasma. However, they noted that animals surviving the endotoxemia had significantly higher Fn levels throughout the observation period, implying that the plasma Fn possessed some biologically real protective function.

The use of Fn as a marker in animal models of inflammation is straightforward. Fn in animal models of inflammation appears to fall into the general classification of an acute phase protein in that: (1) it is synthesized by the hepatocyte, and (2) its synthesis is induced by a variety of inflammatory conditions. Fn, as an acute phase reactant, is unique in that: (1) it has an affinity for a variety of substrates and receptors including macrophages, collagen, fibrin, and heparin, and (2) it is synthesized by a variety of cell types including macrophages, fibroblasts, and endothelial cells. These characteristics give rise to the unique concentration curve exhibited by plasma Fn over the course of inflammation. The mouse model of inflammation described by Dyck and Rogers[85] typifies the plasma Fn concentration curve present in many animal models of inflammation. Mice given a s.c. injection of silver nitrate had depressed levels of plasma Fn 4 h after injection. Within 24 h, plasma Fn levels had increased threefold over uninjected controls. In contrast, another acute phase plasma protein, serum amyloid component, showed no transient depression.

The actual *in vivo* biological activity of Fn is difficult to ascertain. The models of inflammation discussed in this review indicate some level of *in vivo* biological Fn activity. The degree of activity and its overall importance to the animal is a matter which promises to be of continuing interest.

ACKNOWLEDGMENT

We are grateful to Maria Nanfaro for her expert administrative and typing skills in the preparation of this manuscript.

REFERENCES

1. **Horman, H.**, Fibronectin-mediator between cells and connective tissue, *Klin. Wochenschr.*, 60, 1265, 1982.
2. **Weissmann, G.**, Activation of neutrophils and the lesions of rheumatoid arthritis, *J. Lab. Clin. Med.*, 100, 322, 1982.
3. **Akiyama, S. K., Yamada, K. M., and Hayashi, M.**, The structure of fibronectin and its role in cellular adhesion, *J. Supramol. Struct. Cell Biochem.*, 16, 345, 1981.
4. **Norris, D. A., Clark, R. A. F., Swigart, L. M., Huff, J. C., Weston, W. L., and Howell, S. E.**, Fibronectin fragments are chemotactic for human peripheral blood monocytes, *J. Immunol.*, 129, 1612, 1982.
5. **Blumenstock, F. A., Saba, T. M., Roccario, E., Cho, E., and Kaplan, J. E.**, Opsonic fibronectin after trauma and particle injection determined by a peritoneal macrophage monolayer assay, *J. Reticuloendothel. Soc.*, 30, 61, 1981.
6. **Stecher, V. J., Connolly, K. M., and Speight, P. T.**, Fibronectin and macrophages as parameters of disease-modifying antirheumatic activity, *Br. J. Clin. Pract.*, 41 (Suppl. 52), 64, 1987.
7. **Pearson, C. J.**, Experimental joint disease: observations on adjuvant-induced arthritis, *J. Chronic Dis.*, 16, 863, 1963.
8. **Glenn, E. M. and Grey, J.**, Adjuvant-induced polyarthritis in rats: biologic and histologic background, *Am. J. Vet. Res.*, 26, 1180, 1965.
9. **Stecher, V. J., Kaplan, J. E., Connolly, K. M., Mielens, Z., and Saelens, J. K.**, Fibronectin in acute and chronic inflammation, *Arthritis Rheum.*, 12, 394, 1986.
10. **Scott, D. L., Farr, M., Crockson, A. P., and Walton, K. W.**, Synovial fluid and plasma fibronectin levels in rheumatoid arthritis, *Clin. Sci.*, 62, 71, 1981.
11. **Vartio, J., Vaheri, A., Von Essen, R., Isomaki, H., Stenman, S.**, Fibronectin synovial fluid and tissue in rheumatoid arthritis, *Eur. J. Clin. Invest.*, 11, 207, 1981.
12. **Carsons, S., Parenti, D., Lavietes, B., Diamond, H. S., Singer, A., and Boxer, M.**, Plasma fibronectin in systemic lupus erythematosus: relationship to clinical activity, DNA binding and acute phase proteins, *J. Rheumatol.*, 12, 1088, 1985.
13. **Goebel, K. M. and Storck, U.**, Effect of intra-articular orgotein versus a corticosteroid on rheumatoid arthritis of the knee, *Am. J. Med.*, 74, 124, 1983.
14. **Connolly, K., Stecher, V. J., Kaplan, J. E., Mielens, Z., Rostami, H. J., and Saelens, J. K.**, The effect of anti-inflammatory drugs on plasma fibronectin, *J. Rheumatol.*, 12, 758, 1985.
15. **Scherbel, A. L.**, Nonsteroidal anti-inflammatory drugs, *Postgrad. Med.*, 63, 69, 1978.
16. **Scott, D. L., Delamere, J. P., and Walton, K. W.**, The distribution of fibronectin in the pannus in rheumatoid arthritis, *Br. J. Exp. Pathol.*, 62, 362, 1981.
17. **Scott, D. L., Wainwright, A. C., Walton, K. W., and Williamson, N.**, Significance of fibronectin in rheumatoid arthritis and osteoarthrosis, *Ann. Rheum. Dis.*, 40, 142, 1981.
18. **O'Duffy, J. D. and Luthra, H. S.**, Current status of disease modifying drugs in progressive rheumatoid arthritis, *Drugs*, 27, 373, 1984.
19. **Stecher, V. J., Carlson, J. C., Connolly, K. M., and Bailey, D. M.**, Disease modifying antirheumatic drugs, *Med. Res. Rev.*, 5, 371, 1985.
20. **Jessop, D., Wilkins, M., and Young, M. H.**, The effect of anti-rheumatic drugs on the phagocytic activity of synovial macrophages in organ culture, *Ann. Rheum. Dis.*, 41, 632, 1982.
21. **Persellin, R. H. and Ziff, M.**, The effect of gold salts on lysosomal enzymes of the peritoneal macrophage, *Arthritis Rheum.*, 9, 57, 1966.
22. **Mackenzie, A. H.**, Pharmacologic actions of 4-aminoquinoline compounds, *Am. J. Med.*, 75(1A), 5, 1983.
23. **Alitalo, K., Hovi, T., and Vaheri, A.**, Fibronectin is produced by human macrophages, *J. Exp. Med.*, 151, 602, 1980.
24. **Mielens, Z. E., Connolly, K., and Stecher, V. J.**, Effect of disease modifying antirheumatic drugs and nonsteroidal anti-inflammatory drugs upon cellular and fibronectin responses in a pleurisy model, *J. Rheumatol.*, 12, 1083, 1985.
25. **Carsons, S., Mosesson, M. W., and Diamond, H. S.**, Detection and quantitation of fibronectin in synovial fluid from patients with rheumatic disease, *Arthritis Rheum.*, 24, 1261, 1981.
26. **Lu-Steffes, M., Iammartino, A. J., Schmid, F. R., Castor, C. W., Davis, L., Entwistle, R., and Anderson, B.**, Fibronectin in rheumatoid and non-rheumatoid arthritis synovial fluids and in synovial fluid cryoproteins, *Ann. Clin. Lab. Sci.*, 12, 178, 1982.
27. **Connolly, K., Stecher, V. J., Saelens, J. K., and Kaplan, J. E.**, The relationship between plasma fibronectin levels and autoimmune disease in MRL/1 mice, *Proc. Soc. Exp. Biol. Med.*, 180, 149, 1985.
28. **Axelrod, L.**, Glucocortocoids, in *Textbook of Rheumatology*, 3rd ed., Kelly, W. N., Harris, E. D., Ruddy, S., and Sledge, C. G., Eds., W. B. Saunders Co., Philadelphia, 1989, 845.

29. **Amrani, D. L., Falk, M.J., and Mosesson, M. W.,** Synthesis of fibronectin by primary chick hepatocyte cultures, *Fed. Proc.*, 4, 1917, 1983.
30. **Marceau, N., Goyette, R., Valet, J. P., and Deschenes, J.,** The effect of dexamethasone on formation of a fibronectin extracellular matrix by rat hepatocytes in vitro, *Exp. Cell Res.*, 125, 497, 1980.
31. **Baumann, H. and Eldredge, D.,** Dexamethasone increases the synthesis and secretion of a partially active fibronectin in rat hepatoma cells, *J. Cell Biol.*, 95, 29, 1982.
32. **Furcht, L. T., Mosher, D. F., Wendelschafer-Crabb, G., Woodbridge, P. A., and Froidart, J. M.,** Dexamethasone-induced accumulation of a fibronectin matrix in transformed human cells, *Nature (London)*, 277, 393, 1979.
33. **Stecher, V. J., Connolly, K. M., and Snyder, B. W.,** The effect of glucocorticoids on plasma fibronectin levels in normal and arthritic rats, *Proc. Soc. Exp. Biol. Med.*, 182, 301, 1986.
34. **Akayama, S. K. and Amada, K. M.,** Fibronectin in disease, in *Connective Tissue Diseases*, Wagner, B. M., Fleischmajer, R., and Kaufman, N., Eds., Williams & Wilkins, Baltimore, 1983, 55.
35. **Ackerman, N., Tomolonis, A., Miram, L., Kheifets, J., Martinez, S., and Carter, A.,** Three-day pleural inflammation: a new model to detect drug effects on macrophage accumulation, *J. Pharmacol. Exp. Ther.*, 215, 588, 1980.
36. **Stewart, P. B. and Burgen, A. S. V.,** The turnover of fluid in the dog's pleural cavity, *J. Lab. Clin. Med.*, 52, 212, 1958.
37. **Apicella, M. A. and Allen, J. C.,** A physiologic differentiation between delayed and immediate hypersensitivity, *J. Clin. Invest.*, 48, 250, 1969.
38. **Courtice, F. C. and Simmonds, W. J.,** Absorption of fluids from the pleural cavities of rabbits and cats, *J. Physiol.*, 109, 117, 1949.
39. **Richards, P. S. and Saba, T. M.,** Fibronectin levels during intraperitoneal inflammation, *Infect. Immunol.*, 39, 1411, 1983.
40. **Kobayashi, I. and Ziff, M.,** Electron microscopic studies of the cartilage-pannus junction in rheumatoid arthritis, *Arthritis Rheum.*, 18, 475, 1975.
41. **Werb, Z., Banda, M. J., and Jones, P. A.,** Degradation of connective tissue matrices by macrophages. I. Proteolysis of elastin, glycoproteins and collagen by proteinases isolated from macrophages, *J. Exp. Med.*, 152, 1340, 1980.
42. **Schnyder, J. and Baggiolini, M.,** Secretion of lysosomal hydrolases by stimulated and nonstimulated macrophages, *J. Exp. Med.*, 148, 435, 1978.
43. **Burton-Wurster, N. and Lust, G.,** Incorporation of purified plasma fibronectin into explants of articular cartilage from disease-free and osteoarthritic canine joints, *J. Orthop. Res.*, 4, 409, 1986.
44. **Burton-Wurster, N. and Lust, G.,** The fibronectin and water content of articular cartilage explants after partial depletion of proteoglycans, *J. Orthop. Res.*, 4, 437, 1986.
45. **Burton-Wurster, N. and Lust, G.,** Synthesis of fibronectin in normal and osteoarthritic articular cartilage, *Biochem. Biophys. Acta*, 800, 52, 1984.
46. **Burton-Wurster, N. and Lust, G.,** Fibronectin in osteoarthritic canine articular cartilage, *Biochem. Biophys. Res. Commun.*, 109, 1094, 1982.
47. **Burton-Wurster, N., Butler, M., Harter, S., Colombo, C., Quintavalla, J., Swartzendurber, D., Arsenis, C., and Lust, G.,** Presence of fibronectin in articular cartilage in two animal models of osteoarthritis, *J. Rheumatol.*, 13, 175, 1986.
48. **Lust, G., Burton-Wurster, N., and Leipold, H.,** Fibronectin as a marker for osteoarthritis, *J. Rheumatol.*, 14, (Suppl. 14), 28, 1987.
49. **Colombo, C., Butler, M., O'Bryne, E., Hickman, L., Swartzendurber, D., Selwyn, M., and Steinetz, B.,** A new model of osteoarthritis in rabbits. I. Development of knee joint pathology following lateral meniscectomy and section of the fibular collateral and sesamoid ligaments, *Arthritis Rheum.*, 26, 875, 1983.
50. **Consden, R., Doble, A., Glynn, L. E., and Nind, A. P.,** Production of chronic arthritis with ovalbumin. Its retention in the rabbit knee-joint, *Ann. Rheum. Dis.*, 30, 307, 1971.
51. **Gardner, D. L.,** The experimental production of arthritis, *Ann. Rheum. Dis.*, 19, 297, 1960.
52. **Scott, D. L., Almond, T. J., Walton, K. W., and Hunneyball, I. M.,** The role of fibronectin in the pathogenesis of antigen-induced arthritis in the rabbit, *J. Pathol.*, 141, 143, 1983.
53. **Holund, B., Clemmensen, I., and Wanning, M.,** Sequential appearance of fibronectin and collagen fibers in experimental arthritis in rabbits, *Histochemistry*, 80, 39, 1984.
54. **Livne, E., Mark, K., and Silbermann, M.,** Morphologic and cytochemical changes in maturing and osteoarthritic articular cartilage in the temporomandibular joint of mice, *Arthritis Rheum.*, 28, 1027, 1985.
55. **Andrews, B. S., Eisenburg, R. A., Theofilopoulos, A. N., Izui, S., Wilson, C. B., Mc Conahey, P. J., Murphy, E. D., Roths, J. B., and Dixon, F. J.,** Spontaneous murine lupus-like syndromes. Clinical and immunopathological manifestations in several strains, *J. Exp. Med.*, 148, 1198, 1978.
56. **Tsurufuji, S., Yoshino, S., and Ohuchi, K.,** Induction of an allergic air pouch inflammation in rats, *Int. Arch. Allergy Appl. Immunol.*, 69, 189, 1982.

57. **Edwards, T. C. W., Sedgwick, A. D., and Willoughby, D. A.,** The formation of a structure with the features of synovial lining by subcutaneous injection of air: an in vivo tissue culture system, *J. Path.*, 134, 147, 1981.
58. **Scott, D. L., Robinson, M. W., and Yoshino, S.,** Fibronectin in chronic inflammation: studies using the rat air pouch model of chronic allergic inflammation, *Br. J. Exp. Path.*, 66, 519, 1985.
59. **Yoshino, S., Bacon, P. A., Blake, D. R., Scott, D. L., Wainwright, A. C., and Walton, K. W.,** A model of persistent antigen-induced chronic inflammation in the rat air pouch, *Br. J. Exp. Path.*, 65, 201, 1984.
60. **Al Adnani, M. S.,** Concomitant immunohistochemical localization of fibronectin and collagen in schistosome granulomata, *J. Pathol.*, 147, 77, 1985.
61. **Nishimura, M., Asahi, M., Hayashi, M., Takazono, I., Tanaka, Y., Kohda, H., and Urabe, H.,** Extracellular matrix in hepatic granulomas of mice infected with *Schistosoma mansoni, Arch. Pathol. Lab. Med.*, 109, 813, 1985.
62. **Wyler, D. J. and Postlethwaite, A. E.,** Fibroblast stimulation in schistosomiasis. IV. Isolated egg granulomas elaborate a fibroblast chemoattractant in vitro, *J. Immunol.*, 130, 1371, 1983.
63. **Holund, B., Clemmensen, I., Junker, P., and Lyon, H.,** Fibronectin in experimental granulation tissue, *Acta Path. Microbiol. Immunol. Scand.*, (Sect. A), 90, 159, 1982.
64. **Kurkinen, M., Vaheri, D., Roberts, P. J., and Stenman, S.,** Sequential appearance of fibronectin and collagen in experimental granulation tissue, *Lab. Invest.*, 43, 47, 1980.
65. **Grinnell, F., Billingham, R. E., and Burgess, L.,** Distribution of fibronectin during wound healing in vivo, *J. Invest. Dermatol.*, 76, 181, 1981.
66. **Gauperaa, T. and Seljeli, R.,** Plasma fibronectin is sequestered into tissue damaged by inflammation and trauma, *Acta. Chir. Scand.*, 152, 85, 1986.
67. **Clark, R. A. F., Dvorak, H. F., and Colvin, R. D.,** Fibronectin in delayed-type hypersensitivity skin reactions: associations with vessel permeability and endothelial cell activation, *J. Immunol.*, 126, 787, 1981.
68. **Fukuda, Y., Ferrans, V. J., Schoenberger, C. I., Rennard, S. I., and Crystal, R. G.,** Patterns of pulmonary structural remodeling after experimental paraquat toxicity. The morphogenesis of intraalveolar fibrosis, *Am. J. Pathol.*, 118, 452, 1985.
69. **Schoenberger, C. I., Rennard, S. I., Bitterman, P. B., Fukuda, Y., Ferrans, V. J., and Crystal, R. G.,** Paraquat-induced pulmonary fibrosis. Role of the alveolitis in modulating the development of fibrosis, *Am. Rev. Respir. Dis.*, 129, 168, 1984.
70. **Peters, J. H., Ginsberg, M. H., Bohl, B. P., Sklar, L. A., and Cochrane, C. G.,** Intravascular release of intact cellular fibronectin during oxidant-induced injury of the in vitro perfused rabbit lung, *J. Clin. Invest.*, 78, 1596, 1986.
71. **Yamada, K. M. and Kennedy, D. W.,** Fibroblast cellular and plasma fibronectins are similar but not identical, *J. Cell Biol.*, 80, 492, 1979.
72. **Hayashi, M. and Yamada, K. M.,** Differences in domain structure between plasma and cellular fibronectin, *J. Biol. Chem.*, 256, 292, 1981.
73. **Cohler, L. F., Saba, T. M., and Lewis, E. P.,** Lung vascular injury with protease infusion. Relationship to plasma fibronectin, *Ann. Surg.*, 202, 240, 1985.
74. **Lanser, M. E. and Saba, T. M.,** Opsonic fibronectin deficiency and sepsis: cause or effect? *Ann. Surg.*, 195, 340, 1982.
75. **Mc Cafferty, M. H. and Saba, T. M.,** Influence of septic peritonitis on circulating fibronectin immunoglobulin and complement. Relationship to reticuloendothelial phagocytic function, *Adv. Shock. Res.*, 9, 241, 1983.
76. **Velky, T. S., Yang, J. C., and Greenburg, A. G.,** Plasma fibronectin response to sepsis: mobilization or synthesis? *J. Trauma*, 24, 824, 1984.
77. **Kiener, J. L., Saba, T. M., Cho, E., and Blumenstock, F. A.,** Clearance and tissue distribution of fibronectin in septic rats. Relationship to synthetic rate, *Am. J. Physiol.*, 251, R724, 1986.
78. **Saba, T. M. and Cho, E.,** Reticuloendothelial systemic response to operative trauma as influenced by cryoprecipitate or cold-insoluble globulin therapy, *J. Reticuloendothel. Soc.*, 26, 171, 1979.
79. **Kaplan, J. E., Scovill, W. A., Bernard, H., Saba, T. M., and Gray, V.,** Reticuloendothelial phagocytic response to bacterial challenge after traumatic shock, *Circ. Shock*, 4, 1, 1977.
80. **Doran, J. E., Lundsgaard-Hansen, P., and Rubli, E.,** Plasma fibronectin: relevance for anesthesiology and intensive care, *Intensive Care Med.*, 12, 340, 1986.
81. **Grossman, J., Pohlman, T., Koerner, F., and Mosher, D.,** Plasma fibronectin concentration in animal models of sepsis and endotoxemia, *J. Surg. Res.*, 34, 145, 1983.
82. **Loegering, D. J. and Schneidkraut, M. J.,** Effect of endotoxin on alpha-2-opsonic protein activity and reticuloendothelial system phagocytic function, *J. Reticuloendothel. Soc.*, 26, 197, 1979.
83. **Loegering, D. J.,** Humoral factor depletion and reticuloendothelial depression during hemorrhagic shock, *Am. J. Physiol.*, 232, H283, 1977.

84. **Modig, J. and Borg, T.,** Biochemical markers in a porcine model of adult respiratory distress syndrome induced by endotoxemia, *Resuscitation,* 14, 225, 1986.
85. **Dyck, R. F. and Rogers, S. L.,** Fibronectin is an acute phase reactant in mice, *Clin. Invest. Med.,* 8, 148, 1985.

INDEX

A

D

E

F

G

H

I

J

K

L

M

N

O

P

T

U

V

W

Y

Z